Human Factors for Assistance and Automation

hfes europe chapter

Edited by

D. De Waard
F.O. Flemisch
B. Lorenz
H. Oberheid
K.A. Brookhuis

2008, Shaker Publishin

Europe Chapter of the Human Factors and Ergonomics Society

europechapter@hfes-europe.org

http://www.hfes-europe.org

© Copyright Shaker 2008

All rights reserved. No part of this publication may be reproduced, stored in a retrieval system, or transmitted, in any form or by any means, electronic, mechanical, photocopying, recording or otherwise, without the prior permission of the publishers.

Printed in The Netherlands.

D. de Waard, F.O. Flemisch, B. Lorenz, H. Oberheid, and K.A. Brookhuis (Eds.).

Human Factors Issues in Complex System Performance

ISBN 978-90-423-0350-8

Chapter and cover illustrations by Eva Fabriek (http://www.lelijkepoppetjes.nl)

Cover design by Klaas Woudstra

Shaker Publishing BV
St. Maartenslaan 26
6221 AX Maastricht
Tel.: +31 43 3500424
Fax: +31 43 3255090
http://www.shaker.nl

Contents

Preface 7
   Dick de Waard, Frank Flemisch, Bernd Lorenz, Hendrik Oberheid,
   & Karel Brookhuis

KEYNOTE 9

Error and motivation in driving 11
   Oliver Carsten

SURFACE TRANSPORTATION 33

Effects of tree aromas on automobile drivers' fatigue reduction and wakefulness
   assessed with finger-pulse fluctuations 35
   Keisuke Suzuki & Shigeki Harada
Changing lanes with active lanekeeping assistance: a simulator study 49
   Gerrit Schmidt, Kirstin L. R. Talvala, Joshua P. Switkes,
   Miklós Kiss, & J. Christian Gerdes
Speed recommendations during traffic light approach: a comparison of
   different display concepts 63
   Stephan Thoma, Thomas Lindberg, & Gudrun Klinker
Pupillometry as a method for measuring mental workload within a simulated
   driving task 75
   Maximilian Schwalm, Andreas Keinath, & Hubert D. Zimmer
A three-level model of Situation Awareness for driving with in-vehicle devices 89
   Nadja Rauch, Barbara Gradenegger, & Hans-Peter Krüger
Exploring appropriate alarm timing for a driver-adaptive forward collision
   warning system 103
   Genya Abe & Makoto Itoh
Effects of auditory warnings on driving behaviour 117
   Nicola Fricke & Mònica De Filippis
Effects of preactivated mental representations on driving performance 129
   Jessica Seidenstücker & Rainer Höger
The effect of experience, relevance, and interruption duration on drivers'
   mental representation of a traffic situation 141
   Martin R.K. Baumann, Thomas Franke, & Josef F. Krems
Overriding the ACC by keys at the steering wheel: positive effects on driving
   and drivers' acceptance in spite of a more complex ergonomic solution 153
   Ingo Totzke, Véronique Huth, Hans-Peter Krüger, & Klaus Bengler
A model of normal and impaired visual exploration while steering: a way to
   identify assistance needs 165
   Isabelle Milleville, Camilo Charron, Jean-Michel Hoc, & Jean-François Mathé

Advanced Driver Assistance Systems – Impact of psychological variables on the acceptance of modern technologies 179
Stefanie Müller, Heidi Ittner, & Volker Linneweber

Human centred design for informative and assistive technology in transport 193
Annie Pauzié & Anabela Simões

Getting back to basics: using road accident investigation to identify the desirable functionality of longitudinal control systems 203
Nikolaos Gkikas, Julian R. Hill, & John H. Richardson

Simulator training for truck-driving: indications of learning effects 217
Annette Kluge & Dina Burkolter

Analysis of automation in current UK rail signalling systems 229
Nora Balfe, John R. Wilson, Sarah Sharples, & Theresa Clarke

ADAPTIVE AUTOMATION 241

KONVOI: Electronically coupled truck convoys 243
Matthias Wille, Markus Röwenstrunk, & Günter Debus

Automation spectrum, inner / outer compatibility and other potentially useful human factors concepts for assistance and automation 257
Frank Flemisch, Johann Kelsch, Christian Löper, Anna Schieben, & Julian Schindler

Objective and subjective assessment of warning and motor priming assistance devices in car driving 273
Jordan Navarro, Franck Mars, Jean-François Forzy, Myriam El-Jaafari, & Jean Michel Hoc

Adaptive Automation enhances human supervision of multiple uninhabited vehicles 285
Ewart de Visser, Don Horvath, & Raja Parasuraman

A simple minded model for levels of automation 301
Michela Terenzi & Francesco Di Nocera

Inter-individual differences in executive control activity during simulated process control: comparison of performance and physiological patterns 313
Peter Nickel, Adam C. Roberts, Michael H. Roberts, & G. Robert J. Hockey

AVIATION HUMAN FACTORS 325

Modelling the allocation of visual attention using a Hierarchical Segmentation Model in the augmented reality environments for airport control tower 327
Ella Pinska & Charles Tijus

Empowerment of the planning controller in the use of Controller-Pilot Data-Link Communication (CPDLC) 343
Renée Schuen-Medwed, Bernd Lorenz, & Stefan Oze

Designing scenarios: the challenge of a multi-agent context for the investigation of authority distribution in aviation 359
Sonja Straussberger & Guy Boy

Head-Mounted Display – evaluation in simulation and flight trials 373
Sven Schmerwitz, Helmut Többen, Bernd Lorenz, & Bernd Korn

MODELLING AND SIMULATION 387

Everyday mistakes: confidence or cognition? 389
    Robert R.A. van Doorn & Fred R.H. Zijlstra
A model based approach to Cognitive Work Analysis and work process
    design in air traffic control 401
    Christoph Möhlenbrink, Hendrik Oberheid, & Bernd Werther
An integrated model for working environments and rail human factors 415
    Malte Hammerl, Bärbel Jäger, & Karsten Lemmer

HUMAN MACHINE INTERFACE 429

The effects of travel information presentation on driver behaviour 431
    Karel A. Brookhuis, Matthijs Dicke, Rosanne L. Rademaker, Sander R.F.
    Grunnekemeijer, Jorien N.J van Duijn, Maarten Klein Nijenhuis,
    Nynke A. Benedictus, & Elmas Eser
Supporting the localisation of task-relevant information 445
    Martin Groen & Jan Noyes
Virtual Reality in HMI research: attentional and motor performance in 3D-space 457
    Claudia Armbrüster, Marc Wolter, Torsten Kuhlen, Will Spijkers,
    & Brunno Fimm
Evaluation of task-oriented package design to control attention in out-of-box
    experience on internet services 469
    Masaru Miyamoto & Momoko Nakatani
Human factors involved in container terminal ship-to-shore crane operator tasks:
    operator fatigue and performance analysis at Cagliari Port 479
    Gianfranco Fancello, Gianmarco D'Errico, & Paolo Fadda
The different human factor in automation: the developer behind versus the
    operator in action 493
    Hartmut Wandke & Jens Nachtwei

Acknowledgement to reviewers 503

Preface

*Dick de Waard[1,2], Frank Flemisch[3], Bernd Lorenz[4], Hendrik Oberheid[3]
& Karel Brookhuis[1,2]
[1]University of Groningen, [2]Delft University of Technology
The Netherlands
[3]DLR German Aerospace Center
Germany
[4]Eurocontrol CRDS
Hungary*

End of October 2007 the annual conference of the Europe Chapter of the Human Factors and Ergonomics Society was held in Braunschweig, Germany. This publication summarises the content of that meeting, with contributions based on presentations that were given. The book begins with the keynote address "Error and Motivation in Driving" by Oliver Carsten, Professor of Transport Safety at the Institute for Transport Studies (ITS), University of Leeds.

Apart from the keynote, 35 papers are included, grouped into five chapters. Again Eva Fabriek has illustrated all these chapters and the cover of the book. Since she, together with Klaas Woudstra who is responsible for the cover design, started this work three years ago, a more coherent series has emerged that hopefully will continue for many years and will cover all colours of the rainbow.

We are very grateful to all who helped to make the meeting and the book a success. We'd like to thank DLR in Braunschweig who kindly hosted the three-day meeting. We also thank the long list of external reviewers (see page 503), who helped us to evaluate all the manuscripts. Their input is very much appreciated. The conference was also supported by the European Office of Aerospace Research and Development of the USAF, under Award No. FA8655-07-1-5092. We are very grateful for all this support.

In D. de Waard, F.O. Flemisch, B. Lorenz, H. Oberheid, and K.A. Brookhuis (Eds.) (2008), *Human Factors for assistance and automation* (p. 7). Maastricht, the Netherlands: Shaker Publishing.

Keynote

Error and motivation in driving

In D. de Waard, F.O. Flemisch, B. Lorenz, H. Oberheid, and K.A. Brookhuis (Eds.) (2008), *Human Factors for assistance and automation* (p. 9). Maastricht, the Netherlands: Shaker Publishing.

Error and motivation in driving

Oliver Carsten
University of Leeds
UK

Abstract

From in-depth accident studies, the human component in traffic accident causation is clear. Many studies, including the one at Leeds in the late 1980s, found that human error was the overwhelming contribution to accident causation and further found that the major type of error was the mistake. In cruder interpretations, such findings have resulted in a blame culture in which the driver is blamed for his/her errors or (in an only slightly more enlightened version) we are told that more training will improve the driver. But it has long been clear that many road user errors can be designed out of the system, and the Swedish Vision Zero makes the responsibility for safe design and operation of the traffic system on the part of system managers explicit.

As the probability of making a mistake and the seriousness of the consequences of a mistake get gradually designed out of the traffic system, then arguably violations will start to play a proportionately greater role in accident causation. Certainly there is increasing attention to this area, with concerns about a cluster of accidents involving young drivers in the UK and the European Transport Safety Council's focus on enforcement. So are there ways in which we can design out violations?

This paper highlights some recent work which indicates that such an approach is very promising. One study used as an illustration is an investigation into a Forward Collision Warning System which adapted to driving style. Another set of studies is those on Intelligent Speed Adaptation (ISA), a system which strongly discourages the driver from speeding. The paper examines the effect of experience with such a system on attitudes and speeding intention and also examines the role of ISA as an instrument to fulfil drivers' intentions.

Introduction

An invitation to give a keynote address to a conference such as this one, perhaps leads naturally to refection on why one finds the topic area so fascinating. It has to be acknowledged that I am no a human factors scientist by training. I am neither a psychologist nor an engineer, so I do not come from either of the disciplines that normally provide training in human factors. My background is in social science, so I am perhaps a bit of an interloper at an event such as this one.

On the other hand, human factors is by definition an interdisciplinary activity extending nowadays into areas such as cognitive science (including neuroscience),

In D. de Waard, F.O. Flemisch, B. Lorenz, H. Oberheid, and K.A. Brookhuis (Eds.) (2008), *Human Factors for assistance and automation* (pp. 11 - 32). Maastricht, the Netherlands: Shaker Publishing.

physiology and with a strong current of work on social and organisational aspects. In the driving domain, as in others, there is a need to bring together the cognitive and social aspects of the driving task, which traditionally have tended to be dealt with separately. It is interesting but also somewhat worrying to note here that the human factors and (mainly social) traffic psychology research communities are somewhat distinct from each other.

I also have a preference which some might call a bias, for the experimental approach to the study of human factors over the use of questionnaires and other methods. Manipulation — the creation of alternative and novel scenarios or situations — allows us to learn how to make advances in a way that no amount of questionnaire data ever will. Humans are not very reliable predictors of their own behaviour, particularly for situations with which they are unfamiliar. Experiments allow us to build up knowledge incrementally and can be applied not merely to study new treatments but also to test theoretically-based hypotheses and models.

Figure 1. Road scene in Hanoi

My chosen application area, road traffic is fascinating because it is a microcosm of human behaviour. It encompasses all the aspects of human factors — motivation, learning, training, experience, skills, aging, impairment and interaction with the vehicle, in-vehicle systems, other road users and the traffic and road environment. Compared to say commercial aviation, there are a very large number of degrees of freedom in the road traffic environment. This can be illustrated by reference to Figure 1 which shows a fairly typical situation in a less-developed country, in his case Vietnam. It can readily be seen that the situation is somewhat chaotic, although at the same time, because of low speeds, not incredibly unsafe. In road traffic, compared to other transport modes, there is large variability in the population, there is a mix of modes which require formal training and testing and those that do not (e.g. walking and cycling), there are an enormous variety of situations as well as large variation in the quality of infrastructure design, there is a mix of vehicles ranging from bicycles to large trucks and there is a comparative lack of regulation and management, so that most drivers can make individual choices about their behaviour. Even violation is tolerated in a way that would be inconceivable in other modes.

My paper has two interlocking themes. The first is human error and its sources; the second is the application of technology to address error. Humans are flexible but fallible, whereas machines are consistent and can be made to be (generally) reliable. Humans are highly error-prone: "The human brain is full of error", said Steven Rose, Professor of Human Biology at the Open University on a radio discussion programme in 1997. And not just humans — pairs of male deer routinely die because they lock horns with each other in a wrestling competition over a female and are subsequently unable to break loose.

Errors and violations

In a frequently cited study, Parker et al. (1995) emphasised the role of violations (intentional errors) as contributing to accident involvement. The method applied was the use of the Driver Behaviour Questionnaire (DBQ) combined with self-reports on accident involvement for 1656 drivers. The respondents included a specific group of 182 drivers who were known to have been involved in two or more traffic accidents in the previous three years. The results confirmed the distinction drawn in an earlier study (Reason et al., 1990) between errors (deliberate violation of rules), errors or mistakes (incorrect plans) and slips and lapses (correct plans carried out with a minor fault).

Parker et al. (1995) draw an important distinction between the sources of the two major forms of aberrant behaviour. One has a cognitive source; the other can be attributed to social factors:

> Errors may be understood in relation to the cognitive function of the individual. Violations, however, are a social phenomenon and can only be understood in a broader organizational or societal context. Errors can be minimized by retraining, redesign of the human-machine interface, memory aids, better information and the like. Violations should probably be dealt with by attempting to change attitudes, beliefs and norms, and by improving the overall safety culture.

And they conclude that the two types of aberrant behaviours have very different impacts in terms of accident involvement:

A clear link has been established between the self-reported tendency to commit violations and accident involvement. Even after the effects of exposure, age and gender had been partialled out, tendency to commit driving violations proved to be a statistically significant and positive predictor of accident involvement. At the same time, no evidence was discovered for a systematic association between reported error-proneness and accident involvement.

A somewhat contrasting picture emerged from the in-depth study of urban injury accidents carried out in Leeds in the late 1980s (Carsten, Tight and Southwell, 1990). As shown in Table 1, violations in the form of attitude problems constituted only a small proportion of the contributory factors coded as explanations for the events that immediately precipitated the investigated accidents. Unintentional errors of the cognitive type as distinguished by Reason et al., 1990 and Parker et al. (1995) were the overwhelming contribution to accident involvement.

Table 1. Major explanatory factors for driver accident-precipitating errors

Error category	Percentage of contributory factors coded
Perceptual error	16%
Unable to see	12%
Cognitive (judgement) error	12%
Lack of skills	3%
Attitude problem	2%

Thus we can conclude that cognitive and perceptual errors are far more frequent in accident causation. An explanation for the Parker et al. findings may be that, in terms of *individual* differences, inclination to violate is a strong predictor of accident proneness. On the other hand, when looking for example at *between-site* risks, then the cognitive and perceptual problems will tend to predominate. Therefore, from a system perspective it is very rewarding to invest in engineering measures that reduce such errors. We can also conclude that violations are highly risky and likely to be a factor in more serious accidents.

It is interesting to note that speeding can be either a violation or an error, in other words either intentional or unintentional, with unintentional speeding resulting, for example, from failure to perceive a speed limits sign or failure to notice that one's speed has gradually increased. Reason et al. (1990) distinguish between unintentional and intentional violations. They terms the former "erroneous or unintended violations and the latter "deliberate violations". Cognitive factors provide the explanation for erroneous violations, whereas motivational factors lie behind deliberate violations. The prevention of the two kinds of violations may require different approaches. It is clear for example that police enforcement is not the most effective approach to addressing unintentional speeding, even though it can be argued that higher levels of enforcement could have an effect on such violations by arising drives' anxiety levels about being caught and so causing them to devote greater effort to speed compliance. No amount of speed enforcement can help driver detect a speed sign that is obscured by foliage like the one shown (or concealed) on the left-hand side of the road in Figure 2.

Figure 2. Obscured speed sign on UK road

The fascination of new technology

New technology can liberate the system designer from traditional constraints — think of the late-nineteenth-century innovators who were able to exploit the invention of the elevator and steel-frame construction to create the first skyscrapers. With new information and communication technologies, we can create entirely new information and driver assistance systems to aid, monitor and advise the driver and to intervene in vehicle control. The sensors behind these systems can capture information from the road and traffic environment or from in-vehicle databases including digital maps as well as from the driver himself as in the case of eye movement cameras. It has been proposed to create systems that use real-time physiological monitoring of driver brain activity while driving (e.g. Reddy et al., 2007).

Speed and Intelligent Speed Adaptation

One straightforward use of new technology is to overcome the limitations imposed by traditional technologies. Speed management can provide some good examples. Speeds signs were once a new idea and indeed one can find photographs of groups of people gathering to observe the erection of these new-fangled inventions (see Figure 3). The technology of a sign attached to a post which was mounted next to the roadway was innovative then, but is hardly cutting-edge today.

Figure 3. Erection of a 30 mph sign in 1937

Speed management is critical to road safety. Arguably speed is the most important factor in road risk. Without motion, there would of course be no crashes and driving like other forms of motion is not exempt from Newtonian physics in which the energy released in crash goes up with the square of speed. Thus speed is a factor in all crashes and the more severe a crash the more crucial is the role of speed. In other words, speed affects both the risk of crash occurrence and the severity of crash outcome. The review on the relationship between speed and accidents conducted by Finch et al. (1994) concluded that, for every 1 km/h change in the mean speed of traffic on a road, the risk of an injury accident occurring changes by 3%. This relationship can be treated as an approximation — with the actual relationship depending on country, road type, road quality, etc. — but nevertheless, as concluded by Elvik et al. (2004), "it is difficult to think of any other risk factor that has a more powerful impact on accidents or injuries than speed."

In addition to being probably the most important factor in determining crash risk, speed is also central in determining the task demand imposed by driving. Ray Fuller in the latest version of his model of driver behaviour, places speed choice as the principal means through which drivers manipulate the task difficulty of driving and thus their ability to remain in control of the situation (Fuller, 2005). Reduced speed brings with it reduction in workload, greater situation awareness, increased decision time and a reduced probability that error by a driver or other road user will result in a collision.

Speeding can be attributed to both intentional violation and unintended violation. The problem of intentional speeding is not helped by the primitive means with which we exhort compliance, which has not moved very much beyond the messages shown in Figure 4. Nor is it the problem of unintentional speeding helped by the antiquated means used to transmit speed limit information from the roadside to the driver.

Figure 4. Encouragement to drive at a safe speed

New technology does offer an alternative in the form of Intelligent Speed Adaptation, the system whereby speed limit information is brought inside the vehicle. The term ISA was coined by Brookhuis and de Waard (1996) but the concept has been around even longer. The knowledge of the speed limit can be used to display the current speed limit to the driver and to provide warnings when the vehicle is exceeding the speed limit. The information can also be linked to the vehicle engine management system and, if desired the brakes, so as to keep the vehicle below the speed limit when the limit is known. An override to allow the driver to exceed the limit when desired can be provided.

The technology behind ISA is typically the use of an enhanced navigation system, which in addition to all the typical information about road layout and features also incorporates speed limit as a road attribute. There is no need for a driver to enter journey start point or end point into the system; rather the ISA system automatically detects the road on which the vehicle is travelling and hence the speed limit. A special display is provided to show the current speed limit and system status.

Trials of ISA have been carried out in Sweden, the Netherlands, Denmark, Belgium, France, Australia and the UK. Here I will give a flavour of what we found in the UK trials.

The objective of the car trials in the "ISA-UK" project was to investigate how drivers would behave in everyday driving with a car equipped with voluntary ISA. The major focus was naturally on their speed choice — would ISA reduce the amount of speeding and if so by how much, and would it affect their speed choice across the range of speeds or only in terms of curtailing excess speed (i.e. speeds above the limit). Other issues were:

- What would be the acceptance of ISA?
- How would the experience of driving with ISA affect driver attitudes?
- When and where would drivers choose to override the voluntary ISA?
- Would behaviour with and attitudes to ISA vary by type of driver?
- Would compliance with ISA vary by type of road?
- How would drivers assess the impact of ISA on the quality of their driving?

A fleet of twenty cars was converted to provide ISA support, and these vehicles were used in four successive field trials. The participants in the trials drove the converted car on a daily basis for six months; the first month driving without ISA, the next four months driving with ISA, and the final month driving without ISA once again. The first month of driving served as a baseline for comparison with the ISA activated period, and the final month of driving provided the opportunity to identify any impact of experience with ISA on subsequent driving.

The four trials were:

Trial 1: Leeds area with private motorists
Trial 2: Leeds area with fleet motorists
Trial 3: Leicestershire with private motorists
Trial 4: Leicestershire with fleet motorists

The Leeds trial was in a major urban area, although the speed limit data covered the whole of the Leeds Metropolitan District, which includes some outlying rural areas and villages. The Leicestershire area was mainly rural and small-town.

A fleet of 20 Skoda Fabia Elegance 1.4 litre estate was converted for the trials. They were fitted with a voluntary (i.e. overridable) intervening ISA system. During the ISA-active four months of each trial, the ISA system defaulted to providing speed limiting through intervention in the throttle and where needed mild application of the brakes when driving on any road whose speed limit was known. The speed limit database covered all roads in the local area and all the national highways. The driver could override the speed limiting function by depressing the accelerator pedal to its full extent or by pressing an opt-out button on the steering wheel. Speed limiting would be resumed by one of the following: pressing the opt-in button on the steering wheel, bringing the car back below the speed limit or entering a new speed limit zone. A display showed the speed limit (where known) and system status. Changes in speed limit and in system status were signalled by audio alerts. To discourage the driver from demanding too much throttle, a vibrating motor fitted to the accelerator pedal gave tactile feedback when the driver demand exceeded the calculated maximum throttle demand by 40% or more.

Participants for the field trials with private motorists were recruited in response to adverts placed in local newspapers. Participants for the fleet trials were recruited from local organisations — in Leeds from employees of Leeds City Council (LCC), and in Leicestershire from various local authorities as well as a private company. Because one participant had to withdraw during the fourth trial, complete data were only collected for 79 participants.

Within each trial the aim was to balance the number of participants equally across various driver characteristics: male/female, young (25–40) or old (41–60), and intender/non-intender (based on prior intention to speed as defined by an attitudinal questionnaire using the Theory of Planned Behaviour of Ajzen, 1988). It proved impossible to recruit the intended balance of drivers across all the trials. Therefore the final combination of participants was as shown in Table 2.

Table 2. Characteristics of participants

Gender	Age Group	Intention to Speed	Number
Male	23–39	Intender	11
Male	23–39	Non-Intender	8
Male	40–60	Intender	13
Male	40–60	Non-Intender	12
Female	23–39	Intender	6
Female	23–39	Non-Intender	11
Female	40–60	Intender	11
Female	40–60	Non-Intender	7

So did the ISA system affect speed choice? Speeds were compared across trial phase with Phase 1 being the one month baseline with no ISA, Phase 2 the four months of driving with ISA and Phase 3 the final month again with no ISA. The analysis examined speed choice and the propensity to speed by road category, with road category being defined as speed limit. ISA had an impact on the amount of speeding across all speed limits. The only exception was 60 mph roads where, in line with national observations of traffic speeds, there was very little speeding in the before-ISA phase. The typical pattern was for speeding to reduce in Phase 2 as compared to Phase 1, and then for there to be at least a partial return to the baseline behaviour in Phase 3, resulting in the V-shaped pattern which can be seen in Figure 5. This figure shows mean speed within each trial phase. The effect of ISA on speed choice can be observed even more clearly from Figure 6 which shows 85th percentile speed, i.e. the speed below which 85% of the distance travelled was covered. ISA affected high-end speeds much more than it affected average speeds.

The impact of ISA was even more pronounced when looking at the relative amount of speeding in the three phases, as shown in Figure 7. With ISA, there was a reduction in the proportion of distance travelled over the speed limit for all speed limits apart from 20 mph. This reduction was smaller on 60 mph roads. However, the impact of ISA did not generally carry over to driving in Phase 3.

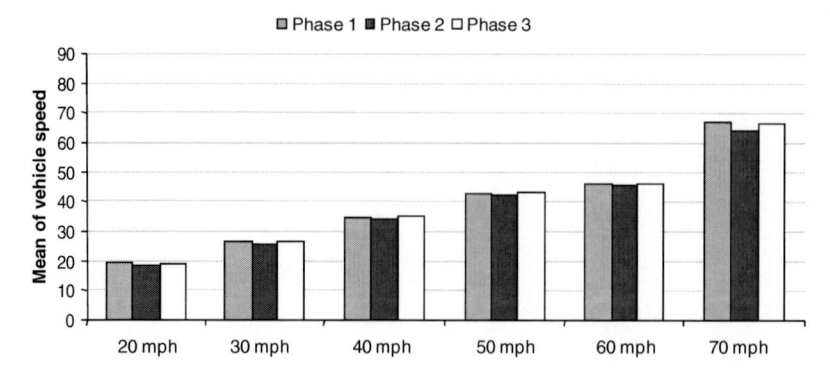

Figure 5. Comparison of mean speed across trial phases

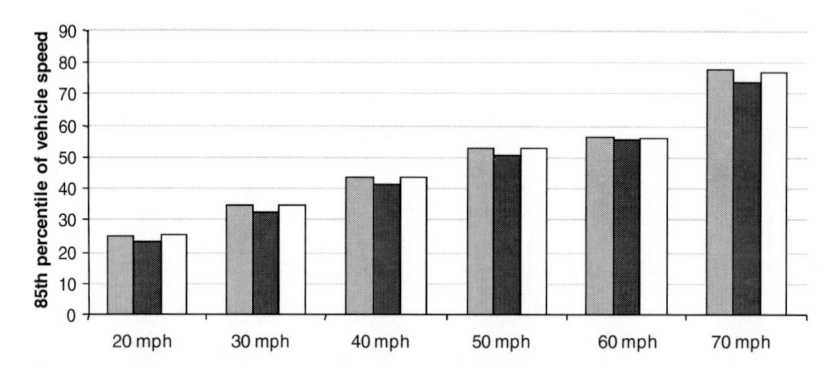

Figure 6. Comparison of 85th percentile speed across trial phases

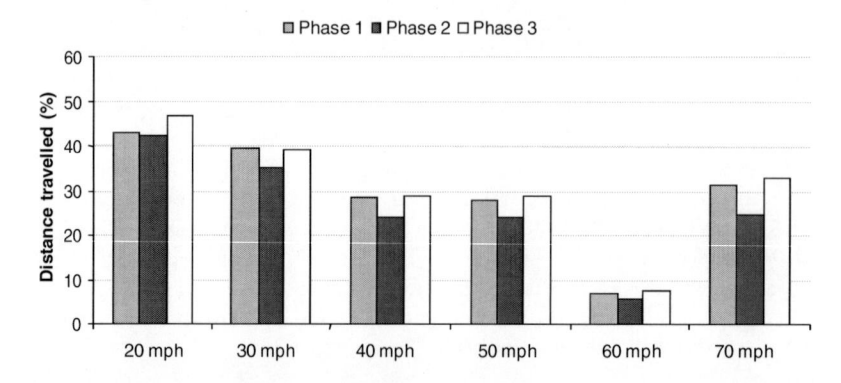

Figure 7. Comparison of percentage of distance travelled over speed limit across trial phases

It is also clear that ISA did not eliminate speeding. This was in part because the drivers could override the ISA system and they chose to do so to a lesser or greater extent. But it was also in part due to the design of the ISA control of vehicle top speed. The ISA system did not cut off acceleration sharply at the speed limit. There was some lag in the reaction of the system, drivers could speed somewhat when going downhill and could also speed for a short while after entering a lower speed zone. But ISA did have a very marked effect on very large exceeding of the speed limit. This can be observed when examining more detailed plots of speed choice.

Such plots are shown in Figures 8 and 9. In each of the plots the proportion of distance travelled within each phase of the trials is shown for 5 mph speed ranges. From Figure 8, for 30 mph (50 km/h) roads, it can be seen that ISA had no impact on the speed of driving at the lower end. However, it had a very pronounced effect on speeds at the top end. Driving with ISA also produced a bulge in the distribution just below and just above the limit. A very similar impact can be seen in Figure 9 for 70 mph (110 km/h) roads, i.e. motorways and high-speed dual carriageways. Here there are even some signs that the impact of ISA, in terms of curtailing very high speeds, persisted after the ISA system was switched off.

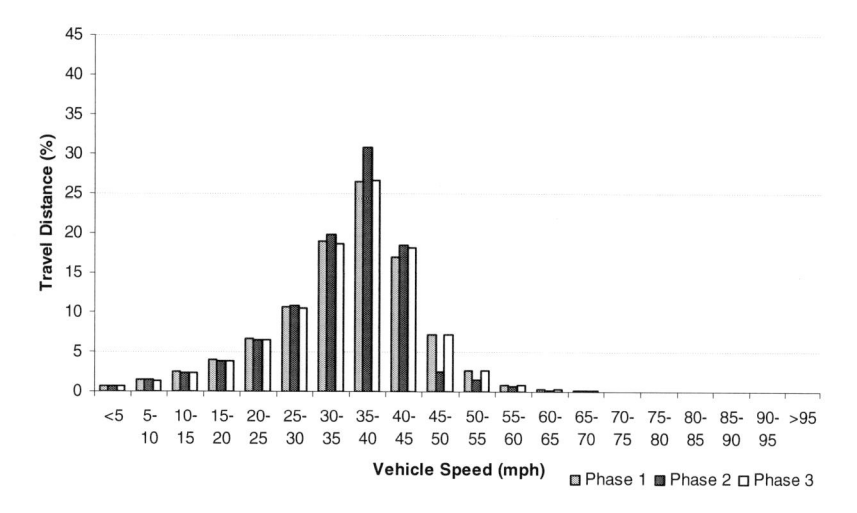

Figure 8. Speed distribution by phase on 30 mph roads

The use of a voluntary ISA system also provides an opportunity to examine where drivers were willing to accept the control of the ISA system and where they chose to override it. The proportion of distance travelled with ISA overridden was highest on 70 mph roads as can be seen in Figure 10. This in general overriding propensity was greatest when surrounding traffic tended to be speeding as is the case on UK motorways.

An important issue with ISA as with other safety systems is whether it only affects the behaviour of the safety-conscious subgroup or whether it affects the behaviour of all drivers including those who are not particularly receptive to safety messages. In the case of this study, the recalcitrant group was the speed intenders. Figure 11, for 70 mph roads, demonstrates that the ISA system had a dramatic effect on the speed choice of speed intenders. Even these drivers chose to override the system only for 20% of this distance travelled on such roads.

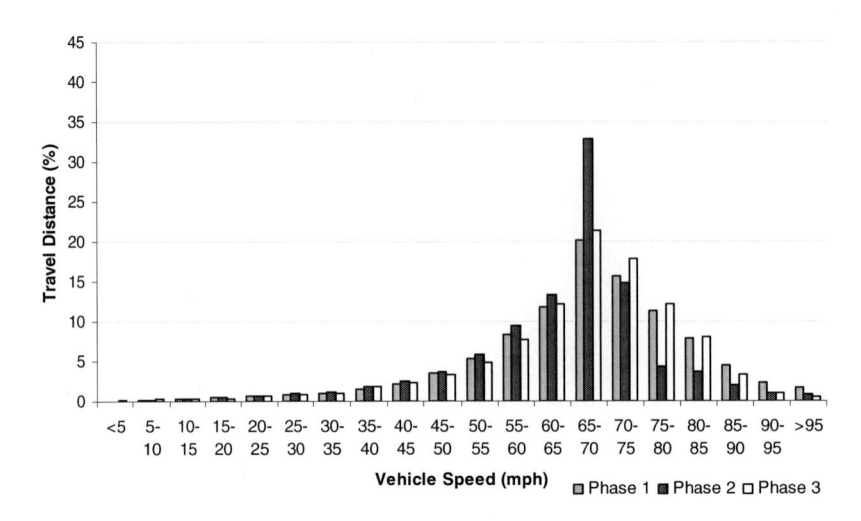

Figure 9. Speed distribution by phase on 70 mph roads

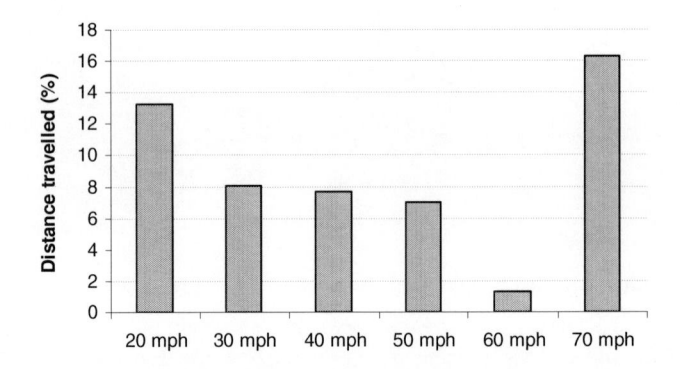

Figure 10. Proportion of distance travelled with ISA when the system was overridden

Figure 12 examines the propensity to override the ISA by trial. There was a striking difference in behaviour between the private motorists and the fleet drivers: private motorists overrode more frequently than fleet drivers on urban roads, while fleet drivers overrode more frequently than private motorists on 70 mph roads. This suggests that the fleet drivers may have been more conscious of the need to comply with the speed limits on urban roads, but felt less compunction about speeding on 70 mph roads. Speeding on high-speed roads also saves more journey time which may be relevant in work-related driving.

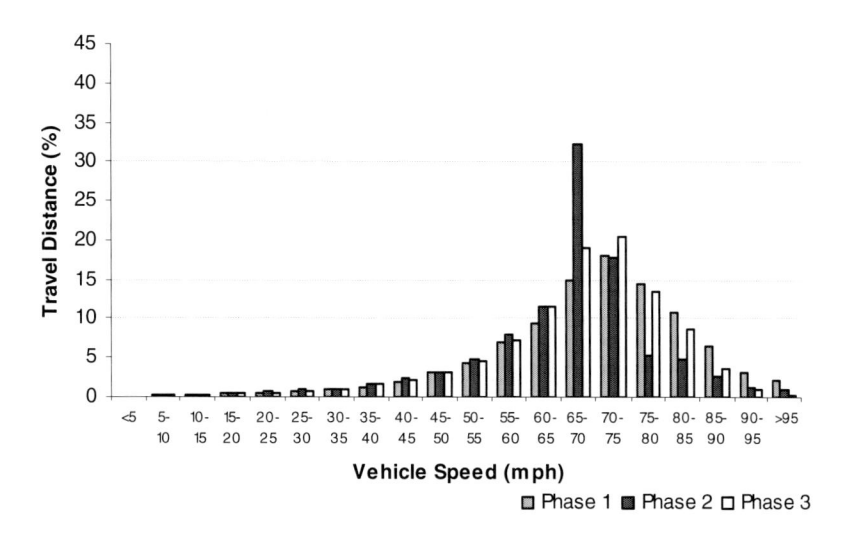

Figure 11. Speed distribution for speed intenders on 70 mph roads

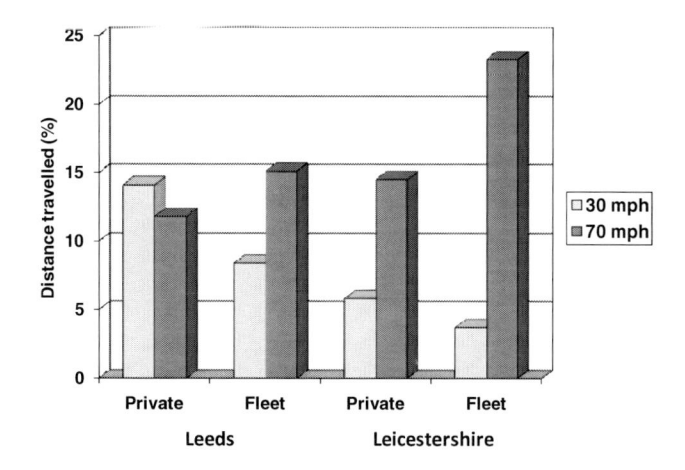

Figure 12. Overriding of ISA by trial

Finally, the log-term trials permitted the tracking of attitudes over time. Intention to speed, as measured by the Theory of Planned Behaviour questionnaires, changed over time. Intention to speed was negative even prior to driving with ISA in Phase 1, but became more negative when driving with ISA in Phase 2 and was most negative following withdrawal of ISA in Phase 3. Thus it can be argued that ISA had a calming effect on attitudes towards speed and that this effect persisted beyond the period of ISA driving.

Changes can also be observed in the correlations between intention and actual speeding behaviour. The correlation between Phase 1 (prior to ISA) intention and behaviour during Phase 2 (with ISA) was 0.24. The correlation between Phase 2 intentions and Phase 3 (after ISA) behaviour was 0.15. This indicates a disjuncture in Phase 3 between intentions and behaviour. The explanation may lie in the fact that withdrawal of ISA made it harder for the drivers to carry out their intentions, in part because it is simply easier not to speed with ISA and in part because not having ISA leads to more unintentional speeding. Thus ISA can truly be termed a driving assistant system — it helps driver to realise their intentions.

The promise of adaptive systems: an experiment with an adaptive Forward Collision Warning system

As stated earlier, with new sensors and new electronic systems, the engineer now has the freedom to create totally new systems. One new freedom is to create adaptive designs that can observe the behaviour of the user and adapt the system response to the "normal" observed behaviour of the individual. Such adaptation may not be desirable — we would not want an ISA system that never pointed out to the habitual speeder that he or she was speeding. But appropriate adaptation that can respond in subtle ways to the observed responses of the driver may increase system acceptance and thus system usage. But of course the added value of such adaptivity must be proven. In the context of the European AIDE (Adaptive Integrated Driver-vehicle InterfacE) project, we sought to create and evaluate a forward collision warning system that would adapt to the driving style of individual drivers. The starting hypothesis was that drivers would not appreciate a system that acted as a kind of "virtual mother-in-law" by warning them when they were already aware that they were closing in on the lead vehicle and were about to brake.

The system

The Leeds Driving Simulator was used for the study. The simulator was based on a complete Rover 216GTi, with all of its driver controls and dashboard instrumentation still fully operational. Forty-five drivers took part (23 males, 22 females) with a mean age of 37.4 (SD = 13.9) years.

The driving simulator was equipped with a virtual FCW system based on the ISO-recognised Stop Distance Algorithm (SDA). The SDA defined a warning distance, based on the difference between the stopping distances of the leading and following vehicles. If the distance between the two vehicles was less than the warning distance, an auditory collision warning alarm was presented to the driver. It is expected that the SDA will be introduced as the main alarm trigger logic in the design of future collision warning systems (Wilson et al., 1997). The SDA was defined as follows:

$$D_w = (V_{driver} \cdot T_{driver}) + \left(V^2_{driver} \Big/ 2 d_{driver} \right) - \left(V^2_{lead} \Big/ 2 d_{lead} \right)$$

Where D_w [m] = warning distance
V_{driver} [m/s] = speed of following (simulator) driver
T_{driver} [s] = the time assumed for a driver to react to an event
d_{driver} [m/s^2] = assumed deceleration of the following vehicle
V_{lead} [m/s] = speed of the leading vehicle

d_{lead} [m/s^2] = assumed deceleration of the lead vehicle during the event

The SDA had three fixed parameters: T_{driver}, d_{driver} and d_{lead}. The real-time speeds of the two vehicles (V_{driver} and V_{lead}) varied as the simulation progressed.

Two different types of FCW were simulated: non-adaptive and adaptive. In the both systems, the fixed deceleration parameters of the SDA (d_{driver} and d_{drone}) were selected as 5.0 m/s^2. The main difference between the two systems was in the use of driver's reaction time (T_{driver}). In the non-adaptive system this was fixed at 1.5s. The adaptive system used an individual driver's brake reaction time, measured in the driving simulator. When the simulator driver encroached on the lead vehicle, following at a distance less than the warning distance, an auditory FCW was presented.

Training of the system

Following simulator familiarisation, adaptive FCW system training was undertaken. For this, participants also found themselves in a repeatable braking scenario designed to allow the measurement of their individual brake reaction time. The virtual environment was rural with alternating 500m straight and curved sections. Drivers were required to maintain 50mph (80km/h) and obliged to follow a lead vehicle. This lead vehicle attempted to maintain a 1s headway in front of the simulator driver using a speed controller. Feedback on participant's speed choice was given in form of a coloured overlay over the computer-generated visual display. If the driving speed was within ±5% of the required 50mph, the overlay disappeared. A scenario in which the lead vehicle braked was choreographed at every other straight section (i.e. approximately every 2km). In the braking scenario, the lead vehicle slowed at a deceleration of 4m/s2 until it reached 5mph (8km/h), continuing at this speed for 10s before accelerating back to 50mph (80km/h). The scenario was only "triggered" if the driving speed (50mph ± 2.5mph) AND the following headway (1s ± 0.05s) were within tolerance, plus the simulator driver did not have the accelerator pedal released. The choreographed simulation ensured that the braking scenario was identical for every participant and that reaction time was measured from accelerator release to brake activation. The braking scenario was repeated six times during system training and at its conclusion the mean brake reaction time for each participant was recorded. Each individual's brake reaction time was used for the parameter $T_{reaction}$ in the SDA of the adaptive FCW system.

Data collection

Since a repeated-measures design was used, participants were required to complete three separate drives during experimental data collection, once with no FCW system, once with the non-adaptive system and once with the adaptive system. FCW system type was counterbalanced across the participants. The participants did not know in which of the two system conditions they were driving.

Each drive was made up of eight, 5km sections of rural road (total road length 40km). Each section was made up of alternating 400m long straight segments and

300m-400m curved segments. The curved segments varied in radius between 400m and 1100m. The posted speed limit of the road was 50mph (80km/h). A car following scenario was introduced early in the drive with the lead vehicle maintaining a fixed 50mph. On-coming vehicles, on average every 4s, made overtaking challenging but if an overtaking attempt was successful, the lead vehicle was "re-cycled" (re-introduced ahead of the simulator driver). Participants were instructed as follows: "… you will be required to follow a lead car. We would like you to follow the lead car as closely as you feel comfortable – just imagine that you are in an absolute hurry to reach the destination."

Each experimental data collection drive consisted of six "expected" braking events, one per each 5km section. During an "expected" event, the lead vehicle also slowed at a deceleration of $4m/s^2$, again reaching 5mph for 10s and accelerating back to the nominal 50mph. "Expected" events were scripted to occur only during five of the eight 400m-long straight segments per section, pre-selected to minimise learning effects. The braking event was "triggered" if simulator driver's headway to the lead vehicle was between 1s and 3s AND the simulator driver had the accelerator pedal depressed by at least 5%. If these constraints had still not been met by the end of the fifth and final pre-selected straight segment, the braking scenario was presented regardless.

After the sixth and final "expected" event had occurred, the lead vehicle left the main roadway by slowing and pulling off the road. Throughout the virtual road network, parked vehicles waited in around 50% of the "lay-bys". Whilst the participant made his/her way unhindered along the seventh 5km section, one of these parked vehicles began to move and pulled onto the roadway, accelerating slowly at $0.5m/s^2$, in front of the simulator driver. To minimise learning effects, this first "unexpected" event occurred randomly during one of the eight 400m-long straight segments situated within the 5km section. This "unexpected" event occurred when the simulator driver had a time-to-contact of 7s to the parked vehicle. On-coming vehicles prevented the simulator driver from simply swerving around the new lead vehicle, thus forcing him/her to brake. The lead vehicle continued to accelerate slowly to the nominal 50mph onto the eight and final 5km section. Using the same logic as for the "expected" braking events (straight segment, 1s-3s headway and accelerator pedal depressed), a second "unexpected" event was activated such that the lead vehicle slowed sharply without showing any brake lights.

Following each experimental session, several questionnaires were administered. Completed questionnaires gave self-reports for:

- FCW alarm timeliness and frequency (rated on a five-point Likert scale anchored at too early/too late and too few/too often.
- Rating Scale of Mental Effort using the FCW (Zijlstra, 1993).
- User acceptance of the FCW (Van der Laan et al., 1997).
- Trust in the FCW (Lee and Moray, 1992).
- Personal factors affected by the FCW (safety, irritation, stress, feeling of being controlled, joy of driving, attentiveness in traffic)

After each experimental session, during which a participant had interacted with one of the two types of FCW system, several questionnaires were administered. The first was on timing and appropriate frequency of the FCW alarms, rated on a five-point Likert scale. The questions were posed as follows:

"What did you think of the timing of the alarms?" (too early – too late).

"What did you think of the frequency of the alarms?" (too few – too often).

Self-reported mental effort was assessed using the Rating Scale of Mental Effort (Zijlstra, 1993). User acceptance was assessed with using the Van der Laan scale (Van der Laan et al., 1997). System trust was assessed by using visual analogue scales over a range from 0-100 (strongly disagree – strongly agree) for a number of questions such as "I trust the system", "The system is reliable" and "I have confidence in the system".

Behavioural results

Two main driver behavioural measures were used. The first was Brake Reaction Time — the time between the illumination of the lead vehicle's brake lights (onset of braking) and the application of some brake pedal effort. The second was Minimum Headway — the minimum value of time headway recorded during the complete event.

The main between subjects factor was individual brake reaction time. The individual brake reaction time groups were split based on the RT recorded during the adaptive system training drive. The sample was normally distributed with a mean score of 0.967s and a median score of 0.937s. Based on this evidence, preferred headway was split into *fast* (< 0.937s) and *slow* ($\geq$ 0.937s).

Figure 13 shows the sample mean brake reaction time from the six expected and two unexpected braking events. The reason for the large difference in brake reaction time is due to the definition of the start of each braking event. Even though the expected braking events were more numerous, the unexpected events suggested a much more powerful effect of system (*expected*: $F(2,86)=2.79$, $p=0.067$; *unexpected*: $F(2,86)=10.7$, $p<0.001$). The greater power demonstrated by the unexpected events was consistent across all the behavioural measures and so only these events are included subsequently.

Figure 14 shows the effect of driver style on brake reaction time. There was no significant main effect or interaction. It can also be observed that there was little difference between the effectiveness of the two systems. The adaptive system, which might have been expected to perform worse because it gave a warning later to those who exhibited short reaction times, in fact performed about as well.

The results in terms of minimum headway were similar, as can be seen from Figure 15. The adaptive system perhaps performed slightly worse, but the differences were not significant.

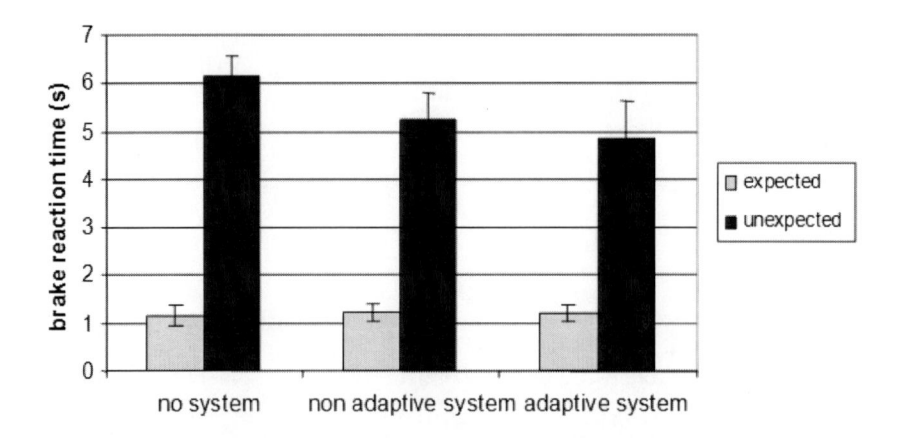

Figure 13. Brake reaction time for expected and unexpected events

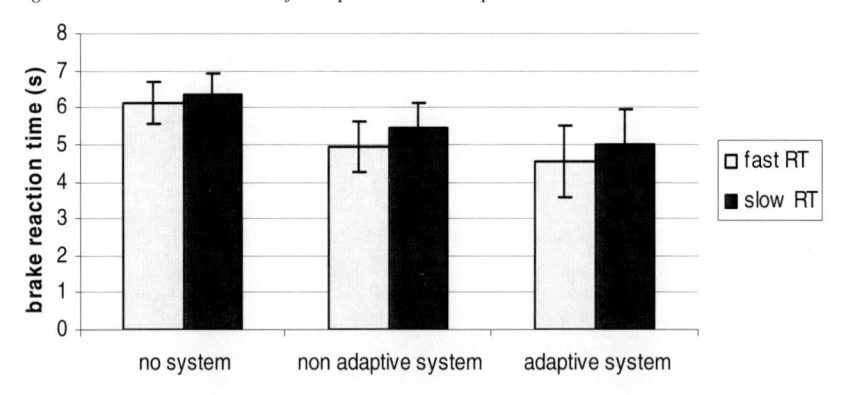

Figure 14. Brake reaction time by driving style

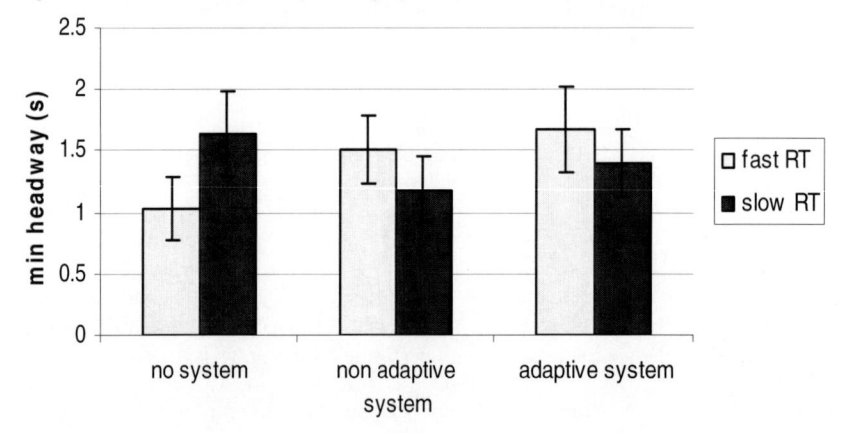

Figure 15. Minimum headway by driving style

Attitudinal results

The results on Rating Scale of Mental Effort indicated no difference between the two types of FCW. In terms of acceptability, participants gave positive feedback on the usefulness of both FCW systems, but rated them less than satisfactory. There was also a significant interaction between system type and driving style for both usefulness, $F(1,43)=10.0$, $p=0.002$, and satisfaction, $F(1,43)=5.83$, $p=0.02$. Slow reactors did not discriminate much between the two systems, whereas fast reactors preferred the adaptive system. The adaptive system generally scored better in terms of the various trust scales (see Figure 16), although differences were not significant. So it can be argued that the adaptive system was more acceptable overall, while being no worse in terms of safety performance.

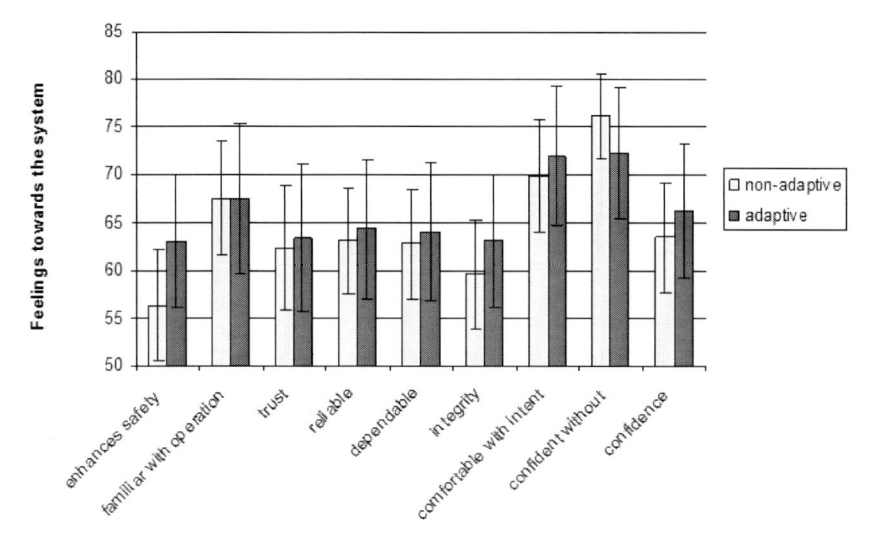

Figure 16. Trust ratings by FCW type

Discussion of FCW results

The adaptive system was clearly less objectively safe in engineering terms in that it gave some drivers a late warning. However that same system was preferred, especially by the drivers with short reaction times. The explanation lies in the fact that these drivers did not receive annoying "false" alarms, in other words alarms that may have been objectively correct but could be perceived by the driver as inappropriate because they were warning about a situation of which the driver was aware. The outcome of creating a system with thresholds that were objectively less "safe" was increased acceptance without any harm to actual safety margins. It can be argued that it would be better not to cater to drivers' preferences, especially when they tend to be unsafe. But a system that does not gain acceptance will not be used and will therefore not be effective.

Conclusions

Motivation is clearly a very important aspect of driving. System designers neglect motivation at their peril. Good design is not just a question of producing usable systems; it is also a question of producing ones that people will use. But that does not mean that we should just pay attention to users' initial prejudices about a system. Drivers may dislike a system such as ISA when they first hear about it, but real-world experience may convert them to acceptance. Actual driving with a system can not only change behaviour but also change motivation — ISA appears to have affected not only speeding behaviour but the intention to speed. Thus short-term acceptance may not predict long-term acceptance or long-term attitudes.

Another message is that we can gain acceptance and thus obtain willingness to use a system by adapting to individuals' attitudes and driving style. Users appear to appreciate systems that fit their own habitual behaviour. Whether over time it is possible to subtly alter system thresholds, so that the drivers who drive with smaller safety margins are encouraged to adopt greater safety margins, remains to be determined.

Finally, there was a recent advertising campaign on the London Underground that stated: "Machines don't care!" The advertisement was promoting personal banking. It is our task to create machines that do care.

References

Ajzen, I. (1988). *Attitudes, personality and behaviour.* Buckingham, UK: Open University Press.

Brookhuis, K.A. & De Waard, D. (1996). *Limiting speed through telematics: towards an Intelligent Speed Adaptor.* Report VK 96-04. Groningen, The Netherlands: Traffic Research Centre, University of Groningen

Carsten, O.M.J., Tight, M.R. & Southwell, M.T. (1990). *Urban accidents: why do they happen?* Basingstoke, UK: AA Foundation for Road Safety Research.

Elvik, R., Christensen, P., & Amundsen, A. (2004). *Speed and road accidents: an evaluation of the Power Model.* TOI Research Report 740/2004. Oslo: TØI: Institute of Transport Economics.

Finch D.J., Kompfner, P., Lockwood, C.R., & Maycock G. (1994*). Speed, speed limits and accidents.* TRL Project Report 58. Crowthorne, UK: Transport Research Laboratory.

Fuller, R. (2005). Towards a general theory of driver behaviour. *Accident Analysis and Prevention, 37,* 461-472.

Lee, J. & Moray, N. (1992). Trust, control strategies, and allocation of function in human-machine systems. *Ergonomics, 35,* 1243-1270.

Parker, D., Reason, J.T., Manstead, A.S.R., & Stradling, S.G. (1995). Driving errors, driving violations and accident involvement. *Ergonomics, 38,* 1036-1048.

Reason, J., Manstead, A., Stradling, S., Baxter, J., & Campbell, K. (1990). Errors and violations on the roads: a real distinction? *Ergonomics, 33,* 1315-1332.

Reddy, B.R., Basir, O.A., & Leat, S.J. (2007). Estimation of driver attention using Visually Evoked Potentials. *Proceedings of the 2007 IEEE Intelligent Vehicles Symposium*, Istanbul, Turkey, June 13-15, 588-593.

Van der Laan, J.D., Heino, A., & De Waard, D. (1997). A simple procedure for the assessment of acceptance of Advanced Transport Telematics. *Transportation Research Part C, 5*, 1-10.

Wilson, T.B., Butler, W., McGehee, D.V., & Dingus, T.A. (1997) Forward-looking collision warning system performance guidelines. SAE *Technical Paper Series 970456*. Warrendale, PA, USA: Society of Automobile Engineers,.

Zijlstra, F.R.H. (1993). *Efficiency in working behaviour: a design approach for modern tools.* PhD Thesis, Delft University of Technology, Delft, The Netherlands.

Surface Transportation

Effects of aroma of trees on fatigue reduction

In D. de Waard, F.O. Flemisch, B. Lorenz, H. Oberheid, and K.A. Brookhuis (Eds.) (2008), *Human Factors for assistance and automation* (p. 33). Maastricht, the Netherlands: Shaker Publishing.

pulse, reported by Miao (2003), Shimizu (2004), and Sano (1985) for quantitative analysis of pulse fluctuation. The effect of Borneol on wakefulness was investigated with analysis of finger blood volume. Although some studies have addressed the effects of aromas on fatigue reduction, few have focused on the effects of aromas on driver behaviour and on vehicle manoeuvres in particular. This study for clarifying the positive effects of aromas on driver behaviour was conducted in a driving simulator.

Method

Apparatus

This study utilized a driving simulator with real-time calculation of vehicle dynamics and real-time presentation of road scenery using personal computers.

LabVIEW supplied by National Instruments Corporation was used for real-time simulation of vehicle dynamics. Digital Loca supplied by i3Lab Corporation was used to construct image databases of winding urban roads and highways. Two different image databases of roads were made: an urban winding road with a radius of curvature of 250 m for analysis of fatigue reduction, and a straight highway for analysis of wakefulness. In the experiment scenario, a preceding vehicle was set up on the road, and the driver was asked to follow it. A pilot study on fatigue reduction conducted before the main experiment indicated that it would take more than three hours for drivers to become fatigued if a straight highway was used. Therefore, a winding road was used so that the driver would become fatigued more quickly.

To analyze fatigue reduction, drivers were asked to keep a specified time headway to the preceding vehicle (exceeding one second) as an additional driving task. A white line indicating one second ahead of the subject's vehicle was drawn on the road in the computer graphics environment of the driving simulator. The drivers were asked to centre the line on the preceding vehicle to keep the time headway longer than one second.

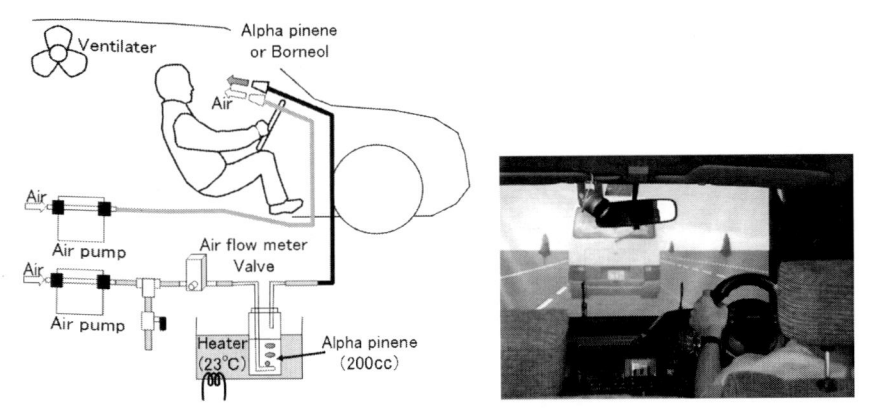

Figure 1. Experiment system and sample image projected on the screen

Effects of tree aromas on automobile drivers' fatigue reduction and wakefulness assessed with finger-pulse fluctuations

Keisuke Suzuki[1] & Shigeki Harada[2]
[1]Daido Institute of Technology
[2]Denso Corporation
Nagoya, Japan

Abstract

The effects of tree aromas on an automobile driver's fatigue level, wakefulness, and active safety (e.g., keeping the appropriate time headway to the preceding vehicle) were investigated using a driving simulator. The fluctuation of the driver's finger pulse was used to analyze fatigue and sleepiness levels. Reaction time to visual stimuli was investigated, and vehicle manoeuvres were assessed with the variation of time margin to the preceding vehicle. It was found that the tree aroma Alpha-pinene minimizes the fluctuation of finger pulse and shortens the driver's reaction time to visual stimuli. Another tree aroma, Borneol, was found to minimize the driver's sleepiness level and thus help prevent swerving of the vehicle. It was concluded that supplying the aroma of trees as an air supplement helps lower a driver's fatigue and sleepiness levels, and thus is beneficial in terms of active safety during car driving.

Introduction

Some analyses regarding the effects of tree aromas on fatigue reduction and wakefulness have been reported by Yatagai (1997, 2000), Sawada (2000), and Homma (2004). For instance, such studies revealed that inhaling Alpha-pinene (a main ingredient of Japanese cedar or cypress) activates the parasympathetic nervous system and relieves mental stress or fatigue; as a result, pulse rate and blood pressure decrease. Additionally, Borneol (a main ingredient of cypress, hemlock, and fir) was found to have an activating effect on the central nervous system and thus to enhance wakefulness.

This study addresses the effects of tree aromas as an air supplement on an automobile driver's fatigue reduction and wakefulness. Focus is on active safety, such as keeping the appropriate time headway to the preceding vehicle to avoid the risk of collision. Driver behaviour was investigated in terms of reaction time to visual stimuli, fatigue level, sleepiness level, and vehicle manoeuvres.

In this study, the effects of Alpha-pinene on reducing mental stress or fatigue caused by continuous long driving was investigated using the Lyapunov exponent of finger

In D. de Waard, F.O. Flemisch, B. Lorenz, H. Oberheid, and K.A. Brookhuis (Eds.) (2008), *Human Factors for assistance and automation* (pp. 35 - 48). Maastricht, the Netherlands: Shaker Publishing.

The actual cabin of a compact car was used in setting up the cabin (capacity of 3.0 m³) of the driving simulator, and red LEDs were installed around the cabin to analyze reaction time to visual stimuli. A ventilator and a valve were also installed. For the without-ventilation condition, the valve was closed and the cabin was kept airproof. Figure 1 presents the experiment system and an example of the image projected on the screen.

State variables for evaluating driver behaviour

To evaluate the effect of an air supplement on fatigue reduction and wakefulness, finger-pulse fluctuation, reaction time to visual stimuli, and vehicle manoeuvre (e.g., time headway and time margin before lane deviation as reported by Godthelp (1984)) were investigated. The following section presents the details of these variables.

(a) Finger-pulse fluctuation
Studies using finger-pulse fluctuation to analyze fatigue level and mental stress have been reported by Miao (2003), Shimizu (2004), and Hashimoto (1998). The present study used the Lyapunov exponent reported by Miao (2003), Shimizu (2004), and Sano (1985).

Shimizu's study on the relationship between mental stress and the Lyapunov exponent can be summarized as follows. When the driver does not feel any mental stress just after the experiment starts, the Lyapunov exponent is small. In this condition, the three-dimensional trajectory of pulse, which can be drawn by transforming the time series data of pulses (Figure 2(a)) into a three-dimensional time delay coordinate (Figure 2(b)), follows the same course, and the fluctuation of the trajectory is relatively small. In this study, the Lyapunov exponent was calculated by applying the Sano-Sawada method (Sano and Sawada, 1985).

The Sano-Sawada method to calculate the Lyapunov exponent can be summarized as follows. The Lyapunov exponent is a state variable for quantifying the instability of trajectories. In the calculation of the Lyapunov exponent, the size of the hyper sphere is a dominant factor for quantifying the fluctuation of trajectories in three dimensions. The size of the hyper sphere folding the trajectories in three dimensions is changed after the specified expansion time (Figure 2(c)). In this study, the expansion time was set at 0.01 seconds. The average regarding logarithm of the pace of expansion in three directions, considering the size of this hyper sphere, was calculated every second. The average of three values in three directions (e_1, e_2, and e_3) was determined as the Lyapunov exponent.

Figure 2(d) illustrates the three-dimensional trajectories of pulse and the Lyapunov exponent in two different mental conditions. One is the trajectory when a driver is relaxed and does not feel any mental stress; the other is a trajectory when the driver feels mental stress in a situation after driving in a driving simulator for 90 minutes. When the driver feels mental stress, the Lyapunov exponent is greater than when the driver feels no mental stress.

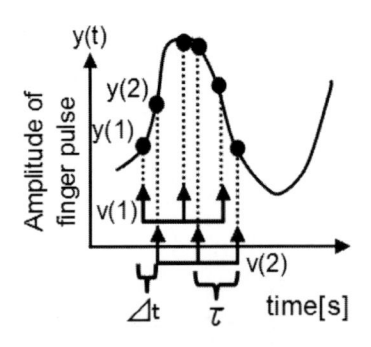

38   Suzuki & Harada

(a) Time series data of pulse.

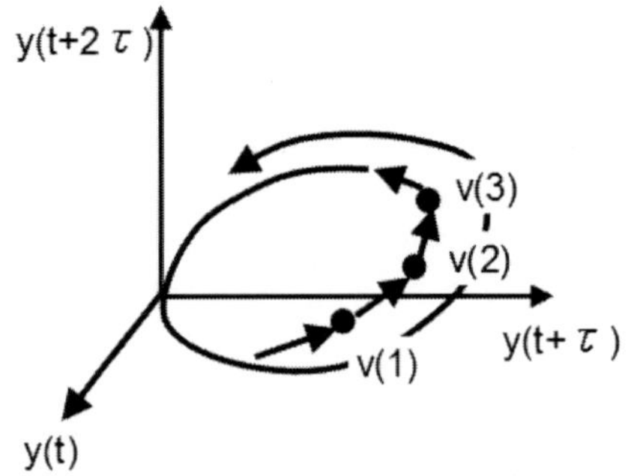

(b) Three-dimensional trajectory of pulse

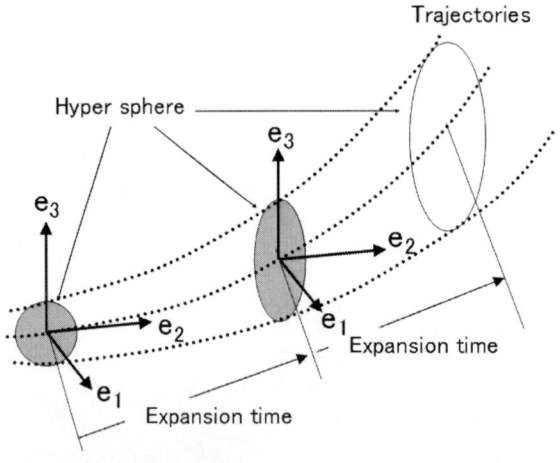

(c) Method for calculating the Lyapunov exponent considering the fluctuation of the size of the hyper sphere including trajectories in three dimensions.

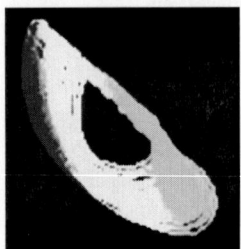

Low fatigue level
Average of the Lyapunov
exponent=2.0
(before driving)

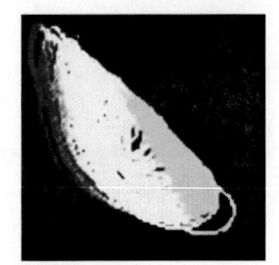

High fatigue level
Average of the Lyapunov
exponent=4.1
(after 90 minutes of driving)

(d) Examples of three-dimensional trajectories in different mental conditions

Figure 2. Procedure of transformation from time series data into three-dimensional trajectories

effects of tree aromas on fatigue 39

To analyze the effect of an air supplement on wakefulness, the amplitude of a time series of finger pulses was investigated every 30 seconds. It is generally accepted that finger-pulse amplitude increases as sleepiness level increases. For instance, an infant's hands become warmer as he/she becomes sleepy because of the increase of blood in his/her hands. In this study, the size of the three-dimensional trajectory (i.e., the averaged distance between each trajectory and the centre of gravity, Figure3(a)) was used to quantitatively estimate the sleepiness level of drivers. Such an investigation is similar to the analysis of the blood flow volume of finger pulse.

For instance, the trajectory becomes larger when the driver's sleepiness level is higher. Figure 3(b) illustrates two different kinds of experiment data. When the sleepiness level is high (right), the trajectory is much farther from the centre of gravity than when the sleepiness level is low (left).

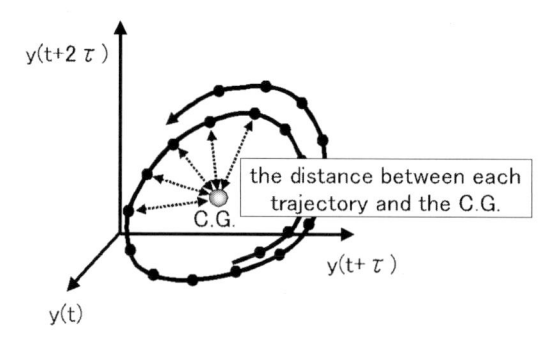

(a) The distance between each trajectory and the center of gravity in the three-dimensional trajectory of pulse

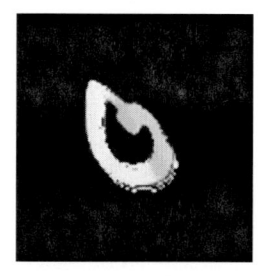

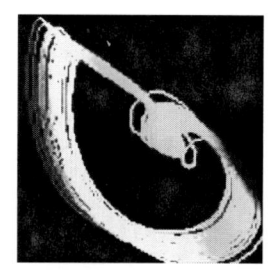

(b) Low sleepiness level   *High sleepiness level*

Figure 3. *Relationship between sleepiness level and the size of three-dimensional trajectories of finger pulse*

(b) Reaction time to visual stimuli
The reaction time to visual stimuli was investigated to quantitatively analyze fatigue and sleepiness levels. It was expected that the driver's reaction time would be prolonged when the fatigue or sleepiness level was high. For this analysis, red LEDs were set up in three positions: in front of the driver's eye point, 45 deg to the right of the centre of the eye point, and 45 deg to the left of the centre of the eye point. The

driver was asked to press the horn button as soon as he/she noticed the lighting of LEDs (the peripheral detection task) during the experiment. The difference between reaction times (i.e., the times between the lighting of an LED and the driver's pressing the horn button) was analyzed before and after the air supplement.

Air supplement

(a) Ingredient
This study analyzed the effect of Alpha-pinene (an aroma of Japanese cedar or cypress) on fatigue level. It has been reported that Alpha-pinene activates the parasympathetic nervous system and thus relieves mental stress or fatigue (Yatagai, 1997). Additionally, this study analyzed the use of Borneol (an aroma of cypress, hemlock, and fir) on sleepiness level. It has been reported that Borneol activates the central nervous system and thus relieves sleepiness (Yatagai, 2000). Each aroma was vaporized with an air flow of 0.002 m^3/min.

(b) Method of supplying the aroma
The olfactory stimulation level is an important factor for this study because the olfactory stimulus is the dominant factor for activating the central nervous system. For instance, if the olfactory stimulation level is high, the effects of aroma on fatigue reduction or wakefulness are large. In this study, three different supply methods were set up and analyzed in order to determine which supply method maintains higher olfactory stimulation levels for the longest period of time.

Continuous supply
The aroma was continuously supplied through the air outlets of the air conditioner in the cabin of the driving simulator when the driver's fatigue or sleepiness level was high (after driving in a driving simulator for 120 minutes). During these times, the ventilation system for the intake of fresh air was turned off.

Intermittent supply
The aroma was intermittently supplied in the cabin of the driving simulator through the air conditioner outlets when the driver's fatigue or sleepiness level was high (after driving in a driving simulator for 120 minutes). The sequence of the air supplement (on/off timing of the system) was optimized in the pilot study of the experiment. The ventilator was shut off when aroma was being supplied. In this study, the cycle of the air supplement was set at 2 minutes, and the duty ratio (i.e., the relative time span of the supplement) was 30% in 2 minutes.

Spot-intermittent supply with ventilation
The aroma was intermittently supplied directly to the face of the driver through the blower fixed on the instrument panel while the driver's fatigue or sleepiness level was high (after driving in a driving simulator for 120 minutes). During these times, the ventilator was activated and fresh air was let in the cabin continuously. The aroma was supplied in an optimized sequence that was identical to the intermittent supply. To keep the driver from perceiving the on/off timing of the air supplement, fresh air without aroma was also supplied to the face of the driver through the blower.

Two types of stimuli presentation were conducted for a period of 30 minutes evaluated with 10 participants: aroma supplied to the driver for a cycle of 2 minutes and 30% duty ratio (supplied for 36 seconds in 2 minutes) with ventilation (spot-intermittent supply with ventilation), and aroma supplied to the driver continuously without ventilation (continuous supply) (Figure 4). In the figure, the rectangular line denoting "On/Off of Air supplement" indicates the operating condition of the air supplement. The top of the rectangle indicates that the supplying system is activated. With intermittent supply of the supplement, the olfactory stimulation level had a rating higher than 3 (Table 1) for a 30-minute period, despite some fluctuations in the air supplement supply. A detailed study of the relationship between the stimulus level of the olfactory nerves and the effect of air supplements is needed; however, it is expected that the aroma supplied intermittently with ventilation will be more effective than the aroma supplied continuously without ventilation. This pilot study for investigating the olfactory stimulation level involved the same participants as the main study of the experiment.

Table 1. Intensity ratings

Rating	Level of stimuli
5	Intensive and unbearable odour
4	Strong odour
3	Odour that is easily sensed
2	Subtle odour
1	Faint odour that is scarcely sensed
0	No odour

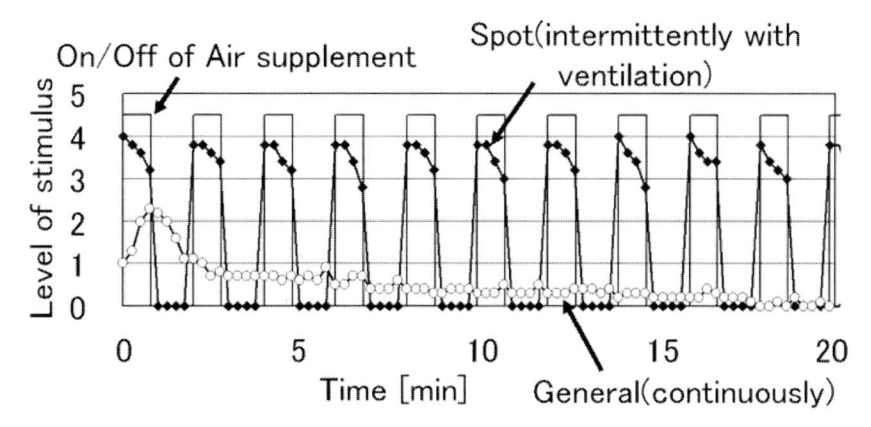

Figure 4. Two types of olfactory stimuli presentation

Participants

The 10 participants were male drivers (average age 29.5 years, SD 3.2) who drive more than 5,000 km per year.

Experiment design

This study used a within-subjects design. Every participant experienced all of the supply methods (i.e., continuous supply, intermittent supply, and spot-intermittent supply with ventilation). The effect of each supply method was evaluated from the data on finger-pulse fluctuation and reaction time to stimuli before and after the aroma supply. Additionally, the condition of not supplying the aroma to the driver was set up as the control condition. The sequence of these conditions was randomized to minimize the learning effect. It took two hours to start the air supplement and three hours to evaluate one supply method per participant, and four days to evaluate all supply methods per participant.

Experimental results

Effect of Alpha-pinene on fatigue level

Three different supply methods were set up. We hypothesized that the aromas would be most effective when the olfactory stimulation level was highest; thus, it was expected that the spot-intermittent supply with ventilation would be the most effective method. To clarify the effectiveness of the supply methods, the effects of Alpha-pinene on each state variable were evaluated statistically with a t-test.

The average of p-value among all subjects (10 subjects) in the t-test between data before and after the air supplement is shown in Table 2. A significant effect on fatigue reduction was verified with spot-intermittent supply with ventilation, as indicated by statistical analysis such as the Lyapunov exponent and reaction time. In the control condition (i.e., when the aroma was not supplied to the drivers), no positive effect was observed. From the results, it was assumed that intermittently supplying the air supplement directly to the face with ventilation reduced fatigue for a long period of time.

Table 2. Effect of Alpha-pinene (p-values)

	Spot-intermittent supply with ventilation	Intermittent supply	Continuous supply	Not supplying (Control condition)
Lyapunov exponent	p= 0.03	p= 0.05	NS	NS
Reaction time / Front	p= 0.04	NS	NS	NS
Reaction time / Side	p= 0.02	p= 0.09	p= 0.10	NS

Figure 5 plots one of the results in a condition of spot-intermittent supply with ventilation regarding the fluctuation of the Lyapunov exponent for quantitative evaluation of fatigue levels. The Lyapunov exponent decreased after onset timing (denoted by 0 min) of the air supplement, compared to that before the air supplement. Spot-intermittent supply with ventilation resulted in the most significant decrease of the Lyapunov exponent after the air supplement.

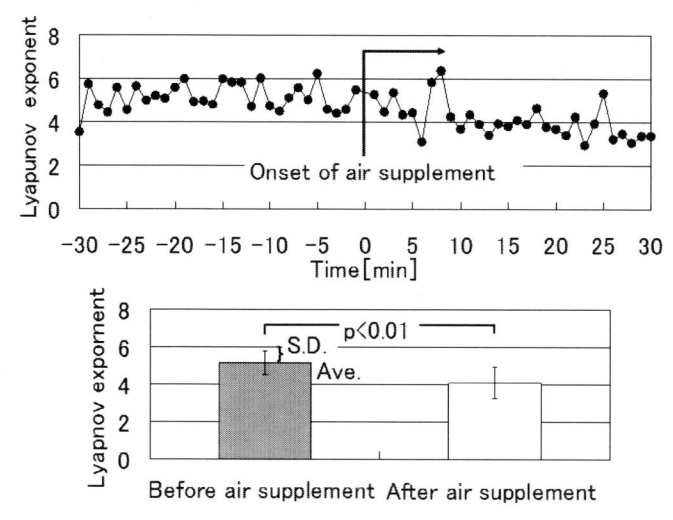

Figure 5. One of the test results regarding the difference of the Lyapunov exponent resulting from the supply of Alpha-pinene (Spot-intermittent supply with ventilation)

Figure 6 plots an example of the effects in a condition of spot-intermittent supply with ventilation on fatigue level as quantitatively evaluated by driver's reaction time to visual stimuli. The reaction time decreased after the air supplement was supplied.

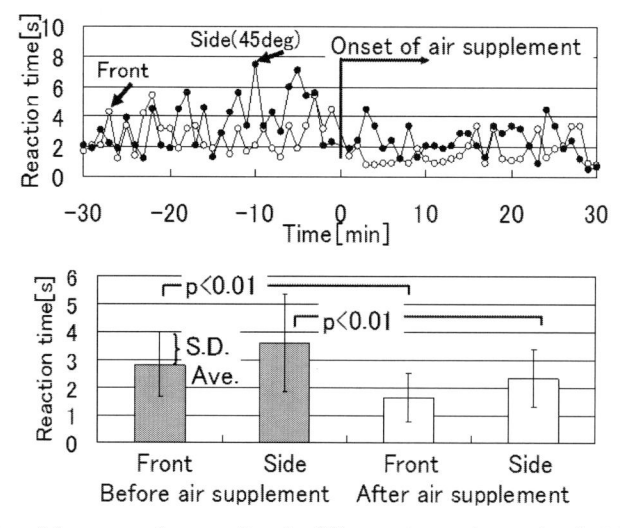

Figure 6. One of the test results regarding the difference in reaction to visual stimuli resulting from the supply of Alpha-pinene (Spot-intermittent supply with ventilation)

Effect of Borneol on wakefulness

To determine which Borneol supply method was most effective, the effects of Borneol on each state variable were evaluated statistically with the method described above. The average of p-value among all subjects (10 subjects) in the t-test between data before and after the air supplement is shown in Table 3. The most significant effect was verified with spot-intermittent supply with ventilation. In the control condition (i.e., when the aroma was not supplied to the drivers), no positive effect was observed.

Table 3. Effects of Borneol

	Spot-intermittent supply with ventilation	Intermittent supply	Continuous supply	Not supplying (Control condition)
Finger-pulse volume	p= 0.04	p= 0.11	NS	NS
Reaction time / Front	p= 0.03	NS	NS	NS
Reaction time / Side	p= 0.04	p= 0.11	NS	NS

Figure 7 presents one of the results regarding the size of the three-dimensional trajectory (i.e., finger-pulse volume) with spot-intermittent supply with ventilation. Though some fluctuations were observed, the finger-pulse volume decreased after onset timing (0 min) of the air supplement, compared to that before the air supplement. In evaluating the effect of the air supplement on wakefulness in terms of the supply method, spot-intermittent supply with ventilation revealed the most remarkable decrease in finger-pulse volume after the air supplement, similar to the results on fatigue reduction.

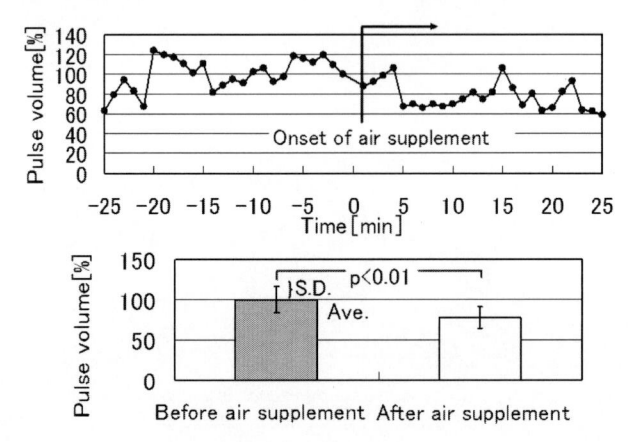

Figure 7. One of the test results regarding the difference in finger-pulse volume after supplying Borneol (Spot-intermittent supply with ventilation)

Figure 8 plots one of the results of spot-intermittent supply with ventilation regarding reaction time to visual stimuli after the air supplement for evaluating sleepiness level. A decreased reaction time was observed after the air supplement, compared to that before the air supplement.

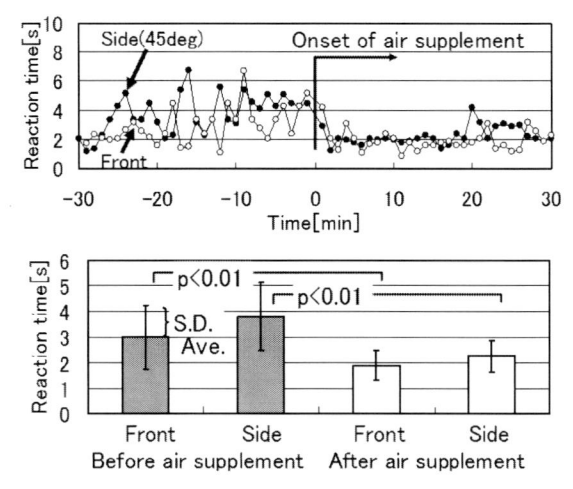

Figure 8. One of the test results regarding the difference in reaction time to visual stimuli after supplying Borneol (Spot-intermittent supply with ventilation)

Effect of air supplement on active safety

The preceding sections discussed the effects of aroma on fatigue and sleepiness levels. This section compares the effect of Alpha-pinene on fatigue reduction with that of the Forward Vehicle Collision Warning System (FVCWS) regarding the frequency distribution of time headway to the preceding vehicle. Watanabe (2001) presented one example of the design concept of the FVCWS to avoid the risk of collision with the preceding vehicle. In this study's design of the FVCWS in a driving simulator, the Stopping Distance Algorithm was applied to control alarm timing.

The state variables of vehicle manoeuvre (e.g., time headway, which can be calculated by dividing the distance to the preceding vehicle by the velocity of the vehicle) were investigated. Further study will be necessary to determine what state variables are appropriate to assess the risk of collision. However, this study compared the frequency distribution of time headway when the aroma was provided to the driver to that when the alarm was provided to the driver, in a situation in which the risk of collision to the preceding vehicle was high.

As mentioned earlier, drivers were asked to maintain time headway exceeding one second to the preceding vehicle. When the driver's fatigue level was high, time frequency with time headway shorter than one second was increased in a pilot study of the experiment. Further study on the relationship between fatigue level and time headway is necessary. However, it can be said that time frequency with time headway shorter than one second increased when the driver's fatigue level was high because drivers were careless in controlling the brake pedal in reaction to the slowing of the preceding vehicle. In the study of fatigue reduction, the driver had to control the brake and gas pedal to keep the specified time headway to the preceding vehicle because the preceding vehicle was given a fluctuating velocity.

The frequency distributions of time headway at every second for a period of 30 minutes before and after supplying Alpha-pinene or activating the warning system were analyzed. Figure 9 plots an example of the investigated data regarding time headway. The focus was mainly on the time frequency when the time headway was shorter than 1.0 second (i.e., the driver had a greater possibility of colliding with the preceding vehicle). The left side of the figure presents decreasing time frequency (i.e., the driver maintained an appropriate distance from the preceding vehicle) with a supply of Alpha-pinene; the right side presents decreasing time frequency with activation of the warning system in the driving simulator.

Future studies should investigate the effects of aroma on active safety in detail. However, this study confirmed that supplying Alpha-pinene has the potential of maintaining time headway for a longer period of time. In terms of active safety to prevent the risk of collision, supplying Alpha-pinene is inferior to using the FVCWS, but it is still useful. Additionally, the effects of supplying Borneol on active safety were analyzed, and it was observed that Borneol reduces swerving of the vehicle and reaction time.

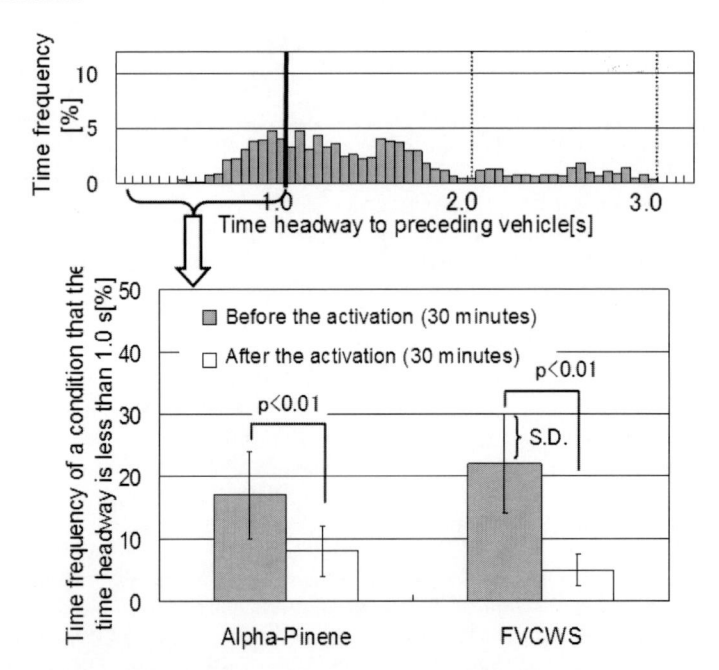

Figure 9. Analysis of the effect of air supplement on active safety in comparison with the Forward Vehicle Collision Warning System

Conclusions

The effect of tree aromas on an automobile driver's fatigue level, sleepiness level, and active safety were investigated with the use of a driving simulator. The results are summarized as follows.

Effect of Alpha-pinene on fatigue reduction

In a quantitative analysis of fatigue level, the Lyapunov exponent of finger pulse was reduced and reaction time to visual stimuli was shortened after supplying Alpha-pinene. A significant difference was observed between before and after supplying Alpha-pinene, indicating that Alpha-pinene reduces fatigue.

Effects of Borneol on wakefulness

In a quantitative analysis of sleepiness level, the size of the three-dimensional trajectory of finger pulse was reduced and reaction time to visual stimuli was shortened after Borneol was supplied. A significant difference was observed between before and after supplying Borneol.

Effect of supply method

Fatigue was remarkably reduced with Alpha-pinene and wakefulness was remarkably increased with Borneol when these aromas were intermittently supplied directly to the face of drivers with ventilation.

Effect of air supplement as an active safety system

The effect of air supplement on active safety was evaluated in comparison with that of the FVCWS. In terms of active safety to prevent the risk of collision, Alpha-pinene is inferior to the FVCWS, but it is still useful. Additionally, it was determined that supplying Borneol helps minimize the driver's swerving of the vehicle or lane deviation, and keeps the driver's reaction time short.

References

Yatagai, M. (1997), Miticidal activities of tree terpenes. *Current Topics in Phychochemistry, 1*, 85-97

Yatagai, M. (2000), Tree aroma and its function of relaxation and rest, *Aroma Research, 1(1)*, 2-7

Sawada, K.,Komaki, R., Yamashita, Y., & Suzuki, Y. (2000), Odor in forest and its physiological effects, *Aroma Research, 1(3)*, 67-71

Homma, S. (2005), Effect of wood land exposure on humans, physical and spiritual-Medical investigation, *Aroma Research, 6(21)*, 47-53

Miao, T., Shimizu, T., & Shimoyama, O. (2003), The use of chaotic dynamics in finger photoplethysmography to assess driver mental work load, *Proceedings of Society of Automotive Engineers of Japan*, #20035065

Shimizu, T., Miao, T., & Shimoyama, O. (2004), Toward measuring driving workload by chaotic dynamics underlying ear photoplethysmo-graphy, *Proceedings of Human Interface symposium*, #2444, 595-598.

Hashimoto, H., & Katayama, T. (1998), Data processing method of finger blood pulse, *Transactions of Society of Automotive Engineers*, #980016 (pp. 33-36)

Sano,M., Sawada,Y. (1985), Measurement of the Lyapnov spectrum from chaotic time series, *Physical Review Letter, 55*, 1082-1085

Godthelp, J., Milgram, J., & Blaauw, G.J. (1984),.The original TLC source would be more appropriate to refer to: The development of a time-related measure to describe driving strategy, *Human Factors, 26*, 257–268.

Japan Automobile Research Institute (1997), Investigation of time headway, *Report of ITS standardization, J-97-1*, 14.

Suzuki, K., Wakazugi, T., & Soma, H. (2002), Designing method of warning timing based on the time criterion for lane departure warning system, *Transactions of the Society of Instrument and Control Engineers, 38,* 567-573.

Watanabe, K., Sakabe, M., Nakasho, T., & Nonaka, M. (2001), Development of Forward Collision Prevention Support System in ASV2, *Proceedings of Society of Automotive Engineers of Japan*, No.58-01, 9-12.

Changing lanes with active lanekeeping assistance: a simulator study

Gerrit Schmidt[1], Kirstin L. R. Talvala[2], Joshua P. Switkes[3], Miklós Kiss[1], & J. Christian Gerdes[2]
[1]Volkswagen AG
Wolfsburg, Germany
[2]Stanford University
Stanford, CA, USA
[3]Volkswagen of America
Palo Alto, CA, USA

Abstract

Active lanekeeping assistance systems are designed to help the driver avoid unintended lane departures. Such a system may also have the additional goal of supporting the driver in changing lanes comfortably and without an unnecessarily large amount of driver effort. Thus, the interaction between the driver and vehicle is important when designing such a system.

This paper presents results from a user study that focused on the driver's ability to change lanes with an active lanekeeping assistance system. Three different systems were tested by each of 20 participants in a fixed-base driving simulator. The assistance from each system was based on the vehicle's position and orientation on the road, and differed only in the way that attempted lane changes were handled. During lane changes one system had an added dependence on time, and the other had an added dependence on the driver's steering rate.

The participants were instructed to make both slow and fast lane changes. Data analysis includes objective parameters such as the steering wheel angle and yaw angle, as well as subjective parameters such as the driver's performance evaluation. The results suggest that an assistance system based on the driver's steering rate is advantageous.

Introduction

Each year, a large number of vehicle fatalities are caused by the vehicle leaving the lane and colliding with a fixed object in the environment. In the United States, this type of accident accounts for approximately one third of vehicle fatalities (NHTSA, 2005). Similar statistics exist worldwide, such as in Germany (Statistisches Bundesamt, 2004). An active lanekeeping system, designed to help the driver avoid unintended lane departures, therefore has the potential to help avoid these accidents.

In D. de Waard, F.O. Flemisch, B. Lorenz, H. Oberheid, and K.A. Brookhuis (Eds.) (2008), *Human Factors for assistance and automation* (pp. 49 - 61). Maastricht, the Netherlands: Shaker Publishing.

Several groups have investigated active lanekeeping systems (see for example, Sato et al., 1998, Suzuki, 2002, Raith et al., 2003, or Morin, 2005) and recently, lane departure alert systems have found their way onto production vehicles (for example, Audi Lane Assist, Citroën AFIL, Honda LKAS, Lexus LKA and LDW, Nissan LKS and LDA).

Although lanekeeping systems are designed to help avoid lane departures, they should also allow the driver to change lanes. This conflicts with the system's primary goal of keeping the vehicle in the lane, and thus presents an interesting challenge. For user acceptance, desired lane changes must be able to be made comfortably and without an unnecessarily large amount of driver effort. For the user to benefit from the system, unintended lane departures must be prevented.

Unfortunately, there are no reliable systems to date that can detect if the driver wants to change lanes, although some promising approaches do exist (McCall et al., 2005). Driving styles differ, since some drivers change lanes very suddenly while others slowly drift over into the new lane. Some drivers use the turn signals, while others do not. Furthermore, lane changes made during normal driving are different than those made in situations where quick evasive manoeuvres are necessary. Yet, in order to design a lanekeeping system that will allow lane changes, the system must be able to differentiate between intended and unintended lane departures.

Since driver's intent cannot easily be directly measured, vehicle position on the road and steering wheel angle rate are used as proxies for the driver's intent to change lanes in this work. The user study presented in this paper focuses on the driver's ability to change lanes with an active lanekeeping assistance system based on these proxies. The underlying lanekeeping system and the algorithm for calculating the force feedback applied to the steering wheel are kept constant. The difference between the system setups is the manner in which lane changes are handled. The assistance from the lanekeeping system is based on the vehicle's position and orientation on the road. During lane changes, one setup has an added dependence on time, and another has an added dependence on the driver's steering rate.

Questions

The study was designed to help answer several questions. The main focus was to investigate how the complexity of the system setup relates to the driver's ability to change lanes. To do this, three different lanekeeping systems of varying complexity were presented to each participant. With each system, the participant was asked to perform two different types of lane changes. An analysis of both participant opinion and driving performance metrics (such as yaw rate and handwheel angle) for both types of lane changes over the three lanekeeping systems was performed to see if there were any noticeable differences between the systems. Training and learning effects over time and across systems were also investigated.

Lanekeeping system description

Basic lanekeeping

The vehicle simulation used in this study is modified to include an experimental lanekeeping assistance system. This system is based on the system of Rossetter et al. (2004). The simulated vehicle is modeled as being equipped with steer-by-wire, which allows the roadwheels to steer independently from the steering wheel. This setup is chosen for experimental purposes, since it allows maximum freedom when designing the assistance system. In this case, the angle of the roadwheels is adjusted to be the sum of the driver's commanded angle and a correction angle from the lanekeeping assistance system. Within each lane, this correction angle is derived from an artificial potential energy function. This potential is a function of lateral error and heading error, described relative to the road centreline. Figure 1 shows a sketch of this potential function around a single lane's centre as well as the definitions used for lateral and heading error.

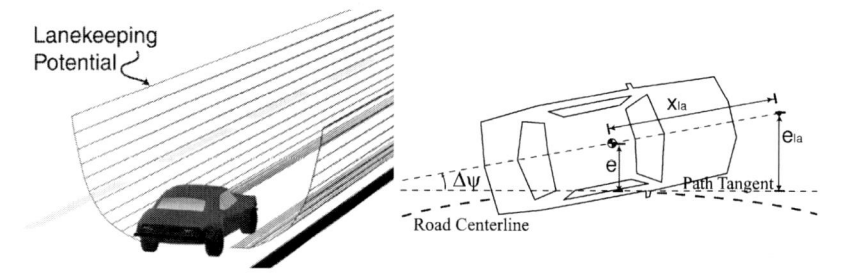

Figure 1. Lanekeeping potential (left) and lateral and heading error definitions (right)

Here x_{la} is a lookahead distance which is multiplied by the heading error, $\Delta\psi$. This is added to the lateral error, e, to form a projected error, e_{la}, which is then multiplied by a gain, k. Thus for a single lane we have a quadratic potential function:

$$V(e_{la}) = k(e_{la})^2 = k(e + x_{la} \sin \Delta\psi)^2$$

In this study the lanekeeping gain and lookahead distance are set to 2000 and 30 m, respectively. These values represent typical values for highway driving, chosen to ensure adequate lanekeeping ability (Rossetter et al., 2004). Graphically the potential function like a parabola, centred at the lane centre as shown in Figure 2 (left), where the lanekeeping assistance force applied to the vehicle is the slope of the potential function.

Clearly this shape of potential function only works properly for a single lane, and adding additional lanes requires modification. In this study, the basic potential function is modified for a two-lane road and then further modified to create a total of three different lanekeeping systems. These three setups are not meant to be the only possible or the best possible, but instead are meant to span the typical types of systems currently in research and development.

Lanekeeping modification for two lanes: "bump" system

To allow the vehicle to change lanes, the potential function is modified to consist of two quadratic functions, with another quadratic "bump" centred on the lane divider line. Around each lane centre, the lanekeeping system works to help keep the vehicle in the lane. Once the vehicle has moved sufficiently far from the lane centre, the assistance decreases to allow the vehicle to change lanes. In this way, the lanekeeping force increases as the vehicle moves from a lane centre towards the lane divider until the vehicle is about 1 m away from the lane divider. At this point, the force decreases and goes to zero as the vehicle passes over the lane divider. The force profile is then mirrored in the other lane. The resistance to lane changing is similar to the existence of a physical hill between the lanes. Figure 2 (centre) shows the Bump system potential function. This system setup is invariant over time and should therefore be learned very quickly by the driver.

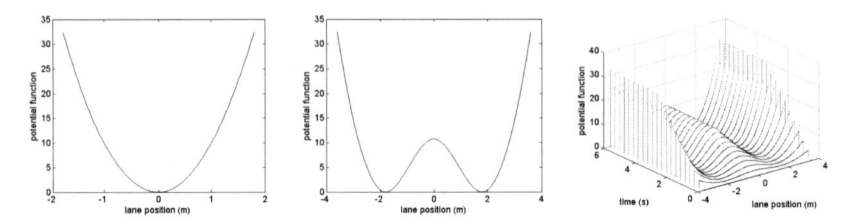

Figure 2. Examples of a single lane potential function (left), Bump system potential function (centre) and Ramp system potential function (right)

Ramp modification when switching lanes: "ramp" system

A possible problem with the Bump system is that the vehicle is pushed into the new lane once it crosses the lane divider line. One solution to this problem is to interpret the presence of the vehicle on the lane divider line as an intention to change lanes. With this interpretation, it makes sense to temporarily deactivate the lanekeeping system until the driver has entered the new lane. At the lane divider line the assistance force is equal to zero. When the vehicle reaches the lane divider line, the lanekeeping potential field gain is set to zero (no change in the assistance force at this point). The system then smoothly increases the gain up to its nominal value over the course of 3 seconds. If the driver is steering towards the new lane centre, the system will reach full lanekeeping strength approximately when the vehicle reaches the new lane centre. On the other hand, if the driver moves only slightly into the new lane, the system will gradually nudge the vehicle towards the new lane centre. Figure 2 (right) shows the ramping behaviour of this system. The potential field is plotted over time, and shown for both lanes. The vehicle crosses the lane divider line (lane position = 0 m) at time 0 seconds. The behaviour of this setup is more complex than the Bump setup as it does vary over time. Learning this system behaviour should therefore be more difficult than in the Bump setup. On the other hand, the system setup is designed to ease the task of changing lanes.

Rate switch to deactivate system: "rate" system

In a similar way to how crossing the lane divider can be interpreted as a desire to change lanes, the speed of handwheel movement can also signal this intention. For this study a rate of 0.63 rad/s at the handwheel was chosen as the threshold, above which the system is deactivated. Thus whenever the handwheel moved faster than this rate, the lanekeeping gain was set to zero for 5 seconds or until the vehicle crossed the lane divider line. At this point, the lanekeeping gain was smoothly increased as in the Ramp system. This system setup has the most complex behaviour of the ones presented here. However, it is designed to try and identify the driver's intentions to change lanes before the driver has crossed into the new lane. This behaviour should consequently inhibit the driver's ability to change lanes the least.

Experimental setup

Simulator description

The driving simulator used is shown in Figure 3. It is a fixed-base simulator consisting of the front two-thirds of a sports coupe. There is no motion base or vibro-acoustic feedback, besides simulated engine and road noise played over the vehicle's original stereo system. In the vehicle cockpit most vehicle interfaces are operational, including the instrument panel, turn signals, pedals (with vacuum assist for good pedal feel), and handwheel force feedback by direct drive DC motor.

Figure 3. Driving simulator (left) and view from driver's seat during the tracking task (right). The road is a straight, two-lane, single direction road

The simulator software is from Systems Technology, Inc. (STI, 2007) and includes a high fidelity nonlinear vehicle model. The chosen model is that of a Ford Taurus. Although the vehicle model is valid at high levels of lateral acceleration, the simulator environment is not. For low levels of lateral acceleration and for the initial reactions to high lateral acceleration manoeuvres, the simulator environment is sufficiently realistic. As will be discussed, the measurements in this study focus on this initial reaction, and do not depend strongly on the ability of the driver to stabilize the vehicle at high lateral accelerations. The software drives three forward facing display channels. Each channel has a resolution of 1024x768 pixels, and is projected onto a screen about 2.7 m on the diagonal. Each screen occupies 45 degrees of the driver's field of view, for a total field of view of 135 degrees. The investigator is seated in the same room, but behind the vehicle such that s/he is not in view of the driver.

Driving task

The basic task for this experiment is highway driving on a straight, two-lane, single direction road. The vehicle is set up with a cruise control function which accelerates to, and maintains, a highway speed of 105 km/h (65 mph). The cruise control function only controls the vehicle's speed and was implemented to ensure that all experiments were conducted at the same speed and not influenced by the participants' ability to maintain a constant driving speed. This constant speed is chosen because the focus of the experiment is on lateral control.

The specific task for each participant is to follow the vehicle in front. This target vehicle is the only other vehicle on the road. Within the experiment, the participants have to perform two types of lane changes: a slow, gradual lane change, and a quick, evasive manoeuvre-type lane change. For simplicity, these are called "normal" and "emergency" lane changes, respectively. In the normal lane change, the driver is instructed to follow the preceding vehicle into the new lane. A target is placed on the screen, as shown in Figure 3 (right), consisting of a black marker which the driver is instructed to maintain in a position between the two rear tires of the preceding vehicle. This task is a continuous pursuit tracking task (McCormick and Sanders, 1982), where the preceding vehicle is the target for the lateral tracking task the driver has to perform. This setup is designed to force a standard lane change that all drivers must follow, so that individual lane changes can be directly compared.

In the emergency lane changes the preceding vehicle suddenly brakes with maximum force, requiring the driver to very quickly change lanes to avoid the preceding vehicle (the driver is instructed not to brake). There is no warning to the driver of the impending emergency, other than active brake lights on the preceding vehicle. The driver is instructed to avoid the preceding vehicle, but is given no other instruction about the quality of the lane change. After an emergency lane change, another vehicle appears in front of the user. After each normal or emergency lane change, the participants have to continue following the preceding vehicle until the vehicle either changes lanes (normal lane change) or suddenly brakes (emergency lane change).

Design of the experiment

Test procedure

Before beginning the test, the participants drive for 7 minutes with no task, to familiarize themselves with the simulator environment. The simulator is then stopped, and the driver is instructed that the main task will be to follow the vehicle in front of them. If the vehicle changes lanes, the participants shall track the vehicle and change lanes as well ("normal" lane change). If the vehicle ahead suddenly brakes, an evasive manoeuvre is necessary and the participants shall avoid the vehicle by quickly changing lanes ("emergency" lane change). The general setup of the two types of lane changes is presented to the driver without any lanekeeping system support. Each user conducted 13 normal lane changes and 7 emergency lane changes in this manner. The time between the start of each lane change was 30 seconds for the first 5 lane changes, and 20 seconds for all other lane changes. Directly after each of the lane changes, the driver was asked to evaluate his/her performance and was asked "How well did you perform in this lane change task?". Participants had to give their ratings on a scale of five categories. Each category contains three steps and the scale ranges from 1 (very weak) to 15 (very well). This type of scale provides high resolution while still being easy to handle for the participant (Heller, 1985). After this run, the driver had to perform the same run three times, once for each of the three lanekeeping system setups. The order of the system setups was randomized for each participant such that at least 3 participants drove each of the 6 possible order combinations. Due to the number of participants, 2 of the possible order combinations were driven by four participants. After each run, the participant completed an additional questionnaire, evaluating their performance on the task, the system behaviour and the setup itself.

Sample

In this study, the behaviour and evaluations of 20 persons between 19 and 53 years old were investigated (m=27.6 years; sd=7.1 years). The participants were recruited among Stanford University students and staff, who reported a mean of 8625 miles driven per year (sd=6857.3). Three participants reported to have experiences with lane departure warning or lanekeeping support systems.

Data processing and statistical analysis

The simulator software calculates and records various vehicle dynamics metrics. Other variables defined for the purpose of the lanekeeping system can also be recorded by the software. Vehicle dynamics metrics such as yaw rate, handwheel angle, and handwheel rate were recorded as a function of time during each experiment. Since the lanekeeping assistance system modifies the roadwheel angle independently from the handwheel, roadwheel angle was also recorded. Parameters not directly computed by the simulator software, such as lateral deviation from the target vehicle, were computed after the study from vehicle position and heading data that was recorded. The time at which the Rate system was triggered was also recorded by the software. The subjects' subjective ratings of the individual lane changes as well as their answers to the post-drive questionnaires were recorded on paper.

The plots that follow show mean values and standard deviations. Due to the experimental design the statistical analysis was performed using T-Tests for paired samples and split-plot ANOVAs (Ross, 2000) including Fisher LSD post-hoc tests. This analysis of variance is used to determine the probability, p, that variation of an independent variable (e.g. system setup) does not affect the considered dependent variable (e.g. lateral deviation). A small p-value indicates a small probability that the two are part of the same distribution. A p-value less than 0.05 is considered

Results

Training and learning effects

Because the users are new to the simulator environment and lanekeeping support, it is useful to examine how their driving behaviour changes over time to identify any learning effects. Figure 4 shows the learning of the test subjects over time and across systems. Here the system order is considered, where "first system" means the first system driven chronologically (and thus includes an even mixture of all three system types). Within each system the lane changes are divided into three blocks, where each block is a third of the total lane changes (emergency and normal combined).

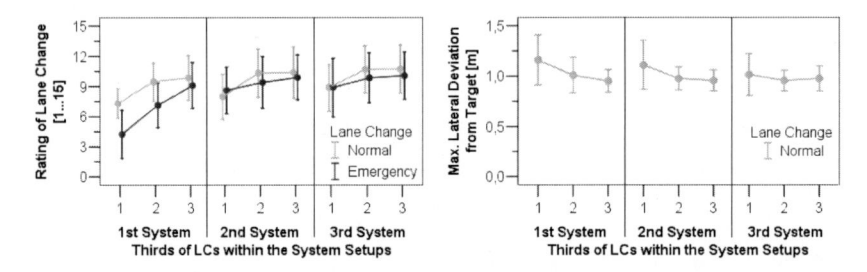

Figure 4. Rating of lane change performance (left) and maximum lateral deviation from tracking target (right) for normal and emergency lane changes. Mean values are calculated for each 33% of performed lane changes per system setup over time

The examination of the rated lane change performance clearly shows an overlay of two training effects over time. First, Figure 4 indicates an improvement over time in the subjects' rating of the lane change performance ($F_{2,38}$=19.420; p_{sys}<0.001; LSD post-hoc Tests: p_{sys1-2}<0.001; p_{sys1-3}<0.001; p_{sys2-3}=0.148) and in the lateral deviation in the tracking task while performing normal lane changes ($F_{2,38}$=5.303; p_{sys}=0.009; LSD post-hoc Tests: p_{sys1-2}=0.116; p_{sys1-3}=0.006; p_{sys2-3}=0.093). Each time the system configuration is changed, the ratings significantly decrease (T-Tests: $p_{1.3-2.1}$=0.002; $p_{2.3-3.1}$=0.002). This drop in subjective evaluation decreases with the number of system setup changes. The tracking performance is only effected by the first change of system setup (T-Tests: $p_{1.3-2.1}$=0.002; $p_{2.3-3.1}$=0.145).

A learning effect can also be observed within each system setup. The subjects' ratings of the lane changes improve from block to block ($F_{2,38}$=85.012; p_{block}<0.001; LSD post-hoc Tests: $p_{block1-2}$<0.001; $p_{block1-3}$<0.001; $p_{block2-3}$<0.001), where each block represents one third of the lane changes. Concerning lateral deviation, the drivers learn from the first to the second block ($F_{2,38}$=16.987; p_{block}<0.001; LSD post-hoc Tests: $p_{block1-2}$<0.001; $p_{block1-3}$<0.001; $p_{block2-3}$=0.303). This indicates that the drivers are able to perform the tracking task better the more time they spend with one system setup. The subjects need to adjust their behaviour strategies and compensate for the system, which they do within the first 33% of lane changes with one system.

Overall, the subjects in this study seem to be able to adapt very quickly to any change in the system setup introduced here. By this, we can replicate general learning effects as shown, for example, for simple mathematical operations (Neves & Anderson, 1981) as well as more complex interaction with driver information systems (Totzke, et al., 2004).

Differences between normal lane changes and evasive emergency manoeuvres

As mentioned, the subjects had to perform two different types of lane changes: normal lane changes where they were tracking a vehicle which slowly changed lanes, and emergency lane changes where an evasive manoeuvre was necessary to avoid a braking vehicle ahead. Figure 5 shows the RMS of yaw rate and handwheel angle with each system setup, showing a much higher yaw rate for emergency lane changes ($F_{1,19}$=203.355; p_{lc}<0.001; η^2=0.915). Obviously, larger vehicle motions are observed in the more extreme manoeuvres. Over time, this difference in yaw rate between the different types of manoeuvres significantly decreases ($F_{2,38}$=6.980; p_{lc*sys}=0.003; η^2=0.269).

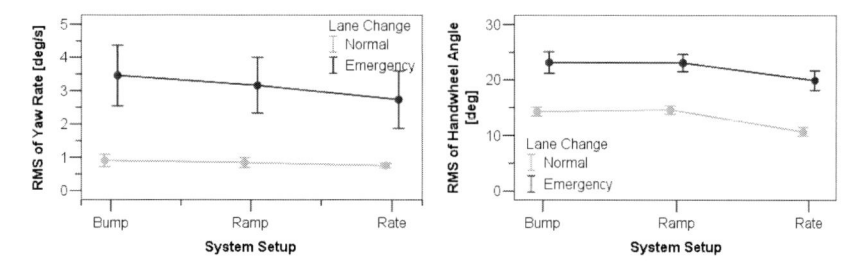

Figure 5. RMSs of yaw rate (left) and handwheel angle (right) within lane change for normal and emergency lane changes with each system setup

The left plot of Figure 6 shows the users' rating of the lane changes. There is no significant difference between the different types of lane changes ($F_{1,19}$=2.855; p_{lc}=0.106). However, as shown in Figure 4, emergency lane changes are rated significantly worse for the first system driven ($F_{2,38}$=7.984; p_{lc*sys}=0.001; η^2=0.296). With further system contact, the difference between the lane changes decreases.

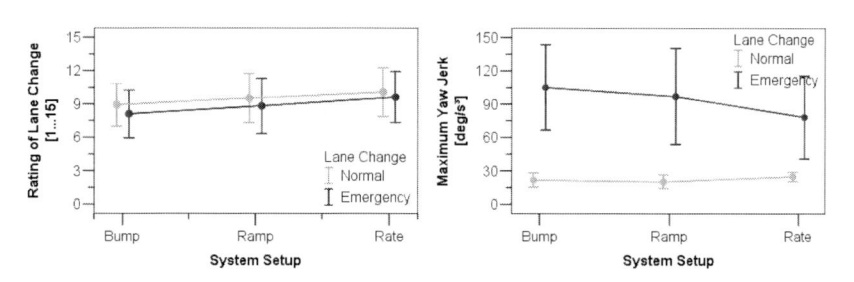

Figure 6. Subjects' performance ratings (left) and maximum yaw jerk within lane change (right) for normal and emergency lane changes with each system setup

Comparison of different system setups

After getting used to the simulator environment in a drive without any lanekeeping assistance system, the subjects had to perform the lane changes with the different lanekeeping systems active. This section focuses on the impact of the system setup on the driving performance and the subjective evaluation. At a first glance, the comparison of the different system setups indicates a preference for the Rate system. Figure 6 shows the evaluation of the lane change performance being advantageous for the Rate system ($F_{2,38}=7.071$; $p_{sys}=0.002$; $\eta^2=0.271$). Within normal lane changes, tracking performance measured by the RMS of the lateral deviation from the target is in tendency better for the rate triggered system ($F_{2,38}=2.745$; $p_{sys}=0.051$; $\eta^2=0.126$). The objective driving performance measures shown in Figure 5 such as the RMSs of yaw rate ($F_{2,38}=11.714$; $p_{sys}<0.001$; $\eta^2=0.381$) and handwheel angle ($F_{2,38}=341.542$; $p_{sys}<0.001$; $\eta^2=0.947$) show a smoother and less dynamic lane change with this setup. As well, observed heading error and handwheel forces are lower.

A closer look does reveal some possible disadvantages of the Rate system. Figure 7 shows the RMSs of roadwheel and handwheel angle rates. The Rate system results in an increased RMS of roadwheel angle rate for normal lane changes as compared to the other system setups ($F_{2,38}=22.051$; $p_{lc*sys}<0.001$; $\eta^2=0.537$). However, the RMS of handwheel angle rate does not support this finding, as it is smaller than for the other setups. The increased roadwheel angle rate must, therefore, be due to the contribution of the lanekeeping system and not due to an input from the driver.

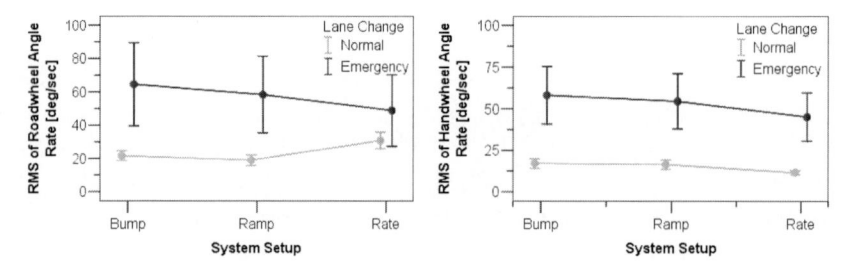

Figure 7. RMSs of roadwheel (left) and handwheel (right) angle rate within lane change for normal and emergency lane changes with each system setup

The maximum yaw jerk (derivative of yaw acceleration) is shown in Figure 6 on the right. The maximum yaw jerk is largest for the Rate system, and supports the findings that the Rate system results in the least smooth motion during normal lane changes. Interestingly, the Rate system results in the smoothest vehicle motion during emergency lane changes ($F_{2,38}=9.746$; $p_{lc*sys}<0.001$; $\eta^2=0.339$). The Rate system setup therefore seems to work very well in emergency lane changes whereas for normal lane changes some difficulties can be observed.

Further exploration of the system behaviour with the Rate setup shows another difference between normal and emergency lane changes. For the emergency lane changes, Figure 8 (left) shows that the rate threshold is always exceeded around one second after the preceding vehicle starts braking. The plot on the right underlines the fact that the handwheel angle at that time is always below 10 degrees and the heading error observed is close to zero. This means that the system was triggered consistently in the very beginning of the lane change. In this scenario, the rate trigger seems to have worked successfully. For the normal lane changes, the data shows the same behaviour in most cases. Nevertheless, in a few cases a different behaviour can be observed as the rate threshold is exceeded much later. At that moment, steering angles are greater than 10 degrees and up to 50 degrees. The higher heading errors for these data points indicate that the lane change is already in full progress.

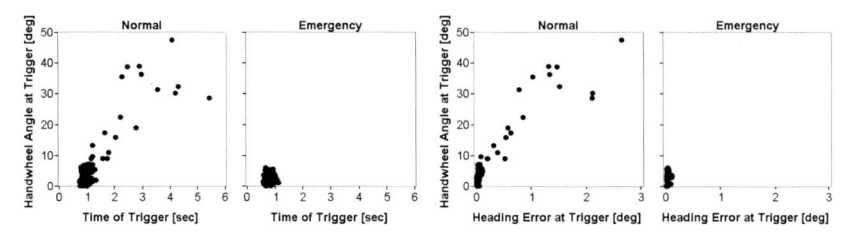

Figure 8. Handwheel angle vs. trigger time point of rate system (left) and vs. corresponding heading error (right) at trigger time point for normal and emergency lane changes with Rate system setup

The Rate system setup, in some cases, is not triggered at the ideal time due to the behaviour of the driver. In those cases, the drivers seem to exceed the steering rate threshold too late. Slowly changing lanes with this setup can lead to a less smooth lane change than with the other setups if the driver expects to trigger the system. The subjective system evaluation after having driven all setups does not clearly identify one system setup as preferable. Nevertheless, 35% of the subjects indicate a preference for the Rate triggered system. Least preferred is the Bump setup with 15% of the ratings, and 25% of the subjects preferred no system.

Conclusions

The participants in this study where presented with three lanekeeping systems: the simple "Bump" system, the time-dependent "Ramp" system, and a more complex, time- and steering rate-dependent "Rate" system. Slow ("normal") and fast ("emergency") lane changes were made with each system active. The participants were able to learn and adapt quickly to all the system setups. The Bump system, which was designed to be the simplest and easiest to predict, resulted in the same learning behaviour as the more complicated Ramp and Rate systems. When driving behaviour is examined, a difference between the systems is evident. The simple Bump system required the driver to use larger steering inputs, and consequently resulted in larger tracking errors in the normal lane changes and larger vehicle motions in the emergency lane changes. This setup is the simplest one, and turned out to be the most difficult to drive with in this experiment.

The subjective evaluations did not show any clear preference among the different systems, however, the Rate system appears to be advantageous in overall examination. In emergency lane changes, the steering rate trigger was exceeded quickly and consistently after the initiation of the manoeuvre. The lanekeeping assistance was temporarily deactivated, and therefore the system did not disturb or inhibit the lane change. During normal lane changes, simulated via the tracking task, the Rate system led to better tracking performance. The subjective ratings of the lane changes support these findings in these measures. The Rate system was not ideal, however, as there were some instances of the rate threshold being exceeded well after the lane change had been initiated. In these cases, the system was triggered when the vehicle had a substantial heading error and the resulting vehicle dynamics were less smooth. For the majority of the normal lane changes in this study, however, the threshold was exceeded sufficiently early during the lane change.

The Rate system could be improved to better support normal lane changes. Including other proxies for driver's intent, such as torque input from the driver, could result in better detection of the driver's intent to change lanes. The results from this study show that the Rate system is not too complex for the driver to quickly learn, and therefore a more complex system could also be acceptable to the driver. Moreover, an adaptive rate threshold could be used, allowing for the system to be customizable for individual drivers.

References

Heller, O. (1985). Hörfeldaudiometrie mit dem Verfahren der Kategorienunterteilung (KU). *Psychologische Beiträge, 27*, 478-493.

McCall, J., Wipf, D., Trivedi, M.M., & Rao, B. (2005). Lane change intent analysis using robust operators and sparse Bayesian learning. In Proceedings of the *Conference on Computer Vision and Pattern Recognition, 2005 IEEE Computer Society, 3,* 59.

McCormick, E.J. & Sanders, M.S. (1982). *Human Factors in Engineering and Design.* New York: McGraw-Hill.

Morin, T. (2005). Tomorrow's smart cars will make drivers better, smarter. *SAE Automotive Engineering International, March 2005*, 108-109.

Neves, D.M. & Anderson, J.R. (1981). Knowledge Compilation: Mechanisms for the Automatization of Cognitive Skills. In J.R. Anderson (Ed.), *Cognitive Skills and their Acquisition* (pp 57-84). Hillsdale, NJ: Erlbaum.

NHTSA (2005). *Traffic safety facts 2004.* (Technical Report). Washington, DC, USA: National Highway Traffic Safety Administration.

Raith, C., Gesele, F., Dick, W., & Mlegler, M. (2003). Vernetzte Produktentwicklung am Beispiel Audi Dynamic Steering. In VDI Gesellschaft Fahrzeug und Verkehrstechnik (Ed.) *Elektronik im Kraftfahrzeug* (pp. 185-205). Düsseldorf: VDI-Verlag.

Ross, S.M. (2000). *Introduction to Probability and Statistics for Engineers and Scientists.* San Diego, CA: Harcourt.

Rossetter, E.J., Switkes, J.P., & Gerdes, J.C. (2004). Experimental Validation of the Potential Field Driver Assistance System. *International Journal of Automotive Technology, 5*, 95-108.

Sato, K., Goto, T., Kubota, Y., Amano, Y., & Fukui, K. (1998). A Study on a Lane Departure Warning System using a Steering Torque as a Warning Signal. In Proceedings of the *International Symposium on Advanced Vehicle Control (AVEC)* (pp. 479-484). Nagoya, Japan.

Statistisches Bundesamt (2004). Fachserie 8, Reihe 7. Wiesbaden, Germany: Statistisches Bundesamt.

STI (2007). Driving Simulator Software. Retrieved September 2007 from http://www.systemstech.com/

Suzuki, K. (2002). Analysis of drivers steering behavior during auditory or haptic warnings in lane departure situations. In Proceedings of the *International Symposium on Advanced Vehicle Control (AVEC)* (pp. 243-248). Nagoya, Japan..

Totzke, I., Krüger, H.-P., Hofmann, M., Meilinger, T., Rauch, N., & Schmidt, G. (2004). Kompetenzerwerb für Informationssysteme - Einfluss des Lernprozesses auf die Interaktion mit Fahrerinformationssystemen. In *FAT-Schriftenreihe Band 184*. Offenbach, Germany: Berthold Druck.

Speed recommendations during traffic light approach: a comparison of different display concepts

Stephan Thoma[1], Thomas Lindberg[1], & Gudrun Klinker[2]
[1]BMW Group Research and Technology, München
[2]TU München
Germany

Abstract

Projects like INVENT[1] or Travolution[2] aim at optimizing traffic flow in cities by linking vehicles and infrastructure elements using infrastructure-to-car communication technologies. Traffic lights equipped with transmitters are able to inform the approaching vehicles about the current and future state of the traffic light control. This information can be used to optimize the speed profile while approaching an intersection. As current vehicles will not perform any speed changes autonomously, an intuitive and easy-to-read human machine interface is needed to present the speed recommendations to the driver.

Three different ways of in-vehicle information presentation were evaluated with 27 participants in a static driving simulator. The first variant is a simple countdown which presents the remaining seconds of the red or the green phase, respectively. The second variant shows the minimum or maximum speed recommendation in a textual manner. The third alternative uses a red and green area on the speed indicator scale.

All three approaches were tested regarding their comprehensibility, acceptance, effectiveness and distraction. The results indicate that the non-textual speed indicator display is preferred by most of the individuals and also has advantages concerning minimization of acceleration and deceleration actions.

Introduction

Traffic flow optimization has always been an important research topic considering growing cities and increasing traffic density. The main motivation behind all measures taken is to reduce CO_2 emissions and to make travelling for drivers more comfortable. This includes expanding the road infrastructure and also improving the throughput of existing traffic routes by establishing intelligent traffic flow management mechanisms. Optimizing traffic light control strategies is not longer a

---

[1] http://www.invent-online.de
[2] http://www.vt.bv.tum.de/uploads/rbraun/Travolution_Plakat_2007-07-09_TUM-VT.pdf

In D. de Waard, F.O. Flemisch, B. Lorenz, H. Oberheid, and K.A. Brookhuis (Eds.) (2008), *Human Factors for assistance and automation* (pp. 63 - 74). Maastricht, the Netherlands: Shaker Publishing.

local problem addressing single intersections but rather a global task involving many intersection points at the same time (e.g. Hewage & Ruwanpura, 2004, and De Oliveira et. al., 2006). Such measures could be even more effective by influencing the drivers' behaviour individually. One possible way to assist the driver is to display information about the traffic light phases while the car is approaching an intersection. Some countries use digital countdown displays or static green wave speed recommendations mounted directly to traffic lights. These displays are designed to help the driver choose an appropriate speed in order to pass the traffic light during the green phase. However, these signals are equal for all drivers approaching the intersection and do not take individual traffic conditions into account (e.g. vehicles driving ahead). Additionally, Kidwai et. al. (2005) state that countdown traffic lights have only little effect on traffic throughput.

Infrastructure-to-car communication can be used to present additional information concerning the current traffic light state to the driver within the car and with respect to his own position and speed. It is important to choose the display type with care as intersections are highly complex situations and additional visual cues may cause distraction.

Concepts

Three different display concepts for traffic light assistance were designed by HMI experts during a brainstorming session and then implemented in a driving simulator environment.

Countdown Human Machine Interface (HMI)

The first one was a simple countdown which displays the remaining time in seconds of the current traffic light phase. The information is shown directly as text and supported by a LED display around it which gives a graphical representation of the number (further referred to as "Countdown HMI"). Figure 1 shows a few snapshots.

Figure 1. Countdown HMI: Remaining seconds in the green phase. For colour versions of the images please visit http://extras.hfes-europe.org

Minimum/Maximum speed HMI

As it might be hard for drivers to deduce an appropriate velocity given only the remaining time, an alternative display type was designed. Using the remaining time in the current phase of the traffic light and the distance to the traffic light, the car computer is able to estimate whether it is possible to pass the traffic light during the green phase and, in that case, calculate a speed recommendation. Basically, the HMI needs to address four scenarios: 1. It is not possible to pass the traffic light during green phase at all. 2. No speed change is necessary to pass the traffic light. 3. Acceleration is required. 4. Deceleration is required.

Taking these facts into account, the second concept was a combination of symbols showing speed recommendations in plain text (referred to as "Min/Max HMI" from now on). Figure 2 shows some examples.

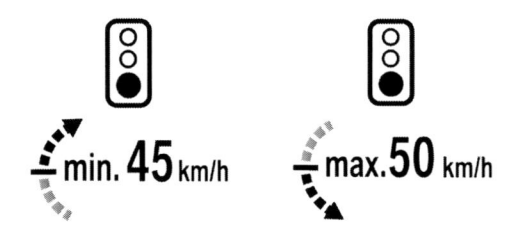

Figure 2. Min/Max speed recommendation HMI

Speed indicator HMI

The "Min/Max HMI" displays the boundary speed which needs to be exceeded in case of a minimum recommendation or must not be exceeded in case of a maximum recommendation. However, the recommendation should rather be a velocity range than a specific speed. Thus a third concept based on Voy et. al. (1981) and Haller (1995) was implemented which makes use of the speed indicator. Coloured areas are overlaid onto the speed indicator: a green area means that travelling in this speed range will allow passing the green traffic light whereas a red area indicates that stopping at the traffic light will be necessary. Figure 3 shows an example where acceleration is required.

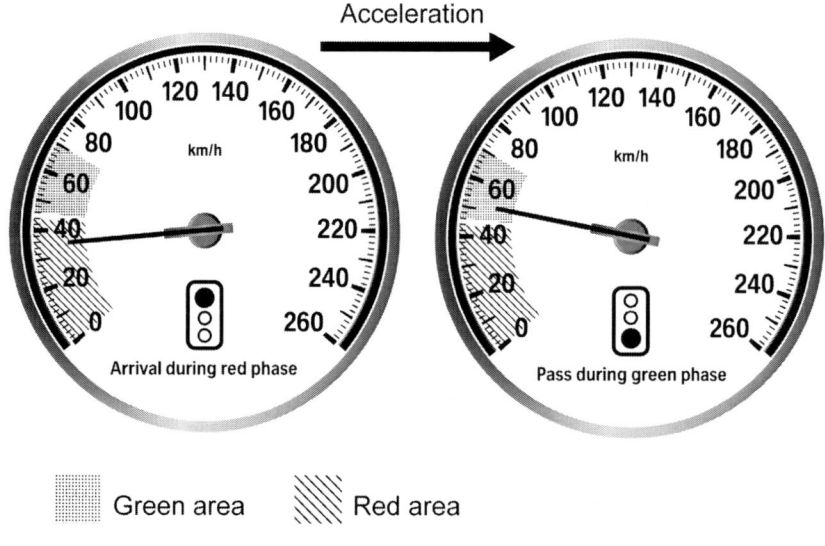

Figure 3. Speed indicator HMI

Since colour-blind persons might have problems to differentiate between the red and green areas on the speed indicator scale, an additional icon was added. This icon shows the expected state of the traffic light on arrival based on present speed. Replacing red and green by different colours is an option, but might weaken the correspondence to the traffic light and thus lead to worse comprehensibility for non-colour-blind persons. Nevertheless, no colour-blind persons were invited for this experiment to avoid an additional disturbance variable in the setup.

Experimental design

The goal of the experiment was to determine which HMI should be implemented in a real prototype car equipped with wireless communication technologies and fully programmable displays.

Comprehensibility during first exposure

Especially for car manufacturers, it is important to use a clear and intuitive information representation within the vehicle as the displays should be comprehensible without reading a manual. Thus participants of the study were asked to pass several traffic lights in the simulator and state their understanding of the HMI during a short interview afterwards. No additional explanation was given in advance. Every participant experienced only one of the three HMI types during the first exposure driving session.

Effectiveness

A traffic light HMI is considered objectively effective if it leads to driving behaviour which is optimal with respect to fuel consumption. Having a realistic consumption model of a real vehicle in the driving simulator is quite complex, as many factors are of influence: environmental conditions, climate controls within the vehicle, engine management system, etc. Therefore, the requirement of minimal fuel consumption was interpreted as follows: the main goal of the HMI is to prevent the driver from performing any unnecessary accelerations or decelerations while approaching the traffic light. As an aggregated value for a single approach, we use the integral of the absolute value of the vehicle's acceleration. Any speed changes - acceleration and deceleration - lead to a greater integral value. Therefore lower values indicate better performance of the HMI. The integration limits were 300m before and 50m after the traffic light.

Acceptance

General data like age, gender and simulator experience was collected prior to the experiment in a demographic questionnaire. During the main part of the experiment, participants experienced each HMI type during a "main session" (details see "Experimental Procedure"). After each main session participants had the opportunity to evaluate the HMI which they just had seen before on a five-point Likert Scale with respect to various aspects (see "Single system evaluation" in Table 1). There was also room for answers to three open questions: Aspects they liked/disliked and suggestions for improvement. After completing all main sessions participants compared the three HMI types on a Likert Scale with respect to distraction and comprehensibility (see "System comparison" in Table 1).

Visual behaviour

During the whole experiment, eye tracking was used to estimate the amount of time that drivers spent reading the additional displays. The drivers' glances were recorded from 300m before to 50m after the traffic light.

Simulator setup

The experiment was conducted in a fixed-base driving simulator equipped with a five-screen projector imaging system (Figure 4). Each screen had a height and width of 2.8 m. The field of view for the participants was approximately 240°. No rear or mirror views were used during this experiment.

Figure 4. Simulator and mock-up setup (In this picture, only three of five projectors are active)

Experimental procedure

The participants were invited to take part in a simulator study without being given any detailed information about the system under investigation. After completing a short training session to familiarize themselves with the driving simulator, each person was confronted with one of the three HMIs. During a short interview after the first two simulator sessions, comprehension of the HMI was tested. A between subjects design was used since an explanation of the first HMI would influence comprehension of all following HMIs dramatically. Although the HMI types "Min/Max" and "Countdown" are predestined for being presented in a head-up display, they were both shown inside the speed indicator to avoid an unintended comparison between head-down and head-up display positions with regard to visual behaviour.

For the four main sessions a within subjects design was used (the three HMI types and a baseline). Prior to every session, the HMI was explained to the subjects and correct comprehension was ensured. Participants drove through a route in an urban environment in the simulator with traffic lights every few hundred meters. The order of the four sessions was randomized. During every main session, subjects passed at least twelve traffic lights with six different scenarios implemented (referred to as factor "scenario"):

1. "Always red": It is not possible to pass the traffic light during green phase at all
2. "Always green": No speed change is necessary to pass the traffic light
3. "+5 km/h": Acceleration by 5 km/h is required
4. "+15 km/h": Acceleration by 15 km/h is required
5. "-5 km/h": Deceleration by 5 km/h is required
6. "-15 km/h": Deceleration by 15 km/h is required

Each scenario was presented twice (factor "recurrence") and in randomized order to the participants during each main session. All subjects were asked to maintain a constant speed of approximately 60km/h when no reasons for changing speed existed (traffic lights or HMI displays). Every main session (except baseline) was followed by a questionnaire prompting (the individuals) for their subjective opinion. After finishing all sessions, the subjects filled out a concluding questionnaire which gave them the opportunity to compare the three HMI types directly.

Participants

Twenty-seven participants, 23 men and four women, took part in the driving simulator experiment; their mean age was 29 years; six persons had prior driving simulator experience; no spectacle wearers were invited due to the eye tracking equipment used. For this first experiment, no persons with colour-blindness were invited.

Results

Comprehensibility during first exposure

The responses collected during the short interviews were categorized: "Instantly understood", "understood after some traffic light passes" and "not understood". It turned out that almost all the subjects understood how the HMI worked and what its purpose was after the first exposure session. This was true for all three HMI types.

Effectiveness

Figure 5 shows velocity profiles of the subjects, the mean value and the standard deviation for the baseline and the speed indicator HMI relatively to the distance to the traffic light in meters. The following scenario is shown: the traffic light is red and turns green while the driver is approaching. The traffic light control is triggered in a way that the maximum speed recommendation presented lies 15 km/h below the current velocity of the vehicle.

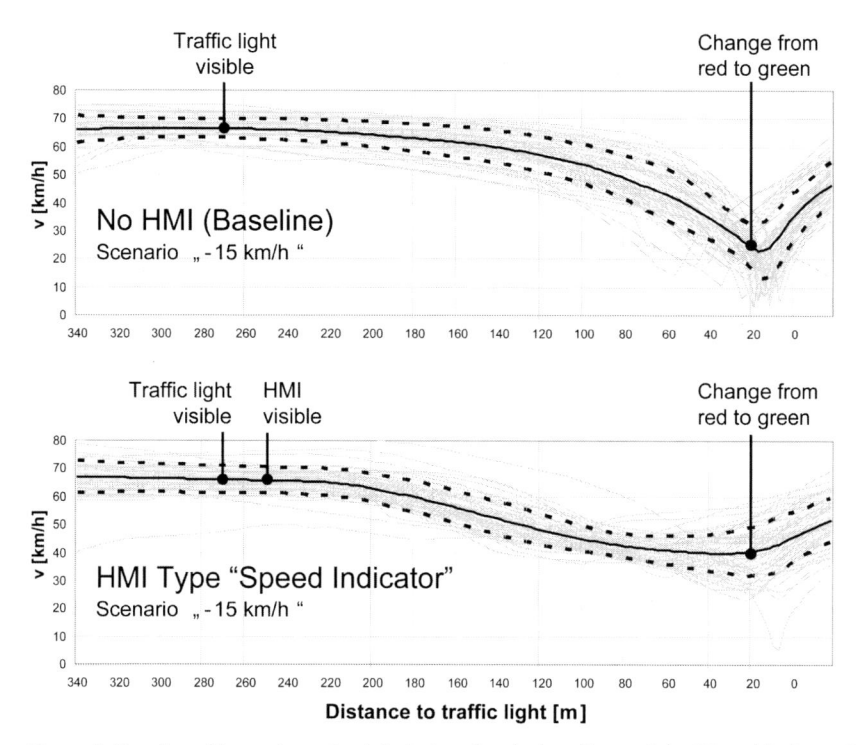

Figure 5. Speed profiles and standard deviations for the baseline and the "Speed Indicator" HMI for scenario "-15 km/h"

It can be seen that the drivers decelerated less and reacted earlier when a speed recommendation was shown. The traffic light became visible between 300 and 250 meters distance from the traffic light, depending on the scenario. The HMI was displayed starting from 250 meters distance to the traffic light. An analysis of variances with repeated measures of the objective acceleration integral measure showed a significant effect for the factor "HMI type" ($F(3, 54)=10.02$, $p<.001$) and "Scenario" ($F(5, 90)=108.60$, $p<.001$). No learning effect could be seen as the recurrence of the scenario for one HMI type did not show a significant effect. Figure 5 shows the results for all HMI types and scenarios.

As expected, scenario "Always green" leads to small and "Always red" to large integral values. As deceleration and acceleration are treated equally by using the absolute value, the scenario "Always red" sums up to approximately 120 km/h: Drivers need to decelerate from 60 km/h to zero and then accelerate again to 60 km/h after the traffic light. The other four scenarios lie somewhere in between these two extremes.

Wilcoxon tests were used to determine significant differences in HMI performance. The speed indicator HMI was significantly more efficient than the baseline and more efficient than the other two HMIs in scenarios "-5 km/h" and "-15 km/h". Even in the scenario "Always green" all HMIs showed some advantage compared to the baseline. Scenarios "+5 km/h" and "+15 km/h" have extremely large standard deviations: As the drivers were not forced to react to the HMI but were instructed to behave as they would in a real world situation, some decided to follow the recommendation – and passed the traffic light, and others did not – and stopped in front of the traffic light. Therefore, some cases in this group tend towards scenario "Always green" and others towards scenario "Always red".

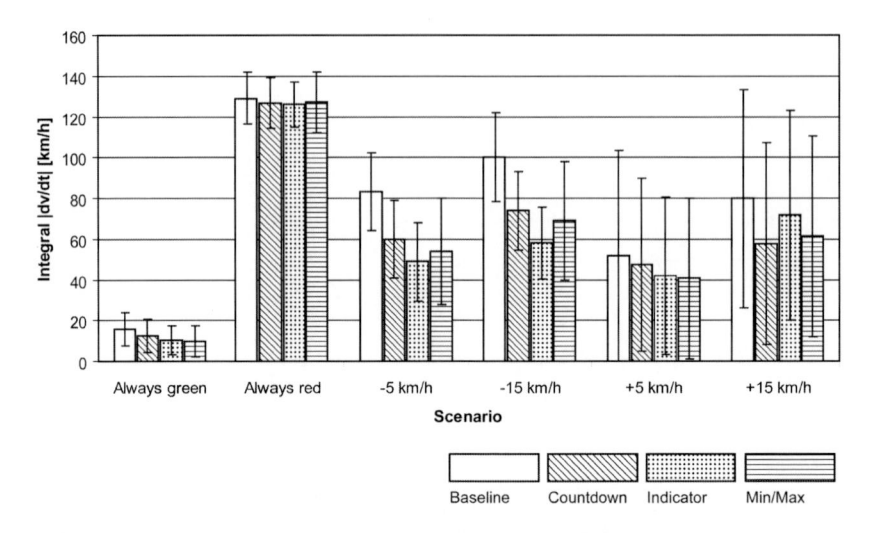

Figure 6. Effectiveness of different HMI types and standard deviation

Generally, high speed variability might also be an indicator for a suboptimal understanding of a certain HMI display. Thus this measure could reflect the comprehension during the driving task (in contrast to the direct evaluation of comprehension during first exposure).

Acceptance

Acceptance was evaluated based on the questionnaires the participants filled out after each main session and after completing all main sessions. Table 1 shows the statements (meaning might vary due to translation from German) and the rating for each HMI type. Apparently almost all the responses are in favour of the speed indicator HMI.

Visual behaviour

Every glance at the assistance display, respectively the speed indicator, was logged from 300m before to 50m after the traffic light.

Figure 7 shows the average glance frequency distribution for one example scenario. It can be seen that most glances are below two seconds and that on average, eight scans are required during one traffic light pass to perceive the HMI display. The countdown HMI produces slightly more but shorter glances than the other two HMI types. The AAM statement of principles (AAM, 2002) for in vehicle information systems states that the 85th percentile of all glances should be below two seconds, which is true for all display variants (Figure 8).

Table 1. Questionnaire statements and number of participants who agreed to them

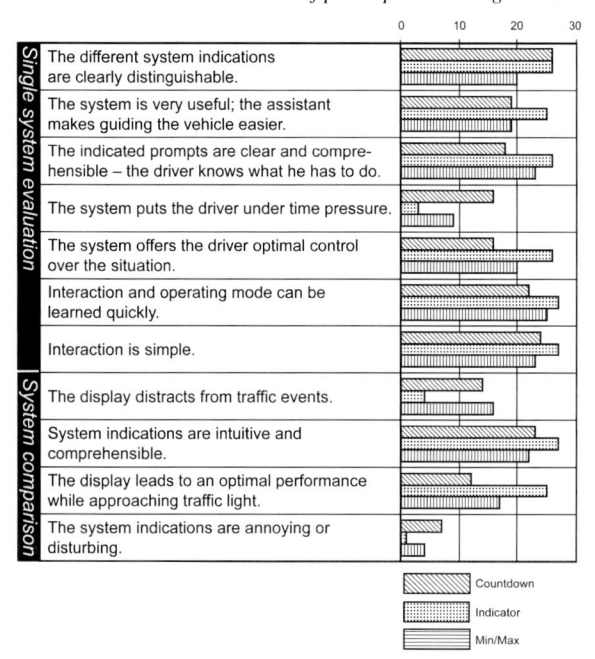

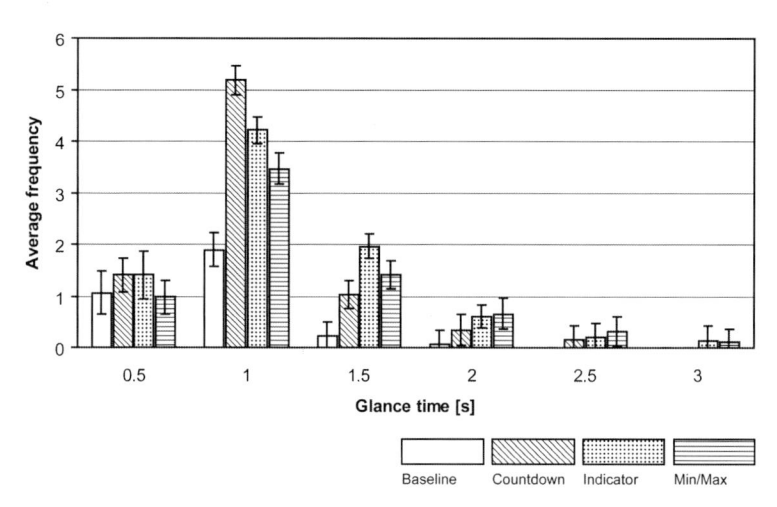

Figure 7. Average glance frequency against glance duration and 95% confidence interval for scenario "-5km/h"

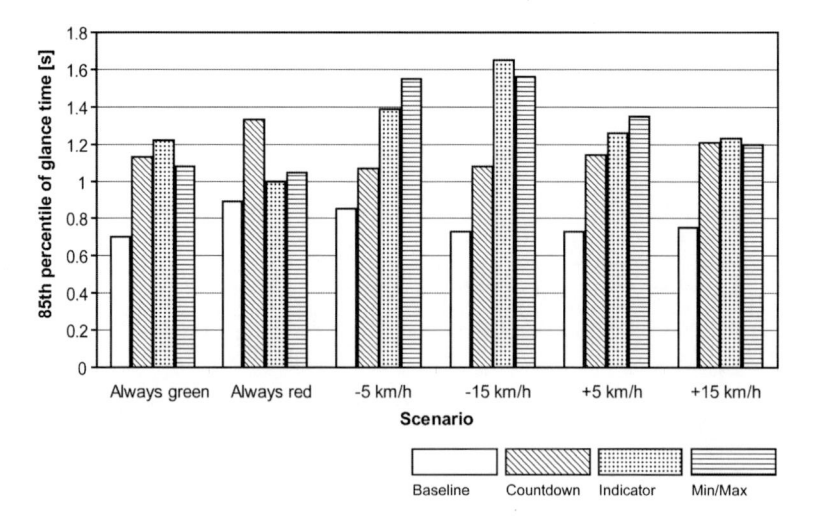

Figure 8. 85th percentile of glance times for all scenarios

Discussion

In this setup, no other traffic was present to allow an experimental setup without any disturbance variables and to assure that the intended scenarios occurred reliably. Although most of the glances were below two seconds, the number of glances was relatively high. It is expected that a natural traffic environment would reduce the number of glances. The countdown HMI is more predictable than the other two HMIs because it does not depend on the vehicle's speed and position and changes in regular time intervals. Therefore, subjects looked at that display less often and for shorter times. In a real world situation, the velocity-based HMIs might be faster to understand as time-based displays require the driver to extrapolate the vehicle's future position. An additional experiment focusing on visual behaviour will be conducted including a more complex and realistic driving course.

Besides usability aspects, many legal issues need to be solved before a system like this could be brought to market. Not giving any recommendations above the local speed limit is a basic requirement. How much of the amber phase is treated as green by the system for example can have a major effect on acceptance, but also has many safety implications on the other hand.

As the experiment was not intended to find any gender effects, the sample was unbalanced. Future studies should include the influence of gender on acceptance and efficiency of the HMI variants.

In order for the driver to benefit from the traffic light HMI, a high penetration grade is desirable. If a traffic light does not transmit its current and future status, no recommendation can be displayed. Image recognition using an on-board camera can detect the current but not the future status of a traffic light. Thus its use is restricted to showing information indicating the existence of a traffic light without being able to give a dedicated speed recommendation. In case of a very high penetration grade, image recognition could avoid confusion for the driver as the vehicle is able to display explanatory information: "There is a traffic light but no speed recommendation can be calculated due to missing radio signal."

Since traffic lights could potentially serve as repeaters for car-to-car communication, network effects can be expected that may accelerate the rollout of wireless-enabled traffic lights (Matheus, 2004).

In summary, it can be stated that the speed indicator HMI has advantages concerning acceptance and efficiency. Therefore, it was implemented in a prototype car. Additional field tests need to show the suitability for daily use under real traffic conditions.

References

Alliance of Automobile Manufacturers. (2002). *Statement of Principles, Criteria and Verification Procedures on Driver Interactions with Advanced In-Vehicle Information and Communication Systems*, Version 2.0 (Report of the Driver Focus-Telematics Working Group). Southfield, USA: Alliance of Automobile Manufacturers

De Oliveira, D., Bazzan, A., Da Silva, B.C., Basso, E. & Nunes, L., Rossetti R., De Oliveira E., Da Silva, R., & Lamb, L. (2006). Reinforcement Learning based Control of Traffic Lights in Non-stationary Environments: A Case Study in a Microscopic Simulator. In *Proceedings of EUMAS'06, Fourth European Workshop on Multi-Agent Systems*. Lisbon, Portugal: Universidade de Lisboa.

Haller, R. (1995). *Anzeigevorrichtung zur Darstellung von Differenzgeschwindigkeiten*. Offenlegungsschrift Patent DE 4325721 A1. München, Germany: Deutsches Patent- und Markenamt.

Hewage, K. & Ruwanpura, J. (2004). Optimization of Traffic Signal Light Timing Using Simulation. In *Proceedings of the 2004 Winter Simulation Conference*, (pp. 1428-1433) Washington, D.C.: IEEE

Kidwai, F.A., Ibrahim, M,R. & Karim, M.R (2005). Traffic Flow Analysis of Digital Count Down Signalized Urban Intersection. *In Proceedings of the Eastern Asia Society for Transportation Studies, Vol. 5* (pp. 1301-1308). Bangkok, Thailand: Eastern Asia Society for Transportation Studies.

Matheus, K., Morich, R., & Lübke, A. (2004). Economic Background of Car-to-Car Communication. In *Proceedings of IMA 2004, Informationssysteme für mobile Anwendungen*. Braunschweig, Germany: Gesamtzentrum für Verkehr Braunschweig e.V. (GZVB)..

Voy, C., Zimdahl, W., Mainka, W., & Stock, F. (1981). *Einrichtung zur Verkehrsführung nach dem Prinzip der grünen Welle. Offenlegungsschrift* Patent DE 3126481 A1. München, Germany: Deutsches Patent- und Markenamt.

Pupillometry as a method for measuring mental workload within a simulated driving task

Maximilian Schwalm[1], Andreas Keinath[1], & Hubert D. Zimmer[2]
[1]BMW Group, München
[2]Saarland University, Saarbrücken
Germany

Abstract

In this contribution a method of pupillometry is discussed to identify high mental demands on the driver in a simulated driving task. A new method allows to identify the effect of mental demand by measuring changes in size of the driver's pupil and to display the actual demand through an index called "Index of Cognitive Activity" (ICA, Marshall et al., 2004). A study will be discussed where a simulated driving task was used in combination with the method of pupillometry. This study shows that the ICA increases in situations with a higher mental demand on the driver when performing lane change manoeuvres or an additional secondary task. Hence the index of cognitive activity seems to be a suitable method for continuously measuring mental demands while driving.

Introduction

The main issues when developing new automotive human machine interaction concepts are to provide an optimum of functionality and to ensure the safe and efficient use of these systems while driving. Additional systems such as driver information and communication systems that can be used while driving may significantly increase the operational demands on the driver. Using these systems while driving, forces the driver to divide his or her attention between the additional system and the driving task itself (De Waard, 1996). If the resulting demand for the driver through such a dual task situation reaches a critical level, the risk of driving errors increases. Therefore when developing new driver-information or driver-communication systems the goal is to design these systems in such a way as to reduce the additional demand on the driver when interacting with these systems to a minimum.

Because of the apparent necessity to identify the amount of mental demand while driving, several different methods have been proposed to identify this effect on a driver. O'Donnell & Eggemeier (1986) specify three groups of workload measures: (1) Subjective measures (or self-report measures, see De Waard, 1996). These measures directly investigate the subjectively perceived mental workload in a given situation (e.g. questionnaires, interviews). (2) Performance measures. Performance

In D. de Waard, F.O. Flemisch, B. Lorenz, H. Oberheid, and K.A. Brookhuis (Eds.) (2008), *Human Factors for assistance and automation* (pp. 75 - 88). Maastricht, the Netherlands: Shaker Publishing.

measures indirectly infer from the performance in a given task to the actual workload of the driver, based on the idea that an increased demand on the driver results in a decrease in performance (e.g. reaction time, accuracy). (3) Physiological measures. These measures use physiological reactions as correlates for the actual demand on the driver (e.g. diameter of the pupil, heart rate).

These three measurement groups that are discussed in the literature each have different methodological advantages as well as disadvantages.

Subjective as well as performance measures have the advantage of a very fast and efficient application during a product development process. The disadvantage of these measures lies in a reduced temporal resolution of the measurement as well as the insensitivity caused by individual compensating strategies while performing a task. Performance measures reflect how a task is performed, but performance may also be protected by the amount of invested effort for a task (Hockey, 1997). Especially when performing a realistic dual task situation such as driving, the driving performance may be deliberately protected by a reduction of the invested effort for the secondary task, leading to a reduced performance in the secondary task but also a reduced demand for the driver.

Physiological measures however offer an objective and direct approach to the measurement of mental workload and do not require an overt response by the subject (De Waard, 1996). Moreover these measures may offer a continuous, interference-free measurement of the actual mental demand with a high temporal resolution. The disadvantage lies in the increased technical effort for the detection of the signal as well as the separation of the relevant to the interfering signals (signal-to-noise ratio, Kramer, 1991 cited in De Waard, 1996).

The current contribution will discuss a specific method of pupillometry as a physiological measurement of mental workload. Especially the objectivity, the high temporal resolution and the online availability of the signal are the main advantages that make the method of pupillometry interesting for an application oriented research in the automotive context.

The method of pupillometry as a physiological measurement

The positive relation between an increase in the pupil diameter and mental demand has been discussed in the literature for decades (Beatty & Lucero-Wagoner, 2000; Just, Carpenter & Miyake, 2003). Empirical evidence could be found for a whole range of tasks such as mental arithmetic (Hess & Polt, 1964; Ahern & Beatty 1979, 1982; Breadshaw, 1968), signal detection tasks (Hakerem & Sutton, 1966 Beatty & Lucero-Wagoner, 1975; Kahnemann & Beatty, 1967), memory tasks (Kahnemann & Beatty, 1966; Graholm, Asarnow, Sarkin & Dykes, 1996) and language processing (Schlurhoff, 1982; Beatty, 1982; Just & Carpenter, 1993). The reason why pupillometry as a measure for mental demand has only been playing a minor role in applied research lies mainly in the fact that the size of the human pupil is dependent on a wide range of different variables such as light density, distance accommodation, heartbeat, breathing frequency, etc, factors that are not directly related to mental demands. The challenge when developing a new measure of mental demand using the size of the human pupil therefore lies in resolving the raw pupil signal from the influence of these other influences.

In previous studies in that the pupillometric measures were used it was attempted to hold irrelevant influences to the size of the pupil constant throughout the data collection. Such an approach, however, drastically limits the application of the method to those laboratory based situations where these variables can indeed be experimentally controlled. But such a procedure could certainly not be applied to realistic and application oriented situations, as for example a driving situation based on the continuously changing environment (lightning condition, distances etc).

An additional problem when measuring the size of the human pupil results from the fact that the absolute size of the human pupil varies between individuals. A comparison of absolute differences in size is therefore not practicable. A meaningful measurement of mental demand through the size of the human pupil should therefore be based on relative rather than absolute changes in size (Marshall, Davis, & Knust, 2004).

The "Index of Cognitive Activity"

In order to cope with these problems, Marshall and colleagues developed the "*Index of Cognitive Activity*" (ICA) (US Patent Marshall, 2000). More detailed discussions of this index can be found in Marshall (2000, 2005, 2007; Marshall, Davis & Knust, 2004). Therefore only a short overview on its rational should be given here instead. The ICA was developed to resolve the previously discussed methodical problems when assessing mental demands through changes in the size of the human pupil. The Index is calculated from high-frequency components of changes in the pupil size while an individual performs a specified task that requires significant cognitive processing. This effect has previously been described as the dilation reflex (Loewenfeld 1993) and is based on the fact that in this case the two muscle groups namely the dilator as well as sphincter muscles that cause the pupil to dilate and to constrict are working together simultaneously. While the dilator muscles are activated (causing the pupil to dilate) the sphincter muscles are inhibited (also causing the pupil to dilate). The result is a brief dilation that is greater than either muscle group could effect, subserved by both sympathetic as well as parasympathetic inputs (Beatty & Lucero-Wagoner, 2000). These changes in the size of the pupil based on the dilation reflex are irregular and sharp, often exhibiting large jumps followed by rapid declines (Marshall, 2000). This response is quite different to the one caused by the light reflex. Therefore the ICA is a measure of relative changes of pupil size that reflects the number of times each second that those large and abrupt increases occur which exceed a specified threshold in the amplitude of the signal. A higher ICA value indicates more changes per second and therefore a higher amount of cognitive activity.

For a more detailed discussion about the calculation of the index from the raw pupil data please refer to the works of Marshall (2000), Marshall, Davis & Knust (2004), Marshall (2007).

The ICA can be computed for short time periods such as single trials in a multi-trial study as well as for long periods of time. While the index is a relative measure of the fast and abrupt changes in the high frequency components of the pupil signal, and as such it is independent of changes in lightning conditions as shown in Marshall (2000) as well as Marshall, Davis, & Knust (2004).

The Index of Cognitive Activity has been shown to be sensitive to different levels of mental load in a whole range of different cognitive tasks. Examples are mental arithmetic and sequence tasks (Marshall, Davis & Knust, 2004), gauge monitoring and the detection of strategic shifts (Marshall, Pleydell-Pearce, 2002; Marshall, 2002), and military tactical decision making (Marshall, 2007). Higher ICA values were observed in situations or with tasks that high demands on cognitive processing. Hence, the index has delivered a suitable measure of cognitive activity in practice. Higher mental load was associated with an increase of the ICA value and the measure should therefore also be appropriate to estimate cognitive load in a driving situation.

Experiment

The aim of the present study is to apply the ICA methodology to the driving situation. It should be used as a method to identify high mental workload situations in the context of a simulated driving task, by measuring the ICA in the course of driving. There are three situations when mental workload is expected to be increased: (1) at the moment the driver is instructed visually via signs to change lanes, work load should increase; (2) if workload is a consequence of processing of the command and not of perceptual processing, the increase should also occur when the command to change is given as an auditory instruction, and (3) when the driver has to perform a secondary task while driving this enhanced load should be visible in a higher ICA value. In order to investigate this dependency we required our participants to perform a simulated driving task and while driving lane change commands were given.

Driving simulation

For the simulated driving task, software called the Lane Change Test (LCT) (Mattes 2003) has been used. The LCT is a PC based driving simulation designed for measuring driver distraction. Drivers have to repeatedly perform lane changes while driving on a simulated roadway that consists of a three-lane road with a total track length of about 3000 m. While participants are instructed to drive with the gas pedal pressed to its maximum, speed is limited to 60 km/h resulting in a total driving duration of about 180 seconds per track. In the original LCT setup the participants are instructed to change lanes according to visual signs situated at the roadside. Each sign indicates the to-be-taken lane (see Figure 1). There are 18 signs along a track and the mean distance between the signs is about 150 m, resulting in a lane change approximately every nine seconds. Distances between signs show minor temporal differences of a few seconds differing between tracks in order to prevent learning effects (ISO Draft TC22/SC13/WG8). In the present study the LCT has additionally been used in a slightly modified form, so that instructions were given via voice instructions through headphones instead of the signs next to the road. In this condition the signs showed no information. Between the manoeuvres the driver has been instructed to keep the lane as precisely as possible.

In the lane change task, driving performance is measured through the deviation between a normative driving path – this is the optimal lane change path prescribed by the model when a change command is presented – and the actual course of the subject along the track averaged over time points. The LCT has been shown to be sensitive to both visual and cognitive distraction while driving (Mattes, 2003, 2006; ISO Draft TC22/SC13/WG8).

Figure 1. The driver's view of the simulated three lane road scene with the signs indicating the correct lane

Secondary task

To manipulate the amount of mental demand for the driver, the LCT is usually combined with different secondary tasks. In the present study the Surrogate Reference Task (SURT) (Mattes, 2006) has been used as a secondary task. The Surrogate Reference Task is a visual search task, where subjects have to identify the position of a target (a white ring) among 50 distractors (white rings with a smaller diameter than the target) on a screen placed next to the LCT screen (distances and viewing angles according to the LCT ISO Draft TC22/SC13/WG8). The area in that the target circle was located had to be marked by a grey bar that could be moved by pressing the cursor keys. The task could be interrupted and started again at any time during execution.

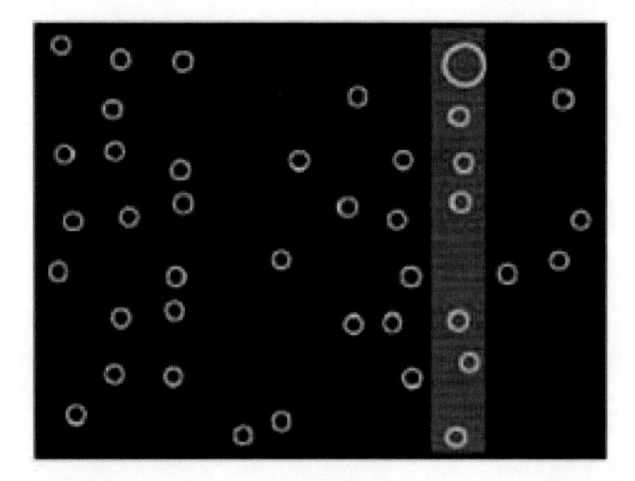

Figure 2. An example of the secondary task display showing the target and distractors in the easy condition

Eye-tracking apparatus

The data reported here was collected using the EyeLink 2 Eye-Tracking System from SR Research, Ltd., with binocular tracking at a sampling rate of 250 Hz. The EyeLink 2 system consists of small video cameras mounted on a lightweight headband. The system offers a pupil size resolution of 0.1% of diameter at an averaged pupil diameter of about 3-7 mm.

Procedure

On arrival at the laboratory the aim of the current study as well as the general procedure was briefly explained to each participant. At the beginning participants were given the opportunity to practice the Lane Change Task while being instructed to perform each of the lane change manoeuvres as fast but as exactly as possible. After having become familiar with the task, participants had to drive one of the tracks until the criteria of a mean deviation less than 1 could be reached (ISO Draft TC22/SC13/WG8).

After calibrating the eye-tracker it was made sure that the quality of the obtained pupil data was sufficient in order to calculate the ICA throughout the whole experimental procedure. Each participant performed each of the three experimental conditions through one complete track on the Lane Change Task: LCT visual, visual plus SURT, LCT auditory. The sequence was interindividually counterbalanced. When performing the LCT plus SURT, participants were instructed not to prioritise one of the tasks but to try to perform as well as possible in both of the tasks.

At the end of each experimental condition the subjectively perceived mental demand was rated by the participants via a German version of the NASA TLX questionnaire (Hart & Staveland, 1988). The NASA TLX includes ratings on six subscales of workload such as mental demands, physical demands, temporal demands, own performance as well as effort and frustration.

Sample

Twenty participants (10 men and 10 women) took part in the current study. The participants ranged in age from 23 to 33 years, with a median age of 26.6. All participants were holding a valid drivers licence.

Results

Driving performance

To determine the performance on the driving task, the mean deviation in the lane change path (determined by the difference between the actual driven path and the optimal path) was calculated for the three experimental conditions (see Figure 3). An analysis of variances (ANOVA) with repeated measures showed a significant main effect for the factor of experimental condition $F(2,28) = 55.9$, $p < .001$. Mean deviation in the dual task condition (LCT plus SURT) was significantly higher than in the two pure driving conditions, $F(1,14) = 70.85$, $p < 0.001$. In contrast, the driving performances in with visual and auditory change commands were not different, $F(1,14) < 1$. Hence, the secondary task caused the intended additional load, but processing the visual signals was not harder than processing the auditory signals although the visual signal puts a strain on the same system as driving whereas the auditory condition did not.

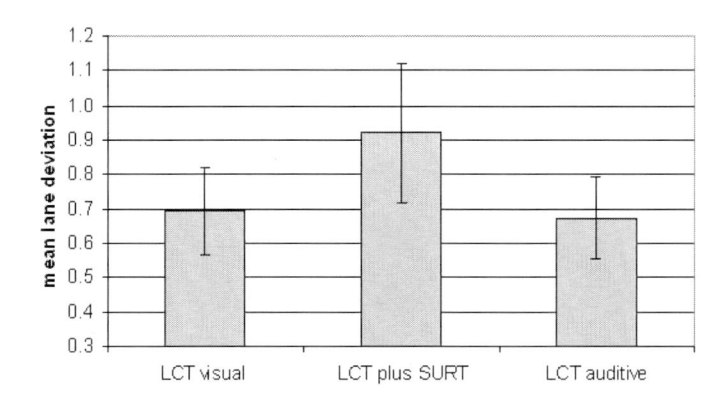

Figure 3. Mean deviation in the lane change path for the condition LCT visual, LCT auditory and LCT plus SURT (error bars indicate +- 1 SD)

Subjective ratings

The subjective ratings of the perceived mental demands as measured by the NASA TLX questionnaire at the end of each experimental condition showed the same pattern of results. Significant differences occurred in the perceived mental demands between the two conditions with and without a secondary task on each of the six dimensions, $F(1, 14) = 46.95$, $p < 0.001$ and no significant difference between the visual and auditory conditions. Here the condition with an additional secondary task was rated to be much more demanding than without such an additional task while driving.

Index of Cognitive Activity

The ICA was calculated for each participant and the period of each experimental condition (see Figure 4). An ANOVA with repeated measures over the three conditions showed a significant main effect for the mean ICA values, $F(2, 28) = 217.3$, $p < 0.001$. A planned contrast revealed a significant increase in the mean ICA values in the LCT plus SURT compared to the two mere driving conditions, $F(1, 14) = 259.17$, $p < .001$. No significant difference was observed between the visual and auditory LCT, $F(1, 14) < 1$. To check the relationship between ICA and task performances we additionally correlation the number of correctly solved tasks (correctly identified targets) and the mean ICA values. Both measures significantly correlated, $r = .55$, $p < .05$. That means, when in the same amount of time more additional tasks were completed while driving, the mental workload of the driver indicated by the ICA increased as well and vice versa.

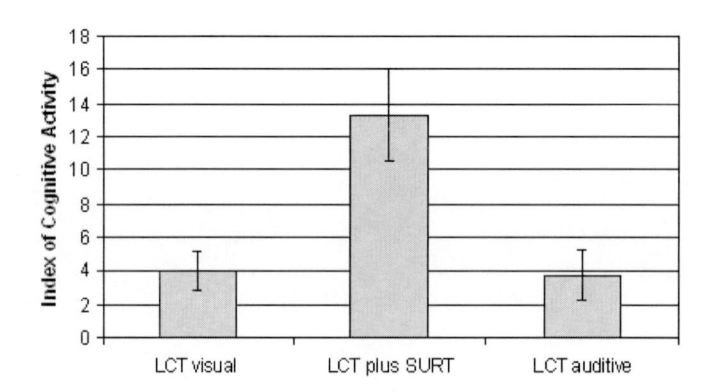

Figure 4. Mean Index of Cognitive Activity (ICA) for the condition LCT visual, LCT audi-tory and LCT plus SURT (error bars indicate +- 1 SD)

The ICA has the advantage that it offers not only a global measure of load, but a continuous measurement of mental workload in a given situation with a high time resolution. We therefore additionally conducted a closer analysis of the changes of ICA in the course of driving. For that purpose we synchronised the starting point of each track and then averaged the ICA values over the 20 participants. Figure 5 depicts the grand averages of ICA during driving separately for the visual and auditory LCT condition. The grey bars indicate the time interval in that participants processed the signal and changed lane. The temporal distribution of ICA revealed a very clear synchrony between the ICA amplitudes and the time periods of the different task components. ICA increased and had a local maximum during any of the 18 lane change manoeuvres (see Figure 5). The local maximum of ICA was more than twice as large (around 10) as during driving without lane change (around 4). In light of the standard deviation of ICA at the group level and of the complete regularity of the effect this can be considered as a highly reliable result.

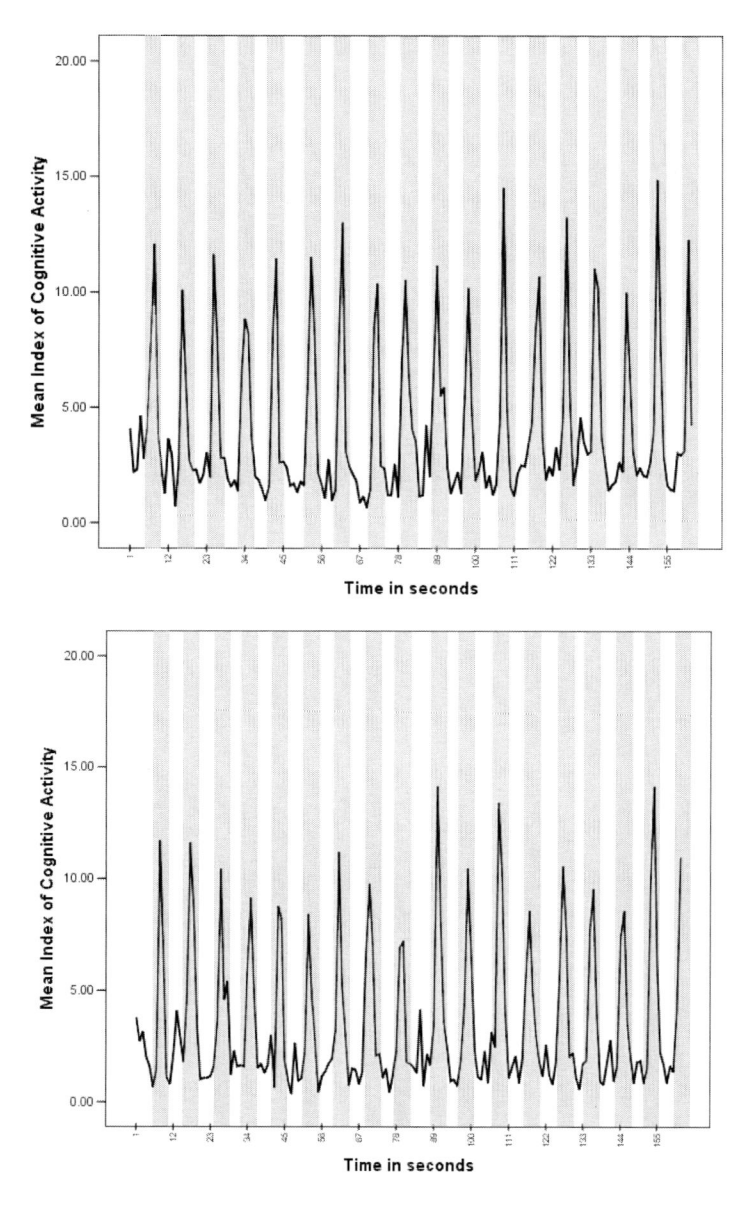

Figure 5. Grand average of ICA in the visual LCT (top image) and in the auditory LCT(bottom image). Grey bars indicate the 18 lane change manoeuvres

Additionally, we analyzed ICA in the LCT plus SURT (see Figure 6). Contrary to out expectation, the due to the secondary task condition generally high ICA values significantly *decreased* around each of the 18 signs. An ANOVA was carried out to compare the mean ICA values during the lane change manoeuvres to the situations where drivers only had to keep the lane while driving. The analysis confirmed this decrease around each of the 18 signs compared to the situations without a sign, $F(1,155) = 92.65; p < .001$.

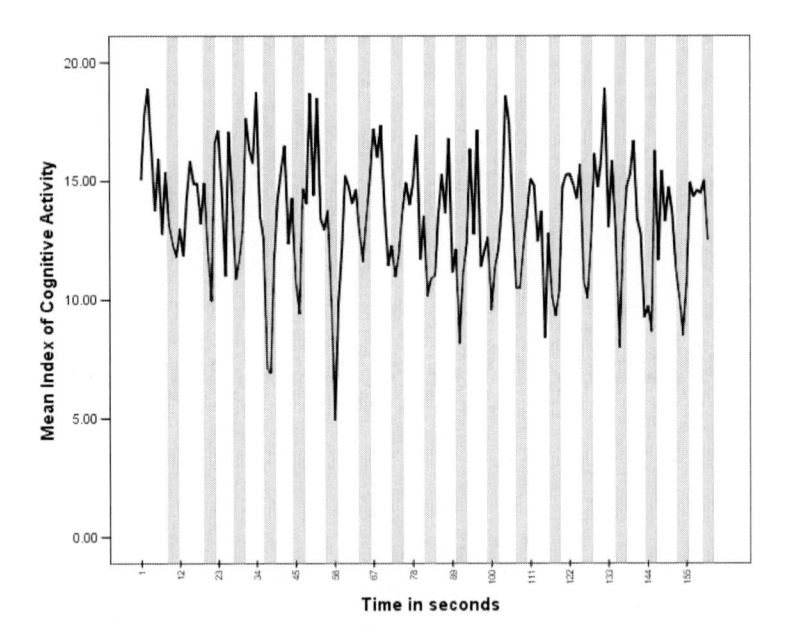

Figure 6. Grand average of ICA in the LCT plus SURT condition. Gray bars indicate lane change manoeuvres

Discussion

The Lane Change Task was developed to manipulate mental demands in driving performance. Correspondingly, we observed the expected results in combination with the secondary task. The mean deviation clearly increased when an additional secondary task had to be performed while driving. The higher demand was also reflected in the subjective measure of mental effort and in the mean ICA. The two instruction modalities did not differ. Obviously, perceptual processing of the simple signs cause only minimal additional costs, so that the driving condition is not impaired by modality-specific interference.

One advantage of ICA is that one can additionally identify the time course of cognitive demands during driving. The ICA dramatically changed during driving and lance change. It significantly increased at the moment the command was presented and the driver had to understand the instruction and had to change lanes. Hence, by the ICA a higher temporal resolution could be reached when measuring mental demands while driving than could ever be obtained by using traditional global performance measures such as the mean deviation from the optimal driving path or subjective ratings. Because this effect was unaffected by the modality of instruction, we conclude that the enhanced load is due to additional load on higher cognitive processes and not perception.

The unexpectedly observed decrease of ICA in the secondary task condition, demonstrates the potential of the ICA for a task analysis and for estimating the on-line demands of a task. When driving together with the secondary task, ICA significantly decreased during the lane change manoeuvres. Such a decrease is counterintuitive. It is plausible, however, when we analyse participants' driving behaviour together with their processing of the secondary task. They stopped working on the SURT when the instruction was given and they focussed on driving. In other words, they switch from a dual task (driving and performing the secondary task) to a single task situation (only driving). This clearly demonstrates that a micro analysis of task is necessary and it is not sufficient to evaluate the task as whole. We have to consider such strategic serial scheduling of two tasks if the discrete characteristic of the secondary task (or the primary task) allows such behaviour. Applied to the driving situation, this follows the idea of a driver who actively manages his workload when performing secondary tasks while driving.

If one follows this interpretation, the ICA has the ability to detect strategic shifts of attention by the driver, as produced by a switch between a dual-task to a single-task situation. ICA would therefore allow insights into the internal strategies of the driver and his workload managing behaviour, independently of the type of system the driver is interacting with. That the ICA is useful for a detection of strategic shifts has previously been shown in experimental settings (Marshall, Pleydell-Pearce, Dickson, 2002, Marshall, 2002, 2007). But this is the first time it has been shown in a driving context.

In summary, the current experimental results show that the "Index of Cognitive Activity" could be a suitable method for measuring mental workload while driving and to detect strategic shifts of attention by the driver. The index could therefore be a valuable instrument when optimising new HMI concepts to the needs of the driver, ensuring a safe and efficient use while driving. One can think on many other applications. Examples are exploring driver learning, the acquisition of new skills, the efficiency of training, the influence on drugs on cognitive load, effects of advanced aging, etc. In all these cases, the "Index of Cognitive Activity" gives us the chance to get insights into the time course of mental workload during driving. This measure therefore is a valuable enrichment of the instruments of cognitive ergonomics.

References

Ahern, S., & Beatty, J. (1979). Pupillary responses during information processing vary with scholastic aptitude test scores. *Science, 205*, 1289-1292.

Ahern, S., & Beatty, J. (1981). Physiological evidence that demand of processing capacity varies with intelligence. In M. Friedman, J. & Das J. & O'Connor, O. (Eds.), *Intelligence and learning* (pp. 121-128). New York: Plenum.

Beatty, J. (1982). Task-evoked pupillary responses, processing load, and the structure of processing resources. *Psychological Bulletin, 91*, 276-292.

Beatty, J., & Kahneman, D. (1966). Pupillary changes in two memory tasks. *Psychonomic Science, 5*, 371- 372.

Beatty, J., & Lucero-Wagoner, B. (1975). *Pupillary measurement of sensory and decision processes in a signal-detection task.* Paper presented at the Meeting of Psychonomic Society, Denver.

Beatty, J., Lucero-Wagoner, B., Cacioppo, J.T., Tassinary, L.G., & Berntson, G.G. (2000). *The pupillary system.* New York, NY, US: Cambridge University Press.

Ben-Nun, Y. (1986). The use of pupillometry in the study of on-line verbal processing: Evidence for depths of processing. *Brain and Language, 28*, 1-11.

Bradshaw, J.L. (1968). Pupil size and problem solving. *The Quarterly Journal of Experimental Psychology, 20*, 116- 122.

De Waard, D. (1996). *The measurement of drivers' mental workload.*PhD Thesis, University of Groningen. Haren, The Netherlands: University of Groningen, Traffic Research Centre.

Hakerem, G.A.D., & Sutton, S. (1966). Pupillary response at visual threshold. *Nature, 212*, 485-486.

Hart, S.G., Staveland, L.E., Hancock, P.A., & Meshkati, N. (1988). *Development of nasa-tlx (task load index): Results of empirical and theoretical research.* Oxford, England: North-Holland.

Hess, E. H., & Polt, J. M. (1964). Pupil size in relation to mental activity during simple problem-solving. *Science, 143*, 1190-1192.

Hockey, G.R.J. (1997). Compensatory control in the regulation of human performance under stress and high workload: A cognitive-energetical framework. *Biological Psychology, 45*, 73-93.

Just, M.A., & Carpenter, P.A. (1993). The intensity dimension of thought: Pupillometric indices of sentence processing. *Canadian Journal of Experimental Psychology, 47*, 310-339.

Just, M.A., Carpenter, P.A., & Miyake, A. (2003). Neuroindices of cognitive workload: Neuroimaging, pupillometric and event-related potential studies of brain work. *Theoretical Issues in Ergonomics Science, 4*, 56-88.

Kahneman, D., & Beatty, J. (1966). Pupil diameter and load on memory. *Science (New York, N.Y.), 154*, 1583-1585.

Kahneman, D., & Beatty, J. (1967). Pupillary responses in a pitch-discrimination task. *Perception & Psychophysics, 2*, 101-105.

Loewenfeld, I.E. (1993). The pupil: Anatomy, Physiology and Clinical Applications, vols. I, II. Ames: Iowa State University Press / Detroit: Wayne State University Press.

Marshall, S.P. (2000). U.S. Patent No. 6,090,051. Washington, DC: U.S. Patent & Trademark Office.

Marshall, S.P. (2002). *The index of cognitive activity: Measuring cognitive workload.* Paper presented at the 7th IEEE Conference on Human Factors and Power Plants, New York.

Marshall, S.P. (2005). *Assesing cognitive engagement and cognitive state from eye metrics.* Paper presented at the 1st International Conference on Augemted Cognition, Las Vegas.

Marshall, S.P. (2007). Identifying cognitive state from eye metrics. *Aviation, Space, And Environmental Medicine, 78,* B165.

Marshall, S.P., Davis, C., & Knust, S. (2004). *The index of cognitive activity: Estimating cognitive effort from pupil dilation.* San Diego, CA: Eyetracking Inc. Technical Report ETI-0401.

Marshall, S.P., Pleydell-Pearce, C., & Dickson, B. (2002). *Integrating psychophysiological measures of cognitive workload and eye movements to detetect strategy shifts.* Paper presented at the 36th Hawaii International conference on System Sciences.

Mattes, S. (2003). *The lane-change-task as a tool for driver distraction evaluation.* Paper presented at the Annual Spring Conference of the GfA/17th Annual Conference of International-Society-for-Occupational-Ergonomics-and-Safety (ISOES).

Mattes, S. (2006). *Messung der Fahrerablenkung in der Fahrzeugentwicklung.* Paper presented at the Vehicle Interaction Summit 3, Fraunhofer IAO, Stuttgart, Germany.

O'Donnell, R.D., Eggemeier, F.T., Boff, K.R., Kaufman, L., & Thomas, J.P. (1986). *Workload assessment methodology.* Oxford, England: John Wiley & Sons.

Schluroff, M. (1982). Pupil responses to grammatical complexity of sentences. *Brain and Language, 17,* 133-145.

Schluroff, M., Zimmermann, T.E., Freeman, R.B., Jr., Hofmeister, K., Lorscheid, T., & Weber, A. (1986). Pupillary responses to syntactic ambiguity of sentences. *Brain and Language, 27,* 322-344.

A three-level model of Situation Awareness for driving with in-vehicle devices

Nadja Rauch, Barbara Gradenegger, & Hans-Peter Krüger
University of Würzburg
Germany

Abstract

Based on theoretical considerations and empirical evidence on Situation Awareness (SA), a three-level model of Situation Awareness in using in-vehicle devices while driving is proposed: SA can be first measured at a "planning level" where the general willingness to additional task performance is influenced by the awareness of associated risk while using in-vehicle devices in different traffic situations and the awareness of own skills and abilities. On a "decision level" SA can be seen in the actual decision for performing a task based on the estimation of current situational demands. During secondary task execution a situationally aware driver monitors the development of the situation via short glances back to the road and adapts his behaviour according to the changed situational demands ("control level"). Evidence for the assumed levels is given by 24 participants completing a 1 hour test course in a motion-base driving simulator containing different complex situations on rural and urban sections. At predetermined points of the route an additional task was offered to the driver (reading numbers from a visual display aloud). Each time he or she had to decide whether the actual situation was suitable for starting a task and for how long it could be executed. The results show evidence for the influence of SA on all the three levels on secondary task performance while driving: Drivers rejected more tasks in demanding situations and adapted eye glance behaviour to situational requirements. Individual differences in compensation strategies were based on drivers' risk perception of using in-vehicle devices while driving.

Theory

Extended research has been done on the effects of dealing with in-vehicle devices while driving. The overall result is that performance of an additional secondary task clearly reduces driving performance and safety. Typical effects are a decrease in lateral control (e.g. Törnros & Bolling, 2005) or delayed reaction times to sudden events (e.g. Strayer & Drews, 2004). On the other hand, compensation strategies in the primary task of driving can be monitored, e.g. reduction of speed (Horberry et al., 2006), an increase of safety margins (Ishida & Matsuura, 2001) or fewer lane changes (Beede & Kass, 2006). An often neglected fact is that drivers are also able to compensate additional workload by specific interaction strategies for dealing with the secondary task. As McCartt et al. (2006) argue: "phone and driving tasks are paced by experimenters, but in the real world drivers decide when and where to use their phones and may adapt their phone use to varying traffic conditions" (p.92).

Results from telephone interviews (Boyle & Vanderwolf, 2005) and video observations (Esbjörnsson & Juhlin, 2003) indicate specific compensation strategies while using a cell phone. Some drivers do not use the phone in a moving vehicle at all; some use it only while they are waiting at a red traffic light, some even stop the car. While calling, drivers tend to drive more slowly, avoid lane changes, choose sections with lower traffic density for calling someone or ask a passenger to do the call. In demanding traffic situations drivers interrupt the conversation, or the remote caller is informed about the environmental conditions to adapt the conversation.

Individual factors like driver age, gender or experience in the interaction with technical systems influence the general risk perception of secondary tasks and may therefore influence the use of such compensation strategies. Younger drivers, for example, rate the use of secondary tasks while driving less risky than older drivers and are more willing to execute those (Lerner & Boyd, 2005).

While performing a secondary task, drivers try to divide the tasks into smaller chunks und look back to the road at adequate intervals (Salvucci & Macuga, 2002). About 30% of the total task duration is spent on road fixations (Victor, 2005). The upper limit for one display fixation is usually about 1.5 s. (e.g. Wierwille et al., 1988). Fixations on the road last about 300 to 700 ms (Rassl, 2004; Schweigert, 2003). Studies with non-visual and visual distraction tasks show that drivers tend to reduce their visual field to the road during secondary task execution (Harbluk et al. 2007). According to situational demands drivers adapt their glance behaviour: in more complex situations drivers increase their attention to the road (Wierwille, 1993) and reduce the remaining time for secondary task performance (Hella, 1987). With high traffic density, drivers decrease display fixations (Rockwell, 1988). In anticipation of higher demands in the driving task, drivers fixate the display less often (Wierwille, 1993).

It is believed that all mentioned compensation strategies can be used as indicators for drivers' Situation Awareness. Only if drivers know what's going on around them, adequately understand the current situation and anticipate future situational demands they can decide whether and how to handle additional demands from a secondary task. The concept of Situation Awareness was originally developed in aviation but has also become more and more important in the domain of driving. The most famous definition is the one by Endsley (1988), who defines Situation Awareness as "the perception of the elements in the environment within a volume of time and space, the comprehension of their meaning and the projection of their status in the near future" (p. 792). In her model she argues that a cognitive representation (schema) of the current and the future situation is formed on the base of knowledge in long-term memory which guides attention to the relevant cues in the environment (Endsley, 1995). Due to the dynamics of the situation, this mental representation has to be continuously updated. These control processes are further addressed in the model of Adams et al. (1995) where this update is understood as a cyclical process, including an active search in the environment for information which may prove or disprove the activated schema and lead to modifications. One mayor drawback especially of the model by Endsley is its broad concept, including a serial information processing model, decision and task execution stages and furthermore allowing influence of concepts like workload, mental model, memory processes etc. on operator performance.

Due to this and also due to different task demands in driving as compared to aviation (e.g. other timely constraints), application of the model and development of concrete study designs and hypotheses in the field of driving is difficult. Therefore a new approach is used, which is especially developed to investigate Situation Awareness while driving with in-vehicle devices. It is assumed, that the self paced use of in-vehicle devices can be used as an indicator for drivers' Situation Awareness. Situationally aware drivers are able to decide in accordance with the demands of the actual driving situation if execution of a secondary task is safely possible or not. This should enable them to avoid higher workload levels in already demanding driving situations. Maintaining Situation Awareness when using in-vehicle devices includes anticipative processes based on accurate perception and comprehension of the situation just before a secondary task is started as well as continuous updates of the situation model via control processes during secondary task execution. In the context of interactions with in-vehicle devices it is argued that Situation Awareness of a driver influences the execution of secondary tasks while driving at three different levels:

- Planning level: On this level, a driver plans general strategies for secondary task interactions, e.g. never executing special secondary tasks in a moving vehicle, waiting or stopping for answering phone calls etc. On this level, an awareness of own skills and abilities, risk awareness and an awareness of specific secondary task demands are necessary.
- Decision level: On this level, an adequate assessment of the situation prior to the start of a secondary task is crucial to decide properly if a secondary task can be performed in the actual situation. This requires the processes of perception, comprehension and anticipation as defined in the model by Endsley (1995). In highly demanding traffic situations the situationally aware driver ignores an additional task or delays the beginning of the task.
- Control level: During secondary task execution the driving task must further be controlled. This requires an adequate monitoring of situation development in order to continuously compare the expected situation development with the actual one. In case of observed differences, the secondary task needs to be interrupted. Also this recurrent situation assessment is a basic process for maintaining Situation Awareness.

To show the influence of Situation Awareness on these three postulated levels, measures have to be developed which indicate situationally adaptive interactions with secondary tasks while driving. This includes measurements that give insights into the anticipation of potential situational demands and decision making processes. The relevant indicators are here, if drivers avoid the execution of secondary tasks in already highly demanding traffic situations and how long they wait with the beginning of a task (indicators on the decision level). The continuous update of the situation during secondary task execution can be seen in the adaptation of eye glance behaviour according to situational demands and in an adequate interruption of the secondary task based on the demands of the situation (indicators on the control level). If the individual risk assessment on the planning level influences secondary task behaviour, this can be seen in correlations between subjective ratings of associated risk with secondary tasks in real traffic and the objective use of secondary tasks (indicators on the planning level). In the end, the general adequacy of situation-dependent decisions can be measured by parameters of driving safety, like driving errors and collisions.

Method

Tasks

To study the effects of SA on the assumed levels of secondary task interaction, a simulator study in a driving simulator with motion system of the WIVW (Würzburg Institute for Traffic Sciences GmbH[*]) was conducted. The drivers were instructed to perform the driving task together with an additional secondary task.

Driving task
The 1.25 h test course consisted of sections with rural roads and urban areas. As an independent variable the criticality of the different driving situations was varied. In general, the course consisted of several non-critical situations like driving on straight sections on rural roads as well as critical driving situations which required a specific driving manoeuvre like braking or avoiding other vehicles. Eight such specific situations, all including a potential conflict, were realised (e.g. a vehicle parking out from a parking zone in front of the participant's vehicle, a pedestrian crossing the road just in front of the participant's vehicle, a broken-down car on a rural road, curvy sections etc.). The other independent variable was the predictability of the critical situations. Each of the situations was realised three times, varying in the salience of environmental cues pointing to the specific conflict (good, medium, hard to predict). A vehicle parking in, for example, can be easily anticipated if the vehicle indicates and plans to stop at a parking area. Less predictive is the situation if the car starts indicating lately. The least predictable situation is if the car indicates lately and stops at a bus station. Due to the limited scope of the paper, this variation of predictability is not further reported in this paper.

Secondary task
In addition, the driver had to perform a secondary task while driving. At predetermined points of the route, the driver was offered the choice to perform an additional task. This offer was given either just before a critical situation (e.g. just after a parked car started to indicate but was still standing) or in a non-critical situation (on road segments between the critical situations). The offer was signalled

---

[*] for more information see www.wivw.de

by a question mark shown in the Head-Up Display on the front scene (see figure 1 left).

Figure 1. Screenshot of the HUD for offering a secondary task (left) and the secondary task (right)

The offer given, the driver had to decide within 3 s whether the situation was suitable for the secondary task or not, according to the situational demands (decision phase). To start the secondary task, the driver had to press a button on the steering wheel. The secondary task was then presented on a visual display located at a lower position on the middle console of the vehicle. If the driver did not make a request within the decision phase, the offer disappeared and he had to wait for the next opportunity to perform a task.

The task itself consisted of sequentially displayed numbers (randomly chosen from 1 to 3) with a fixed frequency of 500 ms per number (see figure 1 right). The driver was expected to read aloud each number just after it appeared on the display. To read the numbers he had to visually focus on the display. It was not possible to perform the secondary task with peripheral vision. As long as the driver held the button depressed, the task continued showing a new number every 500 ms. The task ended as soon as the driver released the button or after a maximum of 5 s.

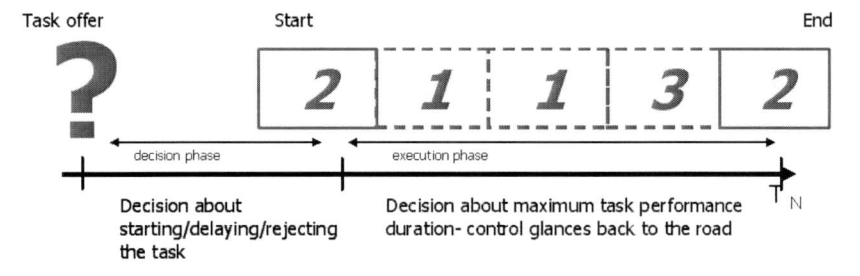

Figure 2. Sequence of task presentation of the secondary task

The driver was instructed to perform the task as long as the situation allowed. He was instructed to interrupt the secondary task and return to the driving task whenever the driving task required full attention. To motivate the driver to perform the secondary task, a reward system was established: for every number read aloud he got a point, for every missed number or serious driving failure (e.g. getting off the road) points were subtracted. The amount of points subtracted was chosen in accordance to severity of the driving failure (e.g. collision with other vehicle -50 points, exceeding the lane -20 points; driving too slowly -10 Points). The driver with the most points at the end of the drive got a reward. Figure 2 shows the sequence of task presentation.

Subjective ratings of using in-vehicle devices in general
After the driving course, the driver had to fill out a questionnaire about his general attitude towards using in-vehicle devices (e.g. radio, CD-player, cell phone) and other activities (e.g. smoking) in real traffic. It was asked to estimate risk and distraction effects of secondary tasks in general (e.g. "I never use in-vehicle devices while driving"; "In-vehicle devices clearly distract me from driving", In-vehicle devices are dangerous for driving"). as well as in specific situations (11 different situations, e.g. overtaking manoeuvres, driving on a motorway) and specific tasks (18 different tasks; e.g. using a cell-phone, a PDA). In addition, the driver had to rate his abilities to adapt secondary task behaviour to the actual situation ("I answer the phone immediately when someone calls me while driving"). For all items a 5-point Likert-scale was used.

Participants

Twenty-four subjects participated in the study: one control-group with a baseline driving-condition (n=8) and one dual task group which performed the additional secondary task while driving (n=16). The mean age of participants was 32.9 years (min. 23, max. 52 years) with 8 female and 16 male drivers. They were all well trained in simulator driving.

Data collection and analysis

100 Hz data recording included for instance parameters of driver action (e.g. brake pedal position), vehicle dynamics (e.g. speed), surrounding traffic (e.g. distance to leading vehicle), parameters of secondary task execution (e.g. button presses) and eye tracking data. The analysed parameters of the secondary task included mean percentage of rejected tasks, mean execution time (for accepted tasks; max. 5 s), mean decision time (only for accepted tasks, max. 3 s), and mean percentage of errors. These errors were monitored and recorded by the investigator. An error was defined as missing one number in the sequence or misspeaking the number. Driving errors were defined by decelerations < -8 m/s^2, critical time gaps < 1 s or critical Time to Collision < 1 s. In addition, the experimenter rated driving behaviour for more complex errors like endangering pedestrians or choosing the wrong lane at intersections.

The eye glance behaviour of the driver was measured by a four-camera eye-tracking system (60 Hz) by SMART-EYE. Raw data were transferred into x-and y-coordinates on the projection screen of the simulator. Absolute accuracy was improved by a 9-point calibration before each test drive. Raw data were then summarised to fixations (algorithm based on Jacob, 1995: standard deviation of measuring points for x- and y-deviations $> 1°$ longer than 100 ms). During secondary task execution, fixations were clustered into roadway and display glances. With this method, only a minimal number of fixations could not be allocated (see figure 3). Calculated parameters given here are "percentage of roadway glances of total execution time" ((number of roadway glances x mean duration of roadway glances)/total execution time of secondary task).

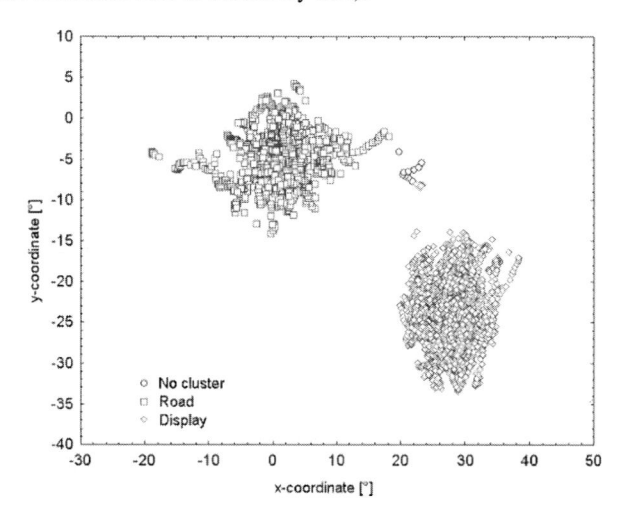

Figure 3. Allocation of roadway and display glances

Results

The analysis of secondary task behaviour showed that drivers adapted their decisions on secondary task interaction according to the situational demands (see figure 4). The percentage of rejected tasks was higher in critical than in non-critical situations (31.6 vs. 11.2%; $F(1,15)=27.4$; $p<.001$). This result is an indicator that drivers anticipated these demanding situations and therefore avoided any additional load due to dual task performance.

Additionally, before starting the secondary task, drivers tried to fully utilize the 3 s phase in order to wait how the situation would develop. This resulted in longer mean decision times until the secondary task was started before critical situations (1.69 s vs. 1.50 s; $F(1,15)= 8.4$; $p=.011$).

Increased mean execution time in non-critical situations showed that drivers were able to detect stable situations where longer performance of a secondary task is less risky. In critical situations they interrupted the task earlier (2.76 s vs. 2.41 s; $F(1,15)=13.8$; $p<.002$). The number of errors in the secondary task was generally very low (only about 5.5% of all executed tasks). The increased percentage of errors in critical situations indicated overload due to the additional task in an already very demanding driving situation (3.8 vs. 8.2 %; $F(1,15)=7.5$; $p=.015$).

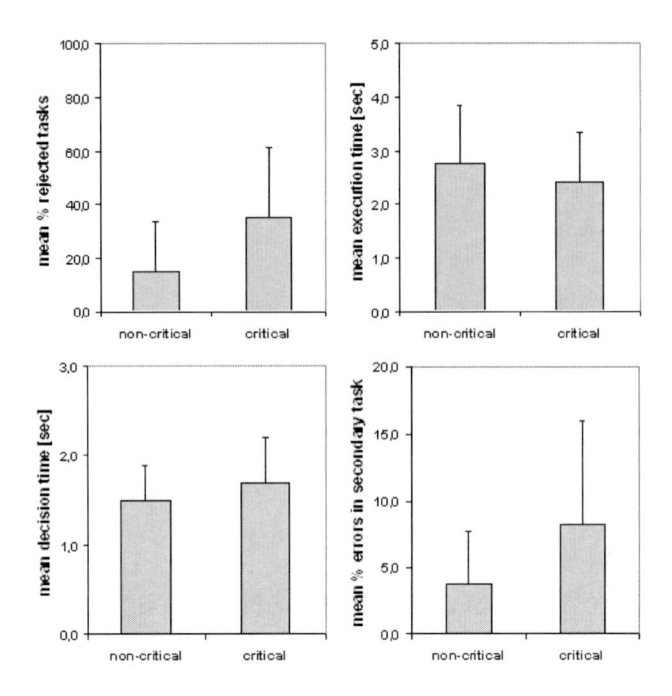

Figure 4. Mean % rejected tasks (upper left), mean execution time (upper right), mean decision time (lower left) and mean % errors in secondary task (lower right), for critical vs. non-critical situations

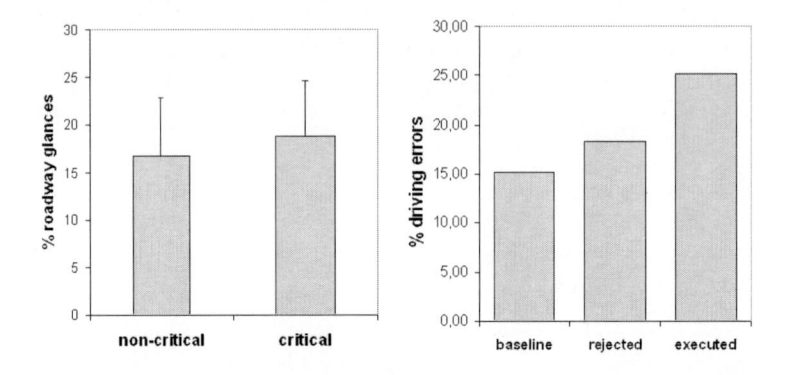

Figure 5. Percentage of roadway glances in non-critical vs. critical situation (left). Percentage of driving errors in driving situations with no secondary task offer (baseline), with rejected task and with secondary task execution, only for critical situations (right)

For the analysis of control behaviour during secondary task execution, drivers' eye fixations were clustered into fixations on the road and fixations on the display. One participant had to be excluded due to insufficient data quality. The results showed that drivers adapted their eye glance behaviour according to situational demands (see figure 5 left): The percentage of roadway glances increased in critical situations (F(1,14)=6.5; p=.023).

To analyse the effects on driving safety, we compared the percentage of driving errors in critical situations (% of all critical situations where a driving error could probably occur) in which no secondary task was offered (baseline condition) to situations in which the task was rejected and situations in which the task was executed (see figure 5 right). If the task was rejected, there was no important increase in driving errors compared to the baseline condition (χ^2=1.89; df=1; p=.170). If the secondary task was executed, there was a significant increase in driving errors (χ^2=4.51;df=1; p=.034). So, rejection of secondary tasks in highly demanding driving situations seems to be an efficient compensation strategy to avoid decreases in driving safety and can be seen as an indicator of having Situation Awareness.

A general result from the analysis of the secondary task behaviour was high interindividual variety in interaction strategies. E.g., for task rejection the number of rejected tasks varied from 15% to 93%. Also, execution time varied highly between drivers who performed the tasks to the maximum of 5 s nearly every time and drivers who executed the tasks only for short times.

To analyse the influences of drivers' general attitude towards the use of in-vehicle devices and other possible distracting activities, items of the questionnaire (general estimation of risk and distraction potential of secondary tasks, estimation of own skills and abilities in executing additional tasks) and parameters of the secondary task were correlated. Correlations were significant in the following cases:

- drivers who rated the distraction effect of additional tasks higher rejected more tasks in the study (r=.38; p=.007)
- drivers who had higher aversion rates against interruptions in general executed the tasks longer (r=.40; p=.005)
- drivers who had higher aversion rates against interruptions in general also showed higher percentages of errors in secondary task (r=.65; p<.000)
- drivers who rated themselves able to adequately interrupt secondary task execution in real traffic executed the task for a shorter time (r=.41, p=.004) and rejected more tasks (r=.41; p=.005)
- drivers who rated themselves as good in estimating situational demands rejected more tasks (r=.36; p=.009) and waited longer with the beginning of the task (r=.34; p=.010).

Discussion and conclusion

The results show influence of Situation Awareness on all the three assumed levels of driving with secondary tasks (see table 1). The high variety between drivers in the number and the duration of secondary task execution independent from the current situation indicate that individual interaction strategies with secondary tasks exist which can be attributed to a higher planning level. On this level the driver decides whether to use secondary tasks while driving at all. Further, correlations between subjective ratings on a questionnaire dealing with the general estimations about using in-vehicle devices and objective secondary task measurements could be found. They indicate that more general attitudes, like individual risk perception and drivers' assessments of their skills and abilities, have an influence on the willingness to execute secondary tasks while driving.

Further it was hypothesised that, on a decision level, the drivers assess the current situation to develop a model of the situation and to anticipate potential conflicts. If drivers act situationally aware this should be seen in adequate decisions according to the situational demands, if, when and how long a secondary task can be executed in the given situation. Parameters of the secondary task, like the number of rejected tasks and decision times until the secondary task is started, show that, in situations with higher expected demands, the drivers either reject a secondary task or delay its beginning.

On the control level, it was assumed that a situationally aware driver continuously monitors situation development during secondary task execution to control for possible changes in the environment. The results show that drivers do this by executing short control glances back to the road and adapt their frequency to situational demands. In anticipation of highly demanding driving situations drivers decide to interrupt the secondary task. This results in shorter execution times in front of critical situations.

Table 1. Three-level model of Situation Awareness for driving with secondary tasks

level	processes	time frame
planning	higher-order planning of interactions with secondary tasks e.g. stopping for executing a task, using of waiting times, execure only low demanding secondary tasks	minutes
decision	decision about the execution of a task according to the actual situational demands (if, how long, when interruption)	seconds
control	monitoring of situation development (during secondary task execution; via short control glances to the driving task)	milliseconds

The adequacy of driver's situation dependent decisions is measured via driving errors in critical situations. The results show that when drivers adequately anticipate the expected high workload of the driving situation and decide therefore to neglect the secondary task offer, they are able to maintain an adequate level of driving safety. The number of driving errors in this condition is comparable to the baseline conditions with no secondary task offers.

The results indicate that drivers are able to use in-vehicle devices according to situational demands. This behaviour can be interpreted as an indicator of drivers' Situation Awareness. It allows the driver to actively react to the constraints of the situation and to avoid overload in already highly demanding traffic situations. These results could only be derived by the special deciding-to-be-distracted approach (term derived from Lerner, 2005) used in the study. With allowing the drivers to decide if, when and how long to execute an offered secondary task their awareness of the current situation can be measured. Not only control processes during secondary task execution but also anticipative processes prior to the beginning of the task can be analyzed with this method. The results give insights into the compensatory behaviour of drivers in using in-vehicle devices in real traffic situations. Due to the use of the multiple different traffic situations, varying in their criticality and predictability, it can be measured if drivers adapt their secondary task behaviour according to the situational demands. The advantage of the proposed three level model of Situation Awareness is that clear hypotheses about the importance of Situation Awareness in the use of in-vehicle devices while driving can be developed and analyzed. The success of anticipation processes on the decision level will mainly be dependent from the driver's experience and the availability of mental models about the driving situation as well as from the availability of environmental cues to understand and correctly interpret the current and future situation. The success of the control processes on the control level during secondary execution will mainly be dependent from the visual and cognitive demands of the secondary task and if the task allows adequate interruptions to frequently return to the driving task. These hypotheses could be analyzed in further studies.

Acknowledgments

The study was conducted within a project financed by the German Association for Research on Automobile Technique and the Federal Highway Research Institute.

References

Adams, M.J., Tenney, Y.J. & Pew, R.W. (1995). Situation awareness and the cognitive management of complex systems. *Human Factors, 37*, 85-104.
Alliance of Automobile Manufacturers Driver Focus-Telematics Working Group (2003). *Statement of Principles, Criteria and Verification Procedures on Driver Interactions with Advanced In-Vehicle Information and Communication Systems.* Retrieved 01.04.2008 from
http://www.umich.edu/~driving/guidelines/AAM_DriverFocus_Guidelines.pdf
Beede, K.E. & Kass, S.J. (2006). Engrossed in conversation: the impact of cell phones on simulated driving performance. *Accident Analysis and Prevention, 38*, 415-421.
Boyle, J.M. & Vanderwolf, P. (2005). *2003 Motor Vehicle Occupant Safety Survey: Volume 4: Crash Injury and Emergency Medical Services Report.* Report No. DOT HS-809-857. National Highway Traffic Safety Administration, Washington, DC.
Endsley, M.R. (1995). Toward a theory of situation awareness in dynamic systems. *Human Factors, 37*, 381-394.
Endsley, M.R., Adams, M.J., Tenney, Y.J. & Pew, R.W. (1995). Situation awareness and the cognitive management of complex systems. *Human Factors, 37*, 85-104.
Esbjörnsson, M. & Juhlin, O. (2003). *Combining mobile phone conversations and driving –Studying a mundane activity in its naturalistic setting.* Paper presented at the ITS World Congress 2003, Madrid, Spain.

Harbluk, J..L., Noy, Y.I., Trbovich, P.L., & Eizenman, M. (2007). An on-road assessment of cognitive distraction: impacts on driver´s visual behavior and braking performance. *Accident Analysis and Prevention, 39*, 372-379.

Hella, F. (1987). Is the analysis of eye movement recording a sufficient criterion for evaluating automobile instrument panel design? In J.A. O'Regan and A. Lévy-Schoen (Eds.), *Eye Movements: from Physiology to Cognition* (pp. 555-561). Amsterdam: Elsevier.

Horberry, T., Anderson, J., Regan, M.A., Triggs, T.J., & Brown, J. (2006). Driver distraction: the effects of concurrent in-vehicle tasks, road environment complexity and age on driving performance. *Accident Analysis and Prevention, 38*, 185-191.

Ishida, T. & Matsuura, T. (2001). The effect of cellular phone use on driving performance. *International Association of Traffic Safety Sciences (IATSS) Research, 2S*, 6-14.

Jacob, R.J.K. (1995). Eye tracking in advanced interface design. In W. Barfield u. T.A. Furness (Eds.). *Virtual environments and advanced interface design*. Oxford: Oxford University Press. Retrieved 01.01.2008 from http://www.cs.tufts.edu/~jacob/papers/barfield.pdf.

Lerner, N. (2005). Deciding to be distracted. In *Proceedings of the third International Driving Symposium on Human Factors in Driver Assessment, Training and Vehicle Design* (pp. 499-506). Rockport, Maine, USA. 27.-30.06.2005.

Lerner, N. & Boyd, S. (2005). *On-Road Study of Willingness to Engage in Distracting Tasks* (No. DOT HS-809-863). Washington, DC: National Highway Traffic Safety Administration.

Lesch, M.F. & Hancock, P.A. (2004). Driving performance during concurrent cell-phone use: are drivers aware of their performance decrements? *Accident Analysis and Prevention, 4*, 471-480.

McCartt, A.T., Hellinga, L.A. & Braitman, K.A. (2006). Cell phones and driving: review of research. *Traffic Injury Prevention, 7*, 89-106.

Rassl, R. (2004). *Ablenkungswirkung tertiärer Aufgaben im PKW-systemergonomische Analyse und Prognose*. Dissertation. München: TU München: Fakultät für Maschinenwesen.

Rockwell, T.H. (1988). Spare visual capacity in driving revisited: new empirical results for an old idea. In A.G. Gale et al. (Ed.), *Visions in Vehicles II* (pp. 317-324). Amsterdam: North Holland Press.

Salvucci, D.D., & Macuga, K.L. (2002). Predicting the effects of cellular-phone dialing on driver performance. *Cognitive Systems Research, 3*, 95-102.

Schweigert, M. (2003). *Fahrerblickverhalten und Nebenaufgaben*. Dissertation. München: TU München, Fakultät für Maschinenwesen.

Strayer, D.L., & Drews, F.A. (2004). Profiles in driver distraction: Effects of cell phone conversations on younger and older drivers. *Human Factors, 46*, 423-428.

Törnros, J.E.B. & Bolling, A.K. (2005). Mobile phone use-effects of handheld and handsfree phones on driving performance, *Accident Analysis and Prevention, 37*, 902-909.

Victor, T.W., Harbluk, J.L. & Engström, J.A. (2005). Sensitivity of eye-movement measures to in-vehicle task difficulty. *Transportation Research F, 8*, 167-190.

Wierwille, W.W. (1993). An initial model of visual sampling of in-car displays and controls. In A.G. Gale, I.D. Brown, C.M. Haslegrave, H.W. Kruysse & S.P. Taylor (Eds.), *Vision in Vehicles IV* (pp.271-280). Amsterdam: New Holland Press.

Wierwille, W.W., Antin, J.F., Dingus, T., & Hulse, M.C. (1988). Visual attentional demand of an in-car navigation display system. In A.G. Gale et al. (Eds.), *Vision in Vehicles II* (pp. 307-316). Amsterdam: North Holland Press.

Exploring appropriate alarm timing for a driver-adaptive forward collision warning system

Genya Abe[1] & Makoto Itoh[2]
[1] Japan Automobile Research Institute, Tsukuba
[2] University of Tsukuba
Japan

Abstract

Determining appropriate alarm timing is important for developing an effective forward collision warning system. Using a high fidelity driving simulator, a driver adaptive alarm that is triggered based on individual driving styles was investigated. Two types of alarm timing were considered in this research: (1) timing such that an alarm is given at the time of ordinary accelerator releasing, and (2) timing such that an alarm is given at the time of ordinary brake implementation. The experimental data for eighteen participants were analysed in order to evaluate the effects of changing alarm timings on driver subjective ratings of trust in an alarm and on driver's timing of releasing the accelerator and implementing the brakes. The results indicate that, overall, both alarm timings are acceptable for the participants from the viewpoint of trust. Driver behaviour was not greatly affected by the presentation of alarms, and drivers avoided collisions effectively. However, for drivers who have particular driving styles, e.g., drivers who tend to implement the brakes very early or very late, simply adapting the alarm timing to brake implementation timing may impair the driver's trust in the alarm. Potential applications of these results include methods for setting alarm timings for driver adaptive forward collision warning systems.

Introduction

The reduction of traffic accidents is an important goal, and rear-end collisions are one of the most common types of accident. Forward Collision Warning Systems (FCWS) may be of great potential benefit to drivers who are not paying sufficient attention while driving, and these systems may reduce the number of traffic accidents (Alm and Nilsson, 2000; Ben-Yaacov et al., 2002).

It is likely that drivers will implement situational recognition, decision-making, and action implementation based on their own intention to avoid risky events. If there is a mismatch between the intended action by drivers and the alarm presentation, then the driver and the FCWS may not agree on a course of action. More specifically, the driver may be annoyed by alarms that do not correspond with their plans to implement collision avoidance actions, resulting in reduced system efficiency (Wheeler et al., 1998).

One possible solution to this mismatch is to individualise alarms to the individual driver by considering the driving characteristics of the driver (Goldman et al., 1995). Consequently, it may be possible to minimise the mismatch between the driver's collision avoidance behaviour and the alarm presentation. In a previous study, the timing with which the individual releases the accelerator and implements the brakes in imminent collision situations was investigated, and the results of the investigation into alarm timings were incorporated in order to assess the driver's response to tailored alarms (Abe & Itoh, 2007). It was found that the individually determined timing based on the mean value of the accelerator release time increased the ratings of trust in alarms more than the timing based on the mean values of braking response. The timing of the braking response time induced a longer response time to application of the brakes. In order to clarify the appropriateness of individualised alarms, it is necessary to assess which timing should be adopted in the process of individual collision avoidance behaviour by considering other timings that have not been estimated previously.

In the present study, two experiments were conducted. In Experiment I, the variation of collision avoidance performance of individual drivers in response to driving conditions was explored. In Experiment II, the influence of driver-adaptive alarms presented at earlier timings of the mean values for releasing the accelerator and implementing the brakes on driver behaviour and driver trust in alarms were investigated. In addition, the appearance of individual differences in response to tailored alarms was considered.

Experiment I

Method

Apparatus
This experiment was undertaken using a driving simulator owned by the Japan Automobile Research Institute. The simulator has six-degree-of-freedom motion and uses complex computer graphics to provide a highly realistic driving environment. The simulated horizontal forward field of view was 50 degrees and the vertical field of view was 35 degrees.

Participants and experimental tasks
Eighteen participants (M = 28.3, SD = 9.0) took part in the experiment. All of the participants were licensed drivers with at least one year of driving experience (Mean=9.2 years, SD=9.3 years), and each participant was given two tasks in the experiment. One was to maintain the speed of a vehicle at a target speed in response to experimental conditions, and the other task was to follow a lead vehicle. The time headway between the lead and following vehicles was controlled at 1.7 seconds.

Experimental design and procedure

For each participant, the experimental condition comprised four sessions with two different lead vehicle decelerations, 0.65g (high time criticality) and 0.39g (low time criticality), and driving speeds, 60 km/h (low speed) and 100 km/h (high speed). Specifically, each participant experienced four sessions with four different combinations of deceleration and driving speed. The order of the experimental conditions was controlled in order to reduce potential effects of the experimental conditions on driver behaviour. Each session involved nine potential collision events with the brake lights of the lead vehicle being illuminated. The participants were instructed to avoid collisions with a lead vehicle using only the brakes. Furthermore, potential collision events were triggered when the drivers travel at speed of 60km/h ±5km/h or 100km/h ±5km/h.

Dependent variables

Two dependent measures describing the collision avoidance behaviour of the driver were recorded:

1. *Braking event to brake onset time*: This is the time period between the braking event of the lead vehicle and application of the brakes.
2. *Braking event to accelerator release time*: This is the time period between the braking event and the release of the accelerator.

Results

Braking event to brake onset time

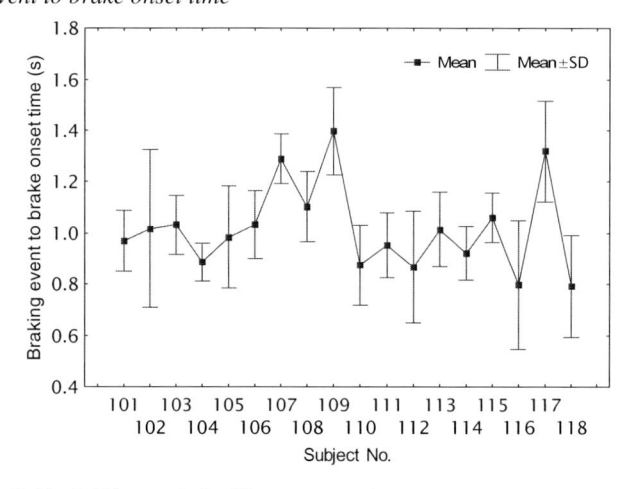

Figure 1. Individual difference in braking response time

A repeated measures ANOVA with the driver and the driving conditions, i.e., the combination of the deceleration of the lead vehicle and the driving speed, revealed a highly significant interaction between the factors, $F(3, 51) = 8.2$, $p < 0.01$. As an example, Figure 1 shows the mean values of the braking event to brake onset time for each driver with a deceleration of the lead vehicle of 0.65g and a driving speed of 60 km/h. As shown in this figure, there may be individual differences in the timing of braking onset for collision avoidance, and the braking strategy may vary in response to the degree of deceleration of the lead vehicle and the driving speed.

Braking event to accelerator release time
A repeated measures ANOVA with the driver and the driving condition, i.e., the combination of the deceleration of the lead vehicle and the driving speed, was performed. There was no significant difference among the driving conditions, $F(3, 51) = 1.806$, NS). As an example, Figure 2 shows the mean values for each driver for a deceleration of 0.65g and a driving speed of 60 km/h. As shown in this figure, there are individual differences in the accelerator release time among the drivers.

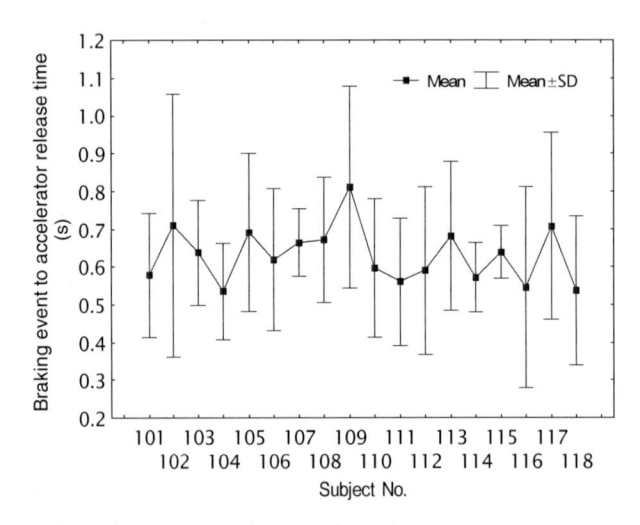

Figure 2. Individual difference in accelerator release time

Experiment II

The purpose of this experiment was to investigate how adaptive-alarm timing may influence the driver response to FCWS. The influences of tailored alarm timings that were determined for the individual on collision avoidance behaviour and driver subjective estimation of alarms were examined.

Method

Apparatus
All materials used in this experiment were the same as those used in Experiment I. In this experiment, a FCWS was introduced with a simple auditory beep that lasted approximately two seconds. All alarms were presented based on the data obtained in Experiment I.

Participants and experimental tasks

In this experiment, the participants were the same as those in Experiment I. For each participant, Experiment II was conducted two weeks after implementing Experiment I. All drivers were required to perform the same tasks as in Experiment I.

Procedures of driver adaptive alarm timing

Driver adaptive alarm timings were determined for each participant based on the data obtained in Experiment I. Specifically, two different alarm timings were used in response to the driving conditions.

- The mean value of braking event to brake onset time -1.282σ (alarm timing Braking_On)
- The mean value of braking event to accelerator release time -1.282σ (alarm timing Accelerator_Off)

Here, σ is the standard deviation of the braking event to brake onset time and braking event to accelerator release time for nine trials for each driving condition obtained in Experiment I, and 1.282σ indicates the 10th percentile from the minimum value for each time. Consequently, each alarm was trigged at an earlier timing compared to the mean value of accelerator release time or braking onset time for each participant. In addition, the mean value and variance were calculated using the natural logarithm of the data and returned to natural number in order to be determined as a specific alarm timing.

Experimental design

Group I: 9 subjects

| Driving condition (0.65g, 100km/h) Alarm timing Braking_On | Driving condition (0.65g, 100km/h) Alarm timing Acclerator_Off | Driving condition (0.39g, 60km/h) Alarm timing Braking_On | Driving condition (0.39g, 60km/h) Alarm timing Acclerator_Off |

Group II .9 subjects

| Driving condition (0.39g, 100km/h) Alarm timing Braking_On | Driving condition (0.39g, 100km/h) Alarm timing Acclerator_Off | Driving condition (0.65g, 60km/h) Alarm timing Braking_On | Driving condition (0.65g, 60km/h) Alarm timing Acclerator_Off |

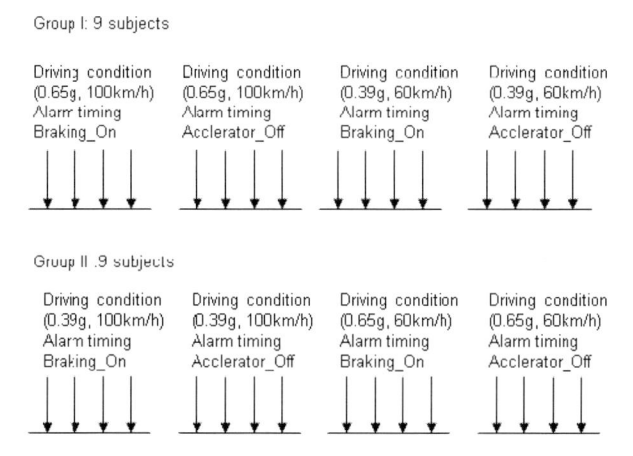

Figure 3. Experimental conditions in response to Groups

The vehicle following condition was the same as in Experiment I. The experiment was a mixed factorial within/between-subject design. Driving conditions, i.e., combinations of the deceleration of the lead vehicle and the driving speed of the following vehicle was the between-subjects variable, and the alarm timing (alarm timing Braking_On and alarm timing Accelerator_Off) was the within-subjects variable. The participants were randomly assigned to one of two groups, Group I and Group II, each of which had nine subjects. The participants who were assigned to Group I experienced a deceleration of the lead vehicle of 0.65g at a driving speed of 100 km/h and a deceleration of the lead vehicle of 0.39g at a driving speed of 60 km/h. The participants who were assigned to Group II experienced a deceleration of the lead vehicle of 0.39g at a driving speed of 100 km/h and a deceleration of the lead vehicle of 0.65g at a driving speed of 60 km/h. Each group included 16 trials with a potential collision event for each trial. The allocation of driving conditions and alarm timings for each trial are shown in Figure 3. For each group, the order of the driving conditions and the alarm timings were partly counterbalanced in response to the number of participants in order to reduce effects of the experimental conditions on driver behaviour.

Dependent variables
Two dependent measures describing the collision avoidance strategy of the driver and three dependent measures related to driver trust in the system were recorded. The two measures of collision avoidance strategy used in Experiment I were also used in this experiment.

1. Braking event to brake onset time
2. Braking event to accelerator release time
3. Perceived alarm timing: An 11-point rating scale was used to record the subjective estimation of alarm timing by the driver, in which 0 indicates 'too late', 5 indicates 'appropriate' and 10 indicates 'too early'.
4. Alarm effectiveness: The estimation by the driver of the alarm effectiveness was recorded using an 11-point rating scale, in which 0 indicates 'not at all' and 10 indicates 'completely'.
5. Trust in the system: The subjective estimation by the driver trust in the systems was obtained using an 11-point rating scale, in which 0 indicates 'no at all' and 10 indicates 'completely'.

All subjective measurements were recorded immediately after the presentation of alarms.

Results

Trust in the system
As an example of the results of driver subjective ratings of alarms, driver trust in the system was considered. Figure 4 shows the mean values of trust ratings in response to alarm timing. Compared to alarm timing Braking_On (based on braking onset), the trust ratings for alarm timing Accelerator_Off (based on accelerator release) were higher, $F(1,17) = 5.4$, $p < 0.05$. However, the value of trust rating for the alarm timing Braking_On itself was not so low. Thus, the trust ratings for both alarm timings may be maintained at reasonable levels. As for other subjective measures, the similar tendencies as obtained in the analyses of trust were observed.

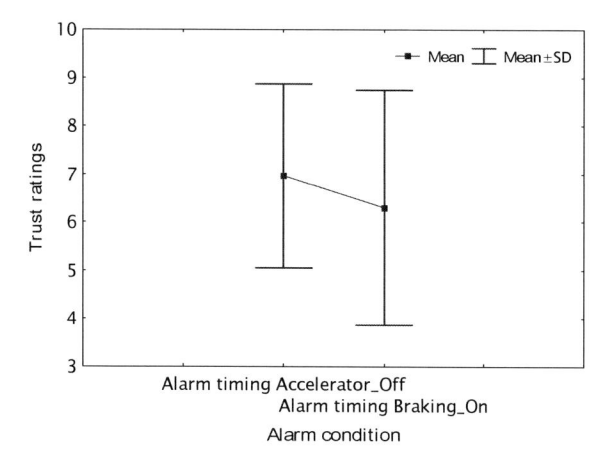

Figure 4. Trust ratings in response to alarm timings

As for individual differences in trust ratings in response to alarm timing, Figures 5 and 6 show the mean values for the individual drivers assigned to Group I and Group II, respectively. As shown in Figure 5, in Group I, there was a significant difference in trust ratings between the alarm timings for Subject Nos. 103 and 117, resulting from decreased trust in alarm timing Braking_On. However, the ratings themselves were not so low (approximately 5). In Group II, there was a significant difference in trust ratings for Subject No. 109, resulting from dramatically decreased trust in alarm timing Braking_On. Moreover, for Subject No. 107 in Group II, in particular, the trust ratings for both alarm timings were relatively low.

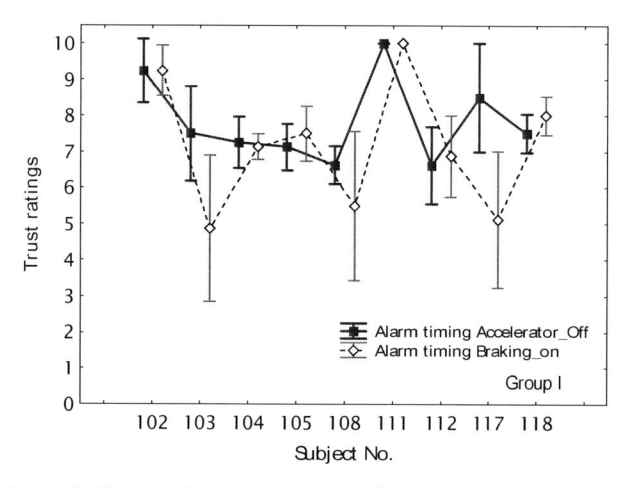

Figure 5. Individual differences in trust ratings regarding alarm timings for Group I

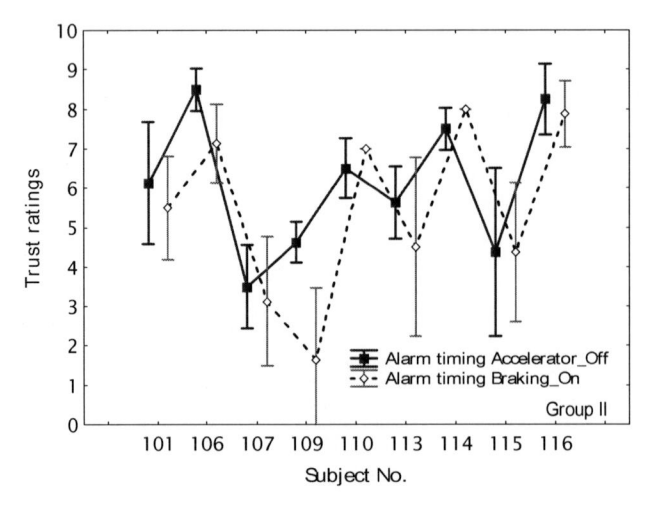

Figure 6. Individual differences in trust ratings regarding alarm timings for Group II

Braking event to accelerator release time

Figure 7 shows the mean values of the braking event to accelerator release time for each alarm timing. Here, the data obtained in Experiment I was considered to provide a comparison of collision avoidance methods with non-assisted driver behaviour. There was no significant difference among the conditions ($F(2, 34) = 1.6$, NS).

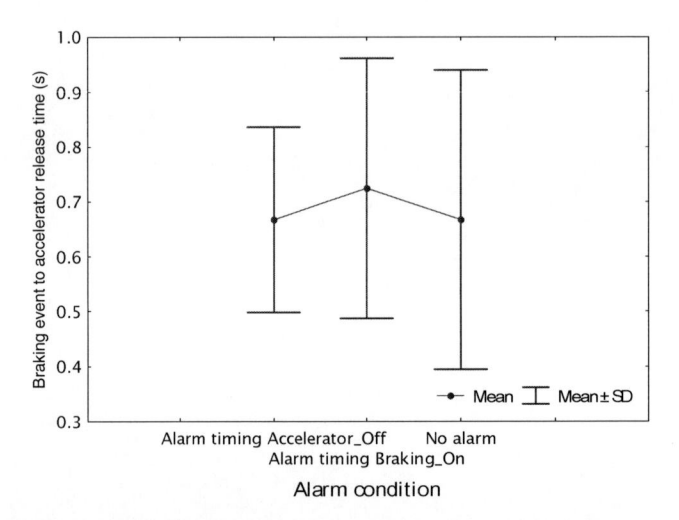

Figure 7. Braking event to accelerator release time in response to alarm timings

Figure 8 shows the frequency distribution for the braking event to accelerator release time for each alarm timing. Alarm timing Accelerator_Off contributed to more consistent accelerator release time in a potential collision situation, compared to other conditions, indicating that the presentation of alarms may induce consistent driver avoidance behaviour in imminent collision situations if the presentation is performed appropriately.

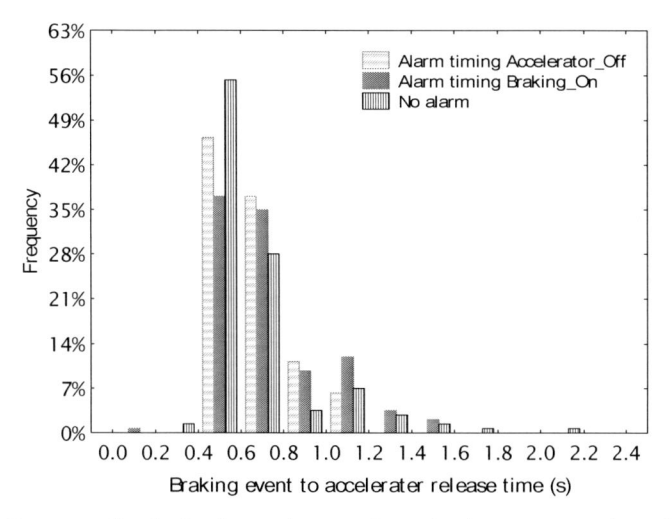

Figure 8. Frequency distribution for accelerator release time in response to alarm timings

Braking event to brake onset time

Figure 9 shows the mean values of the braking event to brake onset time for each alarm timing, indicating that alarms presented using alarm algorithm A induced longer braking response times (F(2,34) = 3.190, p=0.053). Here, the data obtained in Experiment I was considered to provide a comparison of collision avoidance methods with non-assisted driver behaviour. However, for the three conditions, there was not a large difference between the conditions for the values themselves.

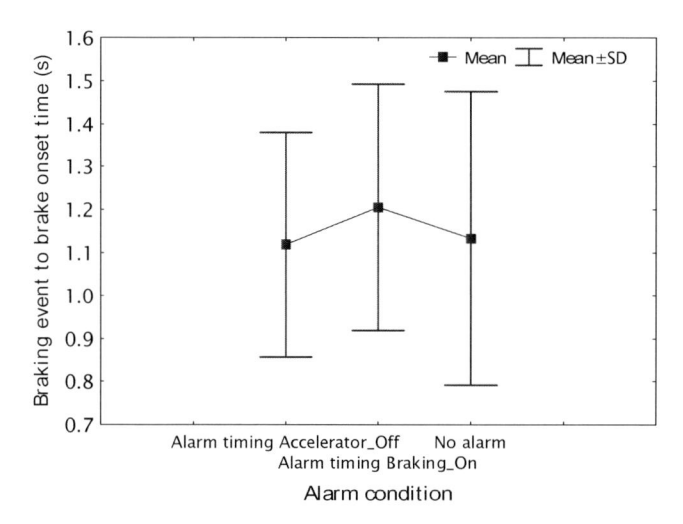

Figure 9. Braking event to brake onset time in response to alarm timings

Individual differences in braking event to brake onset time were considered. Figure 10 shows the mean values for each driver assigned to Group I. For Subject No. 111, the presentation of alarm timing Accelerator_Off induced a longer response time to the brakes. The time difference in braking response due to the presentation of alarms is much longer than the time difference for the other subjects discussed earlier in this section (Figure 9).

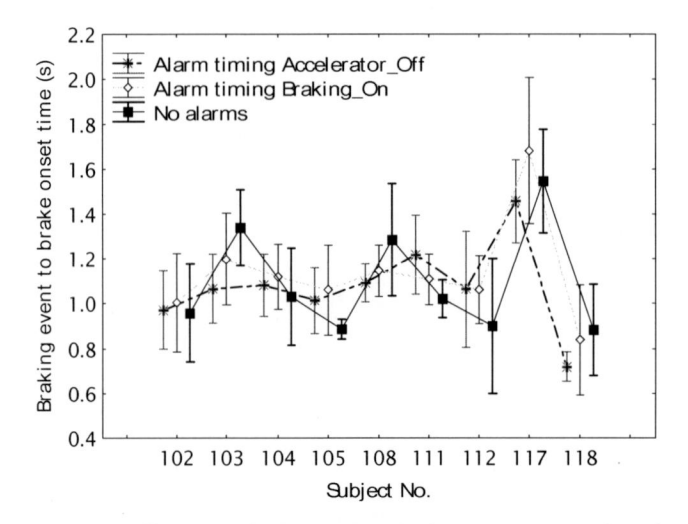

Figure 10. Individual differences in braking event to brake onset time regarding alarm timings

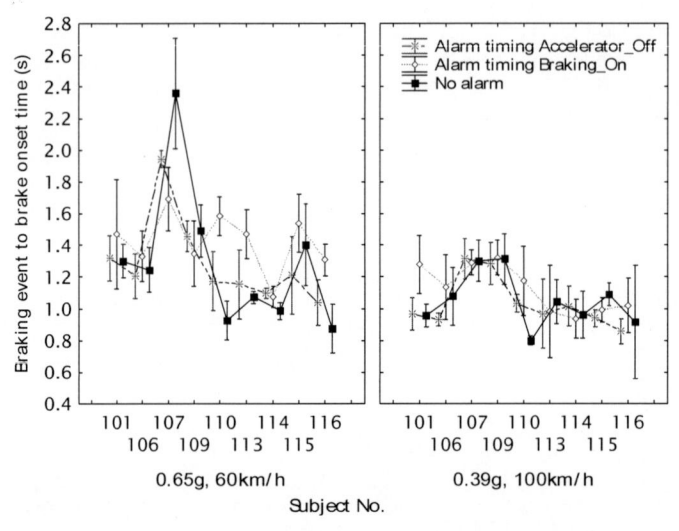

Figure 11. Individual differences in braking event to brake onset time regarding alarm timings and driving conditions

For the drivers who were assigned to Group II (Figure 11), generally, when high deceleration of the lead vehicle (0.65g) occurred, the presentation of alarms did not affect the driver response time to the brakes significantly, as compared to the no-alarm-presentation case. However, when low deceleration of the lead vehicle (0.39g) occurred, alarms induced a longer response time to the brakes for some drivers, as compared to the no-alarm condition. Specifically, for Subject Nos. 110, 113, and 116, alarm timing Braking_On induced longer response times to the brakes compared to the no-alarm condition. The time difference in braking response time due to the presentation of alarms was much longer than the time difference obtained based on the data for the subjects discussed earlier in this section (Figure 9).

These results suggest that from the point of view of individual difference, for some drivers, the presentation of alarms may influence driver behaviour, depending on alarm timing or driving conditions. Therefore, it is necessary to consider whether this issue may impair safe driving and to carefully consider the determination of adaptive alarm timing.

Collision avoidance behaviour and its relation to response to tailored alarms
The alarms triggered in Experiment II were determined based on the timing of collision avoidance for the individual obtained in Experiment I. Therefore, the driver response to alarms would be affected by the characteristic of collision avoidance behaviour for the individual. Here, it is considered that the driver's subjective ratings of trust in alarms may be influenced by the characteristics of the collision avoidance technique.

Figure 12 shows that the time elapsed (x-axis) from the start of deceleration of the lead vehicle to the implementation of the brakes for each driver (y-axis denotes Subject No.) for a deceleration of 0.65g and a driving speed of 60 km/h in Experiment I. For each straight line, the solid line segment indicates the time from a potential collision event to the release of the accelerator and the dotted line segment indicates the time from the release of the accelerator to the implementation of the brakes.

As mentioned earlier, the time until the release of the accelerator and that until the application of brakes differed among drivers. Moreover, as for the swiftness of collision avoidance behaviour for individuals, the similar tendencies were obtained for other driving conditions. In other words, the drivers who exhibited swift/slow responses to the accelerator or to the brakes for a certain driving condition exhibited similar characteristics regarding the response time for other driving conditions. It is suggested that individual driving characteristics may be reflected in a relatively wide range of driving conditions rather than in a particular driving condition.

As shown in Figures 5 and 6, the possibility exists that trust will be decreased even though the alarm timing is determined for the individual. For example, for Subject Nos. 103, 117, and 109, alarm timing Braking_On induced decreased trust ratings, and for Subject Nos. 109 and 117, the braking response times are relatively slow, as shown in Figure 12. Moreover, for Subject No. 107, the trust ratings were relatively low independent of the alarm timings (Figure 5). The elapsed times until the release of the accelerator and until braking for this subject appear to be relatively long, possibly resulting in decreased trust, even though the alarm timing was adjusted for this subject.

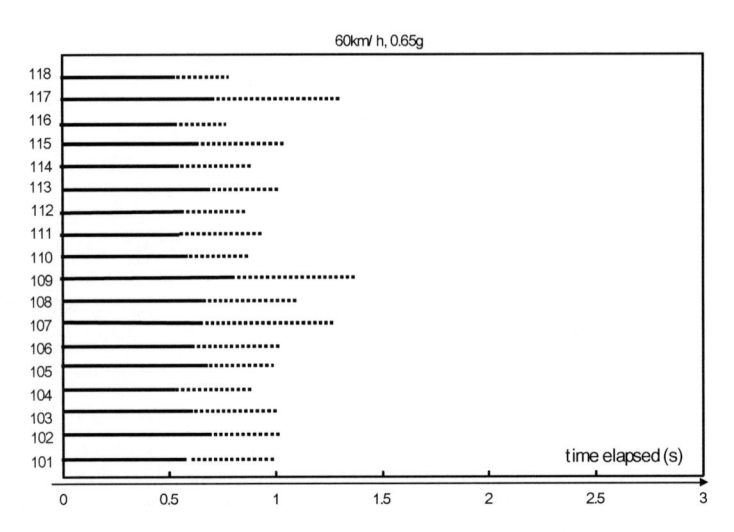

Figure 12. Individual difference in timing of collision avoidance styles

These results indicate that when tailored alarm timing is implemented based on the driving characteristics of the individual, there is a possibility that trust ratings will be impaired for drivers who exhibit particular driving characteristics, i.e., a relatively slow braking response time. Therefore, for these drivers, it may be necessary to consider other factors when determining tailored alarm timing, rather than simply adjusting the timing of collision avoidance behaviour to alarm timing. Specifically how individual differences should be solved will be carefully considered.

Conclusions

The present study focused on the alarm timing for forward collision warning systems and investigated driver response to alarms that were individually determined based on the time for releasing the accelerator and implementing the brakes in imminent collision situations. The results support the following conclusions.

1. Overall, driver adaptive alarms that are triggered at slightly earlier than the time of ordinary accelerator releasing and the time of ordinary brake implementation do not impair subjective estimation of alarms. In addition, driver behaviour is not impaired by the presentation of alarms. Moreover, alarm timing based on accelerator release may contribute to consistent accelerator releasing in imminent collision situations.

2. However, for drivers who exhibit particular driving characteristics, i.e., comparatively slow ordinary accelerator release and ordinary brake implementation, level of trust in alarms may be impaired.

3. Based on individual differences in driver behaviour with respect to alarms, for some drivers, the presentation of adaptive alarms might impair system effectiveness due to changes in driver behaviour, depending on alarm timings and driving conditions. Therefore, it is necessary to carefully consider how changes in driver behaviour due to adaptive alarms may decrease the effectiveness of the FCWS.

References

Abe, G., & Itoh, M. (2007). Effect of alarm timing on trust in driver-adaptive forward collision warning systems. *JARI Research Journal, 28*, 623-626. (in Japanese)

Alm, H., & Nilsson, L. (2000). Incident warning systems and traffic safety: A comparison between the PORTICO and MELYSSA test site systems. *Transportation Human Factors, 2*, 77-93.

Ben-Yaacov, A., Maltz, M., & Shinar, D. (2002). Effects of an in-vehicle collision avoidance warning system on short- and long-term driving performance. *Human Factors, 44*, 335-342.

Goldman, R., Miller, C., Harp, S., & Plocher, T. (1995). Driver-adaptive warning system. *IDEA PROJECT FINAL REPORT Contract ITS-7*. Minneapolis, MN: Honywell Technology Center.

Wheeler, W.A., Campbell, J.L., & Kinghorn, R.A. (1998). Commercial vehicle-specific aspects of intelligent transportation systems. In W. Barfield and T.A. Dingus (Eds.), Human Factors in Intelligent Transportation Systems. (pp. 95-129). New Jersey: LEA.

Effects of auditory warnings on driving behaviour

Nicola Fricke[1] & Mònica De Filippis[2]
[1]Technische Universität Berlin
[2]Deutsches Zentrum für Luft- und Raumfahrt
Braunschweig
Germany

Abstract

Designing crash avoidance warning systems seems a promising starting point for active driver assistance in enhancing driver safety. The aim of this study is to design semantically enriched warnings which not only alert the driver but provide additional information about the critical situation and thereby assist the driver in dealing with the imminent collision. This is implemented by displaying spatial auditory icons (SAIs), which convey information on the identity and the location of the collision object.

In a driving simulator study, several auditory icons were tested in collision scenarios to assess realistic driving behaviour. The main focus of the study was to compare SAIs with spatial tonal warnings and conventional non-spatial tonal warnings. To provide a stable measure for the effects of the different warning systems, a no-warning condition was also implemented. Dependent variables included number of collisions, distance to collision object, reaction time and reaction patterns for braking behaviour.

Results revealed that all warnings prevented a considerable proportion of collisions compared to the no-warning condition. Also, simple, non-spatial tonal warnings evoked strong reactions and prevented collisions more often than SAIs. However, simple tonal warnings led to more reactions before the collision object was even seen, compared to the SAIs. These reactions consisted of stereotypical strong brake-reactions which were initiated before participants knew about the criticality of the situation after having experienced the first scenario. Such premature behaviour could provoke overreactions and be inappropriate in some situations. Overall, the present study suggests that SAIs should be further investigated in comparable realistic situations before concluding whether they can really prevent inappropriate, overhasty reactions and are therefore more useful for collision warnings compared to simple tonal warnings.

Human factor aspects in driver warnings

Increasing safety is one important aspect in human factors research, especially in the driving domain, where many accidents endanger health and life of those involved.

In D. de Waard, F.O. Flemisch, B. Lorenz, H. Oberheid, and K.A. Brookhuis (Eds.) (2008), *Human Factors for assistance and automation* (pp. 117 - 128). Maastricht, the Netherlands: Shaker Publishing.

Driving safety can be enhanced with several approaches. According to Sanders & McCormick (1993), there are three ways to control the risks of a product. Hazards should be designed out, and also guarded against, which could be achieved through road and road-sign design or protective safety systems such as crush zones. Although efforts have been made in these fields, many accidents and physical injuries occur. In 2006 there were a total of 755 113 accidents on German roads (German Federal Office of Statistics, 2006). The third approach to minimizing the hazard of a product or a domain is the use of warning systems. Information is provided about certain states or situations which could result in damage to the driver and others.

Since about 75% of the car crashes in Germany involve other vehicles or pedestrians (German Federal Office of Statistics, 2006) assistance in such collision situations to reduce accidents seems reasonable. A driver can be supported by active driver assistance in which the car takes control of some parts of the driving task, such as an anti-lock braking system, or through informative passive driver assistance, which only supports the driver by processing information, e.g. presentation of warnings.

If the responsibility should remain in the hands of the driver, crash avoidance warnings are one good way of supporting drivers in critical situations and reducing the possibility of accidents. Such warnings can be defined as "an in-vehicle presentation of information alerting the driver to a probable collision situation requiring immediate attention" (Comsis, 1996). The definition emphasises one important difference of collision warning systems to other critical messages inside the car which need attention, e.g. oil-level warning. In contrast to many of those warnings, collison warnings are time-critical and immediate danger will result if no corrective action is taken by the driver in a short period of time. Collision warnings must also lead to appropriate action in the context of the traffic situation so that other collisions can be avoided with following or oncoming traffic. Importantly, such warnings are likely to be rare events over a long period of vehicle usage, which implies that the meaning of a collision warning cannot be learnt by experience. Training the meaning of warnings has been shown to be only effectively achieved in an appropriate amount of time for a small number of warnings which must also be clearly distinguishable (see Patterson, 1982). Unfortunately car manufacturers rarely use tone warnings consistently, but develop different sets for their products. This makes any training difficult, as there would have to be a new extensive and repeated training every time a new car is used, e.g. a rental car. Another possibility would be to describe the warning types in the car manual. However it cannot be assured that drivers read their car manual at all and for this reason only "intuitively" understandable and usable warnings should be used. Therefore, the design of collision warnings has to lead to easily understood signals which produce fast and specific actions.

How to warn?

A good deal of in-car warning signals are simple warning tones, which means that single or grouped frequencies are presented simultaneously (Campbell, 2007). In such applications one universal tone is used in all kinds of situations, for indication of several hazards, and for different information (parking assistance, oil-warning, ...)

or several very similar tones are used which can be easily confused and need to be learnt. The use of various simple tone signals is difficult, because people can only learn by extensive training to differentiate four to six sounds which are clearly temporally and spectrally distinguishable (Patterson, 1982). To differentiate these tonal warnings better, often a visual cue such as an icon is presented simultaneously. This can be a serious problem, because drivers may not notice the visual signal and then cannot distinguish between situations by the tonal warning alone. In addition, looking at a visual signal and processing it is visually distracting and resource binding in the mostly visual task of driving the car. There is a need to design easily understandable warnings with which the driver immediately understands the problem and the information provided. To do this, semantically enriched warnings were created which not only alert the driver but provide additional information about the critical situation and position of the danger and thereby assist the driver in dealing with the imminent collision. As a first approach, information about the nature and the position of the hazard was incorporated.

In general, collision warnings can be communicated by most modalities. So far, visual, auditory, olfactory and haptic warnings have been presented. Nevertheless, auditory information presentation is well suited for warnings because of the omnidirectional nature and the alerting potential. Especially in time-critical situations such as collision situations, where a rapid response is needed, a non-verbal auditory warning should be used. A verbal warning, as speech, is not appropriate in such situations, as it might take too much time for processing and understanding (Haas & Edworthy, 2006). There are two options for non-verbal warnings: a tonal warning, an earcon, which is a combination of synthetic tones in short and rhythmic sequences possibly with varying pitches and intensities (e.g. Brewster, 1993), or an auditory icon. The auditory icon differs from both, a simple tonal warning and an earcon, because it is based on everyday sounds and supposed to be understood easily and quickly, with an inherent meaning (Barrass & Kramer, 1994).

Current guidelines (Campbell, 2007) suggest to either use a simple tonal warning or an auditory icon if an immediate response is required and to use auditory icons in collision warning cases when fast responses are needed. Empirical laboratory studies have also demonstrated that auditory icons are identified more accurately with their meaning compared to earcons (e.g. Lucas, 1994). Even when tested in a driving simulator, they led to accelerated responses in collision situations (e.g. Belz, 1999; Graham, 1999). Based on these findings, it seems reasonable to use auditory icons instead of simple tone warnings for the presentation of the identity of the danger. Can this information reasonably be supplemented by the spatial position of the danger?

In general, humans are able to discriminate spatial positions of auditory signals. This is also true, when spatial tones are presented in a driving context using a stationary vehicle and if the driver is additionally instructed to identify the perceived position (e.g. Tan & Lerner, 1996). Recent studies also provided good results in directing a driver's visual attention towards a critical driving event to the front or the back through spatially presented auditory warnings (e.g. Ho & Spence 2005; Ho et al.,

2006). These studies suggest that spatial auditory icons signalling a critical event such as an impending collision could be beneficial in focussing a driver's attention towards this event. In these studies the driver is instructed to decide whether a collision situation exists after hearing an alarm, and this results in faster reactions. Therefore, spatial auditory icons seem to be a good possibility to meet the requirements of collision warning systems. In the following study spatial auditory icons were investigated in comparison to spatial tones and non-spatial tones in a realistic driving simulator study. These signals not only represented information about whether a dangerous object is in front of or behind the vehicle but further possible locations, and participants were not instructed about the meaning of the warning. Such spatial auditory icons seem to be good combinations to meet many requirements of collision warning systems.

Method

Evaluation of auditory collision warnings has so far either been investigated in laboratory studies with a driving context, simplifying the complexity which exists in a real driving situation, or through simulator studies in which participants did not show spontanoues driving behaviour but had to follow a special instructed task (e.g. Graham 1999; Ho & Spence 2005). However, for a realistic estimation of effects of auditory collision warnings on driving behaviour it is essential to confront drivers with collision situations combined with a continous driving task and present the warnings without instruction about their meaning. This was supposed to display real-life conditions in which drivers do not know which warning system is displaying the warning or in cases where people are not familiar with their car e.g. a rental car. Thus, in this study several spatial auditory icons were tested in collision situations and participants were continously driving without being instructed about the meaning of the warnings before the first encounter.

Material and design

The study was conducted in a fixed-base driving simulator. Audio-rendering was managed through a wave-field-synthesis system with 48 loudspeakers making it possible to place the sound at any position around the car and produce an exact virtual sound source. Localisation ability of humans using this system was tested with 12 participants in a pilot study and results showed that the exact position of a sound source was hard to identify with deviations from 22-50° using free-form answers. For the current study only the discrimination of a frontal, left side, or right side sound source was necessary. The testing showed, that at least this is possible for naïve listeners under single task conditions without driving (accuracy within a 90°-radius around the source was 82-92%). Three auditory icons were used – a car horn, a bicycle bell and a dog bark. These auditory icons had been tested in pilot studies in which participants had to respond to and identify several warning signal candidates. All of the selected signals led to relatively fast reactions (< 1100 ms) and were correctly identified in free association tests for more than 85%. The auditory icons were compared to a conventional tonal warning, which consisted of a 600Hz tone (see Graham, 1999). This tone was also used for the spatial tone warnings, which were presented through the same wave-field-synthesis system as the auditory icons.

All warning signals had a duration of one second and were equalised for subjective loudness.

The design of the study was a 1x4-factorial between-subjects-design which was investigated in six driving scenarios (see table 1). Each participant heard only one type of warning but did every one of the six scenarios in fixed order.

Table 1. Driving scenarios in the simulator study

Scenario-Number	Name	Collision-object	Danger-inducing parameter
1	line-of-sight obstruction I	bike	occlusion: wall
2	traffic jam I	car	occlusion: curve in a tunnel
3	turn	bike	occlusion: truck on a corner
4	line-of-sight obstruction II	dog	occlusion: house
5	parked cars	dog	occlusion: truck parked in a row with other cars
6	traffic jam II	car	bad-weather: heavy rain

The warnings consisted of the spatial auditory icon, the spatial tonal warning, the simple non-spatial tonal warning, and a no-warning condition. The warning scenarios were contructed based on comments made by participants in a pilot study. They had been asked to mention critical driving situations in which warning signals might have helped them to prevent potential collisions. In every scenario the warning occurred before the collision object could be seen to create real critical possible collision situations. Various parameters, like walls, parked cars or poor visibility were used to make it impossible to perceive the different collision objects. In the auditory icon condition three sounds were used: a dog bark, a bicycle bell and a car horn. Presentation of the sounds was depending on the used collision object in each scenario as well as the spatial information given was depending on the position from which the hazard approached and was congruent with it. For example, the scenario traffic jam led to a presentation of the car horn warning (of course only in the auditory icon condition) from the front, in the scenario of the parked cars with a dog running onto the street, the warning was the dog bark presented from front-right. In the tone conditions the tone was always the same.

Recorded data were the number of collisions, brake-reaction time and minimum distance to the collision object. Moreover, participants were presented with a questionnaire to measure acceptance of the warning systems and a structured interview was conducted about the warnings after the driving-study. Results will be reported for each scenario seperately, as the specifics of each scenario make comparisons between the scenarios difficult.

Procedure

Firstly, participants completed a questionnaire concerning demographic factors such as driving history or hearing impairments. Then in the driving simulator they were introduced to the functions and completed a training drive in which they learnt to accelerate, maintain a constant speed and brake to a complete stop. This took about five minutes. Subsequently, the main part of the experiment started and subjects drove five separate sections for about a total of 30 minutes. Four sections included one collision situation and one section consisted of two collision situations. The collision warning was always presented at least three seconds before the actual collision would occur. The single sections took approximately two to eight minutes. Afterwards the participants filled out an acceptance questionnaire and answered a few interview questions regarding the spatial warning stimuli.

Participants

Fourty-six men and 27 women without hearing deficits participated in the study (mean-age: 34.1 years, range 24-40). All participants had a driving licence and more than 80% used their car every day and for more than 10 000 km per annum.

Results

The data were analysed for each scenario separately and the results are reported correspondingly. The most important parameter for data evaluation was the number of collisions during the scenarios.

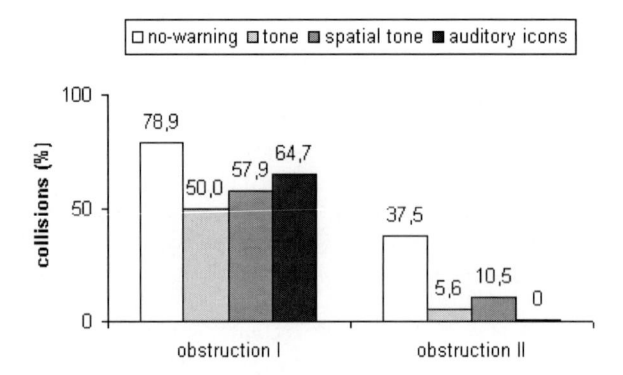

Figure 1. Number of collisions for scenarios obstruction I and obstruction II (in %)

The largest number of collisions occurred in the first scenario (obstruction I), the percentage was especially high for the no-warning condition (see figure 1). The fewest collisions occurred in the simple tone-warning condition. This warning helped to prevent almost 30% of collisions compared to the no-warning condition. Spatial tone warnings and auditory icon warnings lie in-between. In scenario obstruction II most collisions were again found for the no-warning condition, here the auditory icon condition prevented all collisions. No collisions occurred in all other scenarios. Since all collisions took place in these two scenarios, the focus for the others was on brake-reaction times and distance to collision object.

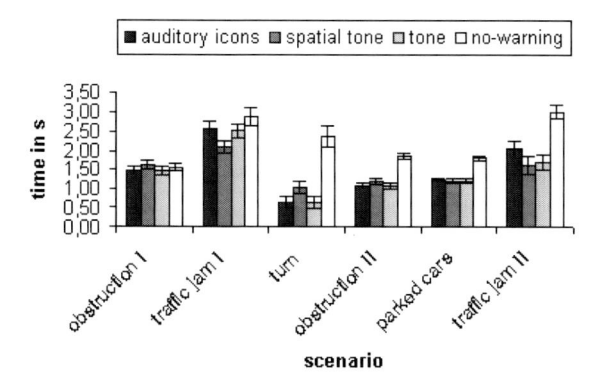

effects of auditory warnings on driving behaviour 123

Figure 2. Mean values and standard errors of brake reaction time (in s)

Multivariate analyses of variance were calculated for each scenario. Considerable differences between the scenarios were found for reaction times as well as distances to collision object (see figures 2 and 3).

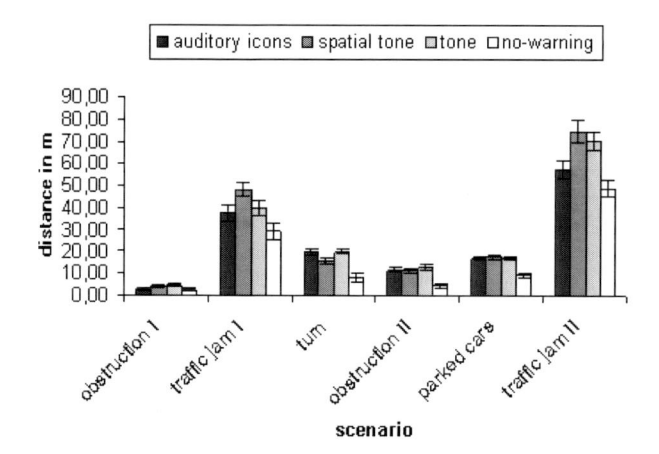

Figure 3. Mean values and standard errors of distance to collision object (in m)

These variations can be explained by the differences in situations of the scenario types, that make any direct comparison very difficult. There is not much differentiation between the warning conditions within one scenario and all measured values are very close together.

Statistically, a main effect for the warning type was found for all scenarios except the first one, there the effect is only a trend (see table 2). Single comparisons showed that mostly from scenario 3 to 6 all warning conditions led to faster reaction times and more distance to the collision object compared to the no-warning condition (see table 2).

Table 2. Main effects and single comparisons for each scenario

	Main effect	Single comparison significant
obstruction I	F= 1.93 (p=.082)	-
traffic jam I	F= 2.50 (p=.027)	spatial tone vs. no-warning (reaction time + distance)
turn	F= 6.25 (p<.001)	all vs. no-warning (reaction time + distance)
obstruction II	F= 8.97 (p<.001)	all vs. no-warning (reaction time + distance)
parked cars	F=10.74 (p<.001)	all vs. no-warning (reaction time + distance)
traffic jam II	F= 5.83 (p<.001)	spatial tone + tone vs. spatial auditory icon and no-warning (distance)

Reaction patterns

Each scenario was constructed so that there was some time between the onset of the warning and the moment at which the collision object could be seen by the driver. Therefore, there were several possible action patterns concerning the order of deceleration and braking behaviour. The possible reaction patterns are displayed in figure 4.

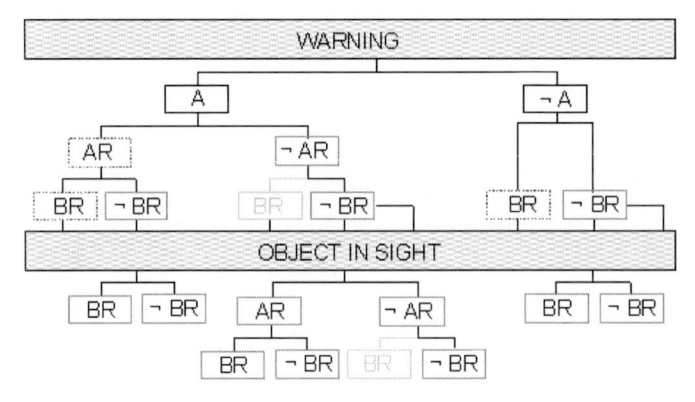

Figure 4. Possible reaction patterns (A = Accelerate; AR = Release Accelerator; BR = Brake, ¬ = No reaction)

Some cases (dashed cuboids, figure 4) represent reaction patterns in which participants braked *before* they could see the collision object. All other cases represent reaction patterns in which participants braked *after* the collision object could be seen. This was measured using markers that showed when the warning was given, the time when the object could be seen and at what time the people reacted. For an analysis of reaction pattern distribution, we categorized all observed reactions into one of the two categories. As can be seen in figure 5, the percentage of reactions *before* the collision object was in sight was larger for spatial tone- and non-spatial tone conditions than for the auditory icons.

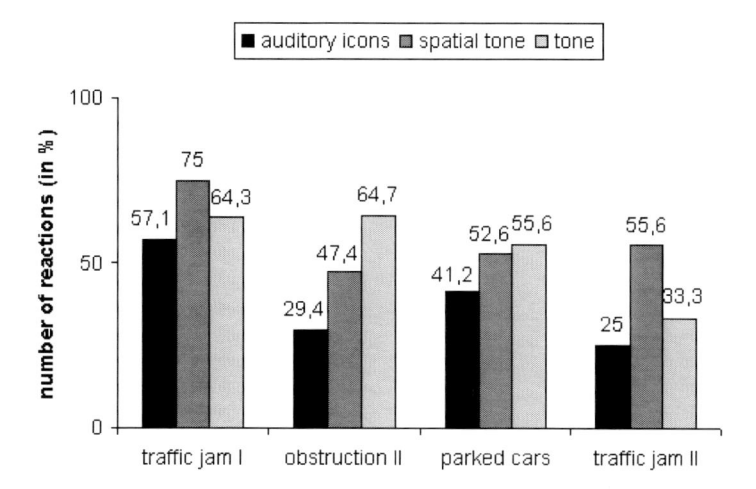

Figure 5. Number of reactions before collision object was in sight (in %)

Especially in the traffic-jam scenarios and in obstruction II, the difference between auditory icons and tone-warnings was about 20% (see figure 5).

Two scenarios are missing in the reaction pattern analysis: scenario obstruction I was left out because all measured reactions occurred after the collision object was seen, and scenario turn was left out because of technical problems in the logfiles.

Interview and acceptance data

After completion of the simulator-driving experiment, participants who were tested in one of the two spatial warning conditions were interviewed about the stimuli. They were asked whether they noticed the spatial presentation of the warnings, and only four out of 34 subjects stated that they consciously perceived spatial manipulated warnings. This is especially interesting since our pilot study had proven good identification accuracy under single task conditions as well as other studies before such as Tan & Lerner (1996) where people responded in the correct quadrant with 80-95% correct. The differences here might be due to the active driving situation in contrast to static situations. All four subjects, who consciously perceived spatial manipulated warnings, had been exposed to auditory icons.

In addition, all participants evaluated the warnings in terms of acceptance. Throughout all warning types, acceptance was high (median: 9; scale ranging from 1 = no improvement, to 15 = great improvement) and did not differ significantly between the conditions.

Discussion

The presented study intended to systematically investigate the effects of auditory icons in a collision warning system in comparison to "normal" simple tone warnings and spatial tone warnings. In a driving simulator study, subjects were confronted with various scenarios in which limited visibility made a collision possible. While performing a dynamic driving task and without any information about the implemented warning system, the participants had to cope with this potentially dangerous situation and correctly interpret the warnings.

Despite the reported advantages of auditory icons (Graham, 1999; Ho et al., 2006), the measurements of dependent variables, such as reaction time and distance to collision object, showed no advantages for auditory icons in comparison to simple, non-spatial tonal warnings. In fact, simple tonal warnings produced the fewest collisions in the first encounter. This first scenario represents the most valid one, because the warning is presented for the first time without any learning effects and is therefore non-predictable and surprising – as is the case in real life situations. This good performance of the non-spatial tone warnings in the first scenario is especially interesting.

Another critical observation was that most participants failed consciously to perceive the spatial characteristics of the respective warnings. This argues against the use of more complex auditory warnings because neither an objective positive effect of spatial warnings nor a consciously perception of them has taken place. In favour of using auditory icons as warnings was the fact that with this kind of warning most reactions were performed after sight of the collision object – which could be interpreted as evidence of such signals leading to orientation reactions without overhasty reflex-like reactions. To support this hypothesis, it would be beneficial to use videotaping for the identification of orienting reactions after warning presentation in all futher studies.

The tonal warnings – the simple tone as well as the spatial tone warning – mostly produced a braking reaction before the actual collision object was seen. Therefore, these signals led to very fast reactions compared to the auditory icons. Although at first glance this seems appropriate for a collision warning system, very quick full braking-action without consideration of the surrounding traffic situation and the collision object might not always be the best reaction. In simple traffic situations, such as pulling out of a parking space, when hazards and risks are clearly distinguishable, fast reflex-like behaviour may represent the optimum reaction in all cases, but collision avoidance in driving situations is much more complex. Braking reaction might not be appropriate in all situations, as rear-end collisions with a following vehicle are also a potential risk.

The interviews after the study showed that participants were quite aware of these effects of the simple tonal warnings. Only participants who were presented with the tone warnings remarked that they noticed that they had shown very strong braking reactions, which might be dangerous if a vehicle were following very closely, for example. No participant in the auditory icon condition mentioned such concerns. Considering the fact that in the present experiment only simple reaction pattern evaluation was possible, this observation has to be further investigated.

Despite the observed differences for the warning conditions, a clear benefit for all collision warnings could be found in the number of collisions avoided in the first scenario. At least 14% of accidents were prevented just by a warning about three seconds before the potential collision was to occur. Thus even for non-predictable situations and completely unknown warnings a clear benefit of a collision warning system could be shown in comparison to the non-assisted driving situation. For future driving simulator studies, we strongly suggest focussing on ecological validity and presenting only one scenario to each person.

In conclusion, the present study revealed limits and opportunities for complex auditory warnings. Evidence was shown that the analysis of simple single dependent variables does not always tell the whole truth about the best solution and more complex analyses like the investigation of reaction patterns might be a promising additional evaluation technique for the effects of warnings on driving behaviour. Moreover, more complex analyses like video or eye tracking analysis should be included in order that more assured conclusions can be drawn for warning design.

References

Barrass, S. & Kramer, G. (1999). Using sonification. *Multimedia Systems, 7,* 23-31.

Belz, S.M. (1997). *A simulator-based investigation of visual, auditory and mixed-modality display of vehicle dynamic state information to commercial motor vehicle operators.* Unpublished Master Thesis, Blacksburg, VA, USA: Virginia Polytechnic Institute and State University.

Brewster, S.A., Wright, P.C., & Edwards, A.D.N. (1993). An evaluation of earcons for use in auditory human-computer interfaces. In S. Ashlund, K. Mullet, A. Henderson, E. Hollnagel, and T. White (Eds.), *Proceedings of InterCHI'93* (pp. 222-227). Amsterdam: ACM Press.

Campbell, J.L., Richard, C.M., Brown, J.L., & McCallum, M. (2007). *Crash Warning System Interfaces: Human Factors Insights and Lessons Learned.* (Report NHTSA Project DOT HS 810 697). Seattle: Battelle Center for Human Performance and Safety. Retrieved September, 14, 2007 from: http://www-nrd.nhtsa.dot.gov/departments/nrd-12/3839/

Comsis (1996). *Preliminary human factors guidelines for crash avoidance warning devices* (Report NHTSA Project DTNH22-91-C-07004). Silver Spring, MD, USA: COMSIS.

Graham, R. (1999). Use of auditory icons as emergency warnings: Evaluation within a vehicle collision avoidance application. *Ergonomics, 42,* 1233-1248.

Haas, E. & Edworthy, J. (2006). An introduction to auditory warnings and alarms. In M.S. Wogalter (Ed.), *Handbook of Warnings* (pp. 189-198). Mahwah, NJ, USA: Erlbaum.

Ho, C. & Spence, C. (2005). Assessing the effectiveness of various auditory cues in capturing a driver's visual attention. *Journal of Experimental Psychology: Applied, 11,* 157-174.

Ho, C., Tan, H.Z., & Spence, C. (2006). The differential effect of vibrotactile and auditory cues on visual spatial attention. *Ergonomics, 49,* 724-738.

German Federal Office of Statistics (2006). Verkehr. Verkehrsunfälle. *Fachserie 8, Reihe 7.*

Lucas, P.A. (1994). An evaluation of the communicative ability of auditory icons and earcons. In G. Kramer & S. Smith (Eds.), *Proceedings of the Second International Conference on Auditory Display, ICAD '94* (121-128). Held November 7-9, 1994, Santa Fe, New Mexico, USA. Retrieved August, 16, 2007 from: http://icad.org/websiteV2.0/Conferences/ICAD94/papers/Lucas.pdf

Patterson R.D. (1982). *Guidelines for auditory warning systems on civil aircraft.* (Report CAA Paper 82017). London: Civil Aviation Authority.

Sanders, M.S. & McCormick E.J. (1993). *Human Factors in Engineering and Design* (7th edition). New York: McGraw–Hill.

Tan, A.K. & Lerner, N.D. (1996). *Acoustic localization of in-vehicle crash avoidance warnings as a cue to hazard direction* (Report NHTSA, DOT HS 808 534). Washington: National Highway Traffic Safety Administration. Retrieved June, 15, 2006 from: http://www.usd.edu/~schieber/materials/in-car-sound-localization.pdf

Effects of preactivated mental representations on driving performance

Jessica Seidenstücker & Rainer Höger
Leuphana University Lüneburg
Lüneburg, Germany

Abstract

This study investigated the distribution of visual attention and driving performance under different conditions of preactivated mental representations. It is propagated that a series of mental concepts is successively activated during driving. Once a concept is activated, reactions to similar objects are facilitated (priming effect). In order to examine to which extent activated concepts influence the behaviour while driving, a driving simulator-study was performed. The difference between the experimental conditions was the existence of a concept-triggering signal: In one version of the traffic scene a premonitory stimulus appeared as a static object (warning sign) and in a second version as a dynamic object (moving pedestrian) before a jaywalker emerged behind a parking bus. The control condition was characterised by the absence of a preactivating stimulus. The task involved a simulator drive while the drivers' eye movements were recorded via a SMI eye-tracker. The main results showed a difference between the trigger and priming conditions respectively in visual search and reaction time. Subjects responded faster (e.g. slowing down the car) if a corresponding traffic object had activated a concept before the emerging event occurred. These findings suggested that mental representations of high relevant aspects which were triggered by salient stimuli facilitate driving actions. Some implications of these results for road traffic design are discussed.

Introduction

Particularly with regard to limited attention an important attribute of the human visual system is the capability to benefit from optical environmental impressions that facilitate a perceptual interpretation and thereby adequate motor reaction. The road traffic often includes corresponding sources of information which could be used to improve the visual information processing (Flowers, 1990). During their driving episodes subjects build up schemata and mental models of certain traffic situations. Aggregated over a longer driving history these aspects constitute the psychological part of the driving experience. In particular, schemata and mental models guide actions and generate a set of expectancies about how the system will behave (Wickens, Lee, Liu & Gordon Becker, 2004). In principle expectancies are the basis of almost any behaviour. To anticipate future events mental mechanisms use past experiences and aggregated knowledge. Even if a subject is unaware of a recent experience, this experience can influence his or her performance (Olson, Roese & Zanna, 1996). In the context of driving, the amount of accessibility of expectancies seems to be very important. The more often a specific fact is presented, the more likely this fact will be utilized to interpret following information (Anderson, 1983). To translate this presumption into the field of traffic psychology, it can be assumed that just the emergence of a certain traffic object increases the subjective probability that this would essentially take place again.

Mental schemata are an important source of expectations and thus both concepts are closely related (Olson, Roese & Zanna, 1996). Schemata and their mental structure can be seen as a set of associations belonging to a specific object (e.g. a pedestrian), an action (e.g. deceleration), or even an event like a specific traffic scene. All those aspects and associations are represented in memory as nodes within a neuronal network (Collins and Loftus, 1975). External stimuli, for example a trigger or a prime, can activate a node and afterwards this activation spreads over the network. In the case of repeated presentations of similar stimuli, a stronger activation of certain nodes can be observed. In that process, the nodes with the strongest activation are those which share analogical characteristics or attributes with the previous stimuli. From this perspective, driving can be classified as a successive activation of concept nodes which were triggered by observable traffic information. Subjects respond faster to corresponding emerging objects if expectancy is built up and thereby a concept node is previously activated. Within the domain of cognitive psychology this effect is called 'priming'. The priming effect can be described by the fact that representations or associations were externally brought in memory just before semantically congruent information occurs. Drivers generate expectancies about targets after perceiving a prime (cp. Posner & Snyder, 1975) and in this survey it is assumed that the following action or response is facilitated because the processing path within the neural network is already activated (Beller, 1971; Höger, 2001). It must be pointed out that the activation of a node declines with time and therefore the priming effect can only be found within a temporary period. Expectancy-based strategies can only operative with long enough time between the presentation of the prime and the target, what implicates that the subject must have enough time to produce expectancies (Posner & Snyder, 1975).

The only requirement to activate a node within the network is that the prime stimuli have to be actively perceived by the subject. The human visual system is not able to process the whole volume of visual stimuli available within the environment. It operates selectively (Posner, 1980). According to Seidenstücker and Höger (2006) the selection of the stimuli depends on their properties. More precisely, there exists an attention attracting hierarchy regarding to specific attributes in traffic related stimuli. Within this hierarchy, the division between dynamic and static objects is very important: living objects attract the most attention (e.g. pedestrians or animals), followed by dynamic objects (e.g. cars), whereas static objects (e.g. signs) have more moderate attention attracting effects. It is assumed that in driving situations, diversified stimuli capture attention and, according to that corresponding mental concept, nodes are activated. If so, the processing of new –but related– objects and the driver's reaction to them should be improved and facilitated (e.g. the act of deceleration). In order to investigate whether these considerations hold, an empirical verification, an experiment was performed.

Method

Participants

Eight female and fifteen male subjects between the ages 20 and 70 years (mean = 39.4 years), participated in the study. All subjects were holders of drivers' licences and had normal visual acuity. For participating in the experiment, subjects were paid a 10 € honorarium.

Stimuli

The stimuli to which the subjects had to respond in the experiment were embedded in a simulated driving task. The scenery consisted of a 3600 m long urban route including signals, signs, buildings, and interactive traffic. To complete the whole simulation scenario a subject drove for approximately 3 minutes. After covering half of the run, the subject is passing a parking bus with a bus stop and an appropriate sign at the roadside. Once the subject were driving, a jaywalker emerged from behind the bus (see Figures 1a and 1b) and crossed the road from right to left.

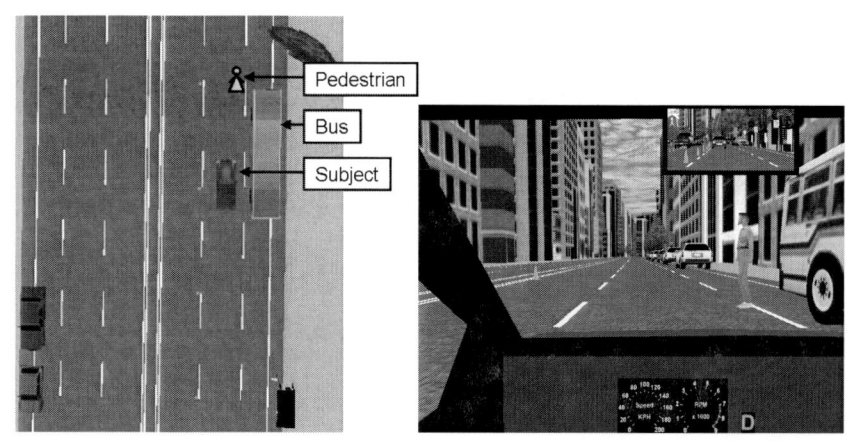

Figure 1a (left). Bird's eye view of the bus scenario. Figure 1b (right). Bus scenario from the subjects' perspective

This pedestrian was triggered relative to the subject's own vehicle, which means he (initially obstructed from the subject's view) would start walking at a speed of 2 m/s once the driver is within 28 m. This hazard conflict is situated on a position where the street was not marked as a crosswalk.

Additionally to the basic (control) version this bus scenario was generated in two further versions:

- Static prime version: A warning sign was set up at the roadside announcing that people might get off a bus and cross the road (see Figure 2a). The distance between this sign and the emerging pedestrian was 105 m.
- Dynamic prime version: A mother and a child cross the road at a safe distance of 86m in front of the jaywalker (see Figure 2b).

Figure 2a (top). Screenshot with priming by a static objects (bus sign). Figure 2b (bottom). Screenshot with priming by a dynamic object (moving pedestrians)

Apparatus

The experiment was conducted using the fixed-based virtual reality *STISIM Drive* Simulator. Subjects operated the simulator using a steering wheel, accelerator and brake pedal. Those controls were mounted in a seat box (including the interior of a real car) and plugged to the control computer. The whole urban scene was projected to a white painted area of 200 cm x 150 cm by a *BENQ DLP*-projector. The participant was seated at a distance of 3.8 meters in front of the video projection. Daylight was slightly reduced to enhance contrast and maintain constant test conditions. All driving scenarios were created with a *Scenario Definition Language (SDL)* that described the assembling of the different objects in the environment (e.g. other vehicles, pedestrians) and traffic characteristics (e.g. markings, signals, curvature of the road, intersections) by text file commands. The *STISIM* program is able to collect all driving activities such as acceleration, braking and steering behaviour.

During the driving task the eye-movements of the subjects were recorded via the *iViewX HED*-system (*SMI*). This head-mounted eye-tracker system records with an accuracy between $0.5° - 1.0°$ and 50/60 Hz sampling rate.

Design and procedure

All participants were tested individually. Initially the subjects had to be introduced to the graphical presentation of the virtual traffic environment as well as to the steering and speed controls via a 6 min familiarization run (including a huge parking site, as well as a rural and an urban highway). The participants' remit was to move straight on the defined route and to act like an ordinary car driver, giving consideration to the general German Highway Code.

After the training the participants were randomly assigned to one of the three experimental conditions. As aforementioned, the different conditions implicate a varied degree of concept activation: (1) No priming (control), (2) priming by static objects (see Figure 2a) and (3) priming by dynamic living objects (see Figure 2b). The factors driving speed, time-dependent acceleration as well as the locations of starting braking behaviour were the dependent variables. In order to ascertain when a certain participant perceived the relevant stimuli, eye-movements were recorded. To derive a synchronisation between the driving behaviour and the eye-movement data, markers were positioned at selected locations within the scenario.

Results

First of all, the course of velocity needed to be analyzed. Therefore velocity profiles for all participants were created. The velocity profiles give both information about the actual speed at the critical sections of the course and temporal aspects of braking behaviour. Finally the velocity profile clearly shows the point where the car comes to the standstill. Within the velocity profiles the appearing moments of the prime stimuli (warning sign, moving pedestrians), the appearing moment of the target (crossing pedestrian) and the moment of target perception were marked. On closer examination it turned out that if a subject moves too slowly towards the bus there are no difficulties in breaking to avoid an accident, because the pedestrian had already crossed the road when the subjects passed the bus. This means that for those subjects the critical event had lost its hazardousness. On this account, 5 of the 23 subjects that drove below the defined speed-threshold of 30 km/h had to be excluded from further analyses. Consequently, a valid sampling of 18 subjects remained. During the roadway arrangement the following moments are of interest:

- the prime stimuli (bus warning sign and mother with child respectively)
- the target stimuli (the moment when the jaywalker emerged behind the bus)
- the moment of target perception.

The beginning of deceleration was defined as the covered distance from the beginning of the course until the point was reached where the speed was reduced with the objective of avoiding a collision with the pedestrian. This kind of measure was taken to create a variable which is largely independent of driving speed. The perception of the emerging pedestrian as well as the static warning sign was defined by a fixation greater than 50 milliseconds. The video-analyzing-program *INTERACT* (Mangold Systems) was used to analyze the eye-movement data and especially the precise fixation position within the time course of the run. This pre-analysis provided a basis to determine important values like 'detection times' and locations of the 'onset of deceleration'.

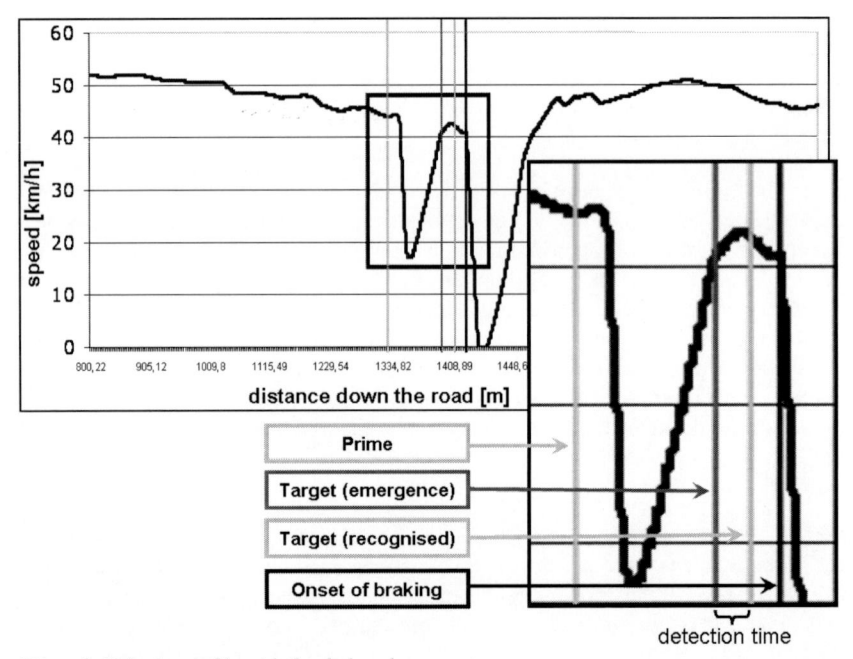

Figure 3. Velocity profile with the deduced parameters

In this context it has to be differentiated between the exact moment when the target emerged behind the bus and the moment when a participant recognizes this target. The variable 'detection time' acts on this differentiation and refers to the elapsed time (in milliseconds) between the emergence of the target and its recognition. In contrast to that, the location of the deceleration is a spatial distance down the road (in meters). Figure 3 represents a prototypical velocity plot with the previously described variables 'detection time' and 'onset of braking'.

In the next step the effects of the three experimental conditions on detection time and deceleration reaction were analyzed. To determine the values of both variables, an analysis of variance was performed. The ANOVA for the dependent variable 'detection time' revealed a highly significant influence of the priming conditions ($F_{2,15}$ = 11.58, p = .001). It turned out that in the condition without any priming the time period between the emergence and the recognition of the target amounted to 489 milliseconds on average. In the condition with the static prime, the mean detection time was 332 milliseconds, whereas in the dynamic priming situation the mean detection time amounted to 99 milliseconds. The results are depicted in Figure 4.

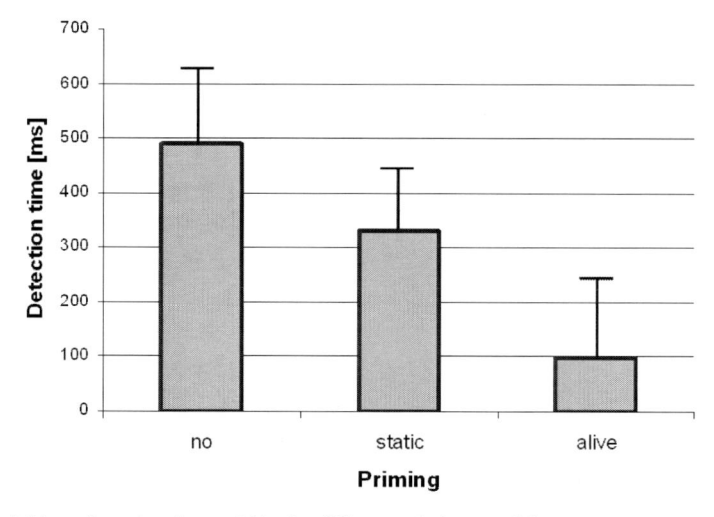

Figure 4. Mean detection times within the different priming conditions

For the variable 'onset of braking' the ANOVA revealed also a highly significant effect of the experimental conditions ($F_{2,15}$ = 13.76, p < .001). Figure 5 shows the average covered distances until the start of decelerating pertaining to the different conditions. In the condition without priming, braking behaviour was initiated after a covered distance of 1433.5 meters on average. In the static priming condition the process of braking started after a covered distance of 1432.5 meters whereas in the dynamic priming situation braking was initiated after 1426 meters.

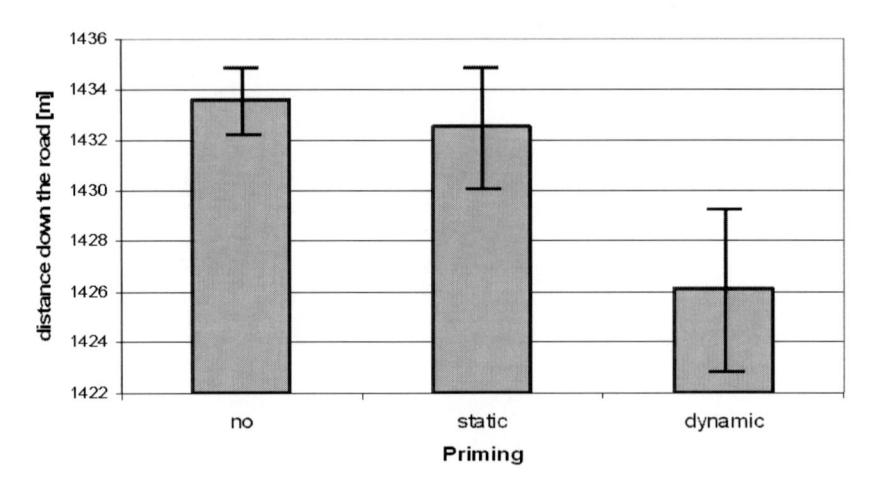

Figure 5. Mean covered distances until braking was initiated

In order to prove the influence of the different priming conditions on the incidence of collisions with the jaywalker, the frequencies of crashes were counted. Because of the exclusion of 5 subjects (who drove to slowly) an unequal distribution of participants over the experimental conditions resulted. Therefore the counted crashes were transformed to percentage values. The proportion of crashes is depicted in Figure 6.

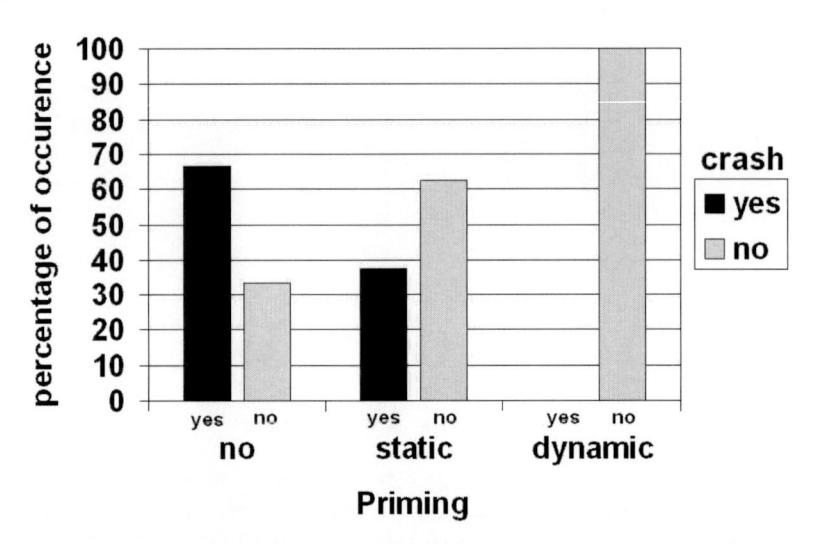

Figure 6. Number of crashes within the different conditions

The relative amount of crashes differs highly significant between the experimental conditions ($\chi^2 = 58.68$, df = 2, p <.001): Proportional most crashes were found in the condition without priming, whereas no crashes emerged in the dynamic priming condition.

Discussion

The study was designed to observe eye-movements and driving behaviour during a simulator drive in a rural traffic scenario in which a jaywalker emerged behind a parked bus under different preactivating conditions. To summarise, the experiment shows the following results:

1. Priming has a strong influence on the perception of critical events
2. Priming influences the reaction time on a target (e.g. the moment of deceleration to prevent a collision)
3. It is different whether the priming situation consists of static or dynamic objects
4. Accident rate depends on the priming condition.

Mental representation of traffic events can be preactivated when external stimuli and associated expectancies acting as a prime. Thereby the speed of perception, action of following and furthermore related stimuli can be improved by such a preactivation of mental concepts. The degree of activation depends on the nature of the presented stimuli (prime): Dynamic stimuli have a greater effect than static stimuli. As shown by previous studies (Seidenstücker & Höger, 2006; 2007), dynamic and living objects attract attention to a bigger magnitude than static, nonliving objects. Concerning the driving situations, the threatening potential of dynamic and particularly living objects is generally much higher than that one of nonmoving, inanimate objects. Potential reasons could be seen in the evolutionary nature of dynamic objects, because those normally have a more unpredictable character and thus implicate a higher potential hazard than static objects. Under critical circumstances the ability to identify participants involved in the traffic scene (e.g. a jaywalker) early enough might be essential to avoid serious accidents involving them. As detected in this experiment, an ability such as this is highly related to the attention attracting properties of the objects available in the road scene. The different nature of traffic objects (living vs. nonliving) influences the degree to which a mental concept is activated by what perception of corresponding objects as well as initiating actions are facilitated and improved.

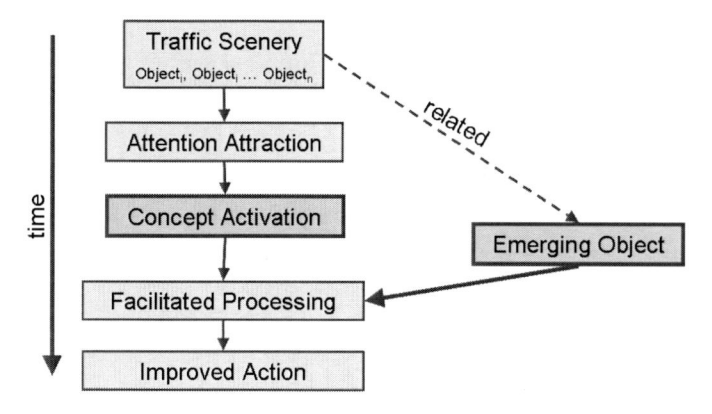

Figure 7. Concept activation model of driving behaviour

Based on the findings, the effects and processes of preactivated mental representations on driving performance are summarized in the model depicted in Figure 7.

Conclusion

Safe traffic behaviour depends on the ability to build up an adequate mental representation of the present traffic scene. Thereby not only the number of details within the mental representation is essential, but more importantly the representation of high relevant traffic aspects like unpredictable living objects (e.g. playing children, thoughtless pedestrians or animals at the roadside).

The simulator experiment clearly shows the role of activated mental concepts while driving. Especially in critical driving situations, preactivated mental concepts reduce response times, e.g. for braking behaviour. These results could be used to review the actual road traffic design. According to the theory of selective attention; the visual perception of the real world is not complete and the road design should integrate more contextual and preactivating information to compensate for this human deficiency. As shown in the present experiment, dynamic objects are good candidates for initiating a preactivation of mental concepts. From this point of view static warning signs should be replaced by signs showing (illusory) motion. One important point is that these signs have to be related to the meaning of the hazardous traffic situation and not only attract attention. So a human-shaped dummy at the roadside should indicate that there exists a danger for a human outside the car (e.g. for a pedestrian or a construction worker). The more road traffic design succeeds in combining attention attraction with content specific information the greater is the benefit for the driver in terms of reducing reaction time and an early initiating of breaking behaviour. Besides more beneficial traffic information might reduce the overall processing load and make room for attention-based resources to relevant traffic aspects.

Acknowledgment

This study is financed by the 'Arbeitsgemeinschaft Innovative Projekte' (AGIP) of the Ministry of Science and Culture, Lower Saxony, Germany. The work would not have been possible without the fundamental support and assistance provided by colleagues (Janina Suhr and Ernst Roidl) and the contribution by the ADAC Nord in Embsen/Germany

References

Anderson, C.A. (1983). Imagination and expectation: The effect of imagining behavioral scripts on personal intentions. *Journal of Personality and Social Psychology, 45*, 293-305.

Collins, A.M. & Loftus E.L. (1975). A spreading-activation theory of semantic processing. *Psychological Review, 82*, 407-428.

Beller, H.K. (1971) "Priming: Effects of advance information on matching," *Journal of Experimental Psychology, 87*, 176-182.

Flowers, J.H. (1990) Priming effects in perceptual classification. *Perception and Psychophysics, 47*, 135-148.

Olson, J.M., Roese, N.J., & Zanna, M.P. (1996). Expectancies. In E.T. Higgins and A.W. Kruglanski (Eds.), *Social psychology: Handbook of basic principles* (pp. 211-238). New York: Guilford.

Posner M.I. & Snyder, C.R.R. (1975). Facilitation and inhibition in the processing of signals. In P.M.A. Rabitt and S. Dornic (Eds.), *Attention and performance V* (pp. 669-682), Amsterdam: North-Holland.

Posner, M. (1980). Orienting of attention. *Quarterly Journal of Experimental Psychology, 32 A*, 3-25.

Höger, R. (2001). *Raumzeitliche Prozesse der visuellen Informationsverarbeitung*. Magdeburg, Germany: Scriptum Verlag.

Seidenstücker, J. & Höger, R. (2006) Gefahrenrelevanz von Objekten im Straßenverkehr. In F. Lösel and D. Bender (Eds.), *45. Kongress der Deutschen Gesellschaft für Psychologie* (p. 331), Lengerich, Germany: Pabst Science Publishers.

Seidenstücker, J. & Höger, R. (2007). Implizite Prozesse der Gefahrenwahrnehmung. In K.F. Wender, S. Mecklenbräuker, G.D. Rey, and T. Wehr (Eds.), *Beiträge zur 49. Tagung experimentell arbeitender Psychologen* (p. 53). Lengerich, Germany: Pabst Science Publishers.

Wickens, C.D., Lee, J.D., Liu, Y., and Gordon Becker, S.E. (2004). An Introduction to Human Factors Engineering. Upper Saddle River, USA: Pearson.

The effect of experience, relevance, and interruption duration on drivers' mental representation of a traffic situation

Martin R.K. Baumann[1], Thomas Franke[2], & Josef F. Krems[2]
[1] DLR German Institute of Aerospace
Braunschweig
[2] TU Chemnitz
Germany

Abstract

While driving a great amount of information has to be attended to, interpreted, and integrated into a coherent representation of the current situation within very limited time constraints. Processing this amount of information exceeds the capacity of working memory (WM). The theory of long-term working memory (LT-WM) describes a mechanism to overcome WM limitations (Ericsson & Kintsch, 1995). According to this theory it is assumed that experienced drivers possess a more differentiated knowledge base of relevant traffic situations allowing them to encode relevant information more reliably than novice drivers. The assumptions were tested in a driving simulator study (n=40). During each drive the participants were interrupted repeatedly and were asked for the number of cars around them. A mixed factorial design was used with experience as between-subjects factor and interruption duration, cuing, and relevance of queried information as within-subject factors. The results indicate that LT-WM might be involved in the construction and maintenance of situation awareness in driving.

Introduction

Performance in dynamic situations is highly influenced by how well the operator knows what is currently going on around him and how well he can predict the development of the situation in the near future. The processes involved in constructing and maintaining such a mental representation of the current situation that forms the basis for the operator's decisions and actions are described in the concept of situation awareness (Endsley, 1995). According to Endsley (1995) situation awareness entails "the perception of the elements in the environment within a volume of time and space, the comprehension of their meaning and the projection of their status in the near future" (p. 36). This concept was successfully applied in the last decade to human factors issues in aviation, nuclear power generation, or military combat systems. Only recently it has been introduced to the analysis of driving behaviour (Baumann & Krems, 2007; Gugerty, 1997; Matthews et al., 2001). This was at least in part driven by concerns that have been raised about the effects of novel driving assistant and information systems. Models of situation awareness in the driving context have been developed to assess the impact of such assistant and information systems on driver's situation awareness and their consequences for driving behaviour and driver's safety (Baumann & Krems, 2007; Matthews et al., 2001). But because of the clear differences between requirements of the piloting or process control tasks and the driving task, such as the dynamics of the situation, it seems necessary to examine in more detail the cognitive processes that underlie situation awareness to determine its applicability to the driving domain and to develop measures of situation awareness while driving (Baumann & Krems, 2007). This paper describes a study that aimed at examining the memory mechanisms that underlie the maintenance of the mental representation of the current situation and factors that influence these maintenance processes.

We view situation awareness as a comprehension process (Baumann & Krems, 2007; Durso et al., 2007) that consists of the construction of a mental representation of the current situation based on the interpretation of perceived environmental information. This construction process involves the activation of knowledge previously acquired by the driver and stored in long-term memory and the integration of this activated knowledge into a coherent situation model that represent the driver's understanding of the current situation and that directs the driver's action planning and decision making. For this construction process to result in a coherent situation model it is necessary that when new information is perceived those parts of the situation model relevant for the interpretation of this new piece of information have to be available in working memory (Fischer & Glanzer, 1986; Glanzer & Nolan, 1986). Because of the amount and the complexity of the information that has to be represented in the situation model, such as the position and speed of other vehicles near to the own vehicle, traffic rules defined by signs, the status of the road surface and so on, it is assumed that the situation model contains much more information than can be kept active in working memory. Therefore a mechanism is necessary that allows to keep this information stored in long-term memory but makes this information reliably available for processing when it becomes relevant.

The theory of long-term working memory (LT-WM) describes a mechanism (Ericsson & Kintsch, 1995) that could provide such a function. According to LT-WM theory "cognitive processes are viewed as a sequence of stable states representing end products of processing [...] acquired memory skills allow these end products to be stored in long-term memory and kept directly accessible by means of retrieval cues in short-term memory" (Ericsson & Kintsch, 1995, p. 211). Such set of retrieval cues are arranged into stable retrieval structures that can have many different forms depending on the domain. In the case of driving such retrieval structures might be based on the huge amount of driving situations an experienced driver encountered. The experienced driver possibly developed many highly differentiated schemata of driving situations that allow him to easily identify many different types of driving situations. Such a mechanism might, for example, at least in part explain why experienced drivers are much better and much faster in identifying dangerous traffic situations than less experienced drivers (Crundall et al., 2003; Crundall & Underwood, 1998; Underwood et al., 2003).

A schema becomes activated when the driver perceives information that is associated with this schema (Norman & Shallice, 1986). If a schema is activated the schema provides the driver with information about what is relevant in this situation and it allows to easily encode current information by connecting it to the schema. This initialized schema is then part of the driver's situation model. For example, the "overtaking on a highway schema" might be activated when the driver being on the right lane perceives a slower vehicle in front. If this schema is activated the driver knows, for example, where to direct attention to: to the left lane behind him to check whether there are faster cars approaching from behind on the target lane or whether there is a car in the blind spot, to the left lane in front to check whether there is enough space to change the lane, to the vehicle in front to estimate its speed relative to the own speed, and so on. The information perceived can then be easily connected with this schema by initializing the relevant slots of the schema. Thereby, information from the current situation is associated with knowledge in long-term memory and represented in a stable form and does not need to be actively kept in working memory. It can be easily accessed and retrieved from long-term memory via the "overtaking schema". Such a mechanism allows that the capacity limitations of working memory can be overcome as task-relevant information that is represented in the driver's situation model can not only be stored in working memory but can also be stored in long-term memory. In this case, information about the current situation kept in working memory and the information connected to this information via this schema establish LT-WM. If then, for example, the driver is shortly distracted during the overtaking manoeuvre by a question from a passenger, leading to the short activation of different schemata, the information previously connected to the overtaking schema is immediately available when the driver focuses on the overtaking manoeuvre again. That is, the driver immediately knows that there was a car approaching from behind rather fast that he or she should let pass before changing to the left lane.

The advantage of applying the theory of LT-WM to situation awareness is that this allows to make specific predictions about the effect different factors should have on the availability of the driver's situation model. These factors are the driver's experience, the relevance of the information, the duration how long information has to be kept available, and whether the driver has available retrieval cues to access information encoded in LT-WM. In the following the key predictions are described that are also addressed in the Results section. A detailed description of all predictions and results can be found in Baumann and Franke (in prep). First, based on the assumption that experienced drivers have available a huge amount of knowledge about traffic situations they should be able to use the LT-WM mechanism to encode driving related information much more efficiently than less experienced drivers can do. This means that they should be able encode more traffic information reliably into long-term memory while driving and therefore should be able to keep this information longer available than less experienced drivers. Second, as the retrieval structures of experienced drivers should be especially adapted to processing driving relevant information LT-WM should support the encoding of relevant traffic information more than that of less relevant information. Therefore it is predicted that the advantage of experienced drivers compared to less experienced drivers in keeping driving relevant information available should be greater when this information is highly relevant than when it is less relevant. Third, as the availability of information encoded in the retrieval structures depends on the presence of retrieval cues in working memory and as experienced drivers should possess more and more differentiated retrieval structures they should be able to take better use of retrieval cues than less experienced drivers.

These predictions were tested in a driving simulator experiment where participants had to drive on a three-lane highway. The simulation was interrupted repeatedly and participants were asked about a crucial aspect of the traffic situation, namely the arrangement of cars around the participant's car. This was done by asking the participant about the number of cars on different locations around the participant's car, for example on the left lane behind the participant. Two groups of drivers were tested: experienced and less experienced drivers. The length of the delay between interruption and recall was either long or short. After the interruption, filled with a working memory task to prevent rehearsal, either a high informative or a low informative retrieval cue was presented. Additionally the relevance of the information that was to be recalled was manipulated. The participant either had to remember the number of cars from a position highly relevant for the current driving manoeuvre or from a position less relevant.

Method

Participants

There were 45 participants in the experiment. Five participants had to be excluded from the sample due to technical problems that lead to missing data. The participants were students of Chemnitz University of Technology. Further participants were recruited from local driving schools and an agency for arranged lifts. All received partial course credit or money for participation.

The sample consisted of 20 male and 20 female participants aged between 18 and 29. There where two groups of driving experience. The 20 "inexperienced" participants were newly licensed drivers with little driving experience with a mean age of 20.32 ($SD = 1.64$) and a lifetime driving experience between 50 and 15785 km ($M = 2619.69$). The 20 "experienced" participants where frequent travellers with a mean age of 26.08 ($SD = 1.73$) and a lifetime driving experience between 83667 and 486558 km ($M = 186896$). They had driven between 10200 and 50000 km during the last year.

Driving scenario and task

The experiment was conducted using STISIM Drive, a medium-fidelity fixed-base driving simulator. The particular setup consisted of a BMW 350i with automatic transmission as driving cab and a projection providing a 135° horizontal field-of-view with screens being 260 cm away from the driver's position. All three rear-view mirrors were projected on the screens in their appropriate position.

The driving scenario consisted of a three-lane highway with a total length of 112 km. There was moderate curvature and sight was reduced to a range of 200 m in front of the participants by simulating misty weather conditions. Traffic density was moderate. Participants were instructed to drive 110 km/h in the middle lane and to react to slower cars in front of them with overtaking where possible. The simulated traffic on the right lane drove slower and the simulated traffic on left lane drove faster than 110 km/h.

While driving the simulation was repeatedly paused with screens being blanked and participants had to switch to a loading WM task of counting backward in threes with a fixed rate of 2 seconds. This interruption lasted for 2 or 20 seconds. After this interruption interval participants had to recall the number of cars on one of four possible locations around the car within a range of 60 m behind or in front of the participant. These locations were the right, middle, or left lane in front or left lane behind the participants. The number of cars in these locations ranged between 0 and 2. The participants were either given a high informative or a low informative retrieval cue at the time of recall (see Figure 1) immediately after the interruption interval. In the high informative cue condition participants were presented with the full traffic scene as it was at the time of the interruption. Only that part of the simulated scene was masked with a green field from which the number of cars had to be remembered. This mask therefore also indicated the area from which to remember the number of cars. As it is not clear which kind of cues drivers might use to retrieve information from LTM the idea of this cue condition was to provide as much information as possible to maximize the probability that the effective retrieval cue was present in the scene. In the low informative cue condition the participants were presented with a picture that contained only the street and the green field to indicate the area to remember the cars from. All objects, such as cars, or trees were removed from the scene to minimize the probability that any effective retrieval cue is present in the scene. After participants finished recall the simulation continued from the moment it was interrupted.

Figure 1. Examples of cuing conditions; left: recall with retrieval cue; right: recall without retrieval cue

Including overtaking manoeuvres as possible reactions to critical events allowed us to classify the traffic on the left lane behind the participant and the traffic directly in front of the participant as relevant, whereas the traffic on the other locations was classified as irrelevant. Accordingly, the locations queried after the interruption were classified into highly relevant and less relevant locations providing us with the opportunity to examine the effect of information relevance on the recall performance.

During the experiment there were a total of 96 interruptions with subsequent recalls where one of these four locations was queried. There were eight additional interruptions where participants had to recall the number of cars in all six possible locations around the car, but the results of these complete recalls will be reported elsewhere (Baumann & Franke, in prep.).

Procedure

After reading the instruction participants were made familiar with the simulator and their driving task. The experiment was divided into two blocks that were separated by a short break. The order of the blocks was balanced across subject. The average time to complete one block was 76 min. Each block was made up of the same 18 event sequences in fixed, pseudo-randomized order. An event sequence was a roadway segment of 2, 3 or 4 km length that included several events and 2 to 4 interruptions with subsequent recalls. In each segment the roadway, the traffic configurations and the specific recall trials were fixed except for the interruption duration.

Design

The resulting design consisted of four factors with two levels each: driving experience (inexperienced vs. experienced) as between-subjects factor, interruption duration (2 s vs. 20 s), retrieval cue (high informative vs. low informative), and relevance of the to be recalled area (irrelevant vs. relevant) as within-subjects factors. The main dependent variable in the experiment was the performance in the recall task after the interruption and the result description will focus on this variable.

Results

For each participant the mean proportion of correct answers in the recall test for each combination of the four within-subject factors was computed. An answer in the recall test was judged correct when the participant recalled the exact number of cars in the queried location. For all statistical analyses reported here the significance level was set to .05. The error bars in the figures denote within-subjects confidence intervals (Masson & Loftus, 2003). The presentation of the results will be focused on effects of driving experience that are directly relevant for the test of LT-WM involvement in situation awareness while driving. The mean proportions were analysed using a mixed factorial ANOVA with the independent variables 2 (driving experience) x 2 (interruption duration) x 2 (cuing condition) x 2 (relevance).

First, given the assumption that experienced drivers can use LT-WM more efficiently to associate new information with knowledge stored on LTM because of their greater and more differentiated knowledge about traffic situations one prediction was that interruption duration should have a smaller effect on recall performance for experienced than for inexperienced drivers. Therefore, it was expected that the interaction between experience and interruption duration should be significant. But this effect was not present, $F(1, 38) < 1$, NS. Instead the ANOVA revealed only a main effect of interruption duration, $F(1, 38) = 113.5$, $p < .001$. This was due to the fact that recall performance after an interruption of 2 sec was much better ($M = .773$) than recall after 20 sec ($M = .627$) independent of the drivers' experience.

Second, based on the assumption that the retrieval structures experienced drivers use to encode new information into LT-WM are adapted to encode task-relevant information it was predicted that the advantage of experienced drivers compared to inexperienced drivers in keeping information available should be greater for highly relevant than for less relevant information. In accordance with this prediction there was a significant interaction between relevance of the information and experience, $F(1, 38) = 4.6$, $p = .038$. This was due to the better recall of experienced drivers for cars from relevant locations than for cars from irrelevant locations, inexperienced drivers did not show any difference in recall performance between cars from relevant and irrelevant locations as can be seen in Figure 2.

The third prediction was that experienced drivers should benefit more from retrieval cues than inexperienced drivers as experienced drivers possess more and more differentiated retrieval structures. But the interaction between cuing and experience was not significant, $F(1, 38) = 1.6$, NS. There was only a significant main effect of cuing, $F(1, 38) = 19.7$, $p < .001$, indicating that a retrieval cue facilitated recall both for experienced and inexperienced participants similarly.

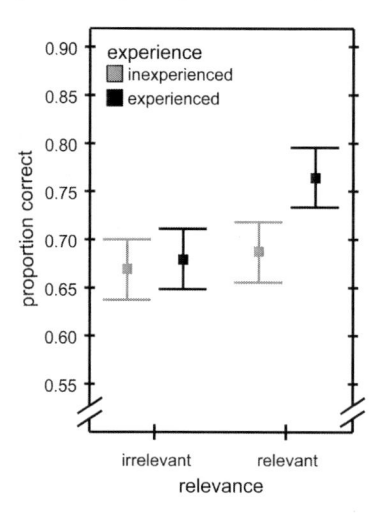

Figure 2. Mean proportion of correctly recalled number of cars as function of driving experience and relevance

But when the effect of relevance was also considered in the three-way interaction of experience, cuing condition, and relevance the ANOVA revealed that the interaction between the cuing condition and experience is at least marginally significant, $F(1, 38) = 2.98$, $p = .093$. Inspection of Figure 3 shows that this is because providing experienced drivers with a retrieval cue facilitates recall both for cars from relevant and irrelevant locations, whereas providing inexperienced drivers with a retrieval cue facilitates only recall of cars from relevant locations. That is, experienced drivers seem to be able to encode more information from the driving situation including both highly and less relevant information. If then provided with the appropriate retrieval cues they can retrieve both relevant and less relevant information. Inexperienced drivers seem to encode mainly information that is relevant according to the current situation interpretation. If provided with retrieval cues these can consequently support only the retrieval of relevant information.

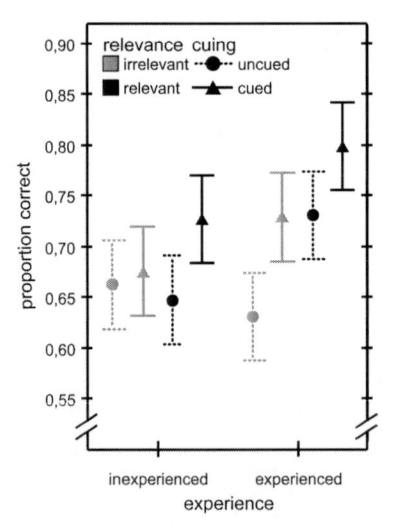

Figure 3. Mean proportion of correctly recalled number of cars as function of driving experience, cuing and relevance

Discussion

The aim of this experiment was to test whether LT-WM is involved in situation awareness while driving. The basic assumption of this theory is that domain experts are able to use their long-term memory to overcome the capacity limitations of working memory when processing task relevant information. That is, because of their experience domain experts possess a huge amount of task knowledge and effective encoding strategies that allow them to reliably associate new task information with knowledge stored in long-term memory. By this, working memory resources are not necessary any more to keep this information available. To retrieve this information into working memory again retrieval cues need to be present in working memory. Such retrieval cues are organized into stable retrieval structures that are adapted to the domain and make it highly probable that relevant information is retrieved when necessary. It was assumed that experienced drivers can use LT-WM to encode traffic situation related information while they construct a mental representation of the current situation – the situation model, whereas inexperienced drivers can use this mechanism to a much smaller extent

To examine this issue a driving simulator experiment was conducted. The simulation was interrupted several times and participants had to recall after varying interruption durations the traffic situation at the time of the interruption. Besides the duration of the interruption the relevance of the queried information in relation to the participants driving task and the retrieval cue provided at recall was manipulated. Two groups of participants were selected for this study: highly experienced and less experienced drivers.

Based on the theoretical framework several predictions about the mental availability of traffic related information after the interruption could be derived. Three were described in detail in this paper as were the results of the empirical tests of these predictions. A complete description of all possible predictions and their empirical tests was beyond the scope of this paper but can be found in Baumann and Franke (in prep.).

The results of this experiment provide some indication that LT-WM is indeed involved in situation awareness processes while driving, but the results are ambiguous. The first prediction was that experienced drivers should be less impaired by the duration of the interruption than the less experienced drivers. This was not the case. Both groups of drivers were affected by the interruption duration in a similar way.

The second prediction was that experienced drivers should be able to encode and therefore retrieve task-relevant information more reliably than task-irrelevant information, whereas less experienced drivers should show a much smaller difference between the retrieval of task-relevant and task-irrelevant. The basis of this prediction is the assumption that the retrieval structures experienced drivers possess are adapted to encode task-relevant information. This prediction was confirmed. Experienced drivers showed a clear effect of relevance of information whereas less experienced drivers did not. The third prediction concerned the effect of retrieval cues on the retrieval of information. Given the assumption that experienced drivers possess more and more differentiated retrieval structures they should be more sensitive to the presence of retrieval cues and should profit more in terms of retrieval accuracy when retrieval cues are presented. The predicted two-way interaction between driving experience and presence of retrieval cues was not significant. But when the effect of relevance was also taken into account it turned out that experienced drivers indeed tend to profit more from retrieval cues as these cues support the retrieval of both highly task-relevant information and less relevant information. Inexperienced drivers profit from retrieval cues only when relevant information is cued, not when less relevant information is cued. This means first that experienced drivers seem to be able to encode more traffic information while driving than inexperienced drivers which is in accordance with results of studies relating the visual scanning behaviour of drivers with their experience (e.g., Crundall et al., 1999). Second, this also is in accordance with the assumption that experienced drivers are able to use retrieval cues more efficiently because they possess more and better differentiated retrieval structures (Baumann & Krems, 2007; Durso et al., 2007).

To summarize, the results on the one hand indicate that retrieval structures as assumed by LT-WM theory are involved in processing traffic information to construct a mental representation of the current traffic situation – the situation model. The interaction of experience and relevance and the interaction between experience, relevance and cuing condition argue for this assumption. But on the other hand there was no interaction between interruption duration and experience as was expected when experienced drivers should be able to use retrieval structures in long-term memory to store results of cognitive processes additional to working memory resources, whereas inexperienced drivers can only rely on working memory resources. One possible explanation is that LT-WM is involved in the construction of the situation model while driving but that the involved retrieval structures are not adapted to store information for a longer period of time. This might be due to the high dynamics of the driving task on the manoeuvre level (Michon, 1985) that make it inefficient to store information about the traffic situation at the manoeuvre level for longer time periods. The information that was to be retrieved in this experiment certainly belonged to the manoeuvre level. The arrangement of cars around one's own car changes rather fast and the representation one has constructed is certainly irrelevant after 20 sec. This contradiction between use of retrieval structures and no long-term availability of encoded information might be resolved by looking at results of Hatano and Osawa (1983), who showed that LT-WM is involved in experts' solving of abacus problems but that the retrieval structures these experts use are not adapted to keep problem related information available after the current problem is solved. That is, after solving an abacus problem experts showed low accuracy in recall of intermediate results of their problem solving process. Hatano and Osawa assumed that the reason is that these experts use a mental image of an abacus to solve problems and that this image overwritten as soon as new problem information is encoded. Therefore only the current problem status is represented in LT-WM, and the states before the current state are lost. The structures drivers use to encode the arrangement of cars near to their own car may be of a similar quality. That is, experienced drivers may possess retrieval structures that have the relevant positions of cars in the current traffic situation as slots. The contents of these slots are continuously updated while driving so that the previous content is lost after an update process occurred. The better performance of experienced drivers in terms of encoding relevant information may then be due to the fact that their retrieval structures better reflect the differences between different types of traffic situations in terms of relevant positions. But the nature of retrieval structures used by experienced drivers to encode traffic information has to be examined in detail in future research activities. The identification of these structures would be an important step in supporting the view of situation awareness as a comprehension process.

References

Baumann, M., & Krems, J.F. (2007). Situation awareness and driving: A cognitive model. In P. C. Cacciabue (Ed.), *Modelling driver behaviour in automotive environments* (pp. 253 - 265). London: Springer.

Baumann, M., & Franke, T. (in prep.) *LT-WM involvement in driver's Situation Awareness.*

Crundall D, Underwood G, & Chapman P. (1999). Driving experience and the functional field of view. *Perception, 28*, 1075-1087.

Crundall, D.E., Chapman, P., Phelps, N., & Underwood, G. (2003). Eye movements and hazard perception in police pursuit and emergency response driving. *Journal of Experimental Psychology: Applied, 9*, 163-174.

Crundall, D.E., & Underwood, G. (1998). Effects of experience and processing demands on visual information acquisition in drivers. *Ergonomics, 41*, 448 - 458.

Durso, F.T., Rawson, K.A., & Girotto, S. (2007). Situation comprehension and situation awareness. In F.T. Durso, R.S. Nickerson, S.T. Dumais, S. Lewandowsky, and T. Perfect (Eds.), *Handbook of applied cognition* (2nd ed., pp. 163-193). Chicester: Wiley.

Endsley, M.R. (1995). Toward a theory of situation awareness in dynamic systems. *Human Factors, 37*, 32-64.

Ericsson, K.A., & Kintsch, W. (1995). Long-term working memory. *Psychological Review, 102*, 211 - 245.

Fischer, B., & Glanzer, M. (1986). Short--term storage and the processing of cohesion during reading. *Quarterly Journal of Experimental Psychology: Human Experimental Psychology, 38A*, 431 - 460.

Glanzer, M., & Nolan, S.D. (1986). Memory mechanisms in text comprehension. In G.H. Bower (Ed.), *The psychology of learning and motivation* (Vol. 20, pp. 275-317). New York: Academic Press.

Gugerty, L.J. (1997). Situation awareness during driving: Explicit and implicit spatial knowledge in dynamic spatial memory. *Journal of Experimental Psychology: Applied, 3*, 42 - 66.

Hatano, G., & Osawa, K. (1983). Digit memory of grand experts in abacus-derived mental calculation. *Cognition, 15*, 95 - 110.

Masson, M.E.J., & Loftus, G.R. (2003). Using confidence intervals for graphically based data interpretation. *Canadian Journal of Experimental Psychology, 57*, 203 - 220.

Matthews, M.L., Bryant, D.J., Webb, R.D.G., & Harbluk, J.L. (2001). Model for situation awareness and driving: Application to analysis and research for intelligent transportation systems. *Transportation Research Record, 1779*, 26 - 32.

Michon, J.A. (1985). A critical view of driver behavior models: What do we know, what should we do? In L. Evans & R.C. Schwing (Eds.), *Human behavior and traffic safety* (pp. 485-524). New York: Plenum Press.

Norman D.A., & Shallice T. (1986). Attention to Action: Willed and Automatic Control of Behavior. In R.J. Davidson, G.E. Schwartz, and D. Shapiro (Eds.) *Consciousness and Self-Regulation.* Volume 4 (pp. 1-18). New York: Plenum Press.

Underwood, G., Chapman, P., & Crundall, D. (2003). Driving experience, attentional focusing, and the recall of recently inspected events. *Transportation Research Part F, 6,* 289-304.

Overriding the ACC by keys at the steering wheel: positive effects on driving and drivers' acceptance in spite of a more complex ergonomic solution

Ingo Totzke[1], Véronique Huth[1], Hans-Peter Krüger[1], & Klaus Bengler[2]
[1]Center for Traffic Sciences, University of Wuerzburg, Germany
[2]BMW Group Research and Technology, Munich, Germany

Abstract

Partial automation of the driving task by driver assistance systems (e.g. ACC) is often criticized for changing the driving task to a monitoring one. Therefore, it is in discussion to offer the driver additional possibilities of control. In this study, a new type of ACC (so-called "ACC-plus-keys") was introduced which allowed the driver to override the ACC temporarily. By pressing additional keys at the steering wheel the driver could choose between different ACC characteristics: a key on the left led to stronger deceleration at shorter distances, a key on the right initiated stronger acceleration. Twenty participants drove a motorway simulator-ride three times: without ACC, with ACC and with ACC-plus-keys. Each ride took about 20 minutes.

The ACC-plus-keys had positive effects on driving and drivers' judgements. Firstly, driving safety of ACC and ACC-plus-keys were comparable, although the additional keys were used frequently. Secondly, driving with ACC-plus-keys resembled manual driving more (e.g. while overtaking). Thirdly, drivers gave higher acceptance ratings for ACC-plus-keys. Thus, ACC-plus-keys is a successful attempt to give the driver control without losing the advantages of this assistance system.

Introduction

Introducing Advanced Driver Assistance Systems (ADAS) into a car implies a change in task requirements for the driver: whereas drivers are responsible for navigation, manoeuvring and stabilization of the car while driving without assistance (e.g. Bernotat, 1970), most ADAS take over defined parts of this task. Though numerous studies report beneficial overall effects of ADAS, it is often criticized that using ADAS might lead to a reduction in driving activity as well: the role of the driver changes from active and non-assisted driving to assisted driving plus monitoring the system, which is typically assumed to be less demanding for the driver. However, monitoring ADAS implies that the driver is no longer "in the loop" as he only has to take over and solve the situation himself in safety-critical situations (e.g. Lee & See, 2004; Muir & Moray, 1996; Parasuraman & Riley, 1997; Parasuraman, Sheridan, & Wickens, 2000). Therefore, monitoring ADAS presents different demands to the driver as compared to manual driving. Accordingly, ADAS are often poorly accepted by the driver even if having high ergonomic quality (e.g. Zwerschke, 2006).

As a rule, ADAS are evaluated on the basis of ergonomic knowledge which focuses on the reduction of the user's workload and the increase of safety. From this perspective, ADAS have to be functional (e.g. consistent, stable, predictable) and usable (learnable, efficient, easy to use; Hancock, Pepe, & Murphy, 2005). Hence, ADAS and other interactive products are usually assessed by parameters of human performance (time and error) and by those related to physiological or psychological effort (Helander & Tham, 2003). Several ergonomic guidelines and standards result from this approach (e.g. EN ISO 15005, 2002; EN ISO 17287, 2003).

Based on theories on the hedonic design of interactive products (e.g. Hancock et al., 2005; Hassenzahl, Platz, Burmester, & Lehner, 2000; Jordan, 2000; Kano, Seraku, Takahashi, & Tsuji, 1984; Norman, 2004), this paper applies this new perspective on developing and evaluating ADAS, including the promotion of positive emotionality through extended interaction with ADAS. This perspective aims at considering hedonic aspects of human-machine interactions so that positive emotions like fun, pleasure or delight are facilitated. Jordan (2000) states, for example, that pleasurability "is not simply a property of a product but of the interaction between a product and a person" (p. 12). Hassenzahl et al. (2000) emphasize accordingly that in making an interactive product "as simple as possible there is a good chance to make it boring as well" (p. 201). This approach, called "hedonomics", assumes that ergonomic aspects of interactive products are the basis for hedonic design (Hancock et al., 2005). The "demand-control-support"-model (Brandtzaeg, Folstad, & Heim, 2003) explains which attributes of interactive products contribute to their hedonic quality. According to this model, interactive products are evaluated most positively if they allow variations in interacting with it ("demand"), feelings of responsibility for and control over the situation ("control"), as well as social interactions with other people ("support").

In the meantime, several approaches deal with hedonic aspects of car design (e.g. aesthetics of car interiors, Karlsson, Aronsson, & Svensson, 2003). However, approaches that consider hedonic aspects of driving itself can only be found sporadically. Knapper and Cropley (1981) emphasize that driving has also to be evaluated concerning arousal, excitement and risk taking. Steg, Vlek, and Slotegraaf (2001) conclude that instrumental motives (like the possibility to drive fast) are relevant for driving. Summala and Naatanen (1988) report that so-called "extra-motives" (e.g. searching for excitement and suspense) contribute to fun while driving. Finally, Beier, Boemak, and Renner (2001) propose that distance keeping and choice of speed are primarily safety-relevant, while accelerating, braking and steering may lead to positive emotions like "fun" or "pleasure" and a feeling of control. This implies that hindering the driver from carrying out basic driving manoeuvres may cause negative emotions like "anger", "annoyance" and "frustration". As a result, the hedonic perspective of driving implies a re-evaluation of task requirements while driving.

However, none of these approaches consider the introduction of ADAS in the car. Theoretical models like the "demand-control-support"-model (Brandtzaeg et al., 2003) have not been developed or transferred yet. One first attempt in this direction may be Battarbee and Mattelmäki (2002), who emphasize that particularly those ADAS that contribute to facilitation of driving or improvement of driving behaviour will be judged positively. Therefore, ADAS that hinder driving or lead to deterioration in driving behaviour compared to driving without ADAS might be rejected.

In the following, it was assumed that considering hedonic aspects in designing ADAS implies an extended view of the classic ergonomic user perspective which might lead to a re-evaluation or even a re-composition of the functionality of existing ADAS. For ACC, a system that controls speed and time headways automatically, this approach might have the following consequences:

1. distance and speed control are to be automated as these aspects of driving are safety-relevant but irrelevant to the positive emotions of the driver and can be regulated by ACC,
2. the regulation of acceleration and deceleration should be accessible to the driver as these aspects of driving are relevant to the emotionality of the driver.

It was expected that drivers would particularly prefer to regulate acceleration or deceleration on their own while driving with ACC if driving with this system differs systematically from driving without ACC. This might be the case while overtaking on motorways, for instance. In this situation, driving with ACC is usually accompanied by losing speed and regulating safe time headways while waiting for an appropriate gap on the left lane. After changing lane, the comfortable acceleration of ACC takes some time to reach the target speed again. In contrast to this, overtaking without ACC can be characterized by approaching the car ahead with the attempt to maintain speed, to find an adequate time gap on the left lane and to accelerate while changing lanes. While overtaking, relatively short but safe time headways can be found.

This difference between driving behaviour while using ACC compared to manual driving may be especially relevant for negative emotions while driving and low acceptance of ACC. Giving the driver the opportunity to regulate deceleration and acceleration in overtaking situations on his own might lead to higher acceptance ratings of ACC and more positive emotions while driving. The driver stays "in the loop" concerning longitudinal control and is not reduced to only monitoring the system. On the other hand, there is no need to offer the driver autonomy of speed and time headway regulation as this is not relevant for the emotional experience of the driver.

In the present study, this re-composition of the functionality of ACC was expected to be achieved with an ACC which offers the driver more interaction opportunities. Therefore, a new type of ACC with additional keys at the steering wheel was introduced by which the driver was able to switch between different ACC characteristics, called "ACC-plus-keys". It was assumed that the ACC-plus-keys would lead to a more convenient overtaking behaviour while driving with ACC. As a consequence, there would be no loss of speed while waiting for an appropriate gap on the left lane, but a strong acceleration right after changing lane. Central issues of this study concerned the impact of the ACC-plus-keys on drivers' experience regarding hedonic aspects of driving as well as the driving behaviour displayed when using the system (particularly while overtaking). For this purpose, driving with ACC-plus-keys was compared to driving without ACC and with a classic ACC, respectively.

Methods

Apparatus

In this study, three experimental conditions were introduced differing in the automation mode: (1) driving without ACC, (2) driving with ACC and (3) driving with ACC-plus-keys.

Firstly, the participants had to drive without ACC (manual control).

Secondly, the participants had to drive with an ACC for speed and headway control (target headway = 1.7 s) developed by the BMW Group. The driver only had to monitor the functioning of the system. Lateral control was and remained manual. Additionally, the driver could override the ACC by (a) pressing the brake pedal to avoid collisions if the ACC was not able to control time headways safely, or by (b) pressing the accelerator to achieve stronger accelerations. Using the brake pedal led to a deactivation of the ACC so that it had to be reactivated manually. The drivers were instructed to use the brake or accelerator pedals only if necessary.

Finally, a new type of ACC was introduced, the so-called "ACC-plus-keys" (see figure 1). In this condition, the ACC took over speed and headway control. Additionally, the driver had the possibility to override the ACC temporarily either by pressing the brake or accelerator (see above) or by using additional keys at the steering wheel. By pressing one of these keys, the driver was able to choose between different ACC characteristics: a key on the left led to stronger deceleration at shorter distances (deceleration was lagged), a key on the right initiated stronger acceleration (the target speed was achieved faster). After releasing the key, the former characteristics of the ACC were reactivated automatically. This means that the usage of the additional keys did not override the ACC totally, but only changed its characteristics. The drivers were instructed to use the keys at the steering wheel as often as they wanted (e.g. while overtaking).

This experiment was carried out in the driving simulator of the Wuerzburg Institute for Traffic Sciences (WIVW, see figure 2 on the left). This simulator with motion platform consists of a 180° front projection system and three rear projections. The driver sits in the front part of a BMW ensconced on a platform which is moved by a 6 degrees-of-freedom motion system. The simulator is controlled by personal computers. Data were sampled with a frequency of 100 Hz. By means of a flexible scripting language developed by the WIVW, scenarios with numerous models for the behaviour of other traffic participants (including vehicles and pedestrians) can be defined easily. Based on the behaviour of the simulation's driver, these are recalculated during simulation. Therefore, it is possible to adapt the behaviour of other traffic participants so that they interact with the simulator car dynamically.

Figure 1. Keys at steering wheel of ACC-plus keys in the simulation.

Figure 2. The simulator of the Würzburg Institute for Traffic Sciences (WIVW; left) and screenshot of driving task (right).

The driving task consisted of a two-lane motorway of 32.5 km length with varying traffic densities (see figure 2 on the right). A speed limit of 140 km/h was set. As a rule, traffic on the right lane drove slower than the speed limit, traffic on the left lane faster. Therefore, drivers had opportunities to overtake regularly. No safety critical situations were introduced explicitly so that it was possible to drive the simulator track without needing to take over manually while driving with ACC or ACC-plus-keys. The participants were instructed to comply with road traffic regulations (e.g. driving on the right lane) and to adhere to the speed limit. The three simulator rides were largely comparable concerning road design and traffic density. The driving task took approximately 20 minutes. Before driving with ACC or ACC-plus-keys, participants were trained to drive with the system on a motorway of 18.5 km length in each case.

Participants and procedure

A total of N = 20 participants (10 male, 10 female) took part in this experiment, ranging in age between 26 and 58 years (M = 39.3, SD = 11.4). All the participants had a minimum driving average of 10.000 km per year. Prior to the experiment, all participants had taken part in an extended simulator training of approx. 3 hours duration (Hoffmann & Buld, 2006) and had experience in driving with ACC in the simulation. As driving with ACC had been practiced extensively in earlier studies, time effects were expected to be minimal but could not, of course, be completely excluded. Eleven participants also had experience with cruise control in real traffic.

In this experiment, the participants drove an approx. 20 min motorway simulator track three times: (1) without ACC, (2) with ACC and (3) with ACC-plus-keys (one-factor within-subject design). In order to control sequence effects, the sequence of the first couple of rides (without ACC and with ACC) was permuted (cross-over design). Driving with ACC-plus-keys was introduced at the end of the session so that all participants had experienced driving with the maximum ("with ACC") and with the minimum ("without ACC") amount of automation before. After each ride and at the end of the session, questionnaires were filled out. The experiment took approximately 2.5 hours.

Table 1. Selected parameters of driving behaviour and drivers' judgements for further analyses

Parameter	Description
Usage of ACC-plus-keys	
Number of pressing the left key	m and sd [number]
Number of pressing the right key	m and sd [number]
Driving behaviour	
Time headway	distribution of headways [in s]
	relative number of critical headways (< 1sec)
Number of lane changes	m and sd [number]
Time gaps while lane change	m and sd [in s]
Velocity	m and sd [in km/h]
Drivers' Judgements	
The usage of the left/right key is useful	m and sd [seven-point scale]
The usage of the keys is distracting	m and sd [seven-point scale]
The usage of the keys is difficult	m and sd [seven-point scale]
Ergonomic Quality (Scale)	m and sd [seven-point scale]
I participated in traffic actively	m and sd [seven-point scale]
I was able to determine my driving style	m and sd [seven-point scale]
I was concentrated while driving	m and sd [seven-point scale]
I used gaps for overtaking	m and sd [seven-point scale]
I did well while driving	m and sd [seven-point scale]
Hedonic Quality (Scale)	m and sd [seven-point scale]

Parameters and analysis

Table 1 lists parameters of the simulation as well as the judgements of the participants which were selected for analysis. The "Ergonomic Quality" and

"Hedonic Quality" scales result from drivers' ratings in a seven-point semantic differential which was adopted from Hassenzahl et al. (2000). In order to analyse the influence of the factor "automation mode" all parameters in Table 1 were tested with a one-factor ANOVA with the within-factor "automation mode" (without ACC vs. with ACC vs. with ACC-plus-keys). As the "Ergonomic Quality" and "Hedonic Quality" was only measured for driving with ACC and ACC-plus-keys, paired-samples t-tests were calculated.

Results

First of all, the usage and the ergonomic evaluation of the controls of the ACC-plus-keys was analysed as a manipulation check. The drivers used the left key to lag the deceleration while waiting for an opportunity to overtake on average 10.3 times (SD = 3.9) and pressed the right key to accelerate more strongly about 13.0 times (SD = 4.9). The drivers rated the introduction of the keys in order to choose between different ACC characteristics as useful (left key: M = 5.50, SD = 1.54; right key: M = 5.70, SD = 1.87). The usage of the keys was judged as non-distracting (M = 2.55., SD = 1.76) and non-difficult (M = 2.15, SD = 1.87). The drivers attributed slightly lower ergonomic qualities (e.g. predictability, comprehensibility) to the ACC-plus-keys itself compared to the traditional ACC (scale "Ergonomic Quality" with ACC: M = 5.92, SD = 1.02; ACC-plus-keys: M = 5.35, SD = 1.34; t(19) = 2.29, p = .034).

As expected, driving with ACC changed driving behaviour concerning time headways compared to driving without ACC (F(2,38) = 39.64, $p < 0.001$). Driving with ACC led to higher mean time headways and lower variations in time headways (without ACC: M = 1.65 s, SD = 0.54 s; with ACC: M = 1.92 s, SD = 0.35; see figure 3). The distribution of time headways for driving with ACC in particular reflects the system functionality of distance control (target headway of 1.7 s). Comparable distributions of time headway can be found for driving without ACC and driving with ACC-plus-keys (ACC-plus-keys: M = 1.70 s, SD = 0.51 s). Thus, though the drivers only had the opportunity to choose between different ACC characteristics by pressing one of the keys at the steering wheel, time headways while driving resembled those generated while driving without ACC.

Also, mean velocity was influenced by different automation modes (see figure 4 on the left; F(2,38) = 45.06, $p < 0.001$): mean velocity was not only higher for driving without ACC, the drivers also exceeded the given speed limit of 140 km/h more frequently (about 56.3% of driving time). In contrast to this, velocity was higher than the speed limit in less than 1% of the driving time with ACC and ACC-plus-keys. This effect can be attributed to the functioning of the ACC. The main difference between driving with ACC and ACC-plus-keys consisted in higher mean accelerations of the latter: compared to ACC-plus-keys, it took longer to reach target speed with ACC due to the lower mean acceleration which could be achieved by the ACC itself (mean acceleration: without ACC: M = 0.13 m/s^2, SD = 0.05 m/s^2; with ACC: M = 0.06 m/s^2, SD = 0.02 m/s^2, with ACC-plus-keys: M = 0.22 m/s^2, SD = 0.16 m/s^2; F(2,38) = 12.68, $p < 0.001$).

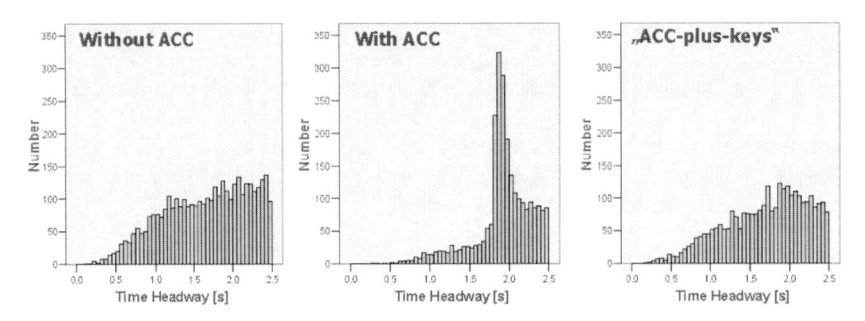

Figure 3. Distribution of time headway while driving (time headways ranging from 0 to 2.5 s; sample rate: 1 Hz; each bar represents 0.05 s).

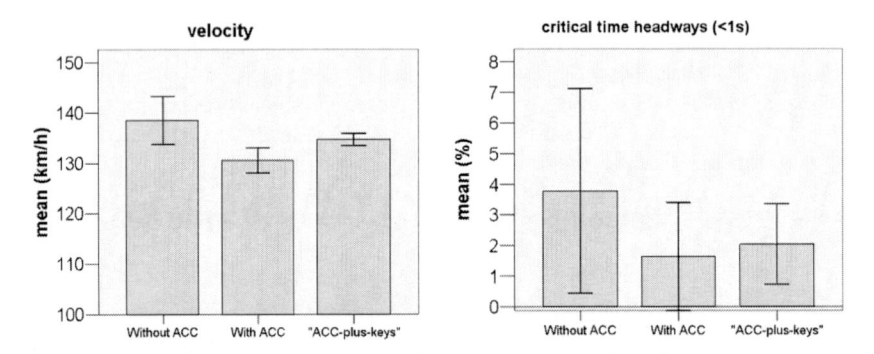

Figure 4. Mean velocity (left) and relative time of critical time headway (number of time headways < 1s, right). Means are displayed with standard deviations.

Nevertheless, driving with ACC-plus-keys did not go along with a higher number of safety-critical situations compared to driving with ACC (see figure 4 on the right): safety-critical time headways (< 1s) with ACC could be detected in 1.64% of driving time (SD = 1.77%), driving with ACC-plus-keys led to a similar number of safety-critical headways (M = 2.04%, SD = 1.31%). Only for driving without ACC did safety-critical headways appear more often (M = 3.78%, SD = 3.34%, F(2,38) = 4.92, p = .013).

It was expected that driving with ACC would lead to fewer lane changes and greater time gaps on the left lane while overtaking compared to driving without ACC. Driving with ACC-plus-keys should have weakened this effect. The results shown in figure 5 confirm this assumption (number of lane changes: F(2,38) = 24.74, p < 0.001, time gaps at moment of lane change: F(2,38) = 10.78, p < 0.001): fewer lane changes with greater time gaps at the moment of lane change were found for driving with ACC compared to driving without ACC. Driving with ACC-plus-keys resembled driving without ACC concerning lane change behaviour.

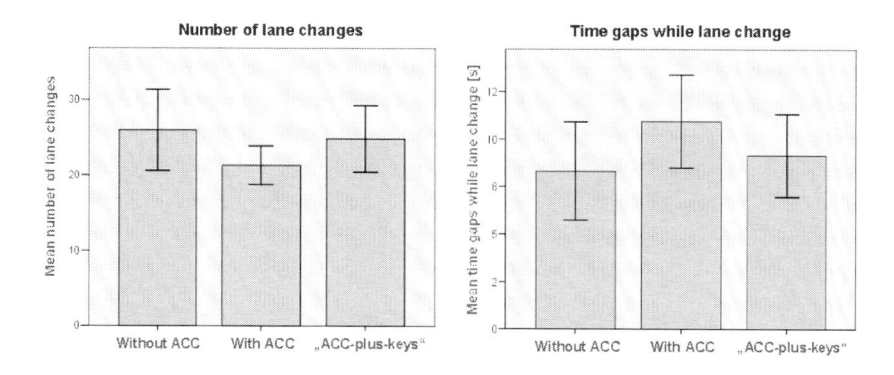

Figure 5. Number of lane changes (left) and mean time gaps at moment of lane change (right). Means are displayed with standard deviations.

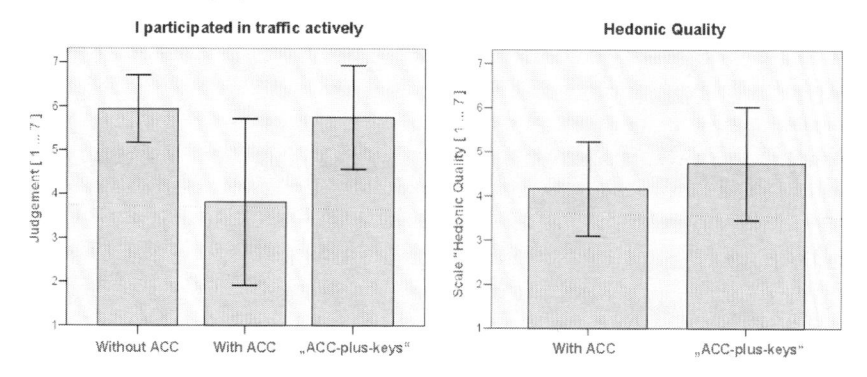

Figure 6. Judgements about perceived activity („I participated in traffic actively", left) and Hedonic Quality of the ACC (right). Judgements were given on seven-point scales. Means are displayed with standard deviations.

In accordance to the results given above, systematic differences in drivers' judgements between driving with ACC on the one hand and driving without ACC or with ACC-plus-keys on the other hand were found. As an example, Figure 6 on the left shows that lower judgements of perceived activity ("I participated in traffic actively") resulted from driving with ACC (F(2,38) = 13.80, p < 0.001). The drivers judged their own activity while driving with ACC as lower compared to driving without ACC or driving with ACC-plus-keys. Similar results resulted for perceived control ("I was able to determine my driving style", F(2,38) = 18.88, p < 0.001), perceived involvement ("I was concentrated while driving", F(2,38) = 8.18, p = .001) and perceived competence ("I used gaps for overtaking", F(2,36) = 3.68, p = .035; "I did well while driving", F(2,38) = 5.51, p = .008). Overall, driving without ACC and with ACC-plus-keys was judged rather similarly concerning central hedonic qualities, whereas the option to change between different ACC characteristics with the ACC-plus-keys was rated as more positive than driving with a classic ACC. Finally, the drivers attributed higher hedonic qualities (e.g.

originality, innovativeness) to the ACC-plus-keys itself compared to the traditional ACC (scale "Hedonic Quality", t(19) = -2.43, p = .025; see Figure 6 on the right).

Conclusions

This paper proposes an extended view of the classic user perspective: the development and evaluation of ADAS should not be restricted to ergonomic aspects of system design, only. It should include the promotion of positive emotionality through the interaction with ADAS as well. Whereas this perspective has already been in consideration in the field of interactive products (e.g. computers, mobile phones) for the past few years, it is new to the design and configuration of ADAS.

The present study confirms previous results that ACC leads to less driving activity and therefore more homogeneous driving behaviour (e.g. waiting for time gaps while overtaking, lower mean acceleration). A lower number of possible driving manoeuvres (e.g. fewer lane changes) and detrimental effects on the hedonic quality of driving (e.g. lower perceived activity, involvement, control and competence while driving) can be found. At the same time, positive implications for safety could be shown for driving with ACC.

Therefore, a new type of ACC, the so-called "ACC-plus-keys", was introduced. While using this system, drivers' experience is characterized by subjective activity, perceived control, pronounced feelings of dynamics and higher driving pleasure. The drivers give higher ratings concerning central hedonic qualities for ACC-plus-keys than for the classic ACC. At the same time, driving behaviour with ACC-plus-keys is more active than driving with ACC and is similar to driving without ACC: drivers overtake frequently and quickly, making use of reduced time headways, rather than with the longer headways associated with ACC. Moreover, driving behaviour resembles more manual driving though the advantages of ACC persist (e.g. lower number of safety-critical time headways).

These results imply that hedonic quality should be taken into account when developing and evaluating ADAS as various positive emotional and behavioural effects result from the interaction with the ACC-plus-keys. However, the introduction of a more complex ergonomic solution with additional controls, like the ACC-plus-keys, may have a negative impact on drivers' workload. Particularly in safety-critical situations for ACC (e.g. abrupt deceleration of car ahead, standing objects on street) the ACC-plus-keys might lead to additional load so that the driver might not be able to handle the situation. Moreover, the ACC-plus-keys might be misused as a new way to interact with the vehicle. Taking it to an extreme, the drivers could use the keys the whole time instead of controlling the car using the traditional pedals or instead of driving with a classic ACC. This might raise new questions concerning the utility of the ACC-plus-keys.

All in all, this experiment represents a first attempt at introducing "hedonomics" in the field of development and evaluation of ADAS without losing the focus on safety aspects. The possibility of transferring this approach to other ADAS (e.g. lane departure warnings, heading control) has to be analysed in further studies.

References

Battarbee, K., & Mattelmäki, T. (2002). Meaningful product relationships. *Proceedings of Design and Emotion Conference*, 01.-03.07.02, Loughborough, England. Taylor and Francis.

Beier, G., Boemak, N., & Renner, G. (2001). Sinn und Sinnlichkeit – psychologische Beiträge zur Fahrzeuggestaltung. In T. Jürgensohn, and K.-P. Timpe (Eds.), *Kraftfahrzeugführung* (pp. 263-284). Berlin: Springer.

Bernotat, R. (1970). Anthropotechnik in der Fahrzeugführung. *Ergonomics, 13*, 353-377.

Brandtzaeg, P.B., Folstad, A., & Heim, J. (2003). Enjoyment: Lessons from Karasek. In M.A. Blythe, A.F.Monk, K. Overbeeke, and P.C. Wright (Eds.), *Funology: from usability to enjoyment* (pp. 55-65). Dordrecht, the Netherlands: Kluwer Academic Publishers.

EN ISO 15005 (2002). *Road vehicles – Ergonomic aspects of transport and control systems - Dialogue management principles and compliance procedures*. Berlin: Beuth-Verlag.

EN ISO 17287 (2003). *Road vehicles – Ergonomic aspects of transport information and control systems - Procedure for assessing suitability for use while driving*. Berlin: Beuth-Verlag.

Hancock, P.A., Pepe, A.A., & Murphy, L.L. (2005). Hedonomics: The power of positive and pleasurable ergonomics. *Ergonomics in Design, Winter 2005*, 8-14.

Hassenzahl, M., Platz, A., Burmester, M., & Lehner, K. (2000). Hedonic and ergonomic quality aspects determine a software's appeal. *Proceedings of the CHI 2000 Conference on Human Factors in Computing* (pp. 201-208), The Hague, the Netherlands. ACM Press.

Helander, M.G., & Tham, M.P. (2003). Hedonomics – affective human factors design. *Ergonomics, 46*, 1269-1272.

Hoffmann, S., & Buld, S. (2006). Darstellung und Evaluation eines Trainings zum Fahren in der Fahrsimulation. VDI-Berichte Nr. 1960. *Integrierte Sicherheit und Fahrerassistenzsysteme* (pp. 113-132). Düsseldorf: VDI-Verlag.

Jordan, P.W. (2000). The four pleasures: A framework for pleasures in design. In P.W. Jordan (Ed.), *Proceedings of the conference on pleasure based human factors design*. Groningen, the Netherlands: Philips Design.

Kano, N., Seraku, N., Takahashi, F., & Tsuji, S. (1984). Attractive quality and must-be quality. *The Journal of the Japanese Society for Quality Control, April 1984*, 39-48.

Karlsson, B.S.A., Aronsson, N., & Svensson, K.A. (2003). Using semantic environment description as a tool to evaluate car interiors. *Ergonomics, 46*, 1408-1422.

Knapper, C.K., & Cropley, A.J. (1981). Social and interpersonal factors in driving. *Progress in Applied Psychology, 1*, 191-220.

Lee, J.D., & See, K.A. (2004). Trust in automation: Designing for appropriate reliance. *Human Factors, 46*, 50-80.

Muir, B. M., & Moray, N. (1996). Trust in automation: 2. Experimental studies of trust and human intervention in a process control simulation. *Ergonomics, 39*, 429–460.

Norman, D.A. (2004). *Emotional design: Why do we love (or hate) everyday things.* New York: Basic Books.

Parasuraman, R., & Riley, V. (1997). Humans and automation: Use, misuse, disuse, abuse. *Human Factors, 39*, 230–253.

Parasuraman, R., Sheridan, T.B., & Wickens, C.D. (2000). A model for types and levels of human interaction with automation. *IEEE Transactions on Systems, Man, and Cybernetics – Part A: Systems and Humans, 30*, 286-297.

Steg, L., Vlek, C., & Slotegraaf, G. (2001). Instrumental-reasoned and symbolic-affective motives for using a motor car. *Transportation Research Part F, 4*, 151-169.

Summala H., & Naatanen, R. (1988). The zero-risk theory and overtaking decisions. In T. Rothengatter, and R. de Bruin (Eds.), *Road user behaviour: Theory and research* (pp. 82–92). Assen, the Netherlands: Van Gorcum.

Zwerschke, S. (2006). Untersuchung zu Bekanntheit, Akzeptanz und Kaufinteresse von Fahrerassistenzsystemen. In VDI-Gesellschaft Fahrzeug- und Verkehrstechnik (Hrsg.), *Integrierte Sicherheit und Fahrerassistenzsysteme* (VDI-Berichte, Nr. 1960, pp. 343-358). Düsseldorf, Germany: VDI-Verlag.

A model of normal and impaired visual exploration while steering: a way to identify assistance needs

Isabelle Milleville[1], Camilo Charron[2], Jean-Michel Hoc[1], & Jean-François Mathé[3]
[1] CNRS and University of Nantes, IRCCyN, Nantes
[2] University of Rennes 2, CRPCC, Rennes
[3] University of Medicine of Nantes, Nantes
France

Abstract

Driving relies mainly on visual landmark processing to determine a car's position and speed on the road. To be efficient, this visual information needs to be explored at the right time during the sensory-motor loop, and is restricted by task constraints. Visual exploration can be impaired for various reasons, including visuo-attentional or cognitive control disturbances in brain-injured people, the elderly, or those experiencing fatigue. The aim of this study was to determine appropriate criteria for visual exploration to develop a model of efficient visual exploration that can be implemented in cars. It may be possible to assist and alert drivers in case of decreasing performance. For this reason, a fixed-base driving simulator and an eye tracker were used to compare the visual exploration of control and brain-injured drivers. The results showed main differences between the two investigated samples: brain injured people had longer mean fixation duration, a reduction in the distance of visual exploration and spent less time in exploring surrounding objects which are not directly useful for the control of vehicle trajectory but are necessary for an awareness of the driving context. Together, these parameters can indicate the necessity of whether or not to intervene. Thus, it is possible to identify the parameters that need to be taken into account to trigger an assistance device and restore driver visuo-attentional and cognitive control capacity.

Introduction

Various driving support systems have been developed over the last ten years. Most of them are devoted to longitudinal control (e.g., Adaptive Cruise Control, ACC), although some of them have dealt with lateral control (e.g., Electronic Stability Program, ESP). Many of these devices warn the driver or partially support steering when a certain threshold is exceeded. These thresholds may be determined by examining the car's behaviour on the road, such as speed, lateral position and Time to Lane Crossing. Alternatively, one can examine driver behaviour, or more specifically, a driver's visual exploration of the road and the surrounding area. Effectively, visual exploration may be a good indicator of the necessity to intervene: indeed, it can provide two invaluable indications about the driver. The first concerns the assurance that the driver actually is looking where he or she is supposed to look to drive safely. The second concerns driver awareness of the situation. This is the reason why this study was aimed at determining appropriate measures to characterize visual exploration in order to develop a model of efficient visual exploration that takes into account attentional and cognitive factors capable of influencing visual exploration. This model could be implemented in cars to detect abnormal visual behaviour and then alert drivers of their diminishing performance.

With regard to visual zones of interest for safe driving, many studies indicate that driving relies on an unconscious detection of numerous visual landmarks, which are then used to determine a car's speed and position on the road. For example, an analysis of eye movements just before a bend indicates that drivers begin to explore bends early on (between 100 m and 30 m before the bend starts, Zikovitz & Harris, 1999). This exploration gives information about the nature of the oncoming bend. Thus, when approaching and negotiating a bend, drivers spend a significant amount of time looking in the vicinity of the tangent point, that is to say the point where the direction of the inside edge line seems to reverse from the drivers' viewpoint (Land & Lee, 1994). Land's studies (1998, 2001) show that there are two distinct regions of information which are useful for guiding the drivers' displacement. The first region is placed approximately 4° below the horizon and corresponds to the position of the tangent point. This region would be perceived by the central vision and drivers devote approximately 80% of their ocular fixations to this point. The second region is placed 7° above the horizon and is perceived by the peripheral vision.

To be efficient, visual information needs to be extracted by an appropriate exploration of the visual scene: this must occur at the right time during the sensory-motor loop and result from an adequate representation of where it is necessary to look (for example: straight ahead and far away) and what must be survey for safe driving (pedestrians, mirror, speedometer, road signs, etc.). Visual exploration may be impaired for various reasons. For example, numerous studies have shown that driving experience has a direct influence on how drivers scan the road ahead. Falkmer and Gregersen (2005) noted that novice drivers fixate more often on relevant traffic cues and in-vehicle objects, compared to experienced drivers. Nevertheless, Underwood, Crundall, and Chapman (2002) reported that as mental load increases, the reverse is observed. In this case, novice drivers fixated less frequently on non-central objects, such as a car's mirror, compared to experienced drivers. Furthermore, novices scanned a road such as a dual carriageway to a lesser extent than did experienced drivers (Crundall & Underwood, 1998). According to Underwood, Chapman, Bowden, and Crundall (2002), novice drivers' inspection of the road ahead is limited, not because they have limited mental resources that are residual to the task of vehicle control, but because they have an impoverished mental model of what is likely to happen on a dual carriageway.

Visual exploration may also be impaired due to attentional disturbance, as may be observed in brain–injured people, the elderly or those experiencing fatigue. The aim of this assertion is not to create confusion between visual exploration and attention. It is known that in many tasks, including car driving, recall performance about objects present in the environment during the task generally reflects the pattern of eye fixations, but viewers do not always recall details about fixated objects and are sometimes able to recall information about objects that were not fixated (Underwood, Chapman et al., 2002). Nevertheless, It is also well known (Jenkin & Harris, 2001) that many aspects of attention lead to a modification of the pattern of visual exploration. For example, in a situation where two dual-tasks require two different visual zones to be simultaneously surveyed, eye fixations alternate between the two zones. However, this alternation diminishes as soon as one or both of the tasks become more attention-demanding. Thus, Wierwille (1993) reported that, when an elderly driver has to share the primary driving task with particularly demanding in-car tasks, this may result in an increased number and duration of in-car single glances. It may thus be envisaged that attentional deficit (normal or pathological) may be identified through a modification of the pattern of visual exploration of the road. It is supposed that attentional impairment will be characterized by a focalization on the steering task with a decrease in the number of visual zones explored, particularly those zones which relate to non-central objects, such as mirrors. This behaviour has to be increased as soon as the steering task becomes more complex (for example, because of internal constraints such as more careful steering or because of the presence of vulnerable passengers in the car) or as the road environment itself becomes more complex (due to an increase in the number of other road users).

Finally, visual exploration may also be influenced by cognitive control modes. In terms of adaptation to dynamic situations, Hoc and Amalberti (2007) have defined cognitive control as the mechanism that enables the individual to bring cognitive activities into play in the correct order and with the appropriate intensity. These authors have also defined two orthogonal dimensions that generate a space where several cognitive control modes can be defined. The first one is the level of abstraction of the data required for control, and roughly contrasts symbolic data (with a need for interpretation before use in action, for example, verbal representations) and subsymbolic data (directly related to action without interpretation, for example, sensations). The second dimension is the origin of data used for control (internal compared to external), and distinguishes between anticipative and reactive control. Their model assumes that the diverse control modes operate together. However, the relative weights of the different modes are seldom equal and their distribution varies in relation to adaptation needs. For example, if routines are not available or are too risky, symbolic control will be preferred over subsymbolic control. In the same way, if the individual has no (internal) model of the situation for anticipation, the control will be reactive (external) instead of anticipative.

For Hoc and Amalberti (2007), adaptation is related to a change in the distribution of control over the modes. Such *cognitive compromise* maintains the situation (the interaction between the individual and the environment) within a domain where the individual thinks it can be mastered. Situation mastery corresponds to the feeling the individual has that he or she can reach an acceptable performance level during a certain period of time. Metaknowledge can play a major role in modifying the weights allocated to the diverse control modes. The cognitive compromise might influence visual exploration. The dimension "level of abstraction of data" is related to the mean fixation duration. Long fixations are interpreted as cues of symbolic processing, which is assumed to take more time than subsymbolic processing. The dimension "origin of the data" is related, in this particular driving task, to the mean fixation distance from the car. Greater distance fixations are interpreted as cues of an anticipative strategy, usually found in car-driving. In normal situations, drivers adopt an anticipative cognitive control mode, devoting much more time to the part of the scene that is further away than to the nearest one where fixation durations are short (Underwood, Chapman, Berger, & Crundall, 2003; Underwood, Chapman, Brocklehurst, Underwood, & Crundall, 2003).

As it was previously said in the introduction, this study was aimed at determining appropriate measures to characterize visual exploration in order to develop a model of efficient visual exploration that takes into account attentional and cognitive factors capable of influencing visual exploration. This model could be implemented in cars to detect abnormal visual behaviour and then alert drivers of their diminishing performance. For this reason, visual exploration of drivers with brain injury was compared with visual exploration of normal drivers. Only brain-injured drivers who presented no particular motor disease and who were characterized only by slight attentional and cognitive impairment were selected to participate in this study (Table 1). This population was chosen because of its particularity to present at the same time both attentional and cognitive impairment. For this reason, an understanding of the visual exploration of this population can serve as a basis to determine visual exploration which may be considered to be normal and that which cannot, thus requiring the intervention of a driving support device. It was supposed that attentional impairments would be characterised by a focalisation on the steering task. For this reason, we defined driving scenarios with various complexities. It was supposed that scenarios complexity will increase focalisation on the steering task. This focalisation was assed thanks to eye position recording. Cognitive impairment was supposed to induce a less anticipative driving style and thus an increase of eye fixations in the nearest part of the road and an increase of speed variability. Finally, it was also supposed to increase fixation duration (du to the increase of symbolic processing).

Method

Participants

Six non-disabled participants served as control (male, aged between 36 and 50 years) in this experiment. All had normal or corrected-to-normal vision. In addition, all participants had driving experience of more than 30,000 km and had held a driving licence for, on average, 23.6 years.

Five drivers with a brain injury due to a trauma (male, aged between 35 and 50 years) also participated in the experiment. These participants were characterized by a slight difficulty in dividing their attention, together with slight cognitive difficulties concerning planning and anticipation (Table 1). All were involved in some professional activity within ESAT (a French institute which helps disabled people through work). All had normal or corrected-to-normal vision. All had driving experience of more than 30,000 km and had passed their licence at least two years before brain injury occurred. All of them were driving again after recovery and had, on average, 11 years' driving experience.

All participants in the experiment were volunteers and were familiarized with the apparatus before the experiment. They were not totally naive about the experience; they were told that the experiment aimed to evaluate whether the simulator was easy for disabled people to handle to assess whether it could be used for steering capacity evaluation. Before and after the experiment, they were comprehensively informed about the experimental objectives. All participants read and signed an agreement to participate in this study. They also signed an authorization for the scientific exploitation of data at the end of the experiment.

Table 1. List of brain injured diseases investigated by means of neuropsychological test battery. -1 and -2 values indicate that performance is one or two standard deviation behind the reference value

TESTS	PARTICIPANTS	CB	SP	JML	DM	HE
D2	Sustained Attention rythm	-2	1	-2	-2	-2
D2	Sustained Attention concentration	1	-1	-2	-2	-2
TMT A	Visual pursuit	-2	1	1	1	-2
TMT B	Mental flexibility	1	1	1	-2	1
STROOP	Selective attention/Cognitive inhibition	1	1	-1	1	1
TEA alerte phasique	Alertness/Attention	1	1	1	1	-1
TEA attention divisée	Shared attention	1	1	-1	-1	1
TEA Go-Nogo	Behavioural inhibition	1	1	-1	-1	-1
TEA Balayage Visuel	Target detection	-1	1	-1	-1	-1
ZOO MAP	Anticipation	1	-1	-2	-2	-2
ZOO MAP	Planning	1	-1	-2	-2	-2
Span direct	Storage in working memory	-2	-2	-1	-1	-1
Span indirect	Manipulation in working memory	-1	-1	-2	-2	-2

Apparatus

The driving simulator software, Sim2 (developed by the MSIS team at the INRETS laboratory) was used, coupled with a fixed-base driving simulator. It was equipped with a manual or automatic gearbox, a steering wheel fitted with force feedback, brake, accelerator and clutch pedals, and a speedometer. The visual scene was projected onto a screen (3.02 m height x 2.28 m width, which corresponds to a visual angle of 80° height and 66° width). The road used corresponded to a simulation of the GIAT test track in Satory (Versailles, France) and was 3.5 km in length. The track is similar to a main road. A mirror was simulated on the visual scene in order to warn participants of other traffic on the road (Figure 1).

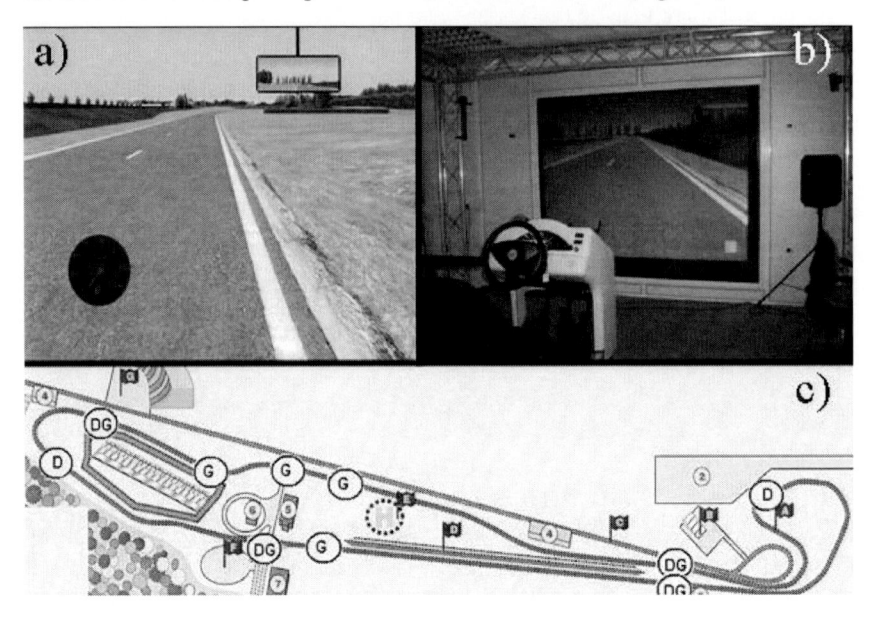

Figure 1. a) Driving simulator software, Sim2, developed by the MSIS team at the INRETS laboratory. b) Fixed-base driving simulator. c) GIAT test track in Satory (Versailles, France, 3.5 km-long)

An eye-tracker, IviewX (SMI), was used to investigate visual exploration. This eye-tracker consists of a hardly invasive, lightweight head-mounted camera that captures images of the subject's eye and field-of-view (Figure 2).

Figure 2. Eye-tracker IviewX (SMI)

Procedure

The experimental session began with a comprehensive presentation of the simulator during which participants were invited to try out the equipment. They were instructed to drive using the automatic gearbox. Participants had to complete a lap on the simulated road to be familiarized with both the simulator and the road. Following this familiarization stage, participants had to complete six laps. Each lap corresponded to a particular driving scenario.

The first two were simple scenarios designed to serve as a basis for further comparisons. The road environment was composed of trees, houses, hangars and hoardings. During the first lap there were no other cars on the road (Basis 1). The second lap was exactly the same, except that there was another car in the lane occupied by the participant (Basis 2). This car was travelling at low speed and was an invitation for the driver to overtake. During these two laps, participants were instructed to drive on the right-hand side of the road (as they are used to doing on French roads) and as they would do on the same road if this was a real-life driving situation.

Two other laps were more complex: pedestrians were added on the road verge, other traffic (cars, motorbikes) was simulated in the opposite lane and a slow-moving car was placed on the lane occupied by the participant as an incentive to overtake (simple incident). During one of these two laps, participants were instructed to adopt a safe driving style (Simple Incident condition with Safe style, SafeSI: "imagine there is a child with you in the car and you must be very careful on the road"). During the other lap they were told to adopt a more sportive driving style (Simple Incident condition with Sportive style or SportiveSI: "imagine you have a very important appointment, for example to find a job, and you are late").

The final two laps were the same, except that when the drivers approached the slow-moving car, they were caught up by a fast-moving car that was visible in the mirror (simulating a complex incident). This car followed the driver for the remainder of the lap and was an invitation for the driver to accelerate. During these two laps, participants were instructed to adopt a safe driving style for one lap (Complex Incident condition with Safe style or SafeCI) and a sportive driving style for the other lap (Complex Incident condition with Sportive style or SportiveCI).

The order of presentation of the four experimental conditions was balanced over participants.

Data recording and analysis

Two types of data were recorded:

- Data that concerned the car's longitudinal behaviour on the road: speed and speed variability.
- Eye fixation: mean duration of eye fixation (automatically stored thanks to BeGaze software developed by SMI), percentage of time spent in particular

visual zones of interest (areas of interest) for the driver (Figure 3). Six zones of interest were defined when driving in straight lines. This definition was based on the function of each of these zones during driving : speedometer (speed control), mirror (anticipation of other road users), surrounding road area (exploration of road environment), near vision as an indicator of on-line control (up to two centre lines demarcations in front of the car), far vision as an indicator of trajectory anticipation (more than four white discontinuous lines) and middle vision as an indicator of difficulties to alternate between near and far vision (between two and four white discontinuous lines). Seven zones of interest were defined in bend negotiation: speedometer, mirror, surrounding road area, near vision (one white discontinuous line), middle vision (between one and two white discontinuous lines), tangent point (large zone surrounding the tangent point known as being useful for trajectory control in bend) and far vision (beyond the tangent point).

As usual, in order to conclude whether a sample effect (δ) is non-null on the basis of an observed effect (d), a Student's t-test of significance was calculated. However, to draw conclusions in terms of population effect sizes, beyond a conclusion in terms of non-null effects, a variant of Bayesian statistical inference (fiducial inference: Lecoutre & Poitevineau, 2005; Rouanet, 1996) was used. On the basis of a maximal *a priori* uncertainty, the technique enables the user to emit a probabilistic judgement on the population effect size. For example, if the observed effect (d) can be considered as large, then a conclusion such as, "there is a high probability (guarantee γ) that the population effect is larger than a notable value" is tried ($P(\delta)>a= \gamma$; shortly $\delta>a$). Conversely, if the observed effect is negligible, the expected conclusion is that, "there is a high probability that the absolute population effect is lower than a negligible value", ($P(|\delta|)<\varepsilon= \gamma$). All fiducial conclusions below will be given with the guarantee $\gamma=.90$. The t- tests will be associated with an observed two-tailed threshold (p).

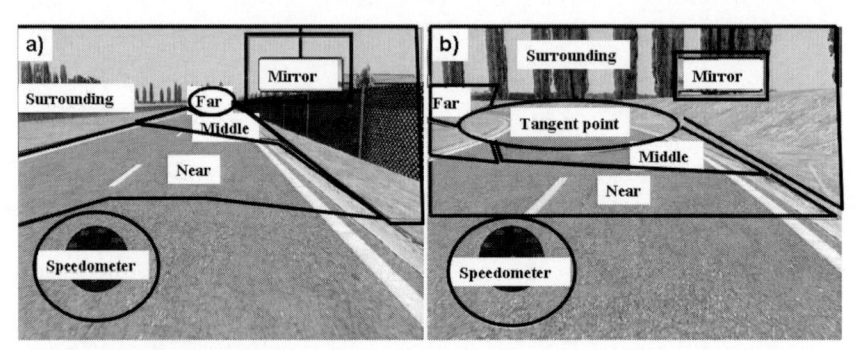

Figure 3. Areas of interest for the analysis of eye fixations: a) straight lines; b) bends

normal and impaired visual exploration while steering 173

Results

Car behaviour on the road

All participants speeded up (Figure 4) in Sportive conditions (d = 2.38 m/s; $t(9) = 5.25$; $p < .001$; $\delta > 1.75$ m/s) and slowed down in the Safe conditions with respect to the two Basis conditions (d = 1.25 m/s; $t(9) = 3.16$; $p < .02$; $\delta > 0.70$ m/s). This result shows that the instructions were effective. In addition, all participants speeded up in the SportiveCI condition with respect to the SportiveSI condition (d = 1.08 m/s; $t(9) = 2.60$; $p < .03$; $\delta > 0.51$ m/s). Thus, in the Sportive conditions, the scenario type was also effective. In the Safe conditions, we can only conclude that, for Brain-injured participants, the speed was higher in CI than in SI (d = 2.86 m/s; $t(5) = 2.72$; $p < .05$; $\delta > 1.32$ m/s). This is compatible with the fact that Brain-injured participants could be more sensitive (more reactive) to speed invitation (CI) than Control participants when the instructions required a safe driving style.

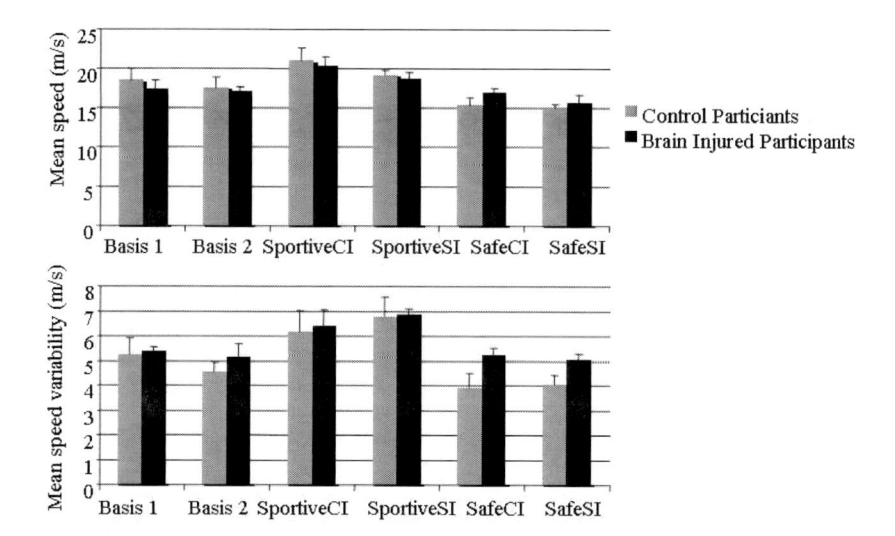

Figure 4. Vehicle parameters (mean speed in m/s, mean speed variability, mean standard deviation in m/s) as a function of experimental conditions. Bars indicate standard errors

All participants increased speed variability (Figure 4) in the Sportive conditions with respect to the two Basis conditions (d = 1.45 m/s; $t(9) = 3.90$; $p < .004$; $\delta > 0.94$ m/s). This result may be due to greater slowing down in bends because of an increase in average speed. However, a reduction in speed variability from the Basis conditions to the Safe conditions is only established for Control participants (d = 0.74 m/s; $t(4) = 2.23$; $p < .09$; $\delta > 0.23$ m/s). On the basis of observed data, speed variability remains constant for Brain-injured participants. They might have been more surprised by hazardous incidents (such as a slow car) or hazardous situations (such as a severe bend or pedestrians on the road verge) that were not compatible with the instruction to adopt a safe driving style. So these participants might have been more inclined to continuously brake and accelerate. This is compatible with more reactive behaviour.

Visual exploration

The mean eye-fixation duration (Figure 5) was different between groups and the experimental conditions, but only on straight segments where the mean duration was higher for Brain-injured participants than for Control participants (d = 98 ms; $t(8) = 2.10$; $p < .07$; $\delta > 33$ ms). No significant difference was observed between participants in bends (d = 2 ms; $t(8) = 0.09$; $p > .93$); however, if it exists, it is smaller than the previous one ($\delta < 30$ ms). This result is in accordance with a more symbolic control for Brain-injured participants. Although we cannot draw a conclusion from this, the observed data showed a stability of the mean eye-fixation duration across the experimental conditions for Control participants. Data also showed some changes for Brain-injured participants who may have had to modify their cognitive compromise to adapt to the various situations.

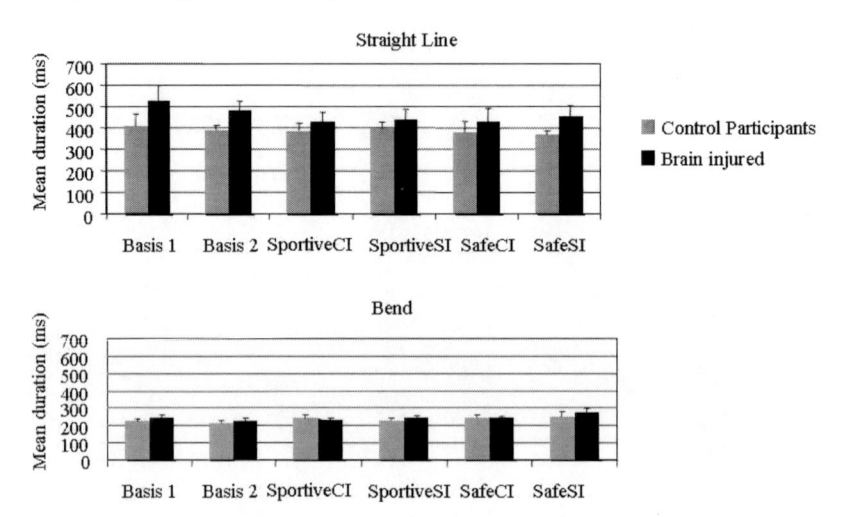

Figure 5. Mean duration of eye fixation in straight lines and bend (ms) as a function of experimental conditions for control and brain-injured participants. Bars indicate standard errors

The percentages of time spent in the particular visual zones of interest were analysed for each driver (Figure 6). On straights segments although the result is marginally significant, we can say that Brain-injured participants spent proportionally more time in the nearest part of the visual scene (d = 27.1%; $t(4) = 1.95$; $p < .13$; $\delta > 5.8\%$) and proportionally less time in the farthest part than did Control participants (d = 30.7%; $t(4) = 1.77$; $p < .16$; $\delta > 4.0\%$). This result is in accordance with less anticipative cognitive control for Brain-injured participants. In the bends, all participants spent most time looking at the tangent point zone (between 75% and 87% of the time). Brain-injured participants devoted a lower percentage of time than Control participants to this zone (d = 12.7%; $t(7) = 2.78$; $p < .03$; $\delta > 6.2\%$). On the basis of the observed data, it seems that this loss was to the profit of the near and middle zones. In addition to the importance of the tangent point zone for trajectory control, this nearer exploration could be interpreted as less anticipative cognitive control.

With regard to the exploration of secondary objects for trajectory control, Brain-injured participants devoted proportionally less time to the mirror, speedometer and other surrounding objects than did Control participants. We can only reach a conclusion for the mirror on straights segments (d = 4.8%; $t(4) = 2.30$; $p < .09$; $\delta > 1.6\%$). This result is in accordance with a focalization of attention on trajectory control for Brain-injured participants.

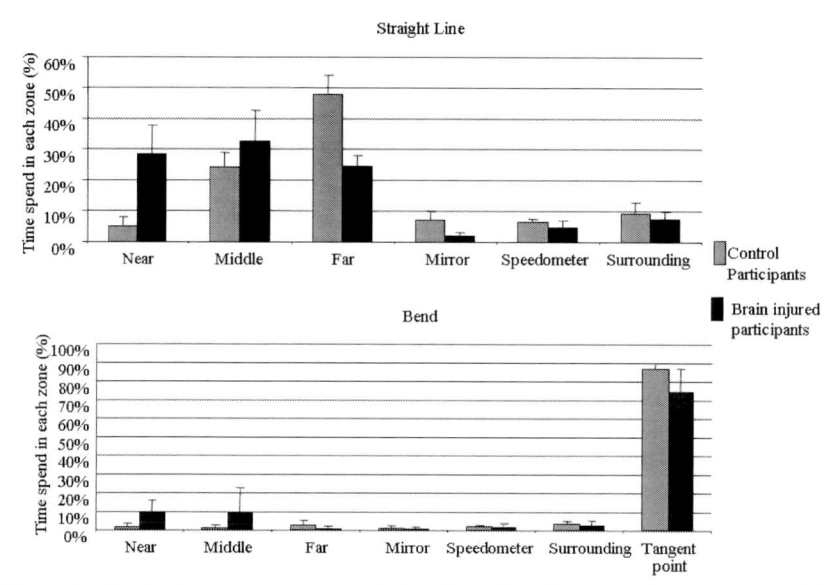

Figure 6. Distributions of percentages of time devoted to the zones of interest on straight segments and in bends (see Fig. 3 for the definitions of the zones). Bars indicate standard errors

Discussion

Cognitive control

This experiment has given some new insight into the cognitive features of brain-injured persons when driving a car after their recovery. From the point of view of cognitive compromise, brain-injured drivers who participated in the present experiment were clearly identified on the two proposed orthogonal dimensions. On the basis of the mean fixation duration, there are indications that their cognitive processes were more symbolic (higher fixation duration) than those of the control participants. On the basis of fixation distance, sensitivity to an incentive to accelerate and variable speed control, their cognitive processes appeared less anticipative (more reactive) than those of control participants. Within the limited framework of this paper we cannot enter into a detailed account of the way that brain-injured participants adapt to instructions and scenarios. However, some observed results, particularly those that concern mean fixation duration, showed some change in cognitive control in order to adapt to instructions and scenarios. Apparently, control participants were able to deal with changes without showing any modification in their cognitive compromise. However, this conclusion is provisional and should be confirmed in the near future.

Attention

Attention is a complementary point of view on cognitive processes and is related to the resources devoted to the main task (trajectory control) and secondary tasks. In the current analysis of this experiment, the well-established difficulty experienced by brain-injured drivers in dividing their attention between several tasks was noted. On the basis of the fixation distance and negligence of peripheral information, and from the point of view of trajectory control, brain-injured participants showed a higher focus on the main task. In comparison with control participants, this was particularly notable for proximate trajectory control. In future work, it may be worth examining the possible relation between this high investment in attention to proximate trajectory control and the symbolic and reactive feature of cognitive control. Symbolic control requires more attentional resources, particularly when applied to proximate control. This kind of cognitive control could explain the attentional focus. However it will be necessary to check whether attentional resources are also reduced.

Implication for steering assistance

The objective of this study was to determine if visual exploration could be a good indicator of driver attentional and cognitive capacity. Previous sections of this paper have shown that at least three aspects of visual exploration can be used to predict driver attentional and cognitive difficulties. Mean fixation duration gives information about the cognitive control level of abstraction: it is longer for people who have difficulties with automated control of vehicle trajectory. In addition, the reduction in the distance of visual exploration is related to a less anticipative cognitive control. Finally, a last parameter is related to attentional load and concerns the amount of time spent in exploring surrounding objects that are not directly useful for the control of vehicle trajectory but are necessary for an awareness of the driving context. Together, these parameters can indicate the necessity of whether or not to intervene.

Results also indicate that visual exploration may bring complementary information about the driver, which is not accessible by the sole means of vehicle parameters. An analysis of visual exploration showed that brain-injured participants tended to reduce the distance they looked away in comparison to control participants. This indicates that brain-injured participants were more likely to be surprised by an unexpected event happening in the far distance. In this case, a device could be designed that stiffens the accelerator pedal when the distance of glance decreases for long periods of time, or something like warning device that indicates when something important must be considered far away.

Conclusion

Within the limited framework of a preliminary study and with small sample sizes, the relevance of a theoretical approach developed for studying adaptation in dynamic situations was explored. We have also looked at a method of analysis of eye movements. More specifically, we brought arguments in favour of the theory and the method for the identification of cognitive differences between brain-injured drivers and control drivers. Of course it is important to keep in mind that it is a pilot study that needs to be confirmed on larger samples of participants. Nevertheless, this attempt is an encouragement to extend the scope of this kind of study. Beyond the limitation of the analysis of car parameters, eye movement analysis is proven to elicit important cognitive features for the design of car-driving assistance systems. The method, when associated to the theory, is capable of extending the sole reference to attention and vigilance towards a richer viewpoint on cognitive control (cognitive compromise). In the future, the same principle needs to be applied to experiments which aim to compare diverse driver populations: for example, young, control, elderly and brain-injured drivers. Some support devices could assist several types of drivers. Finally, with regard to young drivers or brain-injured persons, a developmental perspective should be developed, either during the learning phase or the re-adaptation phase. In the future, our approach could be useful in identifying and explaining such a development.

References

Crundall, D, & Underwood, G. (1998). Effects of experience and processing demands on visual information acquisition in drivers. *Ergonomics, 41*, 448-458.

Falkmer, T., & Gregersen, N.P. (2005). A comparison of eye movement behaviour of inexperienced and experienced drivers in real traffic environments. *Optometry and Vision Science, 82*, 732-739.

Hoc, J.M., & Amalberti, R. (2007). Cognitive control dynamics for reaching a satisficing performance in complex dynamic situations. *Journal of Cognitive Engineering and Decision Making, 1*, 22-55.

Jenkin, M., & Harris, L. (2001). *Vision and attention*. New York: Springer-Verlag,.

Land, M.F. (1998). The visual control of steering. In L.R. Harris and H. Jenkins (Eds.), *Vision and Action* (pp. 163-180). Cambridge, UK: Cambridge University Press.

Land, M.F. (2001). Does steering in car involve perception of the velocity flow field? In J.M. Zanker & J. Zeil (Eds.), *Motion Vision : Computational, Neural and Ecological Constraints* (pp. 227-235). Berlin: Springer.

Land, M.F., & Lee, D.N. (1994). Where we look when we steer. *Letters to Nature, 369*, 742-744.

Lecoutre, B., & Poitevineau, J. (2005). Le logiciel « LePAC ». *La Revue de Modulad, 33* (whole volume). Retrieved 03.03.2008 from http://www.univ-rouen.fr/LMRS/Persopage/Lecoutre/PAC.htm.

Rouanet, H. (1996). Bayesian methods for assessing importance of effects. *Psychological Bulletin, 119*, 149-158.

Underwood, G., Crundall, D., & Chapman, P. (2002). Selective searching while driving: the role of experience in hazard detection and general surveillance. *Ergonomics, 45*, 1-12.

Underwood, G., Chapman, P., Berger, Z., & Crundall, D. (2003). Driving experience, attentional focusing, and the recall of recently inspected events. *Transportation Research Part F, 6*, 289-304.

Underwood, G., Chapman, P., Bowden, K., & Crundall, D. (2002). Visual search while driving: skill and awareness during inspection of the scene. *Transportation Research Part F, 5*, 87-97.

Underwood, G., Chapman, P. Brocklehurst, Underwood, J., & Crundall, D. (2003). Visual attention while driving: sequences of eye fixations made by experienced and novice drivers. *Ergonomics, 46*, 629-646.

Wierwille, W. (1993). Visual and manual demands of in-car controls and displays. In B. Peacock & W. Karwowski (Eds.), *Automotive ergonomics* (pp. 299-320). Washington DC: Taylor and Francis.

Zikovitz, D.C., & Harris, L.R. (1999). Head tilt during driving. *Ergonomics, 42*, 740-746.

Advanced Driver Assistance Systems – Impact of psychological variables on the acceptance of modern technologies

Stefanie Müller, Heidi Ittner, & Volker Linneweber
Otto-von-Guericke-Universität Magdeburg
Magdeburg, Germany

Abstract

Advanced Driver Assistance Systems (ADAS) shall support the human being regarding several driving tasks. The implementation of those systems is assumed to cause a change in the driving process. Therefore the question of acceptance of ADAS arises. This study suggests an own heuristic model to identify psychologically essential variables to explain and enhance the acceptance and therefore adequate usage of ADAS. Two core criteria, willingness to pay for and willingness to use ADAS were defined and examined regarding their relationship with psychological mechanisms. Within the European project Safety Technopro an online survey (N = 7687) has been conducted. The study revealed the significance of cognitive and emotional variables respectively to impact car drivers' potential usage behaviour.

Introduction

Every year there are thousands of people who die in car accidents. According to the European Traffic Safety Council (2007) 39.200 people lost their lives in car accidents all across Europe in 2006. Solely in Germany 1.69 million car accidents were registered from January until September 2007 (Statistisches Bundesamt, 2007). To reduce this huge amount of car accidents electronic safety technologies more and more become an important factor to support the human fulfilling the complex tasks of driving a car. This group of e-safety technologies include Advanced Driver Assistance Systems (ADAS). These provide support for the driver or even take over driving tasks autonomously (Grunenberg, 2003). Consequently, these inventions of modern car technologies lead to relevant psychological changes in the process of driving a car not least because mobility has been and still is of primary meaning in our society (Linneweber & Ittner, 2004).

The crucial point is: how will drivers deal with these changes of the driving process and the consequences of the implementation of new technologies in the car? The proposed benefit of ADAS is to partly take over several driving tasks and facilitate the driving process and increase traffic safety. This might be to navigate the driver via navigation systems in unknown places whereas he or she will not be distracted reading a map and maybe overlook a pedestrian crossing the street. But apart from the help ADAS provide the proper handling of technological systems in the car will also require new additional mental effort (e.g. Walker, Stanton & Young, 2001). This additional effort the car drivers might not be willing to give in. Furthermore, the direct impact of assistance systems in the process of steering the car might lead to acceptance problems of ADAS in the sense that drivers will not be willing to let a technological system take over control over the car (Freymann, 2004). Another important factor concerns the additional financial effort the car drivers are confronted with. The current available ADAS are offered as special equipment that people have to purchase. According to the acceptance of assistance systems this requires to further pay attention to people's willingness to pay for the new systems.

Nevertheless, the human being is primarily responsible for nearly every occurring car accident in road traffic (Gründel, 2005). To reduce the amount of fatalities in Europe the Commission of the European Communities (2001) sets to halve this number until 2010. To contribute, the European project Safety Technopro aims to increase traffic safety through examining and promoting the usage and benefits of e-safety technologies in cars. Safety Technopro is a specific support action founded by the European Commission for Information Society and Media. An important area of expertise of Safety Technopro is that it puts special attention on the human being in the implementation of new technological systems. The project concentrates on the one hand on car drivers who are supposed to use ADAS. On the other hand it focuses professional bodies working in the automotive sector to accurately inform the final end user about the functioning and proper usage of ADAS. Safety Technopro will develop a training system for the professional bodies to foster their status as experts.

The following report exclusively focuses on car drivers and the discussion of the results obtained for this group. This is particularly necessary because the human being dealing with new technological systems has to be primarily focused (Walker et al., 2001) apart from further improving the technology. Moreover it is necessary to look at this subject from a psychological point of view.

The current study aims to centre psychological mechanisms within the process of implementing ADAS. In that sense people's potential usage behaviour is of crucial interest. Established theoretical approaches and their empirical confirmations have been summarized and integrated in the present study. These were explicitly adapted to the subject matter of ADAS and served as the basis for an own heuristic model that has been developed to explain the usage behaviour of ADAS.

Theoretical background

The term "acceptance" has been adapted to various objects of interest but so far it does not follow a unified definition. In order to explain and simplify different forms of understanding of "acceptance", some authors (e.g. Schade, 1999) distinguish between the terms "acceptance" and "acceptability". Hence, acceptability refers to a certain attitude towards an object whereas acceptance is seen as an actual behaviour towards this object (Schade, 1999). The different uses of these terms depend on the research approaches (e.g. Langner & Leiberg, 2002) and a final separation has not yet been completed. The present study defines acceptance as an indicator to show this specific behaviour, namely to use ADAS.

In order to psychologically explain people's acceptance of ADAS a heuristic model is proposed. This heuristic model embraces theoretical assumptions and their implementations in mobility psychology. Figure 1 shows the model summarizing the psychological dimensions that are assumed to influence people's potential usage of ADAS. Due to the fact that this study includes assistance systems currently not available on the market and also referring to the project's conditions requiring quantitative data it was impossible to study or measure the real use of ADAS. Therefore the concept of willingness to show certain behaviour has been adapted to this context. Former empirical studies have shown that the willingness serves as a valid predictor of actual behaviour (Montada & Kals, 2000). Figure 1 demonstrates that two kinds of willingness' are assumed to impact car drivers' usage behaviour. Apart from the car driver's willingness to use certain ADAS the willingness to pay for ADAS was chosen as a second criterion to be analysed. The acquisition of any additional equipment in cars requires car drivers to invest additional money. The additional financial effort is supposed to significantly influence people's acceptance of the in-vehicle technology and is therefore seen as a significant factor to be included in the model. The psychological variables that might determine the two criteria are briefly described below.

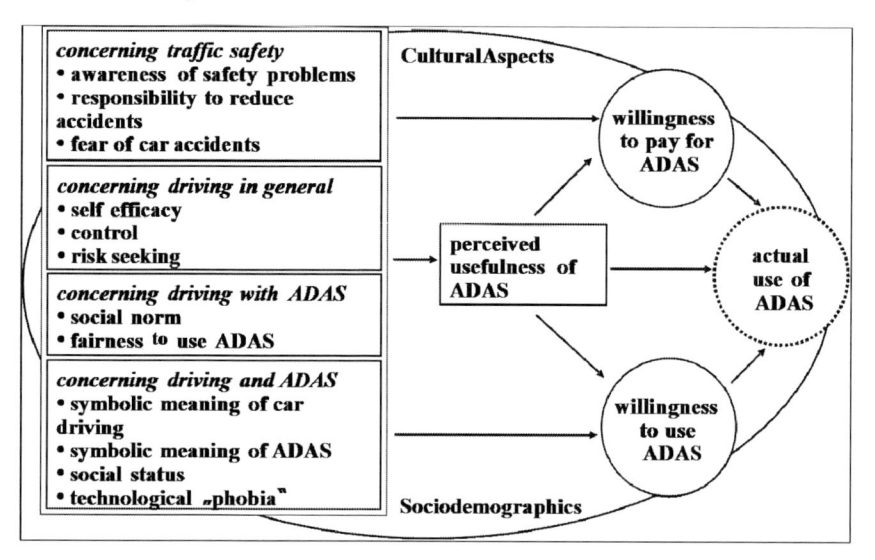

Figure 1. Heuristic model

The two criteria, willingness to pay for ADAS and willingness to use ADAS are expected to be directly influenced by the perceived usefulness of the systems. The perceived usefulness is conceptualized within the Technology Acceptance Model (Davis, 1989, Venkatesh & Davis, 2000). Implemented on the acceptance of modern car technology two aspects, the perceived usefulness and the perceived ease of use are of increment importance for the actual usage of ADAS. Davis (1989) recommends to especially focus the perceived usefulness which refers to the perceived benefit the user of new technology experiences without putting to much additional cognitive effort in because "no amount of ease of use can compensate for a system that does not perform a useful function" (1989, p.333 et seq.). According to this and referring to the current study as a non-experimental one this model assumes that the perceived usefulness of ADAS serves as a crucial predictor for the willingness to use and the willingness to pay for ADAS. The heuristic model furthermore indicates that the perceived usefulness also serves as a kind of mediator between the two criteria and the psychological variables. Figure 1 therefore demonstrates two paths of influences of the psychological mechanisms on the two criteria: one direct one and one indirect one via the perceived usefulness.

The psychological variables were summarized in four specific groups regarding their general issue. Figure 1 shows that these groups concern traffic safety in general, driving in general, driving with ADAS, and the symbolic or emotional dimensions of driving and ADAS. These variables are briefly described in the following and integrate rational as well as emotional aspects. The implementation of "pleasantness and usefulness" has been further successfully achieved e.g. by Van der Laan, Heino & de Waard (1997).

Concerning traffic safety in general the concepts of problem awareness and responsibility are included relying on the model on individual pro-environmental commitments (e.g. Kals, 1996; Montada & Kals, 2000; Ittner, Becker & Kals, 2003). This model reflects the aspects of voluntary engagements and has been applied to different kinds of commitments. Thereby it stresses the role of the necessary awareness of the problem and therefore the need to do something, the perceived own responsibility to engage. Apart from this more cognitive or "rational" point of view that people are only willing to engage in certain behaviour if they immediately see an advantage the model on pro-environmental commitment also focuses emotions as significant factors to influence people's behaviour. Within environmental psychology, especially fear is seen as an important emotional factor (e.g. Hine & Gifford, 1991) to impact human behaviour. Adapted to the current study the fear of being involved in serious car accidents is assumed to influence car drivers' willingness to pay for and to use ADAS.

The heuristic model furthermore includes aspects concerning driving in general. Relying on theoretical as well as empirical research, self efficacy (Bandura, 1982) refers to people's estimation that in a certain situation certain behaviour will be successful. In terms of driving a car people might or might not feel they can handle every occurring situation properly. In addition to that control is a basic human need to perform any specific act relying on one's own performance (e.g. Dörner, 1983). In this context it refers to the act of car driving. This concerns the drivers' personal need for complete control while driving the car instead of being supported by additional technological systems. Also referring to driving in general the heuristic model explicitly refers to risk seeking behaviour respectively risky driving. Risk seeking behaviour regards a very general estimation of people's preference to choose explicitly risky conditions in a certain situation. This is not limited to the area of car driving but refers to the wish to experience risky situations in general.

Concerning driving with ADAS the heuristic model also refers to the theory of planned behaviour (Ajzen, 1991) and its empirical implementation. The theory of planned behaviour implies that individuals form an intention to show certain behaviour. Important determinants of the intention to show this certain behaviour are the attitude regarding the behaviour, the perceived social norm regarding this behaviour and the perceived behavioural control to conduct this behaviour. Bamberg and Schmidt (1993) showed that the TPB provides significant support to forecast individuals' intention to use a specific means of transportation in road traffic. Applied to the acceptance of ADAS the social norm is seen as crucial in accepting assistance systems. The variable refers to what people who are significant for oneself think about paying for and using ADAS. It is assumed that these opinions strongly influence one's own thinking respectively behaviour.

Furthermore, this group of variables also focuses on the role of fairness. The model on pro-environmental commitments (e.g. Kals, 1996; Montada & Kals, 2000; Ittner, Becker & Kals, 2003) conceptualizes fairness as a personal commitment to show a specific behaviour. It refers to the distribution of costs and benefits throughout the society respectively individually relevant social groups. Regarding driving with ADAS the perceived distribution of personal and shared investment and benefit as fair (or not) should impact people's willingness to pay money for assistance systems and use them properly.

The fourth group of psychological variables summarizes so-called symbolic or emotional dimensions of car driving and ADAS. The approach of symbolic meaning of car driving (e.g. Hunecke et al., 2001) illustrates that car use is not only motivated by rational decisions but is also determined to satisfy emotional motivated needs for privacy, fun, autonomy or expressed social status. Interestingly, these additional emotional, symbolic meanings are even much more important for most people than the pure functional aspects of car driving. Hunecke et al. (2001) empirically evaluated the concept of symbolic meaning of car driving. The current heuristic model included and adapted this to the symbolic meaning of ADAS. The following dimensions of possible symbolic or emotional meanings of ADAS were integrated: fun, status, commodity, curiosity, proud, and driving effort. Finally another emotional reaction towards new technology expresses quite more the contrary to the symbolic meanings of ADAS. It refers to the fact that many people do not feel confident in their personal experience with new technologies. They might even feel a kind of fear because they do not know how to handle it properly (or at all) especially in comparison with other people. This might lead to a very sceptical attitude towards this technology and its use. In terms of ADAS the willingness to use and also pay some extra money for car technology is expected to be significantly related to this concept referred to as technological "phobia".

Within Safety Technopro the model has been created that can apply for a variety of ADAS. This implies to use the model to analyse car drivers' acceptance of ADAS that either inform or warn the driver or even take over driving tasks automatically.

An advantage of the model is to be applicable for already available systems but at the same time for ADAS that will be developed in the near future.

Method

An own heuristic model to explain possible influences on car drivers' willingness to pay for assistance systems and the willingness to use those has been developed. The aim was to empirically test this heuristic model.

A Europe wide online study has been conducted. Therefore a questionnaire has been constructed where each psychological variable was operationalised through several items (an item example for each scale is given in Annex I). Participants of the study had to express the degree of their agreement resp. disagreement to every single item on a five-point Likert scale. This questionnaire was translated in 8 languages and posted on web sites of automobile clubs in 12 European countries from March until July 2007. The gathered data were analyzed separately for each country specifically in order to detect similarities or differences between them.

In total 7687 people participated in the online survey. Regarding the gender, there is a clear overrepresentation of male participants (74%) in comparison with female participants (26%). And with respect to the age, there is a slight overrepresentation of middle-aged people, ranging from 31 to 40 years (23%). Interestingly, there is only a very small group of participants aged 18 to 21 (5%). Taken together, nearly half of the sample (46%) is younger than 41 years, whereas the second "half" of the sample (54%) is 41 or older. The following table 1 lists the sample sizes, gender and age for each participating country.

Table 1. Sample sizes and sociodemographic data for all participating European countries

		A	B	CH	D	E	F	GB	HR	I	NL	N	P
Number	N	856	277	117	1750	719	1578	1258	170	480	407	126	130
	%	11	3	1	23	9	20	16	2	6	5	2	2
Age (percentage)	18-21	3	1	3	4	3	5	5	9	4	3	7	6
	22-30	10	13	21	15	24	18	19	38	18	11	20	21
	31-40	16	23	30	17	33	25	25	25	30	23	21	35
	41-50	21	27	26	22	24	15	21	17	21	26	24	25
	51-60	20	20	13	21	11	28	20	11	19	22	23	12
	>60	28	16	9	21	4	8	11	1	8	14	6	2
Gender (percentage)	female	21	19	24	14	23	43	33	28	19	26	15	19
	male	79	81	76	86	77	57	67	72	81	74	85	81

Results

Regarding the willingness not only to use ADAS but also to pay money for them people's general agreement ranged between $M = 2.45$ and 4.13 (scale range: 1-5; 1 = "I don't agree, 5 = "I agree"). The M and SD are shown in Figure 2. All participating European countries are listed. The questionnaire included 9 different assistance systems. Previous to analyzing the data these ADAS were further structured and grouped according to their primary functioning (e.g. Grunenberg, 2003) which was either to warn (e.g. lane departure warning) or to inform (e.g. traffic information system) car drivers. The arrangement in groups of the different systems on the one hand follows methodological aspects. On the other hand the grouping serves to control for the heuristic model to be applicable for any of the ADAS combined according to their primary functioning. The grouping is interesting in a psychological sense because it is assumed that the different levels of intervention of ADAS affect the acceptance of these systems.

Figure 2 separately lists the willingness to pay for warning systems and informing systems. The warning systems included for example the Obstacle Warning System; the informing systems the Navigation System.

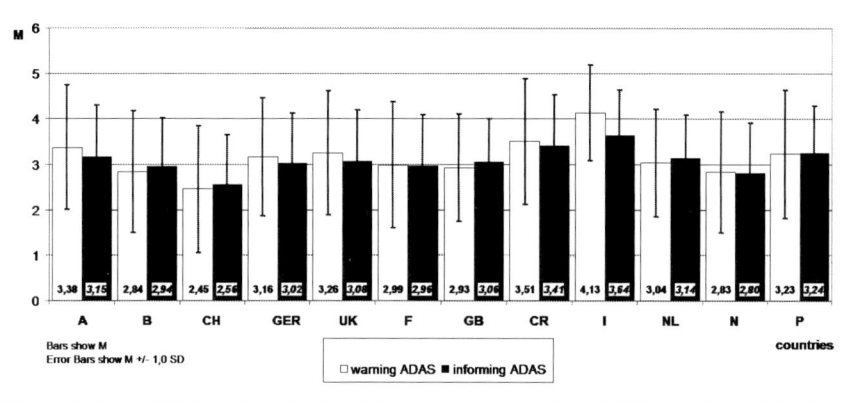

Figure 2. M *and* SD *of the criterion "willingness to pay for ADAS" for all participating European countries*

There is a variance across the different countries concerning this criterion. The Italian sample showed the highest agreement towards a positive attitude of paying money for ADAS. Car drivers from Switzerland mainly disagreed to pay for the systems. The other European participants in general showed an unbiased attitude towards being willing to pay extra money for ADAS. Where participants from Austria, Spain, Croatia, the Netherlands, and Portugal showed values slightly above average (average = 3), the remaining 4 countries slightly stayed below this value. According to the two different system types most car drivers did not differ much between the two groups of assistance systems except for Italy and Austria. These differing results might be due to the fact that "money" has not been further specified in the questionnaire. It was intended to evaluate the car drivers' willingness to accept any additional cost to acquire ADAS. This may have led to varying interpretation in the studies' participants and explain the differing European results.

Regarding the second criterion the results somewhat varied from those for the willingness to pay for ADAS. Figure 3 summarizes the statements exemplarily for the German sample. The exemplary results shown in figure 3 (Germany) and 4 (United Kingdom) have been selected for several reasons. On the one hand both countries have been those with one of the highest amount of car accidents across Europe in 2006 (European Commission, 2008). Moreover in 2006 the highest numbers of injuries caused by car accidents have been found for both these countries. On the other hand the results from Germany and the United Kingdom have been chosen for representational reasons because they are comparable to those found for the other participating European countries in this study.

Concerning the second criterion, willingness to use ADAS, in all participating European countries people expressed an overall agreement. Similar to the willingness to pay for assistance systems the separately listed ADAS in the questionnaire were grouped according to their primary functioning in automatic, warning and informing systems. One system type for each of the three different groups of ADAS is shown in the figure.

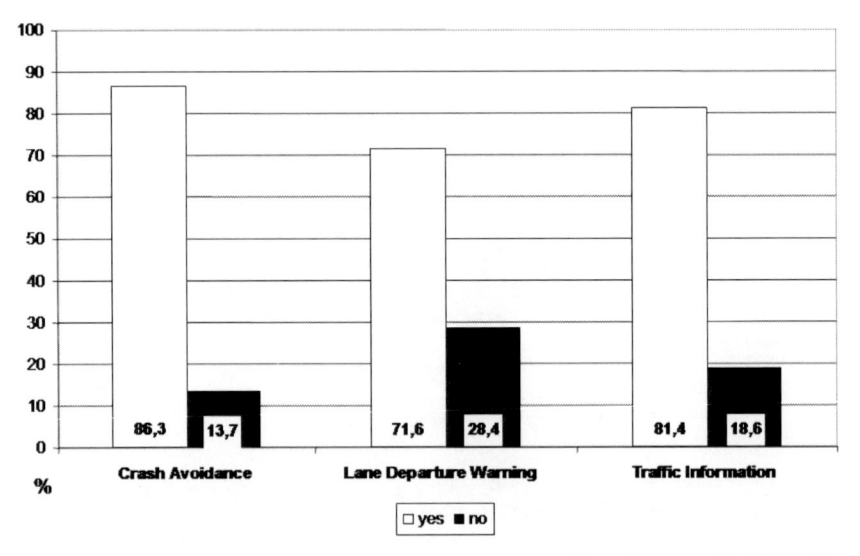

Figure 3. Frequency of the criterion "willingness to use ADAS" in Germany (N=1750). The frequencies are given in percentage

The German participants showed a clear agreement of being willing to use the different ADAS. People's willingness to use did not differ much between the crash avoidance system (automatic system) and the traffic information system (information system). Their willingness to use lane departure warning (warning system) was somewhat lower.

The aim of testing the heuristic model required further multivariate data analyses. To explain the relationship between predictors and criteria and which psychological variables will be most essential stepwise multiple regression analyses were conducted separately for the willingness to pay for and the willingness to use ADAS. Again the different ADAS were grouped according to their primary functioning in automatic, warning, and informing systems. Figure 4 exemplarily shows the multiple regression analysis for the usage of warning systems (e.g. lane departure warning) for the sample from Great Britain.

For the British sample a substantial amount of variance ($R^2 = .31$) in the criterion can be explained through the psychological dimensions including the perceived usefulness ($\beta = .40$). The higher the perceived benefit of warning systems the more likely people are willing to use those systems. According to this the more car drivers are actually aware of current safety problems in road traffic and the more they fear of being involved in serious car accidents the higher is their willingness to use warning ADAS. In addition the higher people want to agree with people who are significant for them the more likely they are willing to make use of warning systems. Furthermore the more car drivers expressed that the usage of ADAS should be a shared issue among all car drivers the higher was their willingness to use assistance systems. Also important is the fact that the people who consider ADAS as being comfortable and exciting also showed a willingness to use ADAS. On the other hand participants who want to keep control while driving are less willing to be supported by an assistance system than those who do not seek absolute control while driving.

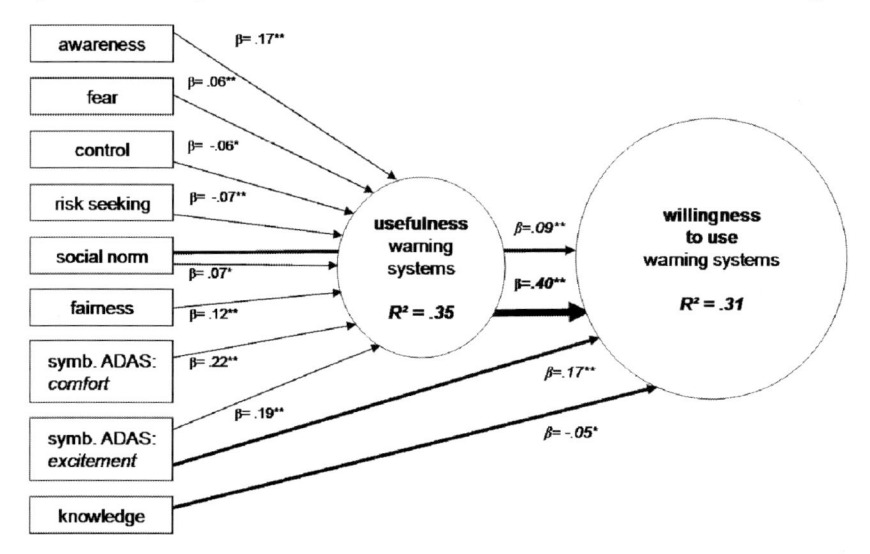

Figure 4. Multiple stepwise regression for criterion "willingness to use automatic ADAS" exemplarily for Great Britain (N=1258)

Figure 4 also indicates that participants who showed a higher awareness that road traffic is problematic and dangerous nowadays more likely perceived assistance systems that might warn them while driving as useful. In addition to this the participating car drivers who also stated that their behaviour is less risky than orientated towards safety also perceived warning ADAS as useful. Furthermore the more likely British participants think of ADAS as symbolizing comfort and excitement the higher is their perceived usefulness of those systems.

Figure 4 exemplarily illustrates the Multiple Regression Analysis (MRA) for Great Britain. Stepwise multiple regression analyses have also been conducted separately for the remaining 11 European samples. The results somewhat differed across the 12 countries. Nevertheless, nearly in every MRA certain psychological variables significantly correlated with the two criteria. The more people are aware of road traffic dangers and the more anxious they are being affected by a car accident the more likely they are willing to use and to pay for ADAS. Furthermore the more car drivers feel responsible to personally contribute to enhance traffic safety the higher is their willingness to use and to pay for ADAS. Additionally, the more car drivers orientate on other people's thinking and behaviour and the higher the car drivers' expectation (or even demand) that the costs of ADAS should be equally shared among all car drivers the more likely they are willing to pay some extra money for assistance systems and to use them. Finally, the higher the participants' meaning of ADAS in a comfort and excitement manner the higher was their willingness to use and to pay for those systems.

Discussion and conclusion

To fully use the safety potential that ADAS might provide and to successfully implement ADAS on the market the car drivers' acceptance of those systems is crucial. Within the European project Safety Technopro a heuristic model was developed to describe and explain potential psychological influences on the usage behaviour regarding ADAS. For the actual behaviour could not be observed in this study the model instead focused two criteria: the willingness to pay for and the willingness to use ADAS. This matter of willingness to show certain behaviour has been found to serve as a valid predictor (Montada & Kals, 2000) for actual behaviour. Apart from the two criteria several psychological variables concerning traffic safety, concerning driving in general, concerning driving with ADAS, and concerning the symbolic meaning of driving and ADAS were implemented in the heuristic model. Furthermore the perceived usefulness (Davis, 1989) of assistance systems was assumed to influence car drivers' willingness to pay for and to use ADAS.

People's willingness to pay money for ADAS varied across the different samples. Interestingly, half of the European sample was rather indecisive to contribute to an additional financial effort. They neither rejected nor agreed being willing to pay money for ADAS. This might be due to the fact that most Advanced Driver Assistance Systems still have to be exclusively added and require extra money investment to the usual prize that has to be paid for a car. Furthermore, also cultural specialties could have caused these different willingness' of paying an extra fee for additional car equipment. The results suggest that people apparently separated between paying for and using assistance systems. Instead the participants' willingness to pay for ADAS their willingness to use those systems revealed a general agreement all across Europe. Therefore it is required to further distinguish between the two kinds of willingness' and separately integrate them in the heuristic model.

To test this heuristic model multiple regression analyses for the willingness to use ADAS has been described. The regression model showed that psychological mechanisms indeed have an impact on the willingness to use ADAS. Accordingly the perceived usefulness (Davis, 1989) of ADAS has proved to be the most important precondition for further willingness to act. Car drivers evaluating a specific assistance system as providing additional benefit will therefore be more willing to make use of.

The most interesting point in this study was to examine the psychological mechanisms that might impact the willingness to use ADAS. The results confirm that cognitive as well as emotional aspects play a significant role in people's evaluation of modern car technology. Referring to that and corresponding to the model on pro-environmental behaviour (e.g. Kals, 1996; Montada & Kals, 2000; Ittner, Becker & Kals, 2003) it appears that the awareness of the current road traffic situation certainly affects people's willingness to use ADAS. This is important for the car industry. The safety benefit ADAS provide has to be explicitly highlighted for example in sales conversations. Car drivers who are aware of road traffic dangers are more willing to use ADAS. This importance of problem awareness for the potential usage of ADAS also implies to actually discuss traffic safety in public to further increase car drivers' awareness on traffic safety and its dangers and therefore strengthen the implementation of assistance systems.

Also relying on the model on pro-environmental behaviour (e.g. Kals, 1996; Montada & Kals, 2000; Ittner, Becker & Kals, 2003) the correlation between fairness and the willingness to use assistance systems must be emphasised. The usage of assistance systems should be intensively promoted through car industry and politicians respectively. As a consequence car drivers would perceive the distribution of ADAS as equally shared among all people. This would include the costs every person will have to effort but also including the safety gain every car driver can contribute to as well as rely on. The results gathered in the current study support that people who perceive the costs of ADAS as fairly shared among all car drivers are also more likely to use those systems.

Another crucial aspect within this study concerns the symbolic meaning of ADAS. These more emotional reactions towards car technology correspond to research on the symbolic meaning of car driving (Hunecke et al., 2001). Therefore people specifically want to fulfil their own personal needs (e.g. fun) when driving a car. This concept was adapted to driving with ADAS. The results show that the willingness to use ADAS goes beyond a pure functional evaluation of technological systems through car drivers. These kinds of emotional reactions towards modern car technology should also be focused when promoting those systems for example in sales conversations. Apart from paying attention to symbolic meanings of yet existing systems this aspect also concerns technicians and designers of upcoming ADAS. The more likely these systems are perceived to fulfil personal needs for comfort and excitement the higher is the willingness to use those systems.

To close, the results gathered in this study provide support for the heuristic model. The impact of psychological mechanisms confirms the assumption that the usage of a modern car technology goes beyond a pure functional evaluation of assistance systems. To describe the relationships between psychological mechanisms and the willingness to pay for and to use ADAS any further additional statistical analyses to test and improve this model are required. Regarding the 12 European countries, the results in this study turned out to be comparable but there were also differences between them. These differences have to be analysed in further detail in order to derive cultural specialties regarding the potential usage of ADAS.

Finally the successful implementation of ADAS involves not only the car drivers but also the car industry, and politicians respectively. The implications of current research on the subject of ADAS including this study might noticeably contribute to help reducing the amount of road fatalities by 2010.

References

Ajzen, I. (1991). The theory of planned behavior. *Organizational Behavior and Human Decision Processes, 50,* 179-211.

Bamberg, S. & Schmidt, P. (1993). Determinanten der Verkehrsmittelwahl –Eine Anwendung der Theorie geplanten Verhaltens. *Zeitschrift für Sozialpsychologie, 24,* 25-37.

Bandura, A. (1982). Self-Efficacy mechanisms in human agency. *American Psychologist, 37,* 122-147.

Commission of the European Communities. WHITE PAPER: European transport policy for 2010: time to decide, retrieved december 2007 from http://ec.europa.eu/transport/white_paper/documents

Davis, F.D. (1989). Perceived usefulness, perceived ease of use, and user acceptance of information technology. *MIS Quart, 13,* 319-339.

Dörner D., Kreuzig,, H.W., Reither, F., & Stäudel, T. (Eds.) (1983), *Lohhausen: Vom Umgang mit Komplexität.* Bern, Switzerland: Huber.

ETCS. (2007). EU road safety plan behind schedule, retrieved 29.11.2007 from http://www.etsc.be/home.php.

European Commission. Transport. Road Safety. retrieved december 2007 from http://ec.europa.eu/transport/roadsafety/road_safety_observatory/care_reports_en.htm

Freymann, J. (2004). *Möglichkeiten und Grenzen von Fahrassistenz- und aktiven Sicherheitssystemen.* BMW Groups Forschung und Technik. Tagungsbericht.

Grunenberg, J. (2003). Tendenzen bei Fahr(er)assistenzsystemen. Seminararbeit Uni Koblenz. Retrieved december 2007 from
http://www.uni-koblenz.de/%7Eagrt/lehre/ss2003/seminar/johannes_grunenberg.pdf.

Gründel, M. (2005). *Fehler und Fehlverhalten als Ursache von Verkehrsunfällen und Konsequenzen für das Unfallvermeidungspotenzial und die Gestaltung von Fahrerassistenzsystemen.* Phd-thesis. Universität Regensburg.

Hine, D.W. & Gifford, R. (1991). Fear appeals, individual differences, and environmental concern. *Journal of Environmental Education, 23,* 36-41.

Hunecke, M., Blöbaum, A., Matthies, E., & Höger, R. (2001). Responsibiliy and environment – Ecological norm orientation and external factors in the domain of travel mode choice behavior. *Environment and Behavior, 33*, 845-867.

Ittner, H., Becker, R., & Kals, E. (2003). Willingness to support traffic policy measures: The role of justice. In J. Schade and B. Schlag (Eds.), *Acceptability of transport pricing strategies* (pp. 249-266). Oxford: Elsevier.

Kals, E. (1996). Are pro-environmental commitments motivated by health concerns or by perceived justice? In L. Montada & M. Lerner (Eds.), *Current societal concerns about justice* (pp. 231-258). New York: Plenum.

Langner, R. & Leiberg, S. (2002). *Determinanz der Akzeptanz von Straßenbenutzungsgebühren*. Bericht zur Forschungsorientierten Vertiefung. Technische Universität Dresden. retrieved december 2007 from http://www.geocities.com/tinkerbellblues/Studium/bericht.pdf

Linneweber, V. & Ittner, H. (2005). Verkehrspsychologie. In W. Heber and T. Rammsayer (Eds.), *Handbuch der Persönlichkeitspsychologie und Differentiellen Psychologie* (pp. 572-579). Göttingen, Germany: Hogrefe.

Montada, L. & Kals, E. (2000). Political implications of psychological research on ecological justice and proenvironmental behavior. *International Journal of Psychology, 35*, 168-176.

Schade, J. (1999). Individuelle Akzeptanz von Straßenbenutzungsentgelten. In B. Schlag (Hrsg.), *Empirische Verkehrspsychologie* (pp. 227-244). Lengerich, Berlin: Pabst Science Publishers.

Statistisches Bundesamt (2007). Pressemitteilungen im Sachgebiet Verkehr, retrieved 29.11.2007 from http://www.destatis.de/jetspeed/portal/cms/Sites/destatis/Internet/DE/Presse/pm/Uebersicht/Verkehr.

Van der Laan, J.D., Heino, A., & De Waard, D. (1997). A simple procedure for the assessment of acceptance of advanced transport telematics. *Transportation Research - Part C: Emerging Technologies, 5*, 1-10.

Venkatesh, V. & Davis, F.D. (2000). A theoretical extension of the technology acceptance model: Four longitudinal field studies. *Management Science 46*, 2, 186-204.

Walker, G. H., Stanton, N.A., & Young, M. S. (2001). Where is computing driving cars? *International Journal of Human-Computer Interaction, 13*, 203-229.

Human centred design for informative and assistive technology in transport

Annie Pauzié[1] & Anabela Simões[2]
[1] INRETS/LESCOT, Bron, France
[2] ISEC, Lisboa, Portugal

Abstract

Information and communication technology offers functions to support the driver by informing about forthcoming critical events, specific weather conditions, distance to the vehicle ahead, directions to follow, speed to respect, functions that support human perception and decision taking. Other functions can take actions to control the vehicle, electronic assistance compensating the shortcomings of human reaction and functional abilities. Informative systems require additional attention from the driver, and the benefit of using available information has to be balanced with potential interference. For automation technologies, assistance systems bring the problem of task assigment between human and machine, as well as the choice of logic used for the management of this control sharing; substitution or co-operation.

In new systems specific care should be devoted to human factors, to avoid misconception of the designers and misuse by the driver. Challenge is to take into account the heterogeneity of functions, of systems, and of driver functional abilities. European research projects concerned with the development of innovative functions, adaptative interface, design guidelines and integrated methodologies are presented.

Introduction

Information and Communication Technology (ICT) in the vehicle allows the development of *informative* functions (i.e., In-Vehicle Information Systems or IVIS) that support the driver. Information can be about about forthcoming critical events, specific weather conditions, distance to a vehicle ahead, directions to follow, or maximum speed to respect, all functions that support human perception and decision taking. On the other hand, *assistive* functions (i.e., Advanced Driver Assistance Systems or ADAS) can take actions to control the vehicle, compensating the shortcomings of human reaction and functional abilities. Implementation of these electronic functions leads to an increased complexity of the Human/Machine Interaction inside the vehicle, a real challenge in terms of human factors and ergonomics. Indeed, these functions can induce deep modifications of the driving task at the operational, tactical or strategical level (Michon, 1985).

This challenge of safe use is also induced by additional factors such as the types of devices, some embedded by car manufacturers, some implemented after marketing some brought by the driver such as nomadic devices.

Finally, the most important challenge in this context is the wide heterogeneity of driver's population in terms of functional abilities, experience, motivation and context of the trip. Devices and their functions will have to be able meet a wide range of users' needs and requirements.

Take for example experience, it has been shown that novice drivers have slow reaction time, and have difficulties in detecting hazardous situations (Deery, 1999). Electronic functions supporting the driver can be expected to be beneficial for this population. Nevertheless, no data are available to identify how the process of building up driving experience can be modified while novice drivers use additional support systems.

Perception, cognition, and motor capacities of elderly drivers in general slow down. This decline however has a large inter-individual variability and is not directly linked to the chronological age. This observation widely studied in the literature induces complexity in the human factors approach and data, as sample of tested drivers are identified according to their chronological age and not according to their functional age. Impact of electronic support for this driver population has to be carefully assessed, taking into account this heterogeneity (Caird 2004; Caird et al., 2005; Vrkljan & Miller-Polgar, 2005). This is a very relevant group and issue, as the number of elderly drivers will increase constantly in the coming years, and mobility of this population has to be preserved for health and well being reasons (Marottoli et al., 2000; Fonda, Wallace and Herzog, 2001). Various studies have indicated that ADAS might be able to provide tailored assistance for elderly driver; efficiency of functions such as Collision Avoidance can be beneficial for all population including elderly, as long as the interface is carefully designed (Kramer, Cassavaugh, Horrey, Becic, & Mayhugh, 2007). Research still needs to be performed to identify how to improve adaptive/compliant interfaces at all levels and how to shorten the training period to enhance safety through electronic support for this population. Motivation to consider this population is also increased by the fact that designing a system fitting elderly needs and requirements will fit better needs and requirements of the whole population of drivers.

Another specific population to take into consideration in this framework are professional drivers, who have specific needs for functions of ITS in their work context. They need communication systems while driving more often than private drivers, as it can be highly important for their job purpose (Harms & Patten, 2001). Nevertheless, a French survey showed that the same work constraint can lead to various strategies according to the companies in terms of in-vehicle design and implementation, and consequences in terms of road safety (Pauzié, 2004, figure 1).

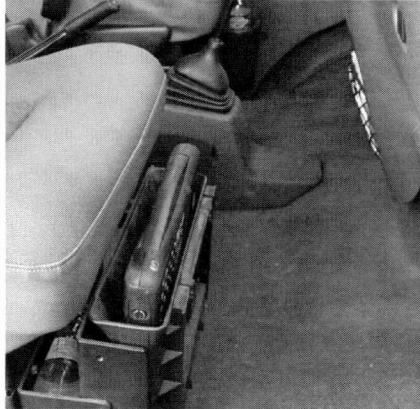

Figure 1. For the same type of job, parcel express delivery, two different company strategies have been observed: on the left, use of the system is possible for the driver while driving; on the right, the system is embedded under the passenger seat, making use while driving impossible

Information and Communication Technology: a chance for the improvement of road safety

The European Community published an explicit document stating the ambitious objective to decrease 50% of the road fatalities by 2010 (White Paper, 2001). The deadline seems close, but, at least, the process of conducting research and development of ICT functions to improve safety has been active for many years, and several projects have already showed the potential of electronic support to limit critical situations happenings, or alleviate seriousness of consequences.

This potential benefit, such as a real-time auditory and visual display guidance, has been demonstrated more than a decade ago (Barham, Alexander & Oxley, 1994). It has been shown on real road experiments that drivers made significantly less driving errors while using this type of guidance system, in comparison with a reference situation, irrespective of their age (Pauzié & Marin-Lamellet, 1989).

Nowadays, more sophisticated functions are studied. For example, in the framework of the European project PReVENT*, active and preventive safety systems have been developed and tested, on different levels from informative to assistive functions. These functons were related to the criticality of the event and the driver's reaction: informing the driver as early as possible, warning him/her if there is no reaction, then actively assisting or ultimately intervening in order to avoid crash, or at least to mitigate its consequences. Development of co-operative system for the prevention of accidents involving vulnerable road users (pedestrians, cyclists and motorcyclists) in

---

* www.prevent-ip.org

complex traffic scenarios is studied in the European project WATCH-OVER[*], with real time detection and relative localization of the vulnerable road users.

To minimize the level of workload for the driver, an integrated adaptative interface has been developed in the European project AIDE[◊] has as objective to manage the set of potential functions available through IVIS, ADAS, and nomadic devices, and to display messages depending on the complexity of the environment,and the driver's attentional resources available.

Information and Communication Technology: a potential decrease in road safety

On the other hand, the use of in-vehicle systems can increase workload and distraction for the driver, especially if functions are not related to the driving task and should be processed in parrallel to driving. A well known example is having a phone conversation while driving (Patten et al., 2004, Matthews, Legg, Charlton, 2003).

Even for systems supporting driving task, informative as well as assistive functions have to be carefully designed to match with driver's functional abilities, requirements, and expectations (Carsten, & Nilsson, 2001). Review of methods for IVIS distraction assessment has been recently conducted (Gelau & Stevens, 2006). Acceptability and usability of automation functions by the driver have to be carefully studied to ensure road safety objective (Hoedemaeker & Brookhuis, 1999; Brookhuis, De Waard, & Janssen, 2001; Griffiths & Gillespie, 2005).

ICT Human centred design and road safety

In-vehicle devices have to be intuitive, self-explanatory, and non intrusive. To reach this goal, the human-centred design approach is relevant at each step of the development: setting up the concept, development of the mock-up and the prototype, implementation of the system, with series of iterations to improve the final result (Pauzié, 2002).

The Network of Excellence HUMANIST (HUMAN centred design for Information Society Technologies[♠]), funded by the European Commission DG InfSo, gathers research activities directly linked to this issue: identification of the driver needs in relation to ITS, evaluation of ITS potential benefits, joint-cognitive models of driver-vehicle-environment for user centred design, impact analysis of ITS on driving behavior, development of innovative methodologies to evaluate ITS safety and usability, driver education and training for ITS use, use of ITS to train and to educate drivers.

Generally, the ergonomic approach for design and evaluation processes aims at :

---

[*] www.watchover-eu.org
[◊] www.aide-eu.org
[♠] www.noehumanist.org

- assisting designers to allow quicker and more efficient design process by setting up ergonomic *guidelines* and criteria, taking into account the wide heterogeneity of drivers' needs and requirements
- evaluating safety for drivers using these devices by setting up *tools, techniques* and *methods*.

Design guidelines

Several research projects in the framework of the European Programme DRIVE have focussed on developing design guidelines for ITS with a human centred design. Examples of these projects are TELAID, VODIS, CEMVOCAS, HARDIE (Ross et al., 1996), TELSCAN (Nicolle & Burnett, 1994). More recently, main recommendations concerning HMI design and implementation of in-vehicle systems have been published by the European Commission in the Official Journal (1999). These recommendations are classically labelled ESoP, for "European Statement of Principles". Following this publication, a European Working Group has been set up in the framework of the eSafety initiative on the specific topic of Human Machine Interaction in order to study this issue, and to allow the follow up of these "Statement of Principles", with a reference to the critical question raised by nomadic devices (see Pauzié, Stevens & Gelau, 2006, for a summary).

Methods for evaluation

Evaluation of in-vehicle systems for improved design can be conducted following various methods of investigation (Cotter & Stevens, 2005). These different methods are suitable for different goals, such as, "to guarantee road safety" or "to generate recommendations for the next generation of improved system". Methods are available for different types of systems such as "informative" or "assistive", for heterogeneous drivers' population such as "elderly" or "novice" (Simoes et al., 2001; Pauzié, 2005).

There are discussions and propositions about relevant tools and methods to investigate the impact of system use according to users' population variability. Typically, parameters that are taken into consideration are related to vehicle control (trajectory deviations as a result of use of the system), drivers' visual strategies (visual demand due to on-board screens) and overall drivers' workload.

Vehicle deviation trajectories can be a good parameter in relation to visual strategies. Unfortunately from an experimental point of view, and fortunately for road safety, this parameter reflects very high and a relatively rare workload situation. In several cases, it can reveal that the driver is on the way to loose control of his vehicle, except in low constraint road context such as motorway driving, where swerving does not constitute such a critical behaviour. So, recording driving deviations is relevant, but certainly not enough to reveal unsafe behavior. Other measurements, e.g. physiology, are necessary to identify more subtle increases in driver's workload (De Waard, 1996, for a review).

Mental workload is multidimensional, and an efficient tool is the NASA-TLX, NASA-Task Load Index, set up by the NASA for the evaluation of pilot's workload. The TLX has been used for decades to evaluate subjective mental workload of operators (Hart & Staveland, 1988; Piechulla, Mayser, Gehrke, Konig, 2003). A modified version of the NASA-TLX, called the DALI for Driving Activity Load Index, has been proposed with the objective to be more adapted to the driving task, and has been tested for mobile phone and navigation functions use while driving (Pauzié & Pachiaudi, 1997).

Recently, an on-the-road experiment has been conducted to assess advantages and limits of the DALI method for the evaluation of driver's mental workload (Pauzié, 2007).

Process for the experimental procedure was the following:

- Set diversified situations varying according to their level of demand: cognitive process (e.g., to memorize a route) and perceptivo-motor process (e.g., to perform a manual action following auditory, visual, or tactile displayed instructions), with or without system use.
- Apply the tool at the end of each of these sessions to gather subjective data in relation to each context.
- Check if the highly task demand session corresponds to the higher values for the tool, and to identify in which way (which DALI factors)

The four tested experimental sessions were presenting the following characteristics:

- to vary according to the level of workload induced on the driver
- to be as realistic as possible in a context of driving task

Two situations with a high task demand:
High (Context + System) *HCS*: While driving, the driver had to run a task according to stimulations emitted by an on-board system. The information to deal with was not related to the task and induced a manual action or a verbal answer. The route to follow was given by a guidance system. The workload was linked to *perceptual processes*, *decision making* and *motor and/or verbal output*

High (Context) *HC*: Before the experimentation started, the driver had to consult a paper map to know the route to follow. Then, he can stop anytime to check again the directions. The workload was linked to the *mental representation* of the route and to *memorization* of it.

Two situations with a low task demand:
Low (Context + System) *LCS***:** The driver had to follow the route according to visual and auditory information given by a guidance system. The workload was linked to *perceptual processes* but the decision making and the mental representation/memorization were lighter than in the previous sessions.

Low (Context) LC: During the route, the experimenter gave clear and on time directions to follow. The workload was linked only with the management of the driving task, without added activity linked to strategical processes.

In Figure 2 the values for each factor and for the global score are displayed for the four experimental sessionsvarying by their level of complexity and induced workload on the driver.

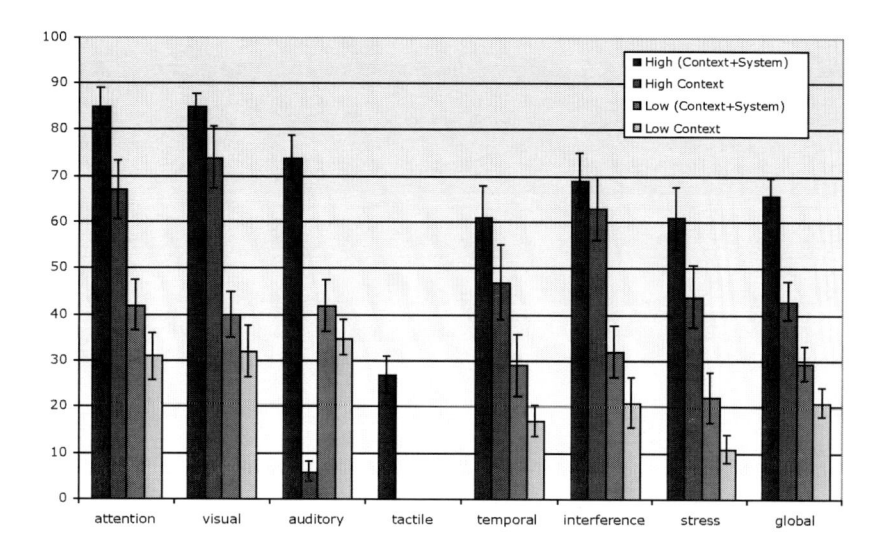

Figure 2. The values of the DALI factors showed significant differences between the four experimental sessions

The classical methods used untill now that have been used to assess an isolated function, will have to be tested in the new context where several functions are made available for the driver: an increased number of functions, and a mix of assistive and informative functions. The issue of hierarchical order and priority of displayed messages will then have to be studied, to get an integrated human machine interaction.

Conclusion

The main objective of most of electronic aids is to improve road safety, by assistance such as lane keeping or alerting or, at least, to increase comfort such as guidance and navigation systems. Nevertheless, system use requires divided attention, with the driver's main task, driving safely, focused on the external road environment. Perceptual and cognitive processes induced by in-car systems have to fit into the general driver's activity, and with road safety critical issues, not to lead to mental/ perceptual overload.

In this framework, goals are to define "what" functions to develop in a user-centred approach to make humans more efficient while using these systems, in addition to "how" to design these functions to be accepted and efficiently used.

This argues for an active participation of the Human Sciences in the various stages starting from system conception to technological development, which all should be centred on the human. Assistance should be designed according to the needs and the capabilities of the human being, and not be technology driven.

Due to the novelty of the topic, there are several European research projects active in this area, with the objective to gather data to support the human centred design of these innovative functions, including a focus on special groups such as the elderly. Further research is needed regarding the emergent human driving tasks of system supervision, system status awareness, and system response representation, particularly in the case of co-operative assistance (Nilsson, 2005).

The success of the deployment of these new products needs, beyond technological research efforts, strong and continuous actions in the scientific research area of human factors and ergonomics, to bring knowledge for a correct usability and acceptability of these systems, and consequently a real improvement of safety and mobility.

References

Barham, P., Alexander, A., & Oxley, P. (1994). What are the benefits and safety implications of route guidance systems for elderly drivers. In *Proceedings from the Seventh International Conference on Road Traffic Monitoring and Control* (pp.137–140), London, UK

Brookhuis, K.A., De Waard, D., & Janssen, W.H. (2001). Behavioural impacts of advanced driver assistance systems: an overview. *European Journal of Transport and Infrastructure Research*, *1*, 245-253.

Caird, J.K. (2004). Intelligent transportation systems (ITS) and older drivers' safety and mobility. *Transportation in an aging society: A decade of experience* (pp. 236–255). Washington, D.C.: National Academy of Sciences, Transportation Research Board.

Caird, J.K., Chisholm, S., Lockhart, J., Vacha, N., Creaser, J.I., Edwards, C., & Hatch, K. (2005). In-vehicle intelligent transportation system (ITS) countermeasures to improve older driver intersection performance. *Transportation Development Centre*, Transport Canada.

Carsten, OMJ & Nilsson L. (2001). Safety assessment of driver assistance systems. *European Journal of Transport and Infrastructure Research*, *1*, 225-243.

COMMISSION RECOMMENDATION of 21 December 1999 on *safe and efficient in-vehicle information and communication systems: A European statement of principles on human machine interface* (notified under document number C(1999) 4786) (Text with EEA relevance) (2000/53/EC), Official Journal n° L 019 du 25/01/2000, p. 0064 – 0068.

Cotter S. & Stevens A. (2005). *Issues related to integrated methodologies*, HUMANIST report, Deliverable E.3, retrieved from www.noehumanist.org.

Deery H.A. (1999). Hazard and Risk Perception among Young Novice Drivers *Journal of Safety Research, 30,* 225-236.

De Waard, D. (1996). *The measurement of drivers' mental workload.* Ph.D. thesis, Haren, the Netherlands: University of Groningen, Traffic Research Centre.

Fonda, S.J., Wallace, R.B. & Herzog, A.R. (2001). Changes in driving patterns and worsening depressive symptoms among older adults', *Journals of Gerontology Series B-Psychological Sciences & Social Sciences, 56B,* 343–351.

Gelau C. & Stevens A. (2006). *Impact of IVIS on driver workload and distraction : review of assessment methods and recent findings,* HUMANIST report, Deliverable D.2/E.2, retrieved from www.noehumanist.org.

Griffiths P.G. & Gillespie R.B. (2005). Sharing control between humans and automation using haptic interface: primary and secondary task performance benefits, *Human Factors, 47,* 574-590.

Harms, L. & Patten, C. (2001). Measuring distraction from navigation instructions in professional drivers—a field study with the peripheral detection task. In *Proceedings from 9th International Conference of Vision in Vehicles,* Brisbane.

Hart, S.G., & Staveland, L.E. (1988). Development of a multi-dimensional workload rating scale NASA-TLX : Results of empirical and theoretical research. In P.A. Hancock and N. Meshkati (Eds.), *Human mental workload* (pp. 239-250). Amsterdam: North Holland Press.

Hoedemaeker, M. &Brookhuis, K.A. (1999). Driving with an adaptive cruise control (ACC). *Transportation Research, Part F, 3,* 95-106.

Kramer A.F., Cassavaugh N., Horrey W.J., Becic E., & Mayhugh J.L. (2007). Influence of age and proximity warning devices on collision avoidance in simulated driving. *Human Factors, 49,* 935-949.

Marottoli, R.A., Mendes de Leon, C.F., Glass, T.A., Williams, C.S., Cooney, L.M., & Berkman, L.F. (2000). Consequences of driving cessation: decreased out-of-home activity levels, *Journal of Gerontology, 55,* S334–S340.

Matthews R., Legg S., & Charlton S. (2003). Distraction, driving, multiple-task, workload, performance. *Accident Analysis and Prevention, 35,* 451-457.

Michon, J. (1985). A critical view of driver behavior models: what do we know, what should we do? In L. Evans and R.C. Schwing (Eds), *Human Behavior and Traffic Safety* (pp. 485- 524). New York: Plenum

Nicolle, C. & Burnett, G. (1994). *Handbook of design guidelines for usability of systems by elderly and disabled travelers,* DRIVE/TELSCAN report. Retrieved 03.02.2008 from http://hermes.civil.auth.gr/telscan/telsc.html

Nilsson L. (2005). Automated driving does not work without the involvement of the driver. In G. Underwood (Ed.) *Traffic and Transport Psychology - Theory and Application* (pp. 293-301). Amsterdam: Elsevier.

Patten, C.J.D, Kircher, A., Östlund, J., & Nilsson, L (2004). Using mobile telephones: cognitive workload and attention resource allocation, *Accident Analysis and Prevention, 36,* 341-350.

Pauzié, A., & Marin-Lamellet, C. (1989). Analysis of aging drivers behaviors navigating with in-vehicle visual display systems, In *Proceedings from Vehicle Navigation & Information Systems Conference* (pp. 61-67), Toronto.

Pauzié, A. & Pachiaudi, G. (1997). Subjective evaluation of the mental workload in the driving context, in T. Rothengatter and E. Carbonell Vaya (Eds.), *Traffic & Transport Psychology: Theory and Application* (pp. 173-182). Oford, UK: Pergamon.

Pauzié, A. (2002). In-vehicle communication systems: the safety aspect, *Injury Prevention Journal, 8*, 0–3.

Pauzié, A. (2004). French survey about the European statement of principles for HMI of in-vehicle systems, *ITS World Congress*, Nagoya, Japan.

Pauzié, A. (2005). Methodologies for driver-centred evaluation of perceptive and cognitive requirements of in-vehicle display, *Human Computer Interaction Congress*, Lawrence Erlbaum Associates (ed.), July 22-27 2005, Las Vegas, Nevada, USA.

Pauzié, A., Stevens, A., & Gelau, C. (2006). *The expansion of ESoP in the framework of the eSafety initiative*, HUMANIST report, Deliverable 4.3, retrieved from www.noehumanist.org.

Pauzié, A. (2007). Evaluation of driver mental workload facing new in-vehicle information and communication technology, in Proceeding from *the 20th Enhanced Safety in Vehicle Conference*, France

Piechulla, W., Mayser, C., Gehrke, H., & Konig, W. (2003). Reducing drivers' mental workload by means of an adaptive man-machine interface, *Transportation Research Part F, 6*, 233-248.

Ross, T., Midtland, K., Fuchs, M., Pauzié, A., Engerts, A., Duncan, B., Vaughan, G., Vernet, M., Peters, H., Burnett, G., & May A. (1996). *HARDIE Design Guidelines Handbook, Human Factors Guidelines for Information Presentation by ATT Systems*, DRIVE/HARDIE Report.

Simões, A., Baligand, B., Bellet, Th., Boudy, J., Bruyas, M-P, Tattegrain-Veste, H., Deleurence, Ph., Guilhon, V., Pachiaudi, G., Carvalhais, J., Forzy, J.F., Lockwood, Ph., Damiani, S., & Opitz, M. (2001). Evaluation of a Vocal Interface Management System, *Transport Research Record, 1759*, 46-51.

Vrkljan, B.H., & Miller-Polgar, J. (2005). Advancements in vehicular technology: potential implications for the older driver, *International Journal of Vehicle Information and Communication Systems, 1*, 88-105.

White Paper. (2001). "European transport policy for 2010: time to decide", COM(2001) 370.

Getting back to basics: using road accident investigation to identify the desirable functionality of longitudinal control systems

Nikolaos Gkikas, Julian R. Hill, & John H. Richardson
Ergonomics & Safety Research Institute
Loughborough, UK

Abstract

ABS (antilock brake system), EBA (emergency brake assist), ACC (adaptive cruise control) and alternative examples of intelligent vehicle control systems aspire to support the driver in controlling the vehicle and alleviate the incidents that would lead to collisions and injuries. This paper considers some requirements for such systems based on a study of accidents occurring in the real-world. While systems are rationally developed in the engineering laboratory, on the test track and through use of simulations, the need for a thorough understanding of the design needs as observed in the real-world of current day accidents is increasingly recognized. This paper overviews the range of data available on the causes of accidents in the UK. A fresh look is taken at some issues relating to braking by specific reference to data from the On-The-Spot (OTS) accident research study in an attempt to consider the necessary functionality of active safety systems pertinent to longitudinal control failures. The road user interactions file from 3024 road accidents in Thames Valley and South Nottinghamshire regions of the UK, as covered by OTS study, were analysed. Significant contributory factors where "failure to stop the vehicle" was identified as the accident precipitating factor were seen to be "following too close", "disobeyed automatic traffic signal", "careless/reckless/in a hurry", "failure to look" and "failure to judge other person's path or speed". On the other hand, where "sudden braking" is identified as the accident precipitating factor, contributory factors included "sudden braking" (as a contributor), distraction, aggressive driving, failure to judge other person's path, "masked road markings", "excessive speed", "following too close", and "road layout". Current systems address some of these issues, while possibly overlooking others; recommendations for future safety engineering designs are made accordingly.

Introduction

The driving task has been described from many different perspectives. Early theories suggested a versatile safe field in which the driver aims to navigate the vehicle (Gibson & Crooks, 1938). More recent work considers time related descriptions (Senders, Kristofferson, Levison, Dietrich, & Ward, 1967; Van Winsum & Brouwer, 1997; Van Winsum, 1998), motivation (Fuller, 1984; Naatanen & Summala, 1976;

In D. de Waard, F.O. Flemisch, B. Lorenz, H. Oberheid, and K.A. Brookhuis (Eds.) (2008), *Human Factors for assistance and automation* (pp. 203 - 216). Maastricht, the Netherlands: Shaker Publishing.

Näätänen & Summala, 1974; Wilde & Murdoch, 1982) and control-based models (Lee & Strayer, 2004; Mcruer, Allen, Weir, & Klein, 1977; Sheridan, 2004; Summala, 1996; Weir & McRuer, 1970). All such models, however, suggest or assume a basic driving task involving longitudinal and lateral control of the vehicle. The initiation of this task for a typical driver-vehicle interface (figure 1) is the operation of the steering wheel, the throttle and the brake pedals, plus the clutch and gear stick (in the case of manual transmission).

Emergency Brake Assist (EBA), Adaptive Cruise Control (ACC), Electronic Stability Control (ESC), Lane Change Support (LCS) and alternative intelligent vehicle control systems claim to support drivers in controlling the vehicle and alleviate the incidents that would lead to collisions and injury (BOSCH, 2000; Society for Automotive Engineers, 1999; *Delphi active safety products for automotive manufacturers.*; *Honda safety - active safety*). Manufacturers' claims have been supported by statistical evidence (Breuer, Faulhaber, Frank, & Gleissner, 2007; Coelingh, Jakobsson, Lind, & Lindman, 2007; Lie, Tingvall, Krafft, & Kullgren, 2006; Page, Foret-Bruno, & Cuny, 2005; Thomas, 2006). However, those studies had to simplify many aspects of an accident event and make assumptions about some factors in order to proceed with data analysis.

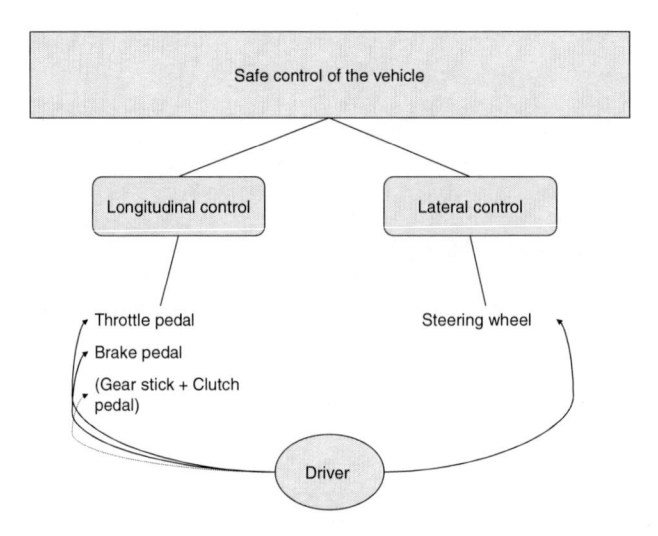

Figure 1. The instruments of basic vehicle control

Within this maze of different systems and component configurations, safety engineers and researchers can focus overly on the means – i.e. the safety system – rather than the problem itself – the accident. Therefore, it is necessary to examine detailed characteristics from accident investigations cases and include findings in the safety design process. This paper makes use of data from the On-the-Spot (OTS) accident investigation study and presents an examination of road user interactions in accidents where one of two specific circumstances had been defined as the initiating (or precipitating) factor in causing an accident, namely a) failure to stop a vehicle and b) sudden braking.

Background - recording the causes of accidents in the British National data

Data on road accidents have been collected since at least 1909 (Hillard, Logan, & Fildes, 2005). However, it was not until 1949 that a nationwide system for accident data collection was introduced, namely STATS19. The original system collected both objective factors (speed limit, time, weather etc.) as well as contributory factors, i.e. the factors which the reporting officer on the accident scene believed had contributed to accident occurrence. The system has been reviewed and improved every five years since its introduction. After some debate about the reliability of the subjective nature of contributory factors, such data ceased to be a national requirement following a review in 1959. However, in 1994 half the country's police forces still used some kind of contributory factors coding in accident data collection (Broughton, 1997).

From then onwards, a continuous process of developing a constant accident data collection system throughout the country commenced with a series of studies and subsequent reports supporting that (Maycock, 1995; Broughton, 1997). Through developing the practical system used by Devon and Cornwall police forces at the time (figure 2) and amalgamating the theoretical model developed in Leeds University during the late 80's (Carsten, Tight, Southwell, & Plows, 1989), with the aggregated experience and practical needs indicated by the police forces, a two-level hierarchy with the following terminology was developed:

Precipitating factors are the failures and manoeuvres that immediately led to the accident.Contributory factors are the causes for these failures and manoeuvres. A recorded contributory factor always relates to a precipitating factor that has already been recorded.

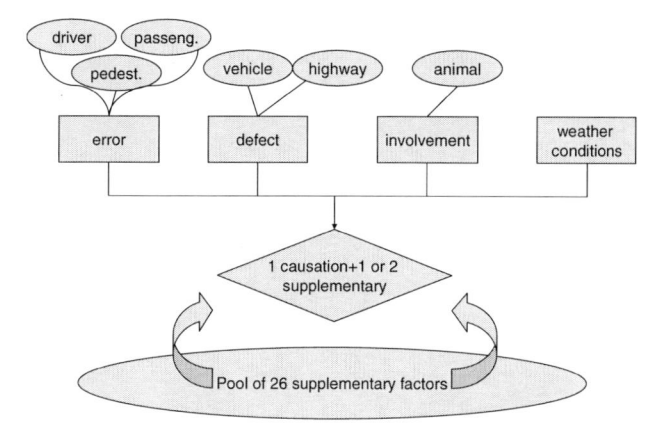

Figure 2. classification system used by Devon & Cornwall police forces

Later consecutive reviews in the year 2000 (Neilson & Condon, 2000) and 2002 (Wilding, 2002) suggested itemised amendments and especially the latter acknowledged the internal "blame machine" of the system, as it tended to lay blame on an individual and was totally inappropriate for accidents were there was contribution from multiple road users. The issues identified in the review in conjunction with the previous paper by Neilson and Condon lead the Department for Transport to commission the Transportation Research Group in Southampton University to go one step further and make suggestions to the Standing Committee on Road Accident Statistics (SCRAS) for the improvement of the contributory system. The subsequent report (Hickford & Hall, 2004), among other recommendations, suggested a revised form for collecting contributory data. However, for ease of use, after consultation with the local authorities and the police, a different layout was adopted by SCRAS. The outcome of that work was the STATS19 contributory factors form now in use, including seventy-six contributory factors and also an option to report "other factor" by text description. The factors are grouped in five main categories: road environment contributed (nine factors); vehicle defects (six factors); driver/rider only (forty-seven factors); pedestrian only (ten factors); and four factors for special codes (stolen vehicle, vehicle in course of crime, emergency vehicle on call, vehicle door open/closed negligently). The driver/rider category is further subdivided into five subcategories: injudicious action, error or reaction, impairment or distraction, behaviour or inexperience and vision affected (by). The reporting officer can select up to six factors from the grid, relevant to the accident. Previously suggested three and four-point scales of confidence are now substituted by a simple two-point scale: the officer indicates for each factor whether he/she considers it "very likely" or just "possible". The system allows for more than one factor to be related to the same road user and for the same factor to be related to more than one road user, if appropriate. This allows the police officer sufficient flexibility to include the necessary details and in a concise manner.

In-depth OTS causation studies

The current UK On-The-Spot accident research study commenced in 2000. Unlike the more traditional retrospective research studies, where accident data are collected several days after an accident occurred, the OTS study offers the ability to collect invaluable data which would otherwise be lost such as vehicle rest position, debris locations, weather conditions, road surface conditions, tyre pressures, temporary changes in the road environment at the time of impact, immediate driver and witness descriptions. Expert research teams attend the scene of road accidents, typically within 20 minutes of the incident occurring to make an in-depth investigation that includes the highway, vehicles and human factors present. In addition to this, it includes data which are collected retrospectively in days or months after the accident (road signs, impact damage on vehicles, road dimensions, injury details, etc.).

The procedure starts with the arrival of the investigation team at the scene of an accident. The serving police officer on the OTS team makes contact with the police officer in charge of the accident scene and briefs him/her about the intended activities of the investigators. After fulfilment of protocols and safety issues, the team makes contact with the people and the various elements involved in the crash. Data are coded in a library of some 200 forms with over 3000 individual variables.

OTS investigators analyse the causes of accidents in detail and record their findings using a suite of causation coding systems. National contributory factors forms (both the current (Hickford & Hall, 2004) and previous (Broughton, 1997) forms, as described above, are routinely coded for all OTS cases according to the same protocols followed by police officers. Thus accident causation is coded in two levels: a precipitating factor and up to six contributory ones.

OTS cases are further analysed to determine more complex descriptions of accident causation in terms of possible interactions between the active road users. A system called "interactions" has been developed to allow analyses and recording of one or more interactions between each road user and his environment to provide a description of events leading to impact at any degree of necessary complexity. All information is held on an anonymous accident database and does not include personal identifying details or other such documents.

Methodology

Accident cases were studied from Phase 2 of the OTS project covering the period from September 2003 to October 2006 and include detailed, disaggregated data from 3024 accidens in the Thames Valley and South Nottinghamshire regions. This study selected accidents where "failure to stop" or "sudden braking" had been coded as the factor initiating the accident sequence. While other precipitating factors are also relevant for the study of longitudinal control failures, those two factors were considered to be of prime interest within the scope of the current study.

It should be noted that "failure to stop" here defines a very specific set of accidents where that is the single, precipitating factor causing the accident. Clearly all accidents are in some way the result of a failure to stop before the collision occurs, but the sub-set under study here represent drivers who were considered to be the predominate, precipitating cause of their accident by failing to stop their vehicle in time. Each "failure to stop" will have been assigned as the precipitating factor following an accident investigation to eliminate other possible precipitating factors, such as for example, the driver travelling too fast, or a pedestrian stepping into the road. This is therefore a set of drivers who were not able to stop for a variety of (personal psychological or other) reasons. There will of course be other drivers who did not stop before collision (all the other drivers in the database). This study, focuses, however, on the unique group for which "failed to stop" was the precipitating factor (together with the additional "sudden braking" group, as explained above). This study does not therefore attempt to consider all possible reasons for drivers failing to stop in time to avoid their accident.

Case selection resulted in 301 cases involving "failure to stop" and 39 cases involving "sudden braking". The study went on to analyse precipitative and causal factors in the context of driver behaviour and longitudinal control of the vehicle. While not explicitly detailed in the results presented, case analyses also focused on the more detailed OTS road-user Interactions coding system, as has been described above. The Interactions file included 1099 interactions in "failure to stop" accidents and 152 interactions in "sudden braking" accidents. The database has been compared against the national data for Great Britain (STATS19) and validated as broadly representative of accidents occurring over Great Britain (Hill, Thomas, Smith, & Byard, 2006).

Results

"Failure to stop"

Failure of a driver or vehicle to stop in time to avoid a collision with another road user or object is identified as the precipitating factor in 301 cases investigated by the OTS team. However these cases were the result of interactions of more than one road users at a time. Browsing through the cases one by one, it is very rare – and naïve – to attribute accidents to a single factor. This is in accordance with experience of accident investigation in high-hazard industry, aerospace and space applications (Kirwan, 1994; Reason, 1990; United States. Columbia Accident Investigation Board., 2003; Whittingham, 2004). Therefore it is necessary to look further into the factors that contributed to the precipitating factor.

Collision types resulting from "failure to stop" are shown in table 1. One might expect junction overshoots and rear-end collisions to predominate, however the OTS cases show a wider variety of collisions. Common collision types associated with such accidents include crossing, merging, turning, and others.

Table 1. Collision type as a result of failure to stop compared to general accident data

	General (all accidents)	Failure to stop
rear-end	20.6	65.3
cornering	12.4	0
lost control or off-road (straight roads)	10.4	0.7
overtaking and lane change	10.4	0
collision with obstruction	3.1	1.4
head on	5.5	1.7
turning versus same direction	3.2	0.9
crossing (no turns)	6.1	15.9
crossing (vehicle turning)	8.5	3.1
merging	3.4	3.3
right turn against	5.5	3.5
manoeuvring	2.6	0
pedestrians crossing road	6.3	1.9
pedestrians other	0.5	0
miscellaneous	0.6	0

Table 1 also makes a comparison with the overall collision-type frequency distribution from the OTS database and that comparison underlines differences in the result of failure to stop in particular collisions. Apart from the widely acknowledged predominance of rear-end collisions (+44.7%= 3 times as common as in general accident data), crossing without turning is particularly common (+9.8%, about 3 times more common), while cornering, overtaking, manoeuvring do not appear at all, and pedestrian crossings are less common (-4.4%, about 3 times less common).

In terms of contributory factors (Table 2) drivers' "too close" car-following strategy is identified as the most common contributory factor, followed by non-adherence to automatic traffic signals and speeding. Cognitive failures – to look and to judge other paths – and inappropriate reactions – sudden braking – are also commonly found in such accidents. Comparing that with the general OTS distribution (table 2), "too close" car-following behaviour is more than three times more frequent as a contributor (+16.81%), non-adherence to automatic traffic signals is more than four times more common in failures to stop (14.06%), while too-fast driving is two and a half times more frequent (+7.33) and psychological parameters (reckless/in hurry) two times more common (+7.30%). On the contrary, non-adherence to give-way signals and more than five times less common factor (-9.03) and failure to judge other paths is somewhat less frequent (-2.71%).

Table 2. Common contributors in failure to stop compared to general accident data

	failure to stop	general (all accident cases)	difference
following too close	22.93	6.12	16.81
disobeyed automatic traffic signal	16.65	2.59	14.06
careless, reckless or in a hurry	13.74	6.44	7.30
travelling too fast for conditions	12.01	4.68	7.33
failed to look properly	6.19	5.15	1.04
exceeded speed limit	4.19	5.10	-0.91
failed to judge other person's path or speed	4.19	6.90	-2.71
sudden braking	2.37	1.01	1.35
special codes: stolen vehicle	2.37	0.73	1.63
slippery road (due to weather)	1.55	2.06	-0.51
disobeyed give-way or stop-sign or markings	1.36	10.40	-9.03
impaired by alcohol	1.36	2.08	-0.71

To make the picture clearer it is necessary to check the type of road users involved in such accidents (table 3). About 80% of road users are car occupants, 3.5% are Light Goods Vehicle (LGV) occupants and 3.1% are Heavy Goods Vehicle occupants. Motorcyclists and bus occupants each constitute about 1% of the road users involved in such accidents. Vulnerable road-users comprise 3% in total, 1.3% are pedestrians and 1.7% are pedal cyclists.

Table 3. The distribution of road user involvement in "failure to stop" accident

- Road users involved in "failure to stop" accidents
  - Car occupants 80.3%
  - LGV occupants 3.5%
  - HGV occupants 3.1%
  - Cyclist 1.7%
  - Pedestrian 1.3%
  - Motorcyclist 0.9%
  - Bus occupant 0.9%

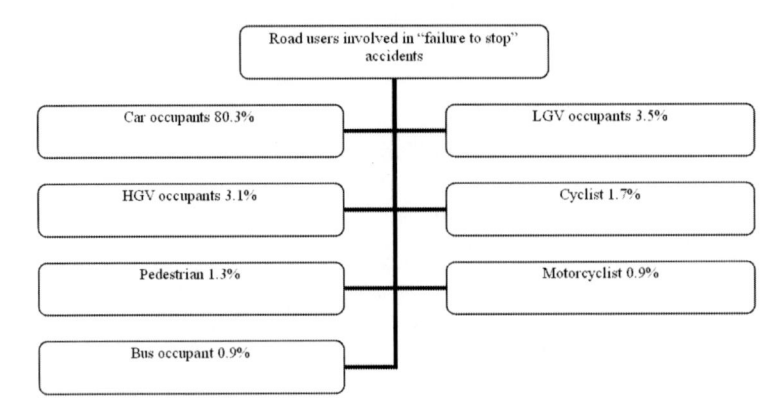

"Sudden braking"

"Inappropriate reaction-sudden braking" is identified as the precipitating factor in 39 cases investigated by the OTS team. The interactions files of those cases include 152 road-user interactions. 73% of those involved are car occupants, while 9.2% are Light Goods Vehicle (LGV) and 6.6% are Heavy Goods Vehicle (HGV) occupants. Bus occupants and cyclists each consist 2.6% of total road users and motorcyclists are 5.9%.

Table 4. The distribution of road users involved in "sudden braking accidents"

- Road users involved in "sudden braking" accidents
  - Car occupants 73%
  - LGV occupants 9.2%
  - HGV occupants 6.6%
  - Bus occupant 2.6%
  - Cyclist 2.6%
  - Motorcyclist 5.9%

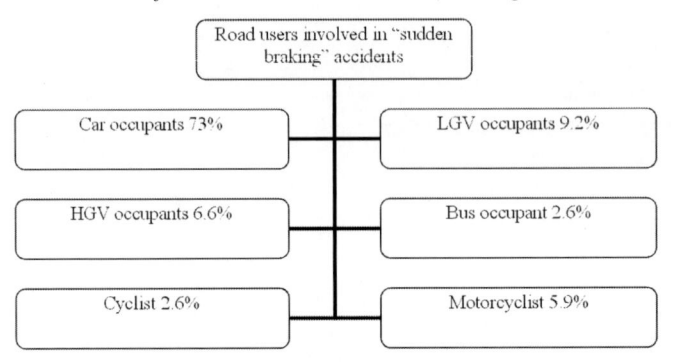

Compared to "failure to stop" cases, there is more frequent involvement of LGVs (about 2.5 times) and HGVs (more than 2 times) and motorcyclists (more than 6 times more common). On the other hand, there have been, as might be expected, no pedestrians involved in this type of accident (compared to 1.3% in failures to stop), and differences below 1% exist in bus occupant and cyclist involvement.

Comparison of collision types in "sudden braking" cases with the general and the "failure to stop" cases reveals some interesting differences (table 5). While the predominance of rear-end collisions is there, as expected, collisions commonly associated with lateral control such as overtaking, cornering and loss of control collisions are initiated by a sudden-braking reaction. Furthermore, miscellaneous collisions (with a trailer mostly) are common results of sudden-braking, unlike other precipitating factors. On the other hand, collisions while crossing and collisions with pedestrians are not found at all in "sudden braking" accidents, unlike "failure to stop" accidents and the database generally.

Table 5. Collision type frequencies in "sudden braking", "failure to stop" and general

	sudden braking	general	failure to stop
rear end	62.5	20.6	65.3
overtaking and lane change	7.9	10.4	0
cornering	7.9	12.4	0
lost control or off road	7.2	10.4	0.7
miscellaneous	5.9	0.6	0
turning versus same direction	3.9	3.2	0.9
head on	2.6	5.5	1.7
collision with obstruction	2	3.1	1.4
crossing(no turns)	0	6.7	15.9
crossing (vehicle turning)	0	8.5	3.1
merging	0	3.3	3.7
right turn against	0	5.5	3.5
manoeuvring	0	2.6	0
pedestrians crossing road	0	6.3	1.9
pedestrians other	0	0.5	0

Examination of the contributory factors in accidents initiated by a sudden-braking reaction indicated a "wave effect" of sudden braking reaction in response to one or more other drivers also braking suddenly ahead to be the most common factor (table 6). Similarly with "failure to stop", close car-following behaviour is a major contributor in this type of accident. Failures of judgement and masked road markings and signs are among the most common contributors as well as distraction, however failure to look properly, junction overshooting and cyclists' intrusions are not common as in "failure to stop" cases.

Discussion

This paper has made a brief overview of the rich and varied range of data describing the causes of road accidents in the UK, from the macroscopic data gathered by the police to microscopic data gathered by the OTS teams. The authors have presented a modest selection of the OTS data of relevance to the study of longitudinal control systems, and with the intention of highlighting the value and importance of real-world data both for a comprehensive understanding of how drivers use these systems and for the most effective safety design.

Table 6. Comparison of contributors in "sudden braking" and general accident data

	sudden braking	general (all cases)	difference
sudden braking	35.53	1	34.53
following too close	24.34	6.1	18.24
failed to judge other person's path or speed	7.89	6.9	0.99
inadequate or masked signs or road markings	4.61	0.4	4.21
careless, reckless or in a hurry	4.61	6.44	-1.84
exceeded speed limit	3.29	5.10	-1.81
road layout (e.g. bend, hill, narrow carriageway)	2.63	6.1	-3.47
travelling too fast for conditions	2.63	4.68	-2.05
distraction outside vehicle	2.63	0.9	1.73
aggressive driving	2.63	0.3	2.33
slippery road (due to weather)	1.97	2.1	-0.13
animal or object in carriageway	1.97	0.5	1.47
junction overshoot	1.97	0.5	1.47
vision affected by road layout (e.g. bend, winding road, hill crest)	1.97	0.4	1.57
cyclist entering road from footway	0.66	0.3	0.36
failed to look properly	0.66	5.2	-4.54

The data presented here are not intended to represent the full range and variety of circumstances where braking plays a role in causing accidents. Rather the intention was to focus on two braking situations as specifically defined in the methodology, involving some (but not all) accidents where the driver fails to stop in time, and others where there was sudden braking. Further work is planned to more fully describe the role of braking in accidents, however the current study raises some notable points for discussion at this stage.

The design of safety systems such as ACC and EBA must focus on the "genetic make up" of the accident scenarios (Hollnagel, 2004), the detailed failure types (Wagenaar & Reason, 1990) and contributory factors if these systems are to best function under the real world conditions involving driver and vehicle interactions. The predominance of rear-end collisions and the contribution of close car-following and speeding imply that ACC, EBA and Intelligent Speed Adaptation (ISA) are all steps towards the right direction. However, there are many other characteristics of the examined accident types that should be considered carefully.

In the case of "failure to stop", a collision when crossing junctions is the second most common collision type recorded. In parallel, the contribution of cognitive failures (to look and to judge) and inappropriate reactions might ideally call for a cognitive/decision making assistant; a system that can track other road vehicles and users, predict their trajectories and provide the driver with a more sophisticated system of prompts to keep him on the "safe path" (as suggested in the very early description of the driving task by Gibson & Crooks in 1938). Certainly, such a system would come with a series of ergonomic issues, including the possibility of unwanted behavioural adaptations to the new system, but possibly no more issues than the problems inherent in any new automated process.

Key contributors to the cause of accidents which – to the knowledge of the authors – has barely been addressed by system developers are the psychological parameters. Recklessness, carelessness etc. are left to the transport authorities to deal with and are sometimes not currently possible to address at all. Future in-vehicle technologies may well offer more dynamic solutions in combination with education and enforcement. Where we currently rely, to a large extent, on road policing to identify drivers in need of behavioural modification, future technologies might help address a wider range of misdemeanours exhibited by the wider driving population and provide instantaneous support to the driving task. Given that all drivers can be and are careless, to a greater or lesser extent, measures able to react to the temporal fluctuation of these parameters therefore offer new and interesting possibilities for accident prevention. Possibly a merge of psychological experience with engineering could offer such solution.

Examination of "sudden braking" cases underlines the strong systemic nature of these accidents. An abrupt reaction of one road user can lead to the abrupt reaction of another user which then initiates an accident (see sudden-braking as contributor). This phenomenon points to the need to account for this systemic nature of accidents when designing safety systems, ensuring multi-vehicle incidents are also considered. Accordingly, current systems should first tune their intervention according to user responses (mostly applicable to EBA) and second, tune their intervention according to other road-user responses (mostly applicable to ACC). Conceptually, ACC systems are highly relevant here as they are intended to assist drivers in close following situations, however system limitations are recognised in consideration of the effectiveness of humans when monitoring the driving task rather than actively operating the vehicle. Such issues will need to be better understood before ACC can be fully developed and deployed as system primarily intended for collision mitigation..

Failures of judgement and inability to see masked road markings are another factor that a visual aid or active decision-aid system as the one described previously could address, while contribution of distraction could be mitigated by a more controlled external environment and an integrated-device manager inside the vehicle (Amditis, Kussmann, Polychronopoulos, Engstrom, & Andreone, 2006).

The low incidence of accidents involving pedestrians was noted, however, the authors wish to examine a more extensive range of accident scenarios within the OTS database before drawing conclusions regarding these road users. Over 5% of users involved in "sudden braking" cases are motorcyclists. These are some of the few single-vehicle cases, where the braking reaction is followed by loss of control with fatal consequences. Although the number of cases was low, the fatality rate suggests the importance of developing intelligent braking systems in two-wheeled motor vehicles.

Acknowledgements

The OTS project is funded by the UK Department for Transport and the Highways Agency. The project would not be possible without help and ongoing support from many individuals, especially including the Chief Constables of Nottinghamshire and Thames Valley Police Forces and their officers. In addition the authors of this paper would like to thank the staff at Loughborough University and TRL Ltd who have helped to establish and carry out this project. The views expressed in this paper belong to the authors and are not necessarily those of the Department for Transport, Highways Agency, Nottinghamshire Police or Thames Valley Police.

References

Amditis, A., Kussmann, H., Polychronopoulos, A., Engstrom, J., & Andreone, L. (2006). System architecture for integrated adaptive HMI solutions. *2006 IEEE Intelligent Vehicles Symposium* (pp. 13-18). Meguro-ku, Japan..

BOSCH (2000). In H. Bauer (Ed.), *BOSCH automotive handbook*. Stuttgart: Robert Bosch GmbH.

Breuer, J. J., Faulhaber, A., Frank, P., & Gleissner, S. (2007). Real world benefits of brake assistance systems. *Proceedings of the 20th Enhanced Safety of Vehicles Conference,* Lyon. (Paper Number 07-0103)

Broughton, J. (1997). A new system for recording contributory factors in road accidents. Proceedings of the Conference Traffic Safety on Two Continents. VTI Konferens 9A, Part 5: Data Collection and Information Systems (pp. 53-70). Lisbon, Portugal

Carsten, O.M.J., Tight, M.R., Southwell, M.T., & Plows, B. (1989). *Urban accidents: Why do they happen*. Basingstoke, Hants: AA foundation for road safety research.

Coelingh, E., Jakobsson, L., Lind, H., & Lindman, M. (2007). Collision warning with auto brake - a real-life safety perspective. *Proceedings of the 20th Enhanced Safety of Vehicles Conference,* Lyon. (Paper Number 07-0450)

Delphi active safety products for automotive manufacturers. Retrieved 12/06/2007, 2007, from http://delphi.com/manufacturers/auto/safety/active/

Fuller, R. (1984). A conceptualization of driving behaviour as threat avoidance. *Ergonomics, 27,* 1139-1155.

Gibson, J.J., & Crooks, L.E. (1938). A theoretical field-analysis of automobile-driving. *American Journal of Psychology, 51,* 453-471.

Girard, Y. (1993). In-depth investigations of accidents: The experience of INRETS at Salon-de Provence. *Safety Evaluation of Traffic Systems, Traffic Conflicts and Other Measures; Proceedings of the International on Theories and Concepts on Traffic Safety (ICTCT) Workshop,* Salzburg.

Hickford, A.J., & Hall, R.D. (2004). *Road safety research report no 43. review of the contributory factors system*. London: Department for Transport [DfT].

Hill, J.R., Thomas, P.D., Smith, M., & Byard, N. (2006). The methodology of on the spot accident investigations in the UK. *Enhanced Safety in Vehicles 2001*, Amsterdam.

Hillard, P.J., Logan, D., & Fildes, B. (2005). The application of systems engineering techniques to the modelling of crash causation. In L. Dorn (Ed.), *Driver behaviour and training* (pp. 407-415). Aldershot: Ashgate,.

Hollnagel, E. (2004). *Barriers and accident prevention*. Aldershot: Ashgate.

Honda safety - active safety. Retrieved 12/06/2007, 2007, from http://corporate.honda.com/safety/details.aspx?id=active_safety

Kirwan, B. (1994). *A guide to practical human reliability assessment*. London: Taylor & Francis.

Lee, J.D., & Strayer, D.L. (2004). Preface to a special section on driver distraction. *Human Factors, 46*, 583-586.

Lie, A., Tingvall, C., Krafft, M., & Kullgren, A. (2006). The effectiveness of electronic stability control (ESC) in reducing real life crashes and injuries. *Traffic Injury Prevention, 7*, 38-43.

Mackay, G.M. (1969). Some features of traffic accidents. *British medical journal, 4* (5686), 799-801.

Maycock, G. (1995). *Contributory factors in accidents; police databases* (Project Report No. PR/SE/083/95). Crowthorne: Transport Research Laboratory.

Mcruer, D.T., Allen, R.W., Weir, D.H., & Klein, R.H. (1977). New results in driver steering control models. *Human factors, 19*, 381-397.

Naatanen, R., & Summala, H. (1976). *Road-user behavior and traffic accidents*. Amsterdam: North-Holland Publishing

Näätänen, R., & Summala, H. (1974/12). A model for the role of motivational factors in drivers' decision-making*. *Accident Analysis & Prevention, 6*, 243-261.

Neilson, I., & Condon, R. (2000). *Desirable improvements in road accident and related data* (Research Paper No. 1). London: Parliamentary Advisory Council for Transport Safety.

Otte, D. (1999). *Description of in-depth investigation team ARU/Medical university Hannover* (No. 1). Hannover: Medical University Hannover.

Page, Y., Foret-Bruno, J. & Cuny, S. *Are expected and observed effectiveness of emergency brake assist in preventing road injury accidents consistent?* Retrieved 02/05/2007, 2007, from http://www-nrd.nhtsa.dot.gov/pdf/nrd-01/esv/esv19/05-0268-O.pdf

Reason, J. (1990). *Human Error*. New York: Cambridge University Press.

Sabey, B.E., & Staughton, G.C. (1975). Interacting roles of road environment, vehicle and road user in accidents. *Fifth International Conference of the International Association for Accident and Traffic Medicine*, London, UK.

Senders, J.W., Kristofferson, A.B., Levison, W.H., Dietrich, C.W., & Ward, J.L. (1967). The attentional demand of automobile driving. *Highway Research Record, 195*, 15-33.

Sheridan, T.B. (2004). Driver distraction from a control theory perspective. *Human Factors, 46*, 587-599.

Society for Automotive Engineers. (1999). ESP electronic stability programme. *Driving safety systems* (pp. 206-242). Germany: Robert Bosch GmbH.

Summala, H. (1996). Accident risk and driver behaviour. *Safety Science, 22*, 103-117.

Thomas, P. (2006). Crash involvement risks of cars with electronic stability control systems in Great Britain. *International Journal of Vehicle Safety, 1*(4), 267-281.

United States. Columbia Accident Investigation Board. (2003). In Godwin R. (Ed.), *Columbia accident investigation board: Report.* Burlington, Ont.: Apogee Books.

Van Winsum, W. (1998). Preferred time headway in car-following and individual differences in perceptual-motor skills. *Perceptual and Motor Skills, 87*, 863-873.

Van Winsum, W., & Brouwer, W. (1997). Time headway in car following and operational performance during unexpected braking. *Perceptual and Motor Skills, 84*, 1247-1257.

Wagenaar, W.A., & Reason, J.T. (1990). Types and tokens in road accident causation. *Ergonomics, 33*, 1365-1375.

Weir, D.H., & McRuer, D.T. (1970). Dynamics of driver vehicle steering control. *Automatica, 6*, 87-98.

Whittingham, R.B. (2004). *The blame machine: Why human error causes accidents.* Oxford: Elsevier Butterworth-Heinemann.

Wilde, G.J.S., & Murdoch, P.A. (1982). Incentive systems for accident-free and violation-free driving in the general population. *Ergonomics, 25*, 879-890.

Wilding, P. (2002). *The 2002 quality review of road accident injury statistics interim report: Extension of timetable.* London: Department for Transport.

Simulator training for truck-driving: indications of learning effects

Annette Kluge & Dina Burkolter
University of St. Gallen,
Switzerland

Abstract

With new European Union regulations mandating training for passenger and freight transport, two experiments tested for conditions under which simulators efficiently develop driving skills. In Pilot Study 1, an experimental ($n = 7$) and a control group ($n = 9$) of truck drivers received initial conventional safety training together at driver's training centre. Using a simulator, both groups then drove a predefined route. The experimental group was presented with three especially dangerous situations along the route and was given constructive feedback; the control group was not. The experimental group outperformed the control group in two of the three situations. In Pilot Study 2, 15 participants were trained in fuel-efficient driving. Those in the experimental group ($n = 8$) then used a simulator to drive each of four different routes twice. Each driver received performance feedback. Each person in the control group ($n = 7$) drove the four routes twice without a simulator and without feedback. The experimental group far outperformed the control group and rated the usefulness of simulator training much higher than the control group did. The results of both studies proved independent of age, experience, attitudes, and motivation.

Introduction

Simulators are widely used in different areas in which it is essential for people to learn to cope with potentially dangerous situations. Flight simulators, for example, have been employed in military training since the end of the 1920s and in training for ground vehicles since the 1950s. Since the 1960s, simulators have had civilian uses as well (Amico & Clymer, 1984). Uhr, Felix, Williams, and Krueger (2004) outline ecological, economic, and didactic reasons for using simulators in driver's education. They point out the low physical risk to the trainee, the opportunity to repeat a situation as often as needed, reduced environmental pollution, high situational controllability, and the possibility of practicing very rare situations. Training on simulators can be effective even when the fidelity of the simulator is low. Participants for whom the simulator's low fidelity poses a problem often perform better than they do on the preceding road test. Experiencing the difficulty seems to become an advantage, much as regularly wearing a weight belt in sports workouts can enhance an athlete's performance in actual competition (the "weight-belt effect"; Dieterich, in press).

In D. de Waard, F.O. Flemisch, B. Lorenz, H. Oberheid, and K.A. Brookhuis (Eds.) (2008), *Human Factors for assistance and automation* (pp. 217 - 227). Maastricht, the Netherlands: Shaker Publishing.

One frequently discussed question is about the transfer of simulation-acquired skills to reality. Parkes and Reed (2005), for example, have shown that drivers whose fuel-efficiency performance has been improved through simulator training are able to transfer those skills to their road performance. And in a study comparing the effectiveness of training on the road and training in a simulator, Uhr et al. (2004) found that participants in both settings learned to avoid an obstacle situated on a road with low adhesion. The results of a three-trial road test indicated a positive transfer of skills from driving in a simulator to driving in reality.

Training in driving simulators has become an issue for professional drivers of motor vehicles in Member States of the European Union (EU) because of EU regulation 2003/59/EG. This new regulation standardizes both basic and continuous driver's training in an effort to assure their quality. As of 2008, for instance, all professional drivers must henceforth document that they have successfully completed basic training. As of 2009, they must also document their participation in continuous training. The training blocks are supposed to address instruction on fuel-efficient driving, safety, and behavior in emergencies. The EU regulation also stipulates that some of the training and driving examination may be conducted in a high-capacity simulator.

But what part should simulator training have in a sequence of different techniques? And how should it be incorporated? We conducted two studies to investigate the conditions under which a simulator can advantageously affect the process of learning and improving skills entailed in driving a truck.

Two pilot studies

Pilot Studies 1 and 2 were designed to compare training strategies that use the simulator for experiential learning. We therefore did not test simulator training against classroom teaching but rather compared two different simulator training procedures.

The underlying assumption was that simulator training makes most sense when used for two different topics. For our purposes, they were taken the curriculum set forth by the EU guideline: (a) emergency and safety training and (b) fuel-efficient driving. We also assumed that training in driving simulators is not in itself supportive, that it requires effective corrective performance feedback from the trainer.

To test for the possible impact of person-related variables (e.g., attitudes toward training, motivation, and perceived usefulness of simulator training), we had all the participants complete pre- and posttraining questionnaires on attitudes and characteristics considered relevant in training research (see Colquitt, LePine, & Noe, 2000, and Parkes & Reed, 2005, for example). The pretraining questionnaire measured

- attitudes toward traffic safety (Iversen & Rundmo, 2004),
- risk behavior in traffic (Iversen & Rundmo, 2004),
- perceived safety climate in the organization (Zohar & Luria, 2005),

- the personality trait "conscientiousness" (Saucier, 1994),
- self-efficacy (Schyns & Collani, 2002),
- motivation (Kluge, 2008), and
- mood (Schimmack, 1997).

The posttraining questionnaire measured

- simulator sickness (Kennedy, Lane, Bernaum, & Lilienthal, 1993),
- attitudes toward computers,
- motivation and interest during training (Kluge, 2008),
- perceived usefulness of the exercises conducted on the driving-practice area, and
- perceived usefulness of the simulator training.

The mobile driving simulator that we used (KMW Progress; see Figure 1) was provided by the Berlin company SiFaT Road Safety GmbH. This type of driving simulator belongs to a new generation of devices that have been improved to reduce the risk of simulator sickness. The simulator consisted of all controls, displays, and pedals of a truck; a visual display presenting a virtual world with its roads and road users; a station for emulating sounds; and a station from which an instructor controls the exercises and the entire simulator.

Figure 1. Mobile simulator: the driver's cab

Pilot Study 1: safety training

The aims of Pilot Study 1 were (a) to assess the added value of simulator training as compared to training received on conventional driving-practice areas and (b) to identify and assess the conditions under which that added value accrues.

Method

Sixteen truck and bus drivers as well as fire fighters (all male) participated in the pilot study. The average age of the participants was 35 years, the youngest being 20 and the oldest being 50. The participants had an average of 9 years of truck-driving experience ($SD = 8.8$) and drove an average of 11,300 km (slightly more than 7,000 miles) a year ($SD = 19,890$ km, or 12,360 miles). There were no significant differences between the groups concerning truck-driving experience ($t = -1.05$, $p > .1$) or average truck-driving per year ($t = 0.66$, $p > .1$). But groups differed in age ($t = -2.00$, $p < .1$), the control group being younger. Participants were recruited through a driving school and through contacts to a fire brigade.

The first phase of training (see Figure 2) consisted of a conventional safety instruction (brake training) conducted at a driver's training centre of the German Automobile Association (ADAC). This phase was the same for all participants. Instruction in brake training consisted of a theoretical section and a section of behind-the-wheel driving on a practice area. In that outdoor facility the participants practiced braking and evasive action on different pavements. Afterward, all participants had a warm-up exercise in the driving simulator. They had to drive through an urban area that confronted with no dangerous situations.

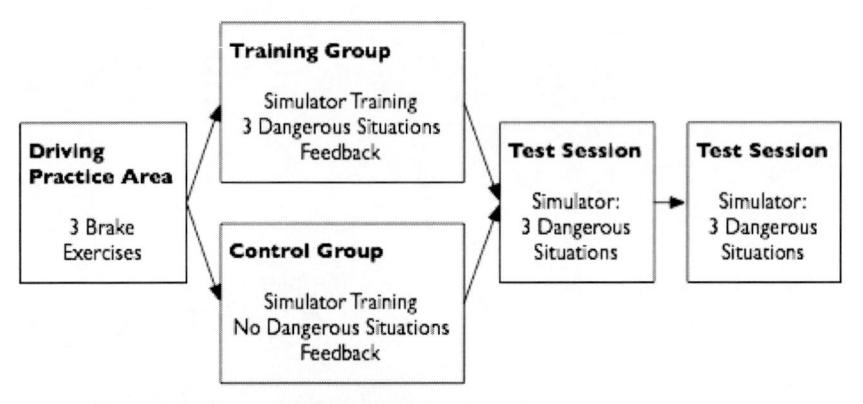

Figure 2. Design of driver's safety training (Pilot Study 1).

In the second phase of training, each participant in the experimental group ($n = 9$) used the driving simulator for approximately one hour to practice responding to three especially dangerous situations and received feedback and suggestions for improvement. The situations were (a) downhill brake failure, (b) the overtaking of a cyclist, and (c) the sudden appearance of a pedestrian emerging from behind a stationary bus. Each participant in the control group likewise spent about an hour in the simulator driving the same virtual route but not confronted with any dangerous situations.

Driving performance was tested after the training session. Using the simulator, each participant twice drove the predetermined virtual route, which presented three dangerous situations. Two performance measures were used:

1. A rating on a scale from 1 (*excellent*) to 6 (*insufficient*) as judged by two instructors who were not involved in the training and who did not know what training the examinee had received.

2. Computer data on braking distance and reaction time for braking. Unfortunately, technical problems prevented the driving simulator from collecting this performance data in Pilot Study 1.

Results

This section presents the results on the effectiveness of the two kinds of training in the driving simulator: one with three dangerous situations, the other without dangerous situations. We also report on the usefulness of the two types of training as perceived by the participants. To attribute training effects to training method, we ensured that there was no significant difference ($p > .1$) between the experimental group and the control group regarding the seven control variables measured before the training sessions (see previous section). The only exception was the mood aspect "calmness" ($t = -2.22, p < .05$).

Analysis of the performance ratings pertaining to the three dangerous situations indicated that the training for dangerous situations was beneficial (see Table 1). The experimental group outperformed the control group on most aspects of the dangerous situations, especially when it came to brake failure on a downhill slope and to the sudden appearance of a passenger emerging from behind a stationary bus.

The participants rated the training on both the driving-practice area and the simulator as useful and very useful, respectively. Their evaluation of the usefulness of the training did not differ significantly (see Table 1). Nor did their interest in the simulator training ($t = 0.05$, *ns*), their training motivation ($t = 0.88$, *ns*), their effort ($t = -0.30$, *ns*), their mood (calm: $t = -0.62$, *ns*; alertness: $t = 0.43$, *ns*), or the perceived possibility of learning transfer ($t = -0.09$, *ns*). However, both groups regarded the simulator training as less useful than the training on the driving-practice area (experimental group: $t = 2.46, p < .05$; control group: $t = 2.88, p < .05$).

The results (for all participants) showed that the demographic variables "age" and "educational achievement" were not correlated with either the perceived usefulness of the simulator training (age: $r_s = -.02$, *ns*; education: $r_s = -.14$, *ns*) or usefulness of the training on the driving-practice area (age: $r_s = -.09$, *ns*; education: $r_s = -.12$, *ns*). By contrast, the number of kilometers the participants annually drove on average was significantly correlated with the perceived usefulness of training on the driving-practice area ($r_s = .47, p < .1$) but not significantly to simulator training ($r_s = .38$, *ns*).

Thus, the greater the number of kilometers the participants drove in a year, the more useful they rated the training on the driving-practice area. Other individual attitudes and characteristics (e.g., perceived safety climate, self-efficacy, and motivation) were not related to performance or the perceived usefulness of training.

Table 1. Descriptive results of the safety training (Pilot Study 1)

	TG M	SD	CG M	SD	Sig. t	p
Dangerous situation a: Downhill[1]						
Overall performance	1.7	1.0	3.0	1.0	2.43	< .05
Speed	2.1	1.4	3.1	1.4	1.42	< .10
Use of retarder	1.8	0.6	2.7	1.4	1.48	< .10
Dangerous situation b: Cyclist[1]						
Overall performance	2.0	0.5	2.0	0.5	-0.04	ns
Overtaking manoeuvre	1.7	0.4	2.1	0.8	1.38	< .10
Speed at overtaking	2.1	1.1	1.6	0.6	-1.13	ns
Dangerous situation c: Bus[1]						
Overall performance	1.9	0.5	2.4	0.5	2.05	< .10
Reaction on pedestrian	1.7	0.4	1.6	0.4	-0.23	ns
Speed	1.8	1.1	3.2	1.1	2.50	< .05
Perceived usefulness[2]						
Perceived usefulness of training on driving practice area	4.9	0.4	4.7	0.7	-0.64	ns
Perceived usefulness of simulator training	3.6	1.3	3.0	1.6	-0.78	ns

[1]Performance rated by instructors according to the German scale of grades: 1 = excellent to 6 = insufficient.
[2]Scale from 1 = not at all useful to 5 = very useful.

The participants did not seem to feel impaired by simulator sickness. On a scale from 1 (*not at all*) to 5 (*very much*), the participants in the experimental group ($M = 1.98$, $SD = 0.83$) and those in the control group ($M = 1.98$, $SD = 0.79$) responded that they did not suffer from simulator sickness ($t = 0.13$, *ns*).

Discussion

The main result of Pilot Study 1 was that the experimental group, which received simulator training for three dangerous situations, outperformed the control group, which did not have such instruction. The experimental group excelled on the performance measures, especially in the downhill-brake-failure situation and the surprise-pedestrian situation. This superiority indicates that the experimental group had learned how to cope with sudden difficult and risky situations especially well—and more effectively than the control group had.

Given the simulator training time of only one hour per participant, the safety training in the simulator proved to be highly efficient. The experimental group benefited from both simulator training and training on the driving-practice area (which both groups completed). Moreover, this training effectivity was independent of individual attitudes and characteristics (e.g., motivation) and the perceived usefulness of training.

The participants regarded both the simulator training and the training on the driving practice area as useful, but the former as less so than the latter. Having had much more experience with training on the driving-practice area than in the simulator by that point, the participants might have had trouble seeing how the acquired skills in the simulator can be transferred to driving in reality.

Pilot Study 2: fuel-efficiency training

As in Pilot Study 1, the aim of Pilot Study 2 was not to study the overall usefulness of driving simulators as a training tool but rather to examine the effect that the instructors' behavior-based feedback had on performance. We also tested for the possible impact of person-related variables by having the participants complete the same pre- and posttraining questionnaires that we had administered to the participants in connection with Pilot Study 1 (see above).

Method

Fifteen volunteer fire-brigade members (including one female) of a fire brigade, each licensed to drive a truck, took part in Pilot Study 2. Their average age was 36 years, the youngest participant being 23 and the oldest being 54. Participants had an average of 9 years of driving experience ($SD = 9.5$) and drove an annual average of 530 km (330 miles) ($SD = 605$). The experimental group and the control group of Pilot Study 2 did not differ significantly in age ($t = -1.83, p < .1$), driving experience ($t = 0.56, p < .1$) or average driving ($t = -1.59, p < .1$).

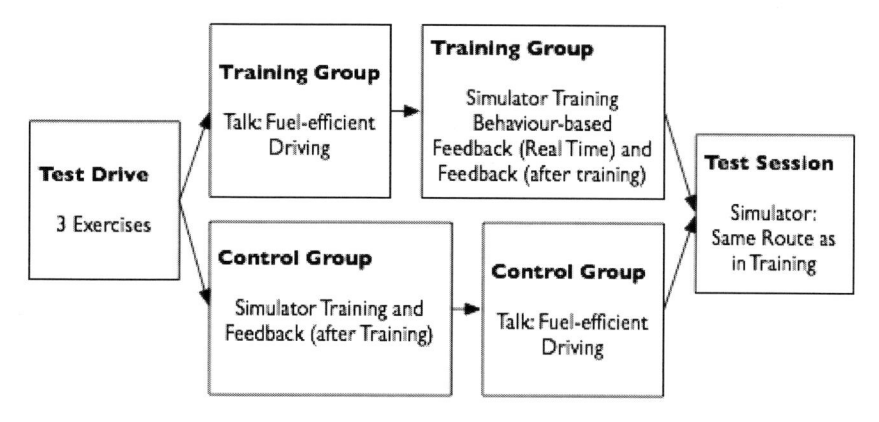

Figure 3. Design of fuel-efficiency training (Pilot Study 2)

All participants first completed an initial test drive (see Figure 3) in which a cyclist had to be passed and the drivers had to turn onto a road that went steeply uphill and then steeply downhill. Participants did not know then that their training was about fuel-efficient driving.

After the test drive, the experimental group ($n = 8$) was informed that they participated in fuel-efficiency training and were given instruction about how to save fuel (e.g., the less braking, the better), how to conserve fuel when driving uphill (e.g., start with full throttle as early as possible) and how to conserve fuel when driving downhill (e.g., employ the retarder). Subsequently, the experimental group completed a training drive in the simulator (the very same one as that used in Pilot Study 1) to apply what they had learned from the presentation (e.g., anticipatory driving to reduce the number of changes between accelerator and brake). While driving, the instructors gave the participants real-time audio feedback on their performance. After the training drive, the participants were given feedback on fuel consumption, the number of changes between accelerator and brake, and the use of the vehicle retarder. In addition, the instructor explained how they could have saved more fuel.

The members of the control group ($n = 7$) drove the same route in the simulator and were requested to limit their fuel consumption as much as possible. Contrary to the presentation received by the experimental group, the control group's instruction about saving fuel was given after the training drive, and no performance feedback was received. However, the control group did receive information about their performance from the simulator data, just as the experimental group did. The training time in the driving simulator was held equal for both groups (approximately one hour).

The posttraining test session was the same for all participants of Pilot Study 2. They drove the identical route for a second time, encountering the same situations as they did during the training drive. Simulator data on fuel consumption, number of accelerator–brake changes, number of uses of the vehicle retarder, braking time, and driving time were recorded. We controlled for the same individual attitudes and characteristics as in Pilot Study 1.

Results

The main difference between the training of the experimental group and that of the control group was that the members of the experimental group received feedback on their performance, whereas the control group did not. The following report of the results focuses on the question of whether this difference affected training outcomes. We also describe relationships between individual attitudes, characteristics, and performance.

As in Pilot Study 1, the experimental group and the control group did not differ significantly in any of the individual attitudes and characteristics measured before the training session. The training received by the experimental group showed benefits for fuel-efficient driving (see Table 2). The experimental group's performance improving in the most relevant criteria. In terms of fuel consumption, the experimental group performed better than the control group, a main effect that was significant ($F = 6.63$, $df = 1.13$, $p < .05$). The experimental group was successful in reducing fuel consumption (from 3.2l to 2.6l), whereas the control group's fuel consumption was even increased (from 3.8l to 4.3l). The interaction between time and training was significant ($F = 9.74$, $df = 1.13$, $p < .01$). The number of the experimental group's accelerator–brake changes declined more (from 54 to 36) than did that of the control group (from 64 to 52). This main effect of training proved to be significant ($F = 7.70$, $df = .13$, $p < .05$), as did the main effect of time ($F = 10.70$, $df = 1.13$, $p < .01$). Furthermore, the experimental group reduced braking time more than the control group did. However, the main effect of training was not significant ($F = 0.63$, $df = 1.13$, ns). The main effect of time with regard to braking ($F = 29.78$, $df = 1.13$, $p < .001$) and the interaction between time and training was significant ($F = 11.79$, $df = 1.13$, $p < .01$). No significant main effects were found for either the number of times the vehicle retarder was used or for driving time.

Table 2. Descriptive pre- and post-training results of driving performance (Pilot Study 2)

Criteria for driving performance	Experimental group Before M (SD)	Experimental group After M (SD)	Control group Before M (SD)	Control group After M (SD)
Fuel consumption (l)	3.2 (0.6)	2.6 (0.4)	3.8 (0.8)	4.3 (1.2)
Number of accelerator-brake changes	54 (8.9)	36 (12.0)	64 (19.9)	52 (6.6)
Number of use of the car retarder	8 (17.0)	39 (38.9)	56 (63.2)	38 (50.1)
Braking time (sec)	44 (12.5)	15 (9.7)	37 (9.7)	30 (11.8)
Driving time per task (sec)	596 (74.0)	588 (48.6)	625 (67.3)	590 (68.6)
Attitudes toward fuel-efficient driving[a]		4.8 (0.4)		4.4 (1.1)
Confidence in ability to apply acquired skills[a]		4.3 (0.7)		2.9 (0.9)
Perceived benefit and transfer[a]		4.2 (0.8)		2.8 (0.9)

[a]The scale ranges from 1 (*Not reasonable or confident at all*) to 5 (*Very reasonable or confident*).

The experimental group and the control group did not differ significantly in their attitudes toward fuel-efficient driving ($t = 1.06$, ns). However, the experimental group was more confident about being able to apply the acquired skills ($M = 4.3$, $SD = 0.70$) than the control group was ($M = 2.9$, $SD = 0.90$). This difference was significant ($t = 3.35$, $p < .001$). The experimental group also rated their benefit from the training significantly higher than the control group did ($t = 3.33$, $p < .001$). The experimental and control in Pilot Study 2 did not differ significantly as regards the posttraining measurements of other individual attitudes and characteristics, such as usefulness of the simulator training, motivation, mood, or interest in simulator training. In terms of the participants' demographic variables, only age was related to the degree to which the training was perceived to be interesting ($r_s = .55$, $p < .05$).

Driving experience (in years) was correlated with the interest to participate again in simulator training for driving ($r_s = -.42$, $p < .1$).

Discussion

The goal of Pilot Study 2 was to assess the effect of behavior-based feedback on performance. The posttraining fuel consumption among the participants of the experimental group, all of whom received behavior-based feedback, significantly declined, whereas the control group's posttraining fuel consumption actually increased. This finding indicates that simulator training alone is not effective and that individual feedback on the driver's performance is needed to improve performance. Additionally, confidence about being able to apply the acquired skills after training was higher in the experimental group than in the control group, as was the perceived benefit of simulator training.

As in Pilot Study 1, simulator training proved to be an efficient learning method, for performance improved after just one hour of training. Simulator training has an additional sociopolitical and ecological advantage in this context: it conserves fuel.

Conclusion

The findings of Pilot Studies 1 and 2 show that simulator training can have useful specific learning effects if the participants receive both performance feedback from instructors and guidance on corrective action to be taken in especially dangerous situations. As a complement to conventional training for truck drivers, simulator training is effective if it has the participants experience and practice situations (such as brake failure on a downhill slope) that cannot be demonstrated in conventional practice areas. In other words, simulator training should be used to train truck drivers for dangerous situations that cannot be practiced in reality.

The findings of these two pilot studies must be considered with two limitations in mind. First, both samples ($N = 16$ and 15, respectively) are relatively small. The research was planned as preparatory work for a larger evaluation of driver's training that uses a simulator, but a surge in Germany's economy just as truck drivers were being sought for the sample groups made it difficult to find volunteers. Many would-be participants were busy with their assignments from dispatchers. Second, technical problems prevented collection of performance data from the simulator and of instructors' ratings in Pilot Study 1.

Acknowledgements

The authors thank Kerstin Schüler for collection of the data during the pilot studies.

References

Amico, G.V., & Clymer, A. B. (1984). Simulator technology—Forty years of progress. In V. Amico & A. B. Clymer (Eds.), *All about simulators* (pp. 176–187). La Jolla, CA: Society for Computer Simulation.

Colquitt, J. A., LePine, J. A., & Noe, R. A. (2000). Toward an integrative theory of training motivation: A meta-analytic path analysis of 20 years of research. *Journal of Applied Psychology, 85,* 678–707.

Dieterich, R. (in press). Simulation, Transfer und Validität. Gemeinsame Probleme zweier Konzepte und ihre Verbindung über das Prinzip der Simulation. [Simulation, transfer, and validity: Shared problems of two concepts and their connection through simulation]. In M. Scherm & C. Roos (Eds.), *Perspektiven der Management-Diagnostik.* Lengerich: Pabst.

EU Regulation 2003/59/EG des Europäischen Parlaments und des Rates vom 15. Juli 2003 [EU Regulation 2003/59/EG] of the European Paliament and the Council of 15 July 2003]. *Amtsblatt der Europäischen Union.* Retrieved October 15, 2007, from http://eur-lex.europa.eu/LexUriServ/site/de/consleg/2003/L/02003L0059-20070101-de.pdf

Iversen, H., & Rundmo, T. (2004). Attitudes towards traffic safety, driving behaviour and accident involvement among the Norwegian public. *Ergonomics, 47,* 555–572.

Kennedy, R. S., Lane, N. E., Bernaum, K. S., & Lilienthal, M. G. (1993). Simulator Sickness Questionnaire: An enhanced method for quantifying simulator sickness. *The International Journal of Aviation Psychology, 3,* 203–220.

Kluge, A. (2008). What you train is what you get? Task requirements and training methods in complex problem-solving. *Computers in Human Behavior, 24,* 284–308.

Parkes, A. M., & Reed, N. (2005). Transfer of fuel-efficient driving technique from the simulator to the road: Steps toward a cost-benefit model for synthetic training. In D. de Waard, K. Brookhuis, and A. Toffetti (Eds.), *Developments in human factors in transportation, design, and evaluation* (pp. 163–176). Maastricht: Shaker Publishing.

Saucier, G. (1994). Mini-Markers: A brief version of Goldberg's unipolar big-five markers. *Journal of Personality Assessment, 63,* 506–516.

Schimmack, U. (1997). Das Berliner-Alltagssprachliche-Stimmungsinventar (BASTI): Ein Vorschlag zur kontentvaliden Erfassung von Stimmungen [The Berlin Everyday Language Inventory of Moods: A proposal for establishing content validity in the indexing of moods.]. *Diagnostica, 43,* 150–173.

Schyns, B., & Collani, G. (2002). A new occupational self-efficacy scale and its relation to personality constructs and organizational variables. *European Journal of Work and Organizational Psychology, 11,* 219–241.

Uhr, M., Felix, D., Williams, B., & Krueger, H. (2004). Transfer of training in a driving simulator. In L. Dorn (Ed.), *Driver behaviour and training* (pp. 330–349). Aldershot: Ashgate.

Zohar, D., & Luria, G. (2005). A multilevel model of safety climate: Cross-level relationships between organization and group-level climates. *Journal of Applied Psychology, 90,* 616–628.

Analysis of automation in current UK rail signalling systems

Nora Balfe[1,2], John R. Wilson[1], Sarah Sharples[1], & Theresa Clarke[2]
[1]University of Nottingham
[2]Network Rail, London
UK

Abstract

The potential to improve performance and efficiency through automation has prompted an aspiration to increase the level of automation in UK rail signalling. This is driving a need to develop new forms of automation both for use with traditional signalling systems and with the introduction of the European Rail Traffic Management System (ERTMS). To support this drive, research is underway to examine and understand both the strengths and the limitations of the current route setting automation system to inform future requirements.

Automation has been a feature of UK rail signalling for many years, but in subtle forms such as the decision support provided by interlocking systems and the augmentation of physical labour in changing signal aspects and point positions. More advanced route setting automation has been introduced over the last 25 years, but the automatic route setting being implemented today has changed very little from the initial configuration introduced.

This paper discusses the methods used to study the current application of automation in the UK rail signalling system and some of the conclusions drawn from these studies.

Automation

Automation may be defined as a device or system that assumes a task previously performed by a human operator (Parasuraman & Riley, 1997). There are many reasons why automation might be introduced into a system. Wickens (1992) lists three main categories of automation purpose; automation which is employed to perform a function that is beyond the capabilities of a human operator, for example performing complex calculations at high speed; automation which performs functions which human operators are poor at, for example, sustained attentiomn, i.e. monitoring a system for a single failure event; and automation which provides assistance to human performance, for example augmenting information on display systems.

In D. de Waard, F.O. Flemisch, B. Lorenz, H. Oberheid, and K.A. Brookhuis (Eds.) (2008), *Human Factors for assistance and automation* (pp. 229 - 241). Maastricht, the Netherlands: Shaker Publishing.

There are a number of perceived benefits of automation; these include a reduction in human error, a saving on labour costs and a reduction in human workload (Bainbridge, 1983; Dekker, 2004; Hollnagel, 2001). Automation certainly contains the potential to bring about a reduction in human error, labour costs and workload but these benefits are not always realised when an automated system is introduced. Human error may be reduced in the task performance; however machines are manufactured, programmed and maintained by humans and an error at any one of these stages may occur which does not become manifest until the operational stage. Automation does not usually replace the human operator in totality; there is often a role for a human supervisor or operator. In addition, there are new job roles associated with the design, manufacture, programming and maintenance of the automated machine and these roles may be more skilled, higher-paid jobs than those the automation is designed to replace. Hence, the saving in labour costs may not be as high as is sometimes perceived. Well designed automation may lead to a reduced workload; however it is often the case that while a reduction in physical workload is achieved, there is an increase in mental workload for the operator. Automation may also lead to peaks and troughs in workload (Woods, 1996) as it reduces workload during periods when workload was already low but becomes a burden during higher workload phases.

Aside from the benefits there are also many problems which automation of a system may bring with it; for example a system with low reliability or a tendency to make errors may induce low operator trust which results in low usage of the automation (Sheridan, 1999). Operators who are not actively involved in the control may suffer from out of loop unfamiliarity (Wickens, 1992) or loss of situation awareness (Endsley, 1996), and this can become a major problem if they are required to take over from the automation, especially during emergency circumstances. Another problem may be the loss of skills on the part of the operator as they are no longer required to use them regularly (Bainbridge, 1983), again this may be an issue during system failures when the operator is required to take over from the automation quickly and effectively.

Automation therefore can be a mixed blessing, and introducing it into a system requires careful analysis, planning and testing to ensure maximum benefit is achieved.

Automation in Rail Signalling

A fundamental method used in this research has been formal and informal discussions with key stakeholders in the industry, including operators, engineers and designers. This has elicited understanding and knowledge of opinions, issues and constraints of both the technical system and the wider socio-technical system.

There is a desire to increase the level of automation in UK rail signalling in order to improve delivery of train services, and this is driving a need to develop new forms of automation and to increase the existing levels of automation on the rail network, both for use with traditional signalling systems and with the introduction of the European

Rail Traffic Management System (ERTMS). Operations staff expect an automated system to reduce staffing requirements and increase efficiency and performance.

Future signalling automation systems are likely to be subject to similar issues and constraints as the current automation. These constraints are often outside of the control of the railway. Weather, mechanical problems, trespassers, engineering works and staff shortages are just some of the potential circumstances which can cause disruption on the railway and although some progress can be made to reduce the occurrence and impact of these situations, it is highly unlikely that they will ever be entirely alleviated. This means that trains will not always run to time, and there will sometimes be blockages of the line which make diversions and cancellations unavoidable. The current automation attempts to deal with some of these situations, but is largely unsuccessful as it does not have sufficiently detailed information on the train and situational characteristics, or sufficient intelligence to fully understand situation as they arise, making it impossible for the automation to develop an optimal solution. Artificial intelligence may never achieve the levels necessary for the machine to rival the operator in dealing with unexpected events, but to achieve the desired reductions in staff and increases in efficiency and performance, future systems must be capable of assisting with the unexpected disturbances as well as the expected.

Rail Signalling

There are three main forms of signalling systems in place on the current UK network; Lever Frame, Entry-Exit panel (NX) and Integrated Electronic Control Centre (IECC), including Automatic Routing System (ARS).

Lever Frame boxes are the earliest type of signalling system still in use today. Each box is equipped with a block bell to communicate with adjoining boxes, and levers are used to change the signal aspects and point end positions. Lever Frames also have a mechanical interlocking system which ensures that the signal and point levers are pulled in the correct order, thus ensuring that an unsafe route is not set.

NX panels, introduced in the 1950s, use electrical circuits to detect train positions and feed this information back to one signaller who can now control a much larger area displayed on a panel. Signals and point ends are mechanised so that they no longer require the operator to pull a heavy lever to move them. The signaller is able to push a button representing a signal at the start of a route and one at the exit of the route and the interlocking will automatically change the signal aspects and point ends appropriate to that route. This technology was instrumental in allowing for an increase in the operator's control area.

The third system, IECC, was introduced in the 1980s. This moved the signalling interface from a panel to computer based (VDU) equipment, but the greatest change was the introduction of ARS.

ARS is an automated system which routes trains according to the working timetable. When a train enters the control area, ARS checks the timetable for a matching train and once found it sets a route in front of the train. This method works well so long as the trains are running to time and there are no obstructions of the line. ARS also incorporates decision resolution algorithms for when trains are not running to time. These come into use when two trains are approaching the same route at the same time and the automation must make a decision as to which train to route first. The algorithms take into account train priorities and the delay of each train, and perform calculations to predict the delay accumulated for each scenario. Hence the automation makes a decision; however, some key information, such as train speed, is currently omitted and so the automation can make some decisions which appear incomprehensible to the operator.

Rail automation model

A rail automation model was developed to illustrate the levels of automation in different generations of signalling systems. The model is primarily descriptive of the current levels of automation, but could also be used to illustrate where there is potential for advances in automation.

This model was based upon the work of Parasuraman, Sheridan, & Wickens (2000) which proposed a model for types and levels of automation based on a simple four stage view of human information processing which breaks the processing down into four functional dimensions; information acquisition, information analysis, decision and action selection, and action implementation. New levels of automation have been generated, as the original levels were found not to adequately describe the levels found in the rail environment, and furthermore the scales developed by Parasuraman et al combined the functional dimensions whereas this model uses a separate five point scale for each of functional dimensions.

Information Acquisition

1	None	Human gathers all information without assistance from computer or technology, using senses for dynamic information and print-outs/publications for static information
2	Low	Human gathers all information but with assistance from ICT (telephone/fax/computer)
3	Med	Computer automatically gathers some information, but human is required to collect other information
4	High	Computer and technology provide the majority of the information to the human
5	Full	Computer gathers all information without any assistance from human

Information Analysis

1	None	Human analyses all information
2	Low	Computer analyses information as it is received and detects conflicts only as they occur.
3	Med	Computer calculates a future prediction based on basic information for the short term (e.g. current trains on the workstation)

| 4 | High | Computer calculates a future prediction based on fuller information (e.g. trains arriving in future, infrastructure state, current situation on other workstations) and highlights potential problems/conflicts over a longer period of time |
| 5 | Full | Computer calculates a long-term future prediction using all relevant data (e.g. up to date information on train speeds, infrastructure state etc.) |

Decision and Action Selection

1	None	Human makes all decisions, without any support
2	Low	Computer provides decision support to the human to help ensure decision is not unsafe
3	Med	Computer performs basic analysis (e.g. first come first serve, run trains to timetable) and leaves perturbed modes to the human
4	High	Computer performs mid-level analysis (e.g. apply set rules to delayed trains) and has basic plans for implementation during perturbed operations
5	Full	Computer makes all decisions under all circumstances using complex algorithms to determine an optimal decision (e.g. based on a high level prediction of the future state, optimal conflict resolution) and provides flexible plans for perturbed operations.

Action Implementation

1	None	Human implements all actions and communications
2	Low	Computer augments human's physical labour (e.g. hydraulic assistance on lever) but human operator performs communications
3	Med	Computer implements physical actions but human is required to perform communications (possibly with assistance from ICT)
4	High	Computer implements physical actions and basic communications but human is required to perform complex or unusual communications
5	Full	Computer implements all actions and communications

Using the scales, each generation of signalling system can be plotted on a graph, as in Figure 1.

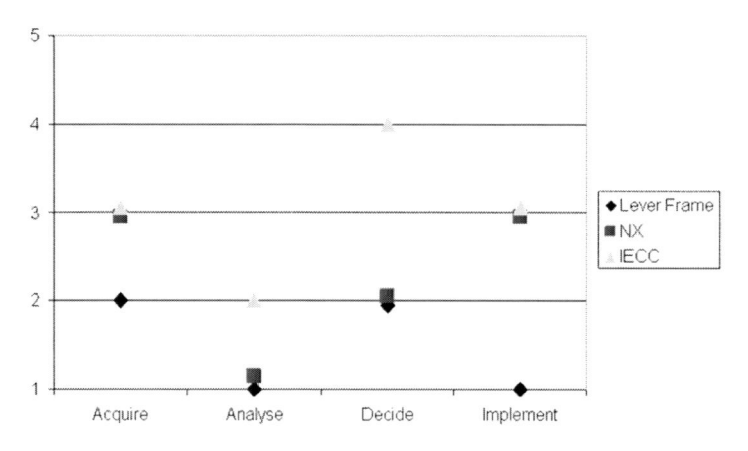

Figure 1. Levels of Automation

As can be seen in the graph, Lever Frame signalling equipment has the least automation, but there is still a low level present in the form of decision support provided by interlocking equipment and assistance from Information and Communications Technology (ICT) in gathering information.

NX panels have similar levels of automation, but offer more support in data gathering as the signaller can no longer physically see the entire track he controls, so information on track occupation is provided by the system. There is also a major leap in terms of action implementation as the physical work of pulling levers to manually move signal aspects and points is removed.

Finally, IECC systems offer the most advanced forms of automation on the UK rail network. However, apart from changing the signalling interface from a panel to a VDU based system there has been no advance in terms of information gathering, and although some form of analysis is now performed by the automation, it is not very sophisticated. The major change in automation has come in the decision and action selection dimension where the automation is now capable of running all timetabled trains under normal conditions and offers some capability to deal with late-running trains and perturbed conditions. The action implementation has not moved on from NX panels.

Principles of automation

A review of literature in the area of general automation was conducted to generate a list of fundamental requirements for automation. Although developed for the rail signalling these principles are intended to be applicable across different domains. The following general requirements for good automation were produced:

- Visible - All relevant dynamic information should be available to the operator so the operator is not required to hold any relevant information in memory (Sarter, Woods, & Billings, 1997; Woods, 1997).
- Observable - The system should provide immediate and effective feedback on system state to support the operator in monitoring and staying ahead of system activities. The feedback should be in a form which reduces the cognitive work required to derive meaning from the data (Christoffersen & Woods, 2001; Dekker, 2004; Sheridan, 1999).
- Understandable - The system should support the operator's development of an accurate mental model by making the underlying logic of the system transparent. The operator should then be able to understand what the system is doing and be able to predict future behaviour (Sheridan, 1999).
- Directable - The operator should be able to instruct the automation easily and efficiently (Christoffersen & Woods, 2001; Dekker, 2004).
- Accountable - The operator should be in command of the automation and the automation should not be allowed to act autonomously (Billings, 1991).
- Reliable - The system must have high reliability in order to maintain trust by the operators (Sheridan, 1999).

- Robust - The system should able to perform under a variety of conditions and should be able to support the operator during both normal and abnormal operational situations (Sheridan, 1999).
- Guard against skill degradation - The system should incorporate a method to guard against skill degradation on behalf of the operators to ensure that safe and efficient operation is within the operator's capacity if the automation fails (Bainbridge, 1983; Dekker, 2004).
- Error resistant - The system should make it difficult for the operator to make an error, and this should be more rigorous than simply asking the operator to confirm their request or verify their entry (Billings, 1991; Dekker, 2004).
- Error tolerant - The system should have the ability to mitigate the effects of an operator error (Billings, 1991).
- These principles will be evaluated using a paired comparisons exercise administered to human factors personnel as well as engineering staff. It is proposed that current and future systems be evaluated against these criteria, and they form the basis of guidelines for engineers and operators involved with specifying future systems.

Observation studies

A series of observation studies were carried out in three IECC signal boxes and signallers' engagement in five basic activities was recorded. The study aimed to identify whether differences exist in individual signallers' use of automation. The objective of the study was to observe signallers' current use of ARS at a basic level and determine what percentage of their time is spent monitoring, planning, controlling, communicating or not actively involved (quiet time) during normal operations. Signallers were observed at their workstation and a coding system was used at time intervals of 5 seconds to capture their activities. The structured observations were undertaken at 6 workstations across 3 different signal boxes with each observation lasting 90 minutes. Three signallers were observed at each workstation. Due to the nature of the live environment opportunity sampling was used and participants were chosen on the basis of availability and willingness to take part in the study. On one occasion the same signaller was observed on two different workstations and this provided an interesting comparison (labelled YS3 and LE1 in Figure 2). All observations took place during the same time of day so that traffic levels, as far as possible, would be comparable. Once the data were gathered a graph was generated showing the distribution of activity for each signaller observed. This graph is shown in Figure 2.

Each bar on the graph describes the distribution of activity for one signaller. Workstations YS and LE were both in York IECC, SF and IL were Liverpool Street IECC, and NK and AS were Ashford IECC. Three signallers were observed at each workstation.

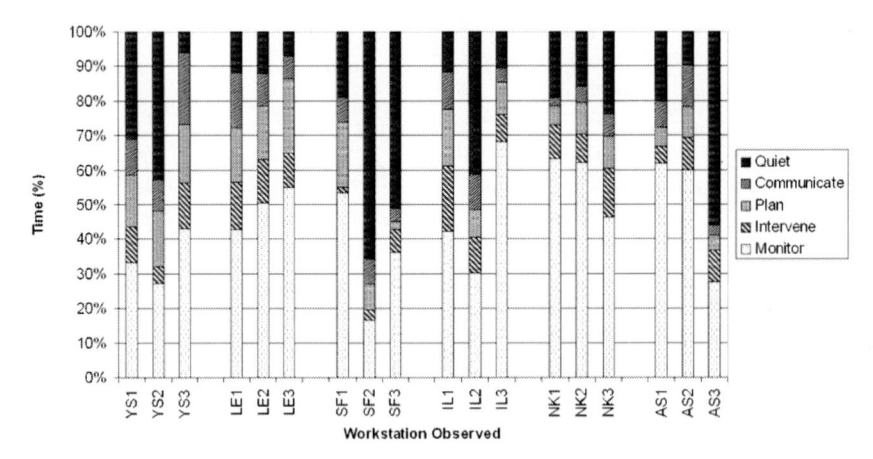

Figure 2. Signaller activity

Discussion

An early finding of the study was that signallers do not appear to maintain consistent levels of monitoring, but vary their monitoring according to events on the workstation. Two main types of monitoring were noted, active monitoring and passive monitoring, although there are likely to be several sub-types within these main categories. Active monitoring usually involved the signaller sitting upright at his workstation often watching a particular spot or scanning the screens looking for a particular item or event. By contrast, during passive monitoring the signaller sat back and appeared to be waiting for his attention to be drawn to something, rather than actively seeking it out. He was also much more likely to be drawn into conversations or to engage in quiet time during a period of passive monitoring. The data gathered in this study did not make it possible to identify what the triggers to switch between monitoring types might be, but a later study using eye-tracking equipment will hopefully address this issue.

From the graph, bearing in mind that traffic levels and demand on the signaller were roughly equivalent on each workstation, it is apparent that signaller activity is driven by the signaller himself rather than by the workstation. The same signaller was observed for YS3 and LE1, and as can be seen in Figure 2, the resulting graphs are very similar, providing further support to the conclusion that the signaller largely drives the activity on the workstation. Furthermore, analysis of the data for each signaller over the observation period revealed that the data remained broadly consistent over time and circumstances on the workstation, barring major events, had a comparatively minor impact on their engagement in each activity.

A variation between intervention levels on each workstation was observed. The traffic patterns were reasonably similar during each observation and SF2 and IL2 both experienced disruption during the observations but still did not achieve the highest level of interventions observed, and this is an indication that each signaller uses a different interaction strategy. It is unclear what the reasons behind the different levels of intervention were, but as discussed above, increased workload due to circumstances on the workstation can be ruled out as the main cause. There are two main possibilities for the difference, some signallers may have intervened more due to distrust of the automation, or some signallers may have had more efficient intervention strategies either using more direct controls or intervening earlier to prevent a problem occurring at a later stage. The data gathered did not allow any analysis of which of these possibilities was more likely, but future studies will gather further data to support a hypothesis.

It can be concluded from this observation study that the use of automation is likely to be driven by the individual. Further research is required to determine what drives these individual strategies, and identifying the factors involved in determining automation usage could provide valuable information to improve future systems.

An analysis of the transfer between activities was performed for all observations and as a result four main groups of activity were identified; monitored interventions, boredom, distraction, and planning updates. Monitored interventions occurred when the data flickered between monitoring and interventions, the signallers were monitoring to decide when to intervene, performing their intervention and then checking the effects of that intervention. Boredom occurred when the data flickered between passive monitoring and quiet time. The signallers were not busy and looked for distractions but they still kept an eye on the workstation. Distraction occurred when the data flickered between active monitoring and quiet time. For example, a conversation may have been going on that they were interested in but at the same time they felt they needed to continue to monitor the workstation. Planning updates occurred when the data flickered between monitoring and planning activities. Signallers were likely to be looking for information on what they could expect to happen in the near future on the workstation.

This data reveals an insight into how signallers use the automation provided to them and gives an indication of the kinds of activities future systems should seek to support.

Trust

A questionnaire was also administered to investigate the signallers' trust in the automation and to determine whether any correlations exist between signallers' trust in the automation and their use of it.

The questionnaire examined trust in the automation on a number of different dimensions adapted from previously developed measures found in the literature (Madsen & Gregor, 2000; Muir, 1994; Muir & Moray, 1996; Rempel, Holmes, & Zanna, 1985; Sheridan, 1999):

- Reliability – in terms of both the mechanical reliability and consistent functioning over time.
- Robustness – the ability to function under a variety of different circumstances.

- Understandability – the ability of the operator to understand what the automation is doing, why it is doing that and how it is doing it
- Competence – the perceived ability of the automation to perform its tasks.
- Explication of intention – the ability of the automation to explicitly give feedback on its intended actions.
- Dependability – the extent to which the automation can be counted on to do its job.
- Personal Attachment – the extent to which operators like to use the automation.
- Predictability – the ability of the operator to predict the actions of the automation.
- Faith – the extent of belief on the part of the operator that the automation will be able to cope with future system states which it may not have yet encountered.

The perceived mechanical reliability of ARS was found to be high, and indeed it is specified to very high levels of reliability as would be expected in a safety critical industry. Respondents rated the robustness of ARS as poor, which was as expected as it is designed to work best during normal operations and signallers are often required to step in under disrupted working. The questionnaire found that although signallers had a reasonable perceived understanding of what ARS is doing and how it does it, they did not have a good understanding of why it makes the decisions it does. Understanding why the automation makes the decisions it does is fundamentally important for predicting and controlling the automation and lack of understanding is likely to impact strongly upon overall trust and use of automation. With respect to the questions on personal attachment, there was a skew towards agreement, indicating that signallers do prefer to use ARS. This has also been noted in unstructured interviews with signalling staff; although they have gripes about the system they will admit that they would not like to work without it.

Conclusion

The ARS system currently implemented in the UK is a powerful tool but there are some issues that should be resolved in future systems. The system is not transparent, or 'observable', making it difficult for signallers to understand and predict future actions of the automation and resulting in unnecessary interventions. The automation is not sufficiently intelligent to fully understand unexpected situations and to develop an optimal solution. Unless huge advances are made in the field of artificial intelligence it is unlikely to rival the expertise of a skilled human operator in coping with unexpected situations. Future automation systems should therefore accept the constraint of the rail network as an open system and not attempt to develop solutions to problems it cannot fully understand. There is capacity for further automation, but within limits and careful analysis and planning is required to achieve maximum benefit.

Further research is underway to elicit more information on problems and issues arising from automation usage, and to examine signallers' monitoring strategies.

References

Bainbridge, L. (1983). Ironies of automation. In J. Rasmussen, K. Duncan and J. Leplat (Eds.), *New Technology and Human Error* (pp. 271-283). Chichester: Wiley.

Billings, C.E. (1991). *Human-centered aircraft automation: A concept and guidelines* (No. NASA Technical Memorandum 103885). Springfield, VA: National Technical Information Service.

Christoffersen, K., & Woods, D.D. (2001). How to make automated systems team players. In E. Salas (Ed.), *Advances in Human Performance and Cognitive Engineering Research* (Vol. 2). New York: JAI Press/Elsevier.

Dekker, S.W.A. (2004). On the other side of promise: What should we automate today? In D. Harris (Ed.), *Human Factors for Civil Flight Deck Design* (pp. 183-198). Aldershot, Hampshire: Ashgate Publishing.

Endsley, M.R. (1996). Automation and situation awareness. In R. Parasuraman and M. Mouloua (Eds.), *Automation and Human Performance: Theory and Applications* (pp. 163-181). Mahwah, NJ: Lawrence Erlbaum.

Hollnagel, E. (2001). *Human-oriented automation strategies.* Paper presented at the World Congress on Safety of Modern Technical Systems, Saarbruecken.

Madsen, M., & Gregor, S. (2000, 6-8 December). *Measuring human-computer trust.* Paper presented at the 11th Australasian Conference on Information Systems, Brisbane.

Muir, B.M. (1994). Trust in automation: Part I. Theoretical issues in the study of trust and human intervention in automated systems. *Ergonomics, 37*, 1905-1922.

Muir, B.M., & Moray, N. (1996). Trust in automation. Part II. Experimental studies of trust and human intervention in a process control simulation. *Ergonomics, 39*, 429-460.

Parasuraman, R., & Riley, V. (1997). Humans and automation: Use, misuse, disuse, abuse. *Human Factors, 39,* 230-253.

Parasuraman, R., Sheridan, T.B., & Wickens, C.D. (2000). A model for types and levels of human interaction with automation. *IEEE Transactions on Systems, Man, and Cybernetics - Part A: Systems and Humans, 30*, 286-297.

Rempel, J.K., Holmes, J.G., & Zanna, M.P. (1985). Trust in close relationships. *Journal of Personality and Social Psychology, 49*, 95-112.

Sarter, N.B., Woods, D.D., & Billings, C.E. (1997). Automation surprises. In G. Salvendy (Ed.), *Handbook of Human Factors and Ergonomics* (2nd ed., pp. 1926-1943). New York: Wiley.

Sheridan, T.B. (1999). Human supervisory control. In A.P. Sage and W.B. Rouse (Eds.), *Handbook of Systems Engineering and Management.* New York: John Wiley & Sons.

Wickens, C.D. (1992). *Engineering Psychology and Human Performance* (2nd ed.). New York: HarperCollins.

Woods, D.D. (1996). Decomposing automation: Apparent simplicity, real complexity. In R. Parasuraman and M. Mouloua (Eds.), *Automation Technology and Human Performance* (pp. 3-17). Hillsdale, NJ: Erlbaum.

Woods, D.D. (1997). Human-centered software agents: Lessons from clumsy automation. In J. Flanagan, T. Huang, P. Jones, and S. Kasif (Eds.), *Human Centered Systems: Information, Interactivity and Intelligence* (pp. 288-293). Washington, DC: National Science Foundation.

Adaptive Automation

Electronicallly coupled truck convoys

KONVOI: Electronically coupled truck convoys

Matthias Wille, Markus Röwenstrunk, & Günter Debus
Rheinisch-Westfälische Technische Hochschule
Aachen, Germany

Abstract

KONVOI is an interdisciplinary project that is being conducted at the RWTH University Aachen. It investigates electronically coupled truck convoys. A convoy consists of up to five vehicles, whereupon the first driver operates manually, while the other vehicles follow fully automatic (lateral and longitudinal guidance). Coupling and decoupling take place while driving on the highway as well as manoeuvres such as lane changes can be executed in the coupled mode. During the automation, the driver is entrusted with the monitoring of the system and must take over the vehicle guidance in some situations (e.g. construction areas). Automation however, can possibly lead to changes in workload and performance, which had to be investigated due to safety concerns. In a simulator experiment, the Institute of Psychology at the RWTH University examined the demand of the driver's mental workload as well as their driving performance in comparing the usage of the KONVOI-system and manual driving. The results in subjective RSME ratings, objective lane keeping (SDLP) and following behaviour (THW) showed no critical changes through the automation.

Introduction

Since the idea of electronically coupled vehicles was introduced with the PROMETHEUS-Project (Progam for European Traffic with Highest Efficiency and Unprecedented Safety) in 1988 and the development of the automated highway system (AHS) in the United States, research on this subject has come a long way. For a couple of years, an interdisciplinary project-group at the Rheinisch-Westfälischen Technischen Hochschule Aachen (RWTH University Aachen), funded by the German Ministry for Education and Research (BMBF) and in association with various partners from industries and businesses, have been working to realize simulator and on-the-road testing of electronically coupled truck convoys. The concept has been established through the EFAS- and MFG-projects, while the testing of the automation is occurring within the KONVOI project.

The KONVOI-system requires the first driver of a convoy of a maximum of 5 vehicles to manually control his truck, while the ones behind are connected through an "electronic drawbar" following fully automatic. It is designed to be used on the highway system (Autobahn), when individual truck drivers occasionally meet and head in the same direction for a long period of time.

In D. de Waard, F.O. Flemisch, B. Lorenz, H. Oberheid, and K.A. Brookhuis (Eds.) (2008), *Human Factors for assistance and automation* (pp. 243 - 256). Maastricht, the Netherlands: Shaker Publishing.

Trucks which have the KONVOI-system installed can send coupling requests through functions on a touch screen as soon as they are within 50 meters away from another equipped vehicle. The addressed driver receives the request and has the choice between accepting and denying it. When accepted, the following truck driver gets informed by the HMI and the automation sets in, closing the gap between the vehicles until a distance of about 10 meters is reached. Decoupling is occurring the same way as the coupling process. First, requests and acceptances are sent before the space between the trucks is automatically being enlarged to 50 meters again. Then a visual and auditory count-down is given to indicate the end of the automatic mode and the driver takes over manually.

The reason for the implementation of the distancing process while decoupling were results from studies by Eick and Debus (2002) as well as Wille (2002). Their data showed that during manual driving, after periods of automatic following with very small headways, gaps to lead vehicles were shorter than normally held. This was interpreted as a distance adaptation towards the spacial circumstances while in the automatic mode. In addition to the KONVOI decoupling process (enlarging the gap to 50 meters before manual takeover), a graphical distance display (see Fairclough, May, & Carter, 1997) was integrated to reduce the possible distance adaptation. The display shows a bar, which is constantly changing in length and colour (green = distance ok, yellow = attention, red = danger, distance too short) according to the current situation.

Another safety feature provided is the possibility for the driver to override the automatic system at any time by pressing the brake- or gas- pedal or moving the steering wheel. Following this, manual control will be immediately given to the driver without the normal decoupling procedure. However, the KONVOI-system is able to handle even difficult and unpredictable traffic situations e.g. a car driving in the gap between the trucks, by enlarging the gap and keeping correct lane position.

The system design combines the advantages of individual traffic (e.g. flexibility) and collective traffic (e.g. safety) through a temporary reduction in space usage and less passing manoeuvres of trucks. Furthermore, it is supposed to improve the efficiency and predictability of travel times for all traffic participants. The last is only one of many advantages but especially important due to predictions that truck traffic in Germany will increase considerably within the next years. Since expanding the infrastructure further is often almost impossible, the KONVOI-system is a reasonable alternative to this problem. Another benefit is a reduction in environmental pollution since the trucks remain in each others slip stream and by doing so use less fuel, which also means fewer emissions.

In this respect, the automatic lateral and longitudinal control of the following trucks enables drivers to move semi freely without their hands on the wheel and feet on the pedals. The idea is that exhausted drivers can relax, prepare paperwork, or entertain themselves by watching TV/DVD, while a fresh driver is steering the first truck manually until someone else is well rested. In case of exhaustion by the current operator, a request to another driver, who will then decouple and manually overtake the KONVOI to become the first truck, will solve the issue. This should reduce workload for all since everyone can take "time-outs". However, driving time is not supposed to be extended. An important reason for this is that at any time requests e.g. to change lanes as well as decoupling of convoy participants can occur and a confirmation of this would be needed by everyone. This also means that operators can not leave the driver's seat or sleep under any circumstances. The driver has to maintain a certain activation level to be able to perceive any changes in the system status.

Obviously the automation relieves the driver from steering and distance control but changes driving to an observational task with different demands and possibly safety issues (see Paragraph below). The Institute of Psychology at the RWTH University Aachen is analysing the specific demands the KONVOI-system brings to the drivers through their subjective workload experience, physiological measures as well as driving performance. A repeated measurement experiment in a driving simulator enables a comparison of travelling with the KONVOI-automation versus driving manually under very realistic traffic conditions for high economic validity.

Automation problems

Automation, whether it substitutes only a singe task or a whole assignment like the entire active driving job in case of the KONVOI-system, changes the obligation of operators but does not replace them (see De Waard, Van der Hulst, Hoedemaeker, & Brookhuis, 1999). As early as 1983, Bainbridge wrote in her article "Ironies of automation" that increase in automation comes with higher observation and control demand for the operator. Manual task vanish while the manipulation as well as error detection of complex systems gains importance (see Endsley, 1996). It has often been discussed that automation reduces the workload or even increases it through less manual but monotone tasks (e.g. Bainbridge, 1983; Endsley, 1996). The theory behind these arguments is that best performances and least workload levels occur when demand is at an optimum level (De Waard, 1996). Underload due to monotony represents suboptimal demands, which often results in boredom and errors (see De Waard, 1996; Hoyos & Kastner, 1986). The KONVOI-automation could lead to the same results as studies of long lasting system controlling (vigilance tasks), which show decreases in performance while workload and stress increases (e.g. Billings, 1988; Wickens, 1992; Wiener & Curry, 1980). This problem may be intensified with deficit in situational awareness, especially when automation fails and manual control has to be taken (Wickens & Hollands, 2000).

Situational awareness is "the perception of the elements in the environment within a volume of time and space, the comprehension of their meaning and the projection of their status in the near future" (Endsley, 1988, p. 1). With diminished situational awareness, the ability to perceive and interpret critical factors as well as anticipate incidents is at a lower level so that decisions and behaviour can be flawed. Automation can have a negative effect on situational awareness through suboptimal workload, monotony and also over reliance on automation (see Endsley, 1996; Sarter, Woods, & Billings, 1997). Operators, who truly trust their driving automation, might not feel any necessity to perceive the traffic situation nor check on any system status.

Furthermore, automation reduces the active experience manual driving provides, adding up to less awareness and the out-of-the-loop performance problem (Endsley, 1996). It was found that people often manage automatic systems effectively under normal circumstances but make wrong decisions and behave inappropriately when unexpected conditions or emergency situations arise. For example, Wickens (1992) and Endsley (1996) state that catastrophic plane crashes and traffic accidents happen because operators are not fully integrated in the active loop of the automatic system and therefore perform incorrectly.

It is crucial that any out-of-loop performance problems do not occur within the KONVOI-project. Even in allegedly harmless situations, drivers have to remain situational aware. Coupling and decoupling of any convoy participants as well as lane changes can happen at any time and therefore always requires the observation of the surrounding traffic (e.g. mirror checking) and having to respond in the form of pressing buttons. However, during the automatic mode, operators can take "time outs" because the KONVOI-system remains in control and adjusts the driving even in unpredictable situations, such as a car driving in the gap between trucks. The status change will be communicated to the driver; occasionally manual overtaking will be requested but not executed until a confirmation is received.

Study

Since the KONVOI-automation changes the driving task from manual control to system observation, which possibly results in differences in situational awareness, workload and performance, evaluation in a simulator is necessary. The Institute of Psychology at the RWTH University Aachen is analysing the specific demands the KONVOI-system brings to the drivers by conducting a repeated measurement experiment in a driving simulator. It enables a comparison of travelling with the KONVOI-automation versus driving manually under very realistic traffic conditions for high economic validity. The rating scale of mental effort (RSME) from Zijlstra (1993) is used to measure the subjective workload, while heart rate and electro dermal activity broaden the assessment. Additionally, driving performance is controlled through lane keeping (standard deviation of lateral position) and following behaviour (time headway). The purpose of the study is to ensure that there will not be any risks and safety issues to the truck drivers themselves or to other traffic participants before introducing the coupling system to on-the-road testing and the market.

Method

Participants

The co-operating transport companies Offergeld Logistik, Hammer, and Ewalds Cargo Care provided N=20 professional truck drivers with ages 29 to 58 years (Mean = 41 years, SD = 6.55 years). Their driving experience ranged from 3 to 31 years (Mean = 15.2 years, SD = 8.4 years), with an annual kilometres of driving from 5000 to 175000 (Mean = 124 000 km, SD = 45 500 km). Prior to the experiment, each driver has handled the specific simulator to rule out simulator sickness as well as for training purposes. The latter reason is important so as to prevent any unwanted training effects that could disturb the actual study (automation effects) (see e.g. Hoffmann & Buld, 2006). Every participant drove the simulation randomly -once with the KONVOI-system and another time manually- so that subject comparison could be drawn.

Simulator

The KONVOI-study was conducted at the RWTH University Aachen InDriveS driving simulator, a fully equipped original MAN truck driver's cab. The vehicle control console included all instruments and functions of a standard factory truck with automatic transmission. Only the radio, fuel indicator, and electrical window lifts were not functional due to irrelevance and control purposes of the experiment. Two beamers projected the traffic simulation onto a 7.30 x 2.80 meter ($\approx$ 21 m²) screen providing a 180° display window. The on-coming traffic could be observed through TFT monitors on each side of the truck functioning as rear view mirrors. Two extra LCD displays were installed inside the cabin. One showed the front view of the first vehicle in the convoy; the second was a touch screen to control the KONVOI-functions (HMI). The drags of the steering wheel as well as the gas and break pedal were close to reality, while the maximum speed was set to 90km/h. A stereo sound system simulated the engine noise according to the speed.

All necessary information about the cabin was transported through a CAN-Bus communication interface to the computers, which in turn control the simulation. At the control desk, different monitors showed the traffic route, vehicle information and allowed their regulation. Each automobile could be instructed to accelerated, break or change lanes in order to perform various manoeuvres. Without any intervention, all vehicles were in a dynamic mode, with specific driving behaviour, some going faster and follow closer than others. This dynamic traffic was simulated by the PELOPS-system, which produces a very realistic driving behaviour and therefore high economic validity (see www.pelops.de). Unfortunately, there was no motion system installed, therefore sensation derived from curves, slopes, acceleration and breaking could not be fully experienced.

Simulation route

The InDriveS simulator produced a traffic scene of a four lane highway (Autobahn), divided by a double crash barrier (standard in Germany). Additionally, emergency lanes in each direction with a single crash barrier to the outer limit of the road were projected.

The route and the landscape were exact copies of the topology of Autobahn A1 from Bliesheim to Blankenheim in Nordrheinwestfalen, Germany. Road signs, bridges, trees, noise barrier and other specific characteristics created a very realistic traffic scenario (see figure 1).

The length of a single course was about 40km but at the end, the participants drove into a tunnel, which was the beginning of the next loop. This was repeated three times so that four loops were driven with a total distance of 160km, taking roughly two hours. Unfortunately, the duration of the simulations could not be extended any further due to availability restrictions of the drivers.

Figure 1a (left) external view of the simulated highway, 1b(right) view of the test-driver

Design

The study is based on a repeated measure plan, with the independent variable being the driving mode. All participants drove in a randomized order, once without automation and on another day with the KONVOI-system (within-subject-design). Additionally, the KONVOI-drive was divided into three coupling conditions.

Independent variable:
Within -Subject-Design:
    IV 1: Condition (KONVOI-drive vs. manual-drive)
Coupling conditions at the KONVOI-drive:
    IV 2: Coupling demonstration (Loop 1)
    IV 3: KONVOI-leading (Loop 2)
    IV 4: KONVOI-following (Loop 3-4)

The order of the loops were always identical because coupling demonstrations at the beginning, followed by leading the convoy and a long automatic following period, presented an increase in automation level and seemed to be most logical. This design inhibited carry-over-effects from one condition to the other best.

Furthermore, the experiment is characterized by a variety of dependent variables which measure the experienced workload, physiological parameters as well as driving performance.

Dependent variables:
    DV 1: RSME (rating scale mental effort) from Zijlstra (1993)
    DV 2: Heart rate (HR)
    DV 3: Heart rate variability (HR var)
    DV 4: Electro dermal activity (EDA)
    DV 5: Standard deviation of lateral position (SDLP)

DV 6: Mean Time headway (THW) *
DV 7: Minimal Time headway (THW) *
DV 8: Time to collision (TTC)

* In this study, the overall distance and time headway are not appropriate measures to discuss truck drivers' behaviour because PELOPS vehicles were often entering the right lane shortly after passing but with much higher speed than the trucks. In those cases, small THW were not reflecting critical encounters in contrast to short distances while front vehicles were slower. Therefore THW was taken into consideration only when approaching another automobile (TTC > 0).

Unfortunately, heart rate variability, electro dermal activity and time to collision are not yet analyzed and therefore cannot be reported at this time.

Procedure

Since all drivers participated in a pre experimental test drive, there was only a short 30 minute introduction of the simulator specifics and rehearsal of manoeuvres with the KONVOI-system prior to the main experiment. Nevertheless, it was announced that questions could be asked at any time during the driving test. The general instruction was to follow another truck with free choice of speed and distance like they ordinarily would. Following this, the procedure was depending on the condition of driving (KONVOI vs. manual), while the measurement phases and also the manoeuvres were relatively at the same kilometres markers of the route.

Manual-drive

The first 7 km were reserved for the driver to get adjusted, followed by a measurement phase from kilometre 8 to 13 as baseline data. The RSME rating was asked at kilometre 17 and 22 after lane changes. The remaining distance (km 23-37) of the first loop and the beginning of the second (km 1-13) was driven as a plain following task with two additional measuring phases. Another measuring phase was included in the second loop at the end as well as RSME ratings before and after a lane change around kilometre 14. The third loop was composed the same way as the second, while in the last loop, the RSME was rated only once (km 16) and no lane change took place.

KONVOI-drive

The testing with the KONVOI-system included the same route and length of the loops. The measurement phases, manoeuvres and RSME ratings were also arranged to be identical so that a comparison with manual-drives could be possible. However, technical and driver related coupling problems led to some variations in locations and numbers of couplings, manoeuvres and RSME ratings.

During the first loop, coupling demonstrations (short periods of automatic following) were executed around km 18. At the beginning of the second loop, the test participant had to pass the truck in front and let him couple from behind. The driver was in the position of the lead vehicle of the convoy for roughly 20 km, followed by manual driving within the measurement phase. From loop three until the 24th km of loop four, the driver was automatically following the lead truck. Immediately and 10 kilometres after the approximately 50 minutes lasting period of coupling, measurements were taken.

Data analysis

The data from the InDriveS simulator comprised of 46 different variables which were registered with 100 Hz per second, and were further reduced to 10 Hz. Now the measurement phase and manoeuvres were coded with numbers to be able to choose and calculate specific one. These steps were done separately for 20 manual as well as 20 KONVOI data, followed by matching of the files to a single data matrix. This matrix has all variables for each participant itemized so that the comparisons of averages and significance test could be performed.

Results

Subjective measure

The rating scale mental effort (RSME) from Zijlstra was used to measure the subjective workload. First, a baseline was conducted through 6 imaginable traffic scenes, averaging a score of 47.4. In specific, a 4 hour drive on a highway with much traffic was rated with a mean of 50.13 (SD = 22.7). The overall average of either condition in the simulation drive on a highway with normal traffic is as expected less demanding than the imagined 4h with much traffic (Manual Mean = 29.6, SD = 23.6; KONVOI Mean = 27.1, SD = 24.8).

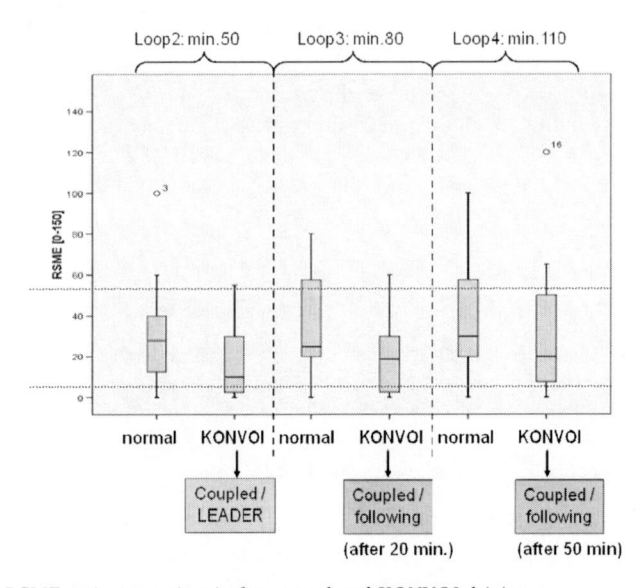

Figure 2. RSME-rating over time in for normal and KONVOI driving

In loop 2 the subjective workload is higher for normal manual driving compared to the KONVOI-condition while leading the convoy, however not significantly (normal driving Mean = 27.6, SD = 24.2 vs. KONVOI Mean = 18.2, SD = 17.0) (see figure 2 – a Boxplot to show the statistical range of distribution).

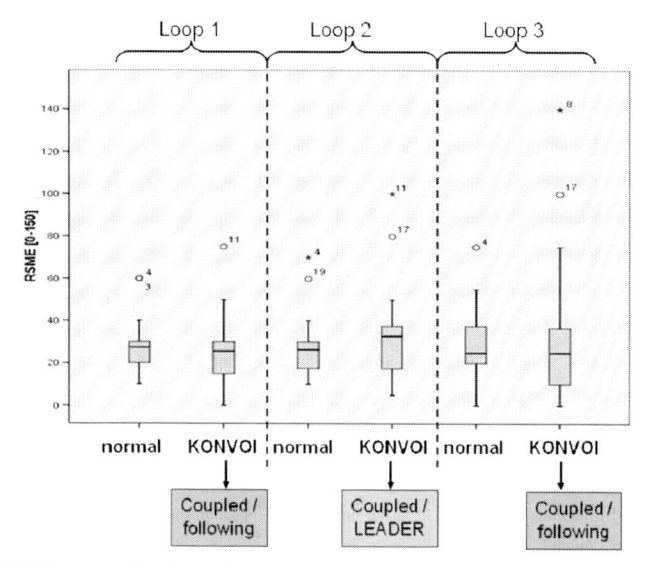

Figure 3. RSME-rating after lane changes for normal and KONVOI driving

After a total of 80 minutes simulated driving, the RSME rating while driving normally is with a mean of 33 (SD = 27.5) significantly higher ($\alpha=.04$) then in the KONVOI-condition where drivers have been couples for the last 20 minutes (Mean=18.6 SD = 18.1). In the same manner, are the scores in loop 4 after a total of 110 minutes driving, lower in the condition with automatic following (normal Mean = 41.3, SD = 31.7; KONVOI Mean = 31.7; SD = 32.2).

Additionally, it was found that there are no significant differences between RSME scores for lane changes while manually or automatic driving. The mean rating for coupled lane change during the demonstration in loop 1 is 25.6 (KONVOI SD = 20) compared to a slightly higher score of 27.4 (normal SD = 15.7) for normal steering. In loop 2 however, the same manoeuvre but leading the coupled convoy contributes to more experienced workload than the manual drive but not significantly (KONVOI Mean 32.8, SD = 25.0; normal Mean 26.4, SD = 17.9). Also the RSME rating for the lane change while coupled following in loop 3 is to some extend greater in automatic mode (KONVOI Mean 33.2, SD = 36.8; normal Mean 27.6, SD = 19.0).

Physiological measures

Baseline Measurement of the heart rate was carried out before simulations were started and showed a mean heart rate of about 73 beats per minute for KONVOI-condition and 74 beats per minute for the normal-driving-condition. An individual correlation of $R^2=.73$ between both measurements was found. Furthermore, the differences in heart rate were compared to the individual baselines to be able to calculate contrasts. The comparisons of heart rates show no significant differences between normal and KONVOI driving. However, while being coupled for a long period of 50 minutes and afterwards while driving manually again, heart rates tend to be somewhat higher in KONVOI compared to normal driving (see figure 4).

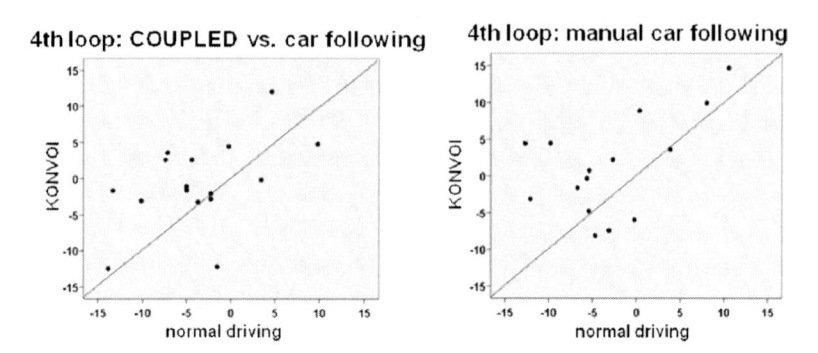

Figure 4. Heart rate difference from baseline, while coupled for 50 minutes (left) and afterwards while driving manually (right)

Driving performance

The standard deviation of lateral position (SDLP) shows the performance in lane keeping but is also an indicator for objective workload. At the beginning, baseline SDLP for normal driving is 17.0 cm (SD = 4.6) while the average is 20.3 cm (SD = 6.8) for the KONVOI-condition, but these slight group difference do not become significant.

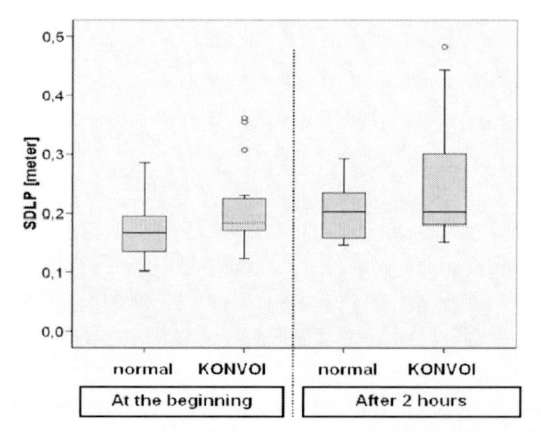

Figure 5. SDLP at the beginning and after 2 hours of driving for normal and KONVOI driving

After two hours of driving, the SDLP tends to be higher with α = .076 for the KONVOI-drivers in contrast to the normal drivers (Mean = 24.5 cm, SD = 9.8 cm vs. Mean = 20.3 cm, SD = 4.5 cm) (see figure 5). However, directly at the manual takeover, the lane keeping is always absolutely normal (the first coupling demonstration: Mean SDLP of 18.1 cm SD = 8.1 cm; after 50 minute of being coupled: Mean = 19.8 cm, SD = 9.5 cm) (see figure 6).

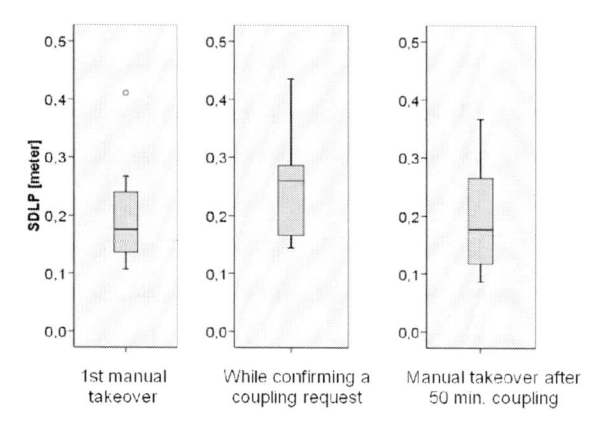

Figure 6. SDLP at manual takeovers for normal and KONVOI driving

During the coupling request while leading the convoy, the average is 25.3 cm (SD = 8.3 cm), somewhat elevated as shown by the graph in the middle of figure 6.

Following distance towards other vehicles is a safety relevant driving performance, measured through time headway (THW). The mean THW while approaching a vehicle is at the beginning of the simulation-drives for the manual condition (normal) 4 s (SD = 2.4) versus 3 s (SD = 1.2) for the KONVOI-condition.

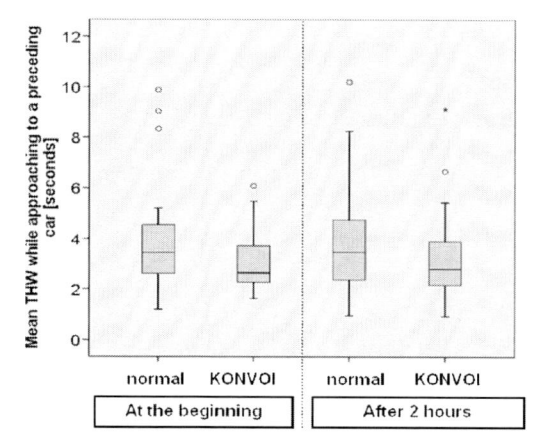

Figure 7. Mean THW at the beginning and after 2 hours of driving for normal and KONVOI driving

After 2 hours, the mean THW is just as the first comparison not significantly different for normal (Mean = 4.0 s, SD = 2.5) (range 1.0-10.2 s) and KONVOI-driving (coupled following) (Mean = 3.4 s, SD = 2.0 s) (range 0.9-9.1 s). Also the minimal THW did not show any considerable dissimilarity and are within a normal range of 0.35 to 4 s.

Discussion

Since the automation through the KONVOI-system possibly brings reduction in workload because active vehicle control is not necessary but at the same time could increase stress due to monotony and underload, the results are satisfying. The RSME ratings show no significant differences in subjectively experienced workload while using the KONVOI-system. Even new ways to execute manoeuvres such as lane changes do not lead to excessive stress for the driver. All mean RSME ratings are not much higher or respectively lower then the total baseline average of 47.4, which is comprised of differing traffic scenarios.

Heart rate measurements show no significant differences between KONVOI and normal driving. A slightly higher heart rate was observed while being in automatic mode for a long period of 50 minutes and during lane change manoeuvres. While driving manually again, rates were higher compared to the normal driving condition. But these differences are not large in amount and could be based on the unfamiliar situation of the KONVOI-condition.

Performance in lane keeping (SDLP), as a driving parameter and indicator for workload, is in general not affected by electronical coupling. Nevertheless, operating the KONVOI-touch-screen can decrease the steering behaviour in some situations. An easier to manipulate control panel is currently being planed. Additionally, after a total driving duration of 2 hours, including 50 minutes of consecutive automatic following, a tendency in higher SDLP is given for KONVOI drivers. This could mean a somewhat increase in workload but again the unfamiliarity with the system and operating procedure might have an effect as well. However, all average SDLP remain below 36 cm found in a car-simulator by Reed and Green (1999). They argued that in simulators without motion systems, like the one used for this study, higher variances are possible due to the missing of kinaesthetic feedback.

Another concern in regards to the coupling system was the short following distances of roughly 10 meters behind leading trucks. Eick and Debus (2002) and also Wille (2002) found an adaptation towards smaller gaps to lead vehicles after such a coupling, which are risks to safety. The results of this study show however, that the HMI feedback (see Fairclough, May & Carter, 1997) works and distances remain within a normal range of 1.0 to 3.0 s (see Fuller, 1981; Fuller, 1984; Van Winsum & Heino, 1996). The average THW for both conditions is longer than the German law with 2.1 s requirements (§ 4 Abs. 3 StVO) and therefore not relevant for safety concerns. Also the minimal THW are within a range found to be normal in simulators by Skottke (2007).

In general the simulator evaluation of the KONVOI-system does not show any critical results, which could have indicated that drastic changes have to be made within the project. It is however necessary to test the system on-the road as well as for longer durations to rule out any possible risks in technology and human behaviour. This also has the potential to show further positive contributions the KONVOI-system can bring to the driver, the overall infrastructure, and economy.

References

Bainbridge, L. (1983). *Ironies of automation*. Automatica, *19*, 775-779.

Billings, C.E. (1988). Towards human centered automation. In S.D. Norman and H.W. Orlady (Eds.), *Flight deck automation: Promises and realities* (pp. 167-190). Moffet Field, CA: NASA-Ames Research Center.

De Waard, D. (1996). *The Measurement of Drivers' Mental Workload*. Haren, The Netherlands: The Traffic Research Centre VSC, University of Groningen.

De Waard, D., Van der Hulst, M., Hoedemaeker, M., & Brookhuis, K.A. (1999). Driver behavior in an emergency situation in the Automated Highway System. *Transportation Human Factors, 1*, 67-82.

Eick, E.-M. & Debus, G. (2002). Situational Awareness und Abstandsverhalten im automatischen Kolonnenverkehr: riskante Fehlanpassung? In M. Grandt and K.-P. Gärtner (Eds.), *Sitaution Awareness in der Fahrzeug- und Prozessführung*. DGLR-Bericht 2002-04. Bonn, Germany: DGLR.

Endsley, M.R. (1996). Automation and situation awareness. In R. Parasuraman and M. Mouloua (Eds.), *Automation and human performance: Theory and applications* (pp. 163-181). Mahwah, NJ: Lawrence Erlbaum.

Endsley, M.R. (1988a). Design and evaluation for sitaution awareness enhancement. *In Proceedings of the Human Factors Society 32nd Annual Meeting* (pp. 97-101). Santa Monica, CA: Human Factors Society.

Fairclough, S.H., May, A.J., & Carter, C. (1997). The effect of time headway feedback on following behaviour. *Accident Analysis and Prevention, 29*, 387-397.

Fuller, R.G. (1981). Determinants of time headway adopted by truck drivers. *Ergonomics, 24*, 463-474.

Fuller, R.G. (1984). Prolonged driving in convoy: the truck driver experience. *Accident analysis & prevention, 16*, 371-382.

Hoffmann, S. & Buld, S. (2006) Darstellung und Evaluation eines Trainings zum Fahren in der Fahrsimulation. In VDI-Gesellschaft Fahrzeug- und Verkehrstechnik (Hrsg.), *Integrierte Sicherheit und Fahrerassistenzsysteme* (VDI-Berichte, Nr. 1960, S. 113-132). Düsseldorf: VDI-Verlag.

Hoyos, C.G. & Kastner, M. (1986). *Belastung und Beanspruchung von Karftfahrern. Unfall- und Sicherheitsforschung Straßenverkehr*. Hrsg: Bundesanstalt für Straßenwesen, Heft 59. Bremerhaven: Wirtschaftsverlag NW.

Reed, M.P., & Green, P.A. (1999). Comparison of driving performance on-road and in a low-cost simulator using a concurrent telephone dailing task. *Ergonomics, 42*, 1015-1037.

Sarter, N.B., Woods, D.D., & Billings, C.E. (1997). Automation Surprises. In G. Salvendy (Ed.), *Handbook of Human Factors and Ergonomics, 2nd ed.* (pp. 1926-1943). New York, NY: Wiley.

Skottke, E.-M. (2007). *Automatisierter Kolonnenverkehr und adaptiertes Fahrverhalten - Untersuchungen des Abstandsverhaltens zur Bewertung möglicher künftiger Verkehrsszenarien.* MSc thesis, not published, Rheinisch-Westfälische Technische Hochschule Aachen.

Van Winsum, W. & Heino, A. (1996). Choice of time-headway in car-following and the role of time-to-collision information in braking. *Ergonomics, 39,* 579-592.

Wickens, C.D. (1992). *Engineering Psychology and Human Performance.* (2nd ed.). New York: Harper Collins.

Wickens, C., & Hollands, J. (2000). *Engineering psychology and human performance.* Upper Saddle River, NJ: Prentice-Hall Inc.

Wiener, E.L. & Curry, R.E. (1980). Flight deck automation: Promises and problems. *Ergonomics, 23,* 995-1011.

Wille, M. (2002). *Abstandsadaptation in elektronisch gekoppelten Fahrzeugkolonnen.* MSc thesis, not published, Rheinisch-Westfälische Technische Hochschule Aachen.

Zijlstra, F.R. (1993). *Efficiency in work behavior. A design approach for modern tools.* PhD thesis, Delft University of Technology. Delft, The Netherlands: Delft University Press.

Automation spectrum, inner / outer compatibility and other potentially useful human factors concepts for assistance and automation

Frank Flemisch, Johann Kelsch, Christian Löper, Anna Schieben,
& Julian Schindler
DLR German Aerospace Centre
Braunschweig, Germany

Abstract

Enabled by scientific, technological and societal progress, and pulled by human demands, more and more aspects of our life can be assisted or automated. One example is the transportation domain, where in the sky commercial aircraft are highly automated, and on the roads a gradual revolution takes place towards assisted, highly automated or fully automated cars and trucks.

Assistance and automation can have benefits such as higher safety, lower workload, or a fascination of use. Assistance and automation can also come with downsides, especially regarding the interplay between human and technology (e.g., Bainbridge, 1983; Billings, 1997; Norman, 1990; Sarter & Woods, 1995a). In parallel to the technological progress, the science of human factors has to be continuously developed such that it can help to handle the technological complexity without adding new complexity (e.g., Hollnagel, 2007).

In this overview article, some fundamental human factors issues for assistance and automation that the authors found useful in their daily work are briefly sketched. Some examples are described how those concepts could be used in the development of assistance and automation systems. While the article deals especially with assistance and automation in vehicles, the underlying concepts might also be useful in other domains.

From levels of automation to automation spectrum

Sometimes the terms "assistance" and "automation" are used as if they are clearly distinct or even opposite poles. In addition, some technologically brilliant developments (Dickmanns, 2002; Parent, 2007; Thrun et al., 2006) might suggest that fully automated vehicles are the "natural" follower of manually controlled vehicles and the unavoidable future. The challenge of automation is more complex, there might be solutions between assistance and automation. Which concepts could help to structure the discussion about automation issues?

In D. de Waard, F.O. Flemisch, B. Lorenz, H. Oberheid, and K.A. Brookhuis (Eds.) (2008), *Human Factors for assistance and automation* (pp. 257 - 272). Maastricht, the Netherlands: Shaker Publishing.

In science, there are already examples extending the common dual approach of manually controlled system vs. full automation. Sheridan and Verplank (1978) for example, expand the binary "either/or"-perspective on automation ("the computer decides everything, acts autonomously, ignoring the human" and "the computer offers no assistance") with eight more levels (e.g., "computer informs the human only if asked" or "computer suggests one alternative" etc.), and open up the discussion about a multi-dimensional design space of automation. Billings (1997) extends the automation concept such that, in addition to control automation and management automation (automation of complex management issues like navigation supporting by, e.g., flight management systems), the provision of information is already automation ("information automation"). Parasuraman et al. (2000) assign the various processing stages of automation (information acquisition, information analysis, decision selection and action implementation) as the second dimension to the continuous "levels of automation".

While the approach of Parasuraman et al. (2000) is quite helpful, essential aspects of automation might be efficiently communicated with a one-dimensional spectrum of continuous automation degrees (figure 1). This spectrum indicates the involvement of human and automation in the control of the human-machine system.

Figure 1. Involvement in system control as a spectrum between fully manual and fully automated

In this continuous spectrum different regions can be identified as assisted, semi-automated, highly-automated and fully automated control (figure 2). Control over the system might be transferred from the human operator to the automation and vice versa for all automation levels. Throughout the complete spectrum of assistance and automation similar principles of the transition might be applicable.

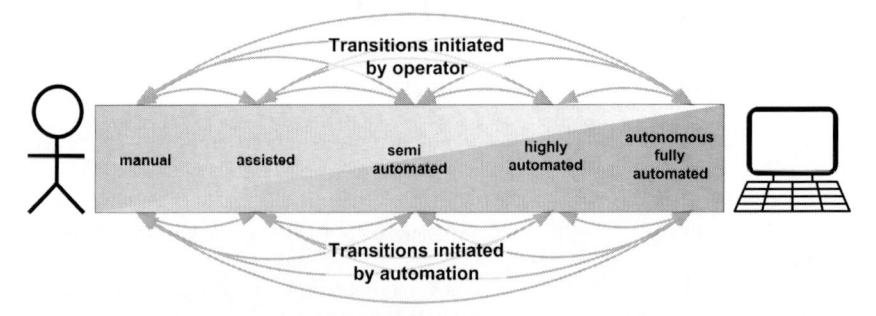

Figure 2. Automation spectrum, automation regions and transitions

This simple "map" of the automation spectrum can be used, e.g., for describing two different but related aspects:

a) The level of involvement of human and machine in the control of a human-machine system for a specific moment ("We are driving highly automated right now.").
b) The automation capabilities of a specific vehicle ("A Boeing 777 is a highly automated vehicle.").

At the beginning of the 21st century some automation subsystems in cars that influence the control of the vehicle are often called assistant systems, like Adaptive Cruise Control ("ACC") or Manoeuvring Aids for Low Speed Operation (MALSO, "Park Assistant"). For the benefit of cross-utilization of research and development efforts, the automation spectrum described here includes assistance as part of the global research and development effort of automation. Both ACC and MALSO Systems control either the longitudinal or lateral axis of the vehicles completely, and can therefore be assigned to in the region of semi-automation.

With "highly automated vehicles", research efforts in the car and truck community are addressed that go beyond semi-automated vehicles, but actively involve the driver in the control task, and link those efforts with the development in aviation, where highly automated aircraft with flight management systems have been in use already for decades.

A soft classification description for vehicle classes "Semi-automated vehicle" and "Highly automated vehicles" could be: A semi-automated vehicle has automation capabilities that allow to automate about half of the control of the movement (e.g. either lateral or longitudinal control.) Highly automated vehicles have automation capabilities higher than semi-automated up to fully automated control of the movement, where a human is usually actively involved in the control of the vehicle.

At the beginning of the 21st century, examples for highly automated vehicles are modern aircraft like the Boeing 737-400 to 777 or Airbus 320 to 380. While in 2007 highly automated cars and trucks are mainly a matter of research (e.g., Holzmann, 2006), some Japanese cars on the Japanese and UK market equipped with both ACC and LKAS (Lane Keeping Assistant System) already cross the border from semi- to highly automated vehicles. An example of a fully automated vehicle is Cybercars (Parent, 2007), where the user only communicates the destination, and from then on is a passenger.

The automation spectrum described above offers a strongly simplified perspective on human-machine systems. To design human-machine interaction in detail, especially regarding time related aspects, more precise perspectives are necessary. Some of those perspectives can be described in the Unified Modelling Language (UML), which can be expanded towards human-automation issues. As an example, figure 3 shows a sequence diagram for the transition of control from an "Automated Highway System" (AHS) to a human operator (Bloomfield et al., 1998, diagram by DLR). The diagram shows the sequence of interaction between the AHS system, covering the automation system, the vehicle and the infrastructure, and the human operator. Longitudinal and lateral control (grey columns) is transferred to the driver, after a short visual message of the system, as soon as the driver actuates the accelerator or brake pedal and steering wheel.

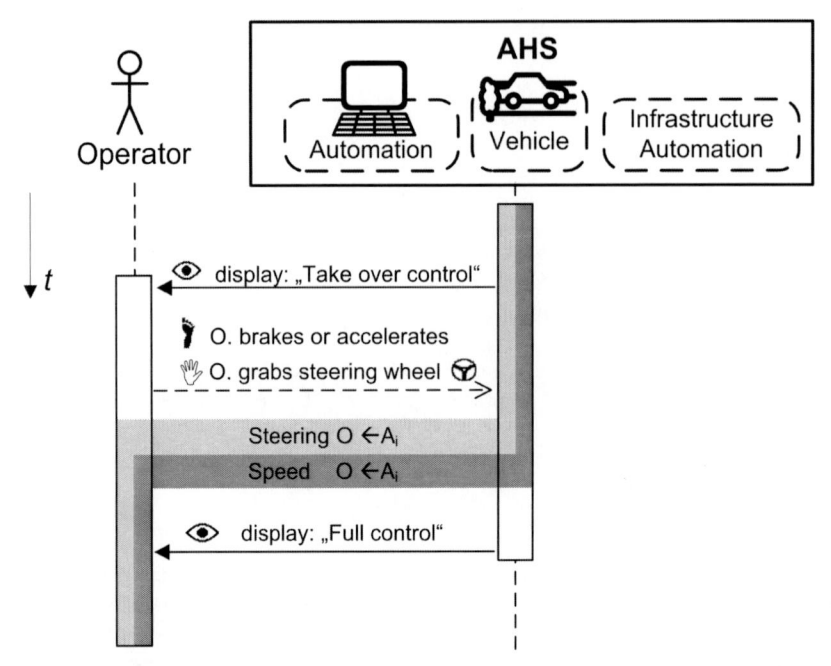

Figure 3. UML-based sequence diagram illustrating the interaction between operator and vehicle during a transition of control from the Automated Highway System (AHS) back to the human operator

Like the sequence diagram in figure 3, each of the diagrams presented in this article offers a specific perspective. Only together these perspectives open up the chance to sufficiently map, in width and depth, the territory of human-machine co-operation.

From "either/or"-automation to shared and co-operative control

The common approach for designing automated systems is to build up an automation, which perceives the environment and provides feedback or smaller control actions. The human, who also perceives and intervenes, is connected over a more or less compatible human-machine interface. When the human switches the automation on, he often leaves the control loop for this particular task. Human and automation both act on the vehicle as two relatively independent sub-systems in an "either/or"-relationship (figure 4).

The more the research and development community explores automation beyond assistance towards semi- and highly automated vehicles, the more important it becomes to think beyond classical control and thoroughly investigate relationships between the human and the automation beyond an "either/or"-relationship. Christoffersen and Woods (2002), for example, suggest designing an automated system as a team player, allowing a fluid and co-operative interaction. Of particular importance are the observability and directability of the co-operative automation. Schutte (1999) proposes using complementary automation or "complemation" in design, where complemation uses technology to fill in the gaps in the human's skills rather than replacing those skills. Miller and Parasuraman (2007) introduce a concept of a flexible automation. In this concept the operator can delegate tasks to the automation like in a human-human interaction so that the automation is adaptable to the specific needs of the human. The research of Griffiths and Gillespie (2005) deals with the exploration of a haptic control interface for vehicle guidance. In this context they use the term "shared control" to describe that the driver as well as the automation can have control over the vehicle at the same time.

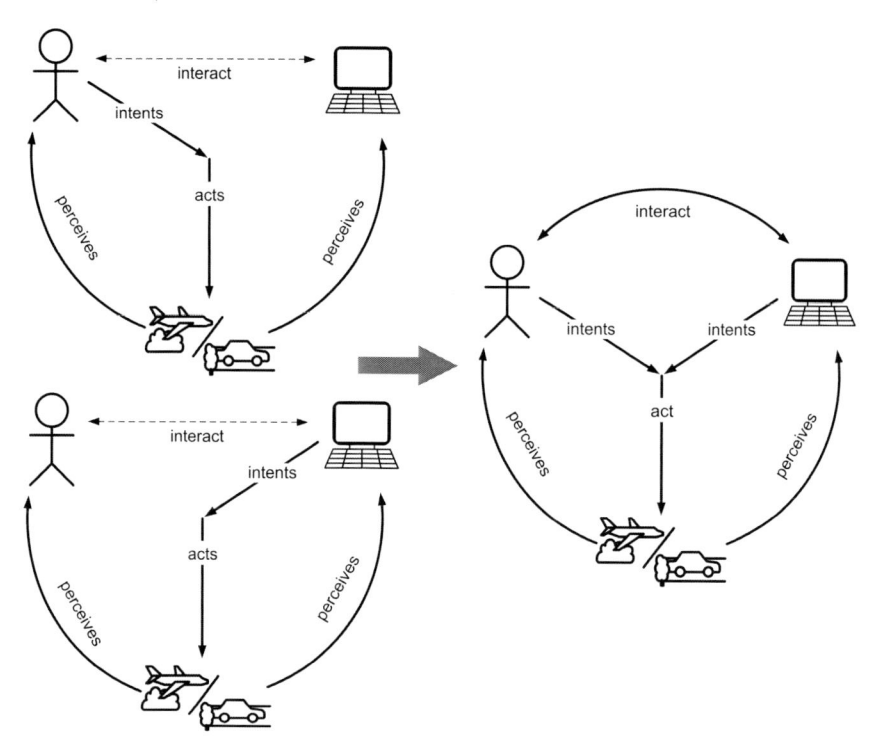

Figure 4. From classical "either/or"-automation to shared control and co-operative control

A concept that includes shared control but goes a step further can be described as co-operative control. Co-operation can be understood as working jointly towards the same goal. In 2008, cooperation in the context of vehicles is mainly used for the cooperation between vehicles. Co-operation can also be applied to the cooperation between operator and the automation, as hinted already by Onken (2002) and described for military systems by Schulte et al. (2006). For vehicle control cooperation this means that the functions which are needed to steer a vehicle are handled together and that the automation actively supports a harmonization of control strategies of both actors (automation and driver) towards a common control strategy. To enable this, the inner and outer design of the automation has to be compatible with the human and a continuous interaction has to be established. Intentions for actions are matched via a corresponding human-machine-interface and a joint action implementation can take place. Co-operative control should make the automation responsive to the driver's intentions and gives the driver the opportunity to optimize his own strategy. An example how co-operative control can be achieved is described further down.

From mental model to compatibility

Automation and assistance systems are additional subsystems in human-machine systems that could add additional complexity, especially when the level of automation is varying. Knowing what the automation does, why it does this and what it will do in the future is crucial for a successful interplay between operator and automation. Generalizing Sarter and Woods' (1995b) concept of mode awareness, this build-up of situation awareness (Endsley, 1995) about the automation can be called automation awareness. A simple system analysis of the information flow in the human-machine system shows that in order to gain and maintain situation awareness, there has to be a sufficient representation, a mental model of the automation inside the operator.

The term mental model has been used in different contexts since it was first mentioned by Craik (1943). In the context of system design and usability Norman (1983) describes mental models as follows: "In interacting with the environment, with others, and with the artefacts of technology, people form internal, mental models of themselves and of the things with which they are interacting. These models provide predictive and explanatory power for understanding the interaction" (p. 7). Therefore, usability seems to be strongly linked to the quality of the matching of the user's mental model of a system and the system functionality.

When humans interact with humans, each of the interaction partners seems to have a mental model of the other partner, whose neurological representation is, for example, described in the compelling concept of mirror cells (Rizzolatti et al., 1996). Applying this thought to human-machine interaction, a human operator builds up a mental model of the whole human-machine system and its relationships (figure 5). Similarly, the automation has to have implicit or explicit representations of its relationship with the vehicle and the environment, in order to perform its task.

The next logical step is to discuss whether a "mental" model of the automation about the operator also makes sense and is feasible (figure 5). In human-computer interaction there are already some approaches to provide the computer with information about the user. This information can be made available in an explicit way for example by user profiles or in an implicit way by the computer itself, analysing the users past behaviour (e.g., Allen, 1997).

To build up a "mental" model of an operator is not a trivial step: humans can be rather complex, especially regarding emergent effects that are much more difficult to model than deterministic effects. Moreover, if we want to use this "mental" model as human factors concepts for assistance and automation 263

a basis for building up an adaptivity of the automation this can add an additional complexity to the overall system and to the mental model of the operator, especially when the adaptivity of the automation meets human adaptivity. However fruitful such approaches will be in the future to add a "mental" model of the operator to the automation, there is a clear priority: good design of artefacts should at first enable humans to build up and maintain a sufficient mental model of the artefact and its use, before it might enable the artefact to build up a "mental" model of the human.

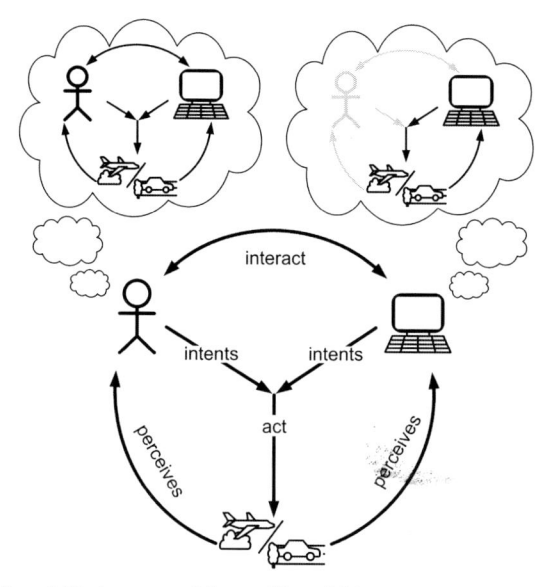

Figure 5. Mental model in human and "mental" model in automation

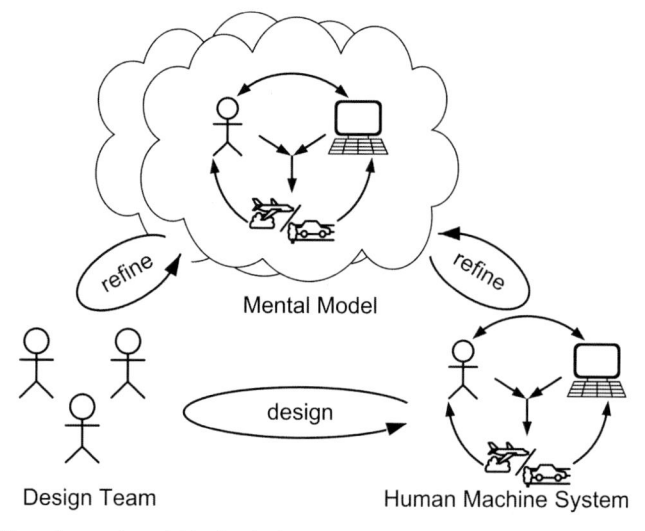

Figure 6. Shared mental model in the design process

If we go back one step further and look beyond the human-machine systems, mental models are also a useful concept in the design process: good design does usually not fall out of the sky, but derives from the hard work of many people in the design process. A key challenge here is to develop, refine and communicate a common and clear "picture", a shared mental model of the future human-machine system early enough in the design process (figure 6). Many shortcomings in the design seem to be related to shortcomings and discrepancies in the mental models shared within the design team, and with future users of the human-machine system. An example later in the article will illustrate how the build-up of a shared mental model can be supported with a "seed crystal".

Back to the human-machine systems: Norman (1988) defines the terms "Gulf of Execution" and "Gulf of Evaluation" and illustrates that the fitting of the mental representation of the user and the physical components and functionality of a system is essential for good usability. The term "Gulf of Execution" describes how well a "system provides actions that correspond with the intention of a person", whereas the term "Gulf of Evaluation" focuses on feedback issues. It is defined as "the amount of effort a person must exert to interpret the physical state of the system and to determine how well the expectations and intentions have been met" (p. 51). For a more detailed discussion on gulfs and distance between humans and machine see Schutte (2000). While "Gulfs" are stressing the distance between humans and machine, the concept of "compatibility" described in the following paragraph stresses the necessary matching of humans and machines.

Inner and outer human-machine compatibility

Compatibility is a quality describing the fit or match between two entities. Human-automation compatibility is a subset of human-machine compatibility and specifies how easy it is for the user to interact and understand the actions of the automation in each situation. Bubb (1993) describes compatibility as the effort a human needs to recode the meaning of different information. High compatibility leads to a reduction of recoding effort. He differentiates between "outer" compatibility as the correct fit of the interfaces between human, machine and reality, and "inner" compatibility as the fit of the operators' mental (inner) model with the perceived information via the operated machine. Based on Bubb's definition, the notion of compatibility can be broadened:

Outer human-machine compatibility describes the fit of the outer borders of the human (such as their eyes, ears and hands) with the outer border of the machine (the hardware interface). This means for example that the machine only uses signals for interaction that are in the range of human sensors. Moreover, the inceptors of the machine, e.g., buttons and levers, should be designed in an ergonomic way. Many issues of outer compatibility are addressed in the field of classical ergonomics.

Inner human-machine compatibility can be defined as the match or fitting of the inner subsystems of the human with the inner subsystems of the machine. Coll and Coll (1989) for example describe a cognitive match, which can be reached by "making the system operate and interact with the user in a manner which parallels the flow of the user's own thought processes" (p. 227). This cognitive match is essential for usability. A cognitive match or cognitive compatibility is one part of the inner compatibility, but not the only one: other parts of inner compatibility are emotional compatibility, and compatibility of values and ethics, concepts that have been hinted by science fiction (e.g., Asimov, 1950), but still have to be explored in the science of human factors.

Compatibility does not necessarily mean similarity or even equality: in the same way as a power outlet and a power plug are different, but compatible, humans and machines can be different, but should also be compatible. The concept of human-machine compatibility described so far offers a well defined boundary between inner and outer compatibility, i.e., the outer border of the human and the machine, but also stresses that inner and outer compatibility belong together inseparably: overall human-machine compatibility is the product of inner and outer compatibility. Sufficient compatibility between humans and machines can only be reached, if there is enough outer AND inner compatibility.

Automation roles

Thinking about mental models and inner or outer compatibility can help us design human-machine systems, but might not be sufficient, especially when machines get more complex. In human-human relationships, additional concepts have proved to be helpful, and might cautiously be applied to human-machine relationships.

In literature, there are already concepts dealing with the role-sharing between human and automation within a human-machine system and the design of this role-relationship. Billings (1997) pointed out that "responsibility" and "authority" are two very important characteristics in the design of a human-centred automation. He discussed the problem of "limitations on pilot authority" and pointed out two types of such limits as part of a special design space: "hard limits" and "soft limits". A role includes responsibility and authority but might cover also more aspects: Linton (1979) defines the social role as an entirety of all "cultural models" attributed to the given status (e.g., mother, boss). This includes expectations dependent on the social system, values, patterns of activity and behaviour. A social actor has to rise to these requirements according to his position. Altogether, a role is a multi-dimensional construct, in which parts of it can be dependent on each other. Similar to the automation spectrum described above this multi-dimensional construct can be mapped to a one-dimensional role-spectrum (figure 7).

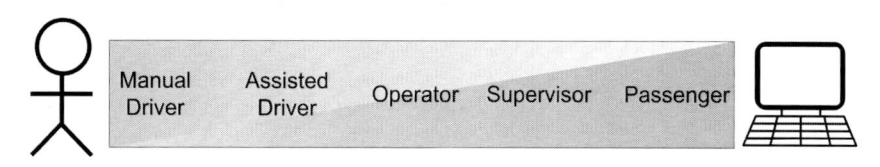

Figure 7. Potential role spectrum in vehicle assistance and automation

While the potential of role concepts applied to automation still waits to be exploited, one way to create explicit roles is the use of metaphors in system design. A metaphor serves both the designer and the user to understand the possible diversity of roles and of the related complex system easier and quicker.

Example: co-operative, manoeuvre-based automation and arbitration

The following section gives a brief overview of the concept of arbitration and co-operative, manoeuvre-based automation, which implement the common concepts of co-operative control and compatibility.

What happens if human and automation, both intervening in the vehicle control, have different perceptions of the situation or different intentions? An example would be a road fork, where the human wants to turn to the left and the automation right. Such conflicts between human and automation have to be resolved. The human-machine system must achieve a stable state, in which a clear and safe action can be executed. Time is often a critical factor. Griffiths and Gillespie (2005) already speak about the "collaborative mode of interaction"(p. 575) and a need for negotiation between human and automation suggesting a kind of human-machine haptic negotiation on the same control interface. To achieve this, a concept of "arbitration" can be helpful, i.e., a fast negotiation between human and machine about co-operative actuator access with the aim of reaching a "joint will" and a "joint action" (Kelsch, 2006). The concept of arbitration uses dialogue rules and psychological conflict solving approaches as described for verbal and non-verbal human-human and human-animal communication. Arbitration can be enabled implicitly by an appropriate design of the automation and the interface, or explicitly with an arbiter, a specialized subsystem of a co-operative automation that moderates the negotiation between human and automation and if necessary makes an equitable decision in a time-critical situation (figure 8).

To enable the described arbitration process the underlying automation has to be co-operative. Which specifics are needed for the design of the co-operative automation? The automation has to generate action suggestions and rate these suggestions. Within the discussion process with the human via the interaction the action intentions may be modified or revaluated. Finally the common action intention has to be prepared for execution. Since there is now a common intention about what to do, the proportion of control can be dynamically distributed between human and automation, a common action implementation can be executed.

To facilitate the co-operative handling of vehicle control an automation structure is needed which allows the discussion and the generation of a joint driving strategy. To achieve this, the concept of inner compatibility described above is employed. Figure 8 shows one way how cognitive compatibility (part of the inner compatibility) of human and automation can be increased. For the human information processing some aspects of the models of Donges and Naab (1996), Endsley (1995), Parasuraman et al. (2000) and Rasmussen (1986) were combined and simplified. This basic structure can also serve as the basic "cognitive" structure for the automation. Firstly, the automation module "perception" generates and process sensor data about the environment. After this, the module "situation assessment" builds up a situation representation which serves as basis for the following processes. Intended actions are generated on four levels in continuous communication with the operator.

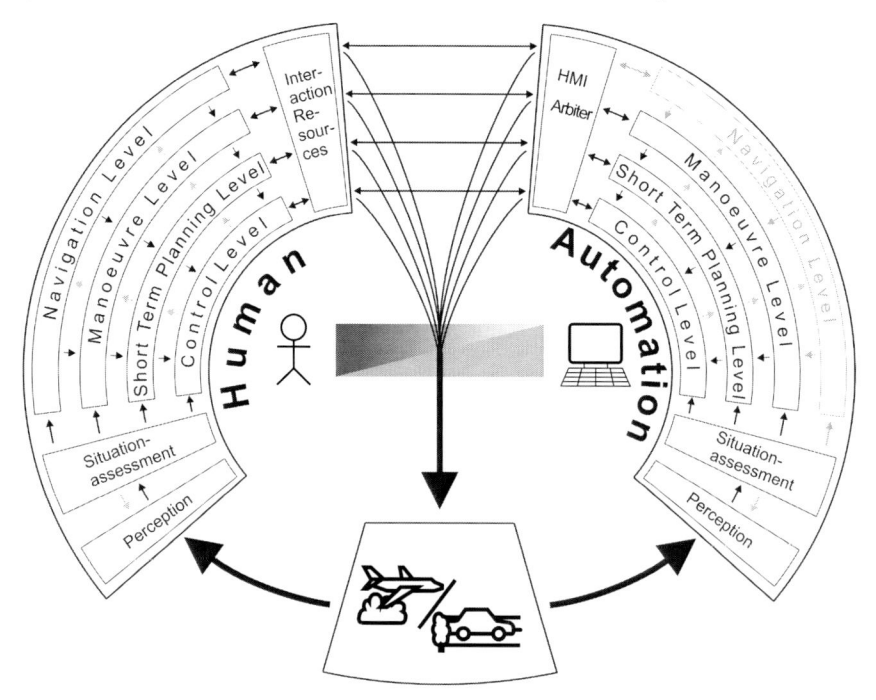

Figure 8. Co-operative automation and arbitration

The navigation level is used for planning a route for the vehicle through the road network to reach a certain destination. The next lower level is structured in manoeuvres, time and space relationships that are also meaningful to the operators / drivers. Manoeuvres are, for example, "follow right lane!" or "overtake!". The short term planning level provides a trajectory for the vehicle movement for a short period of time. Based on this trajectory, the control level generates control actions that are fed to an active interface and are combined with the user's actions.

Based on the experience with co-operative control so far, it seems to be essential that the loop between human and automation is closed and maintained on all four levels simultaneously. This closing should be in a way that allows the human to fluidly change his focus to one particular level without loosing track of the other three. This also opens up the option of a "Fluid Automation", a special form of adaptive automation, where the automation "flows" into those levels that are currently not in the focus of the human (P.C. Schutte & F.O. Flemisch, personal communication, November, 2002).

There is a good chance that a structure of the automation similar to the operator's understanding of the task, combined with an explicit arbitration, leads to a higher inner and outer compatibility and therefore to a better interaction and co-operation between human and automation. It is important to keep in mind, that even if the internal structure of human and a co-operative automation might look similar, e.g. in figure 8, and even if there are co-operative design metaphors like an "electronic co-pilot", this does not necessarily mean that the automation has to be human like. Capabilities and implementation can be vastly different, as long as human and automation are compatible, as shown in the next chapter.

Example: H-Metaphor, a design metaphor for highly automated vehicles

An example where all of the concepts described above come together is the H-Metaphor. A metaphor applies a source (e.g., a natural example) to a target (e.g., a technical artefact), creating something new. An example for a design metaphor is the desktop metaphor, where the concept of an office desk is applied to the surface of a computer operating system, creating a "computer desktop". Another example in the domain of vehicle automation is the H-Metaphor, where the concept of horseback riding/horse carriage driving is applied to the haptic-multimodal interaction (H-Mode) with highly automated vehicles (Flemisch et al., 2003; figure 9). One potential benefit of a design metaphor is that it provides an easy to communicate seed crystal for a shared mental model between the members of a design team and the operators of the designed system.

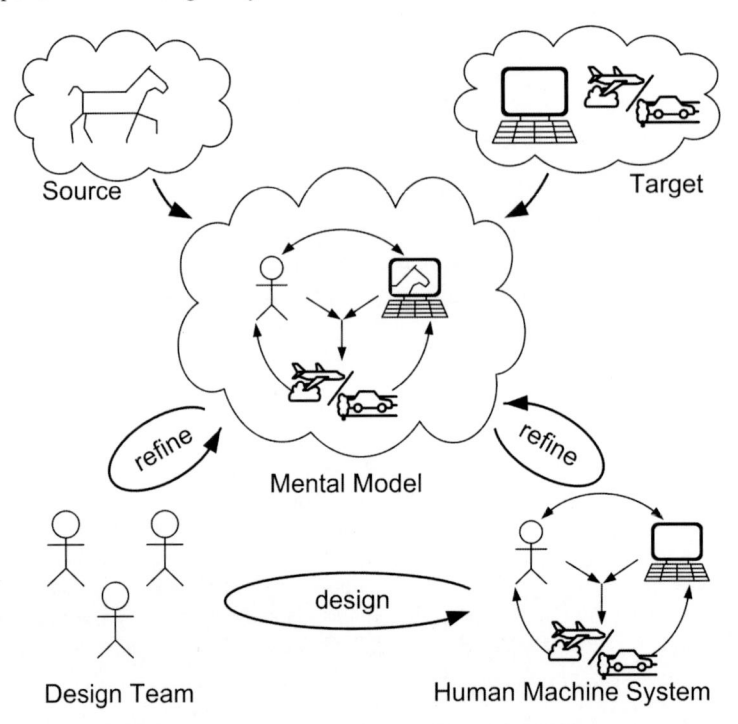

Figure 9. Design Metaphor as technique to create shared mental models (Example H-metaphor)

The H-Metaphor also describes levels of automation (tight rein / loose rein), cooperative control with a mix of continuous and discrete interaction, the transitions and the general role of the operator (figure 10).

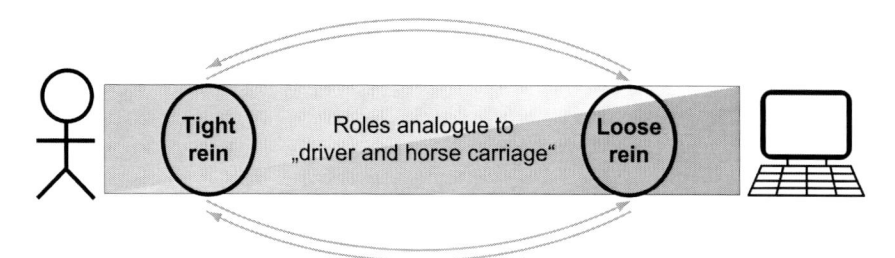

Figure 10. Automation and role spectrum described by the H-Metaphor

Arbitration, as sketched in the last chapter, has been implemented as a fast haptic-multimodal negotiation between human and machine, similar to the communication between human and horse.

The H-Metaphor has been applied to wheelchairs (Tahboub, 2001), to aircraft (Goodrich et al., 2006) and to cars (Flemisch et al., 2007), with a far reaching goal to develop a universal, haptic-multimodal language (H-Mode) for the interaction between humans and highly automated vehicles at all.

Assistance and automation: a risk, a challenge and a chance, also for human factors

At the beginning of the 21st century, technology pushes strongly towards more complex assistance and automation. This is a challenge, a risk and a chance for all of us, and especially for human factors. On the one hand, if human factors would only use the mindset and methods of yesterday to solve the problems of today, it would inadvertently contribute to the complexity of tomorrow and would be in a strong dilemma, as Hollnagel (2007) puts it. On the other hand, if and only if the human factors community continuously develops appropriate mindsets and methods in close coupling with solving the problems of today, there is a realistic chance that human factors can help to handle the complexity of tomorrow, and can make a difference.

References

Allen, R.B. (1997). Mental Models and User Models. In M. Helander, T.K. Landauer, and P. Prabuh (Eds.), *Handbook of Human-Computer Interaction* (pp. 49-63). Amsterdam: Elsevier.
Asimov, I. (1950). *I, Robot*. New York: Gnome Press.
Bainbridge, L. (1983). Ironies of Automation. *Automatica, 19*, 775-779.
Billings, C.E. (1997). *Aviation automation: The search for a human-centered approach*. Mahwah: Lawrence Erlbaum Associates.

Bloomfield, J.R., Levitan, A.L., Grant, A.R., Brown, T.L., & Hankey, J.M. (1998). *Driving performance after an extended period of travel in an automated highway system* (Report No. FHWA-RD-98-051). Georgetown Pike: U.S. Federal Highway Administration.

Bubb, H. (1993). Systemergonomie. In H. Schmidtke (Ed.), *Ergonomie*. München: Carl Hanser.

Christoffersen, K., & Woods, D.D. (2002). How to make automated systems team players. In E. Salas (Ed.), *Advance in human performance and cognitive engineering research: Automation* (Vol. 2, pp. 1-11). Amsterdam: Elsevier Science.

Coll, R., & Coll, J.H. (1989). Cognitive Match Interface Design, A Base Concept for Guiding the Development of User Friendly Computer Application Packages. *Journal of Medical Systems, 13*, 227-235.

Craik, K.J.W. (1943). *The Nature of Explanation*. Cambridge: Cambridge University Press.

Dickmanns, E.D. (2002). Vision for ground vehicles: History and prospects. *International Journal of Vehicle Autonomous Systems, 1*(1), 1-44.

Donges, E., & Naab, K. (1996). Regelsysteme zur Fahrzeugführung und -stabilisierung in der Automobiltechnik. *at - Automatisierungstechnik, 44*(5), 226-236.

Endsley, M.R. (1995). Toward a theory of situation awareness in dynamic systems. *Human Factors, 37*, 32-64.

Flemisch, F.O., Adams, C.A., Conway, S.R., Goodrich, K.H., Palmer, M.T., & Schutte, P.C. (2003). *The H-Metaphor as a Guideline for Vehicle Automation and Interaction* (Report No. NASA/TM—2003-212672). Hampton: NASA, Langley Research Center.

Flemisch, F.O., Kelsch, J., Schieben, A., Schindler, J., Löper, C., & Schomerus, J. (2007). *Prospective Engineering of Vehicle Automation with Design Metaphors: Intermediate report from the H-Mode Projects*. Paper presented at the 7. Berliner Werkstatt Mensch-Maschine-Systeme, Berlin.

Goodrich, K.H., Schutte, P.C., Flemisch, F.O., & Williams, R.A. (2006). *A Design and Interaction Concept for Aircraft with Variable Autonomy: Application of the H-Mode*. Paper presented at the 25th Digital Avionics Systems Conference, Portland, USA.

Griffiths, P., & Gillespie, R.B. (2005). Sharing Control Between Human and Automation Using Haptic Interface: Primary and Secondary Task Performance Benefits. *Human Factors, 47*, 574-590.

Hollnagel, E. (2007). *Modelling human-system interaction: From input-output to coping with complexity*. Paper presented at the 7. Berliner Werkstatt Mensch-Maschine-Systeme, Berlin.

Holzmann, F. (2006) Adaptive cooperation between driver and assistant system to improve road safety; Dissertation; Lausanne: Swiss Federal Institute of Technology

Linton, R. (1979). *Mensch, Kultur, Gesellschaft*. Stuttgart: Hippokrates-Verlag.

Kelsch, J., Flemisch, F.O., Löper, C., Schieben, A., & Schindler, J. (2006). *Links oder rechts, schneller oder langsamer? Grundlegende Fragestellungen beim Cognitive Systems Engineering von hochautomatisierter Fahrzeugführung.* Paper presented at the DGLR Fachauschusssitzung Anthropotechnik, Karlsruhe.

Miller, C.A., & Parasuraman, R. (2007). Designing for Flexible Interaction Between Humans and Automation: Delegation Interfaces for Supervisory Control. *Human Factors, 49*, 57-75.

Norman, D.A. (1983). Some Observation on Mental Models. In D. Genter & A. L. Stevens (Eds.), *Mental Models.* Hillsdale: Lawrence Erlbaum Association.

Norman, D.A. (1988). *The Psychology of Everyday Things.* New York: Basic Books.

Norman, D.A. (1990). The 'problem' with automation: Inappropriate feedback and interaction not 'over-automation'. In D.E. Broadbent, J. Reason, and A. Baddeley (Eds.), *Human Factors in Hazardous Situations* (pp. 137-145). Oxford: Clarendon Press.

Onken, R. (2002). Human Process Control and Automation - still compatible concepts? In B.-B. Burys and C. Wittenberg (Eds.), *From Muscles to Music: A Festschrift to Celebrate the 60th Birthday of Gunnar Johannsen* (pp. 75-87). Kassel University Press.

Parasuraman, R., Sheridan, T. B., & Wickens, C.D. (2000). A Model for Types and Levels of Human Interaction with Automation. *IEEE Transaction on Systems, Man, and Cybernetics, 30,* 286-297.

Parent, M. (2007). Advanced Urban Transport: Automation Is on the Way. *Intelligent Systems, 22*(2), 9-11.

Rasmussen, J. (1986). *Information processing and human-machine interaction: An approach to cognitive engineering.* New York: North-Holland.

Rizzolatti, G.L.F., Gallese, V., & Fogassi, L. (1996). Premotor Cortex and the Recognition of Motor Actions. *Cognitive Brain Research, 3,* 131-141.

Sarter, N.B., & Woods, D.D. (1995a). *'Strong, Silent and Out of the Loop:' Properties of Advanced (Cockpit) Automation and their Impact on Human-Automation Interaction* (Tech. Rep. CSEL 95-TR-01). Columbus: Cognitive Systems Engineering Laboratory, Ohio State University.

Sarter, N.B., & Woods, D.D. (1995b). "How in the world did we ever get into that mode?" Mode Error and Awareness in Supervisory Control. *Human Factors, 37,* 5-19.

Schulte, A. (2006). Co-operating Cognitive Machines - An Automation Approach to Improve Situation Awareness in Distributed Work Systems. NATO RTO System Concepts and Integration (SCI) Panel Workshop on Tactical Decision Making and Situational Awareness for Defense against Terrorism. Turin, Italy.

Schutte, P.C. (1999). Complemation: An Alternative to Automation. *Journal of Information Technology Impact, 1,* 113-118.

Schutte, P.C. (2000). Distance: A Metric for Flight Deck Design an-d Function Allocation. *Human Factors and Ergonomics Society Annual Meeting Proccedings, 44,* 85-88.

Sheridan, T.B., & Verplank, W.L. (1978). *Human and Computer Control of Undersea Teleoperators* (Technical Report). Cambridge, MA: MIT Man-Machine Systems Laboratory.

Tahboub, K.A. (2001). A Semi-Autonomous Reactive Control Architecture. *Journal of Intelligent and Robotic Systems, 32*, 445-459.
Thrun, S., Montemerlo, M., Dahlkamp, H., Stavens, D., Aron, A., Diebel, J., et al. (2006). Stanley: The Robot that Won the DARPA Grand Challenge. *Journal of Field Robotics, 23*, 661-692.

Objective and subjective assessment of warning and motor priming assistance devices in car driving

Jordan Navarro[1], Franck Mars[1], Jean-François Forzy[2], Myriam El-Jaafari[2], & Jean Michel Hoc[1]
[1]IRCCyN, CNRS & University of Nantes
[2]Renault, Guyancourt
France

Abstract

This paper deals with moderately intrusive driving assistance devices that intervene when lane departure is imminent. A previous simulator study (Navarro et al., 2007) showed that motor priming devices were more effective in assisting drivers than other lane departure warning systems. Such motor priming devices prompt the driver to take action by means of an asymmetric steering wheel vibration. This current experiment is aimed at gaining a deeper understanding of motor priming mechanisms. It raises the question of whether it is more efficient because it provides motor cuing, because it provides directional information on the steering wheel, or because the haptic modality elicits a faster response from the driver. In addition, subjective data were used to assess drivers' acceptance of the assistance devices. Results confirm that motor priming devices are more effective than auditory and vibratory warning devices during recovery manoeuvres. Neither the site of stimulation, nor the modality used for conveying information, appeared to play a significant role in the results. Interestingly, subjective data showed that drivers globally preferred auditory warning devices to motor priming devices. These results support the hypothesis that motor priming devices directly intervene at the motor level, in contrast to more traditional warning systems that act at the level of situation diagnosis.

Introduction

A significant number of all road accidents can be linked to lane departure. Bar and Page (2002) have estimated that accidents following an unintended lane departure represent about 40 percent of all crashes and about 70 percent of all road fatalities. In order to reduce the number of such accidents, various types of assistance devices are being investigated. These are expected to help drivers maintain a safe position in their lane. They rank from a simple warning when the vehicle is about to leave its lane to a complete delegation of lateral control. In all cases, cooperation between the driver and the automation will take place (Hoc, 2001). Hoc and Blosseville (2003) put forward a four-level classification system in order to categorize types of car-driving assistance within the framework of human-machine cooperation. All the driving assistance devices assessed in this study belong to the "mutual control" category, in the sense that they react to driver behaviour when the systems detect an imminent lane departure.

The current study follows on from work carried out by Navarro et al. (2007), which assessed several assistance devices belonging to the mutual control category. In particular, it focused on the evaluation of a new type of assistance called "motor priming". Such a device triggers asymmetric steering wheel vibrations when the car is about to cross one of the lane edge lines. More precisely the device triggers alternating steering wheel motion. The first movement of the steering wheel is directed toward the road centre (side of correction), with a stronger torque and speed than the one in the direction of the side of lane departure. The aim is to provide directional information on the steering wheel without correcting the vehicle's trajectory. In this way, the device intervenes at the motor level, preactivating the corrective gesture at the proprioceptive level, without actually performing it in the place of the driver. The motor priming device was compared to more traditional warning devices, such as a simple steering wheel vibration or a sound indicating the side of lane departure. The benefits of all assistance devices were measured during lane departures which were generated by occluding the driving scene at specific locations. Results showed that all driving assistance devices improved recovery manoeuvres in comparison to a condition without assistance. In all cases presented, the drivers spent less time in a dangerous lateral position; however, the benefits were significantly greater with the motor priming device. This was due to an improved action on the steering wheel when the corrective manoeuvre was initiated. The results gathered by Navarro et al. (2007) support the idea that motor priming not only improves situation diagnosis, in the same way as warning systems, but also provides a motor cue to the effectors of steering control, i.e. the hands.

The main objective of this current experiment was to further investigate the determinants of benefits associated with the motor priming approach. For this, a progressive method was used which compared assistance devices which were increasingly different from the motor priming mode. The aim was to assess the relative contribution of the different characteristics which define the motor priming mode to the observed benefits on recovery manoeuvres.

The first step was to compare the motor priming mode to a lateralized vibratory warning delivered on the steering wheel. Both devices are identical (i.e. they both provide directional information to the hands by means of the haptic modality), with the exception of the motor prompt which characterizes motor priming. This comparison will enable the specific role of the motor incentive in the improvement of recovery manoeuvre. The hypothesis is that the effect of the motor priming mode at the action level mainly resides in that part of the stimulation.

Using the steering wheel to stimulate the effector of the manoeuvre may also result in faster responses from the driver. To determine the effect of the localization of the stimulus, a comparison was made between a lateralized vibratory warning on the steering wheel and a lateralized vibratory warning on the seat. Both devices gave directional information via the haptic modality, but at different locations.

Finally, the simple fact of using the haptic modality may explain some of the benefits associated with motor priming (Sklar & Sarter, 1999; van Erp & van Veen, 2004). In an attempt to isolate this dimension, the lateralized vibratory warning on the seat was compared to a lateralized warning sound.

A secondary objective of the current experiment was to assess drivers' acceptance of all the driving assistance devices in parallel with their objective effects on steering behaviour. Drivers' judgments may not favour an automation device acting on the steering wheel, even if it does not interfere with the control of the vehicle (Lefeuvre *et al.*, 2004). It may be especially true for motor priming, due to the motor prompt. Related to this question, the combination of motor priming with a lateralized auditory warning was also studied. Navarro *et al.* (2007) did not observe any difference between this kind of combination and the unimodal motor priming mode when steering behaviour was analysed, but there may exist a difference in terms of acceptability. An auditory warning which mimics the sound of rumble strips was thought to be more acceptable because situation diagnosis is known to be based on the matching of the perception of an event and the previous knowledge of similar events (Wickens & Hollands, 2000). Applied to lateral control in driving assistance devices, rumble strip noise refers to well-known situations and can therefore be expected to be more acceptable. Hence, the combination of motor priming and an auditory warning of this type may form an optimal compromise between efficiency (brought about by motor priming) and acceptability (brought about by auditory warning).

Method

Participants

Four women and sixteen men 34 years of age on average (from 23 to 52 years old) were volunteered to take part in this experiment. Driving experience ranged from 4 to 35 years (16 years on the average). All of them had normal or corrected-to-normal vision. None experienced motion sickness.

Simulator

The experiment was carried out on a high-fidelity moving-base simulator (Cards2, developed by Renault's Technical Simulation Centre). The simulator constituted of a car's cockpit placed on a six-degree-of-freedom platform. The visual scene was projected onto 3 screens with 150° of visual angle. The simulator was kitted with the same equipment found in a real car, including a manual gearbox, force feedback steering wheel, pedals for brakes, accelerator and clutch, and a speedometer. The simulation was generated using the simulation software SCANeR© II (Oktal). The visual database represented a two-lane secondary road of 3.9 km in length.

Driving assistance devices

Six experimental conditions were compared in this study. All assistance devices were brought into action each time the vehicle moved more than 85 cm from the lane centre. They remained active as long as the vehicle position exceeded this threshold.

- Auditory warning (AW): a rumble strip noise was played by one of the loudspeakers placed in the doors of the simulator, in the direction of lane departure.
- Seat vibratory warning (SVW): two vibrators were used, one placed in the seat and the other in the back of the seat, in the direction of lane departure.
- Wheel vibratory warning (WVW): the vibration was delivered by one of two vibrators placed in the upper part of the steering wheel ("ten-to-two" position), in the direction of lane departure.
- Motor priming (MP): a triangular asymmetric steering wheel oscillation was generated by means of two opposite impulses of different strength. The torque applied on the steering wheel was of 2 Nm was when it moved toward the road centre and of 0.5 Nm when it moved in the direction of lane departure. The period of the command signal was of 0.3 s.
- Auditory and motor priming (AMP): the AW and the MP devices were combined.
- Control condition without assistance (WA).

Reading task

Lane departures were provoked by means of a distraction task that consisted of reading a succession of words displayed through a monitor placed on the dashboard (the position usually occupied by a car radio). While driving, participants were instructed to read aloud as many words as possible. During that task, the vehicle trajectory was changed slightly (drivers were unaware of this change) in order to take the car in one direction or the other. The reading task stopped (no more words on the monitor) when the vehicle reached the lateral position that triggered the driving assistance. In order to avoid too much predictability of the consequences of the distraction task, not all episodes led to lane departure.

Procedure

The study lasted about 90 minutes and consisted of 10 laps, each 3.9 km in length. Each of the five assistance devices was assessed over the course of one lap. Laps with assistance were alternated with laps without assistance. The order of presentation of the different assistance devices was fully counterbalanced between drivers. After each lap, drivers were briefly asked about the device they had just experienced.

Drivers were instructed to drive in the right-hand lane, respect speed limits and keep both hands on the steering wheel in a position close to the "ten-to-two" position. Lane departure situations were provoked both in bends and straight lines. The location of these critical situations changed depending on the lap. Traffic in the opposite lane was present at a rate of approximately four vehicles per kilometre and at a speed of 50 km/h. However, the traffic was arranged in such a way that the drivers never had to take into account a potential risk of collision.

After the driving test, post-experimental interviews were conducted and drivers were asked to rank the assistance devices in order of preference (without *ex-aequo*).

Data analysis

To assess drivers' performance, several variables were analysed, from the moment lane departure was imminent to the moment the car returned to a normal position in the lane. The main variable was the time spent by drivers outside the safety envelope of 85 cm from the lane centre. This will be referred as the duration of lateral excursion. Steering reaction times were computed to test drivers' reactivity after lane departure. This variable corresponds to the time between the end of the reading task and the drivers' first action on the steering wheel. The maximum rate of steering wheel acceleration once the recovery manoeuvre was engaged was also calculated. This variable represents the strength of the steering reaction.

Each assistance device (AW, SVW, WVW, MP, and AMP) was compared to control condition (WA) by means of a Student's *t-test*. In order to compare the effect of the driving assistance devices, the data obtained in the control condition (WA) were subtracted trial by trial from the data obtained with assistance devices. ANOVAs were carried out on these data sets. Newman-Keuls tests were used for post-hoc comparisons. The level of significance of $p<0.05$ was used in all tests.

The order of preference given by the participants was compared across assistance devices by means of a Friedman test.

Results

Duration of lateral excursion

The duration of lateral excursion in conditions without assistance (WA) was, on average, 2.79 seconds in straight lines and 3.30 seconds in bends. In straight lines, WVW, MP and AMP significantly reduced the duration of lateral excursion (Figure 1: WVW: $t(19)=2.31$, $p<.05$, d (size of the observed effect compared to WA) = 12%; MP: $t(19)=3.3$, $p<.01$, d=20%; AMP: $t(19)=2.81$, $p<.05$, d=15%). The effects of AW and SVW were not significant. In bends, only the effect of MP and AMP reached statistical significance (MP: $t(19)=3.38$, $p<.01$, d=21%; AMP: $t(19)=2.81$, $p<.05$, d=20%).

The ANOVA performed on the difference between conditions with assistance and WA revealed a significant effect of the assistance devices ($F(4,60)=10.04$; $p<.001$), no significant difference between bends and straight lines ($F(1,15)=.05$, ns) and no significant interaction between both variables ($F(4,60)=1.29$; ns). Post-hoc analysis indicated that MP and AMP significantly differed from all other devices (p<.05) and did not differ one from the other. There were also no significant differences between AW, WVW and SVW.

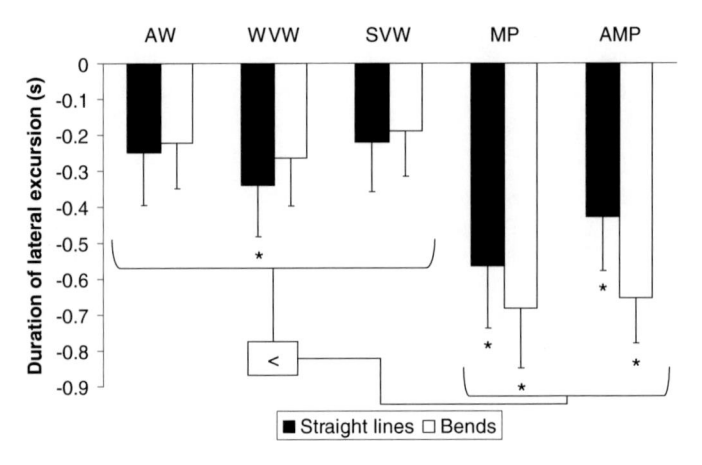

Figure 1. Effects of driving assistance devices on the duration of lateral excursion relative to the control condition. Stars represent significant differences compared to the control condition (0 on the figure). Error bars represent one standard error

Maximum rate of steering wheel acceleration

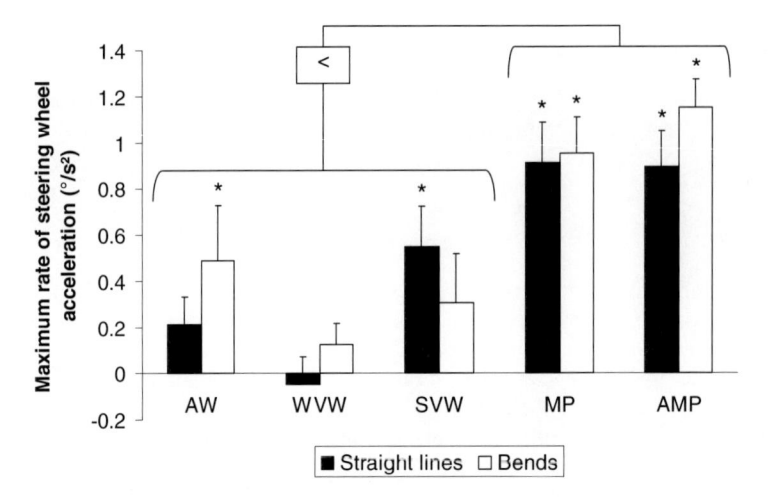

Figure 2. Effects of driving assistance devices on the maximum rate of steering wheel acceleration relative to the control condition. Stars represent significant differences compared to the control condition (0 on the figure). Error bars represent one standard error

The maximum rate of steering wheel acceleration in WA was, on average, 1.58°/s² in straight lines and 1.60°/s² in bends. In straight lines, the maximum rate of steering wheel acceleration significantly increased with SVW, MP and AMP, but not with WVW and AW (Figure 2: SVW: $t(19)=2.2$, $p<.05$, d (size of the observed effect compared to WA) = 35%; MP: $t(19)=4.87$, $p<.001$, d=58%; AMP: $t(19)=4.78$, $p<.001$, d=57%). In bends, the maximum rate of steering wheel acceleration significantly increased with AW, MP and AMP, but not with WVW and SVW (AW: $t(19)=2.26$, $p<.05$, d=30%; MP: $t(19)=6.4$, $p<.001$, d=59%; AMP: $t(19)=9.6$, $p<.001$, d=72%).

The ANOVA performed on the difference between conditions with assistance and WA revealed a significant effect of the assistance devices ($F(4,60)=20.45$; $p<.001$), no significant difference between bends and straight lines ($F(1,15)=0.34$, ns) and no significant interaction between both variables ($F(4,60)=1.79$; ns). Post-hoc analysis indicated that MP and AMP significantly differed from all other devices ($p<.05$) and did not differ one from the other. There were also no significant differences between AW, WVW and SVW.

Steering reaction time

The steering reaction times in WA were, on average, 0.469 s in straight lines and 0.419 s in bends. In straight lines, none of the assistance devices significantly changed steering reaction times compared to the control condition (Figure 3). In bends, only WVW significantly reduced steering reaction times ($t(19)=2.7$, $p<.05$).

The ANOVA performed on the difference between conditions with assistance and WA revealed a significant effect of the assistance devices ($F(4,60)=3.34$, $p<.05$), no significant difference between bends and straight lines ($F(1,15)=0.07$, ns) and no significant interaction between both variables ($F(4,60)=0.66$; ns). Post-hoc analysis showed the effect of assistance devices was mainly due to the WVW condition, but no difference reached statistical significance.

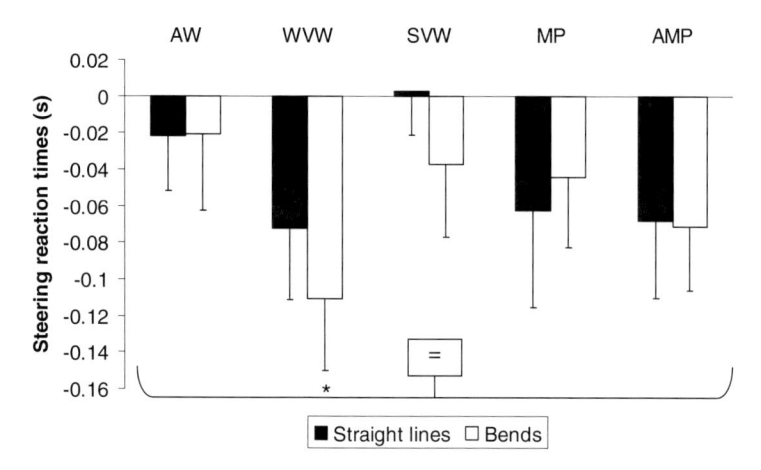

Figure 3. Effects of driving assistance devices on steering reaction time relative to the control condition. Stars represent significant differences compared to the control condition (0 on the figure). Error bars represent one standard error

Subjective data: ranking

Figure 4 presents the distribution of the ranks of preference assigned to all driving assistance devices, from the most acceptable (AW: mean rank = 2.39) to the least acceptable (MP: mean rank = 3.83). WVW (mean rank = 2.83), AMP (mean rank = 2.94) and SVW (mean rank = 3) gave rise to intermediate results. A Friedman test did not reveal a significant effect of driving assistance on the ranks. However, an analysis of contents tends to confirm the contrast between AW (favourable attitude) and MP (unfavourable attitude), with AMP giving rise to mixed feelings. Details about those subjective assessments can be found in El-Jaafari *et al.* (submitted).

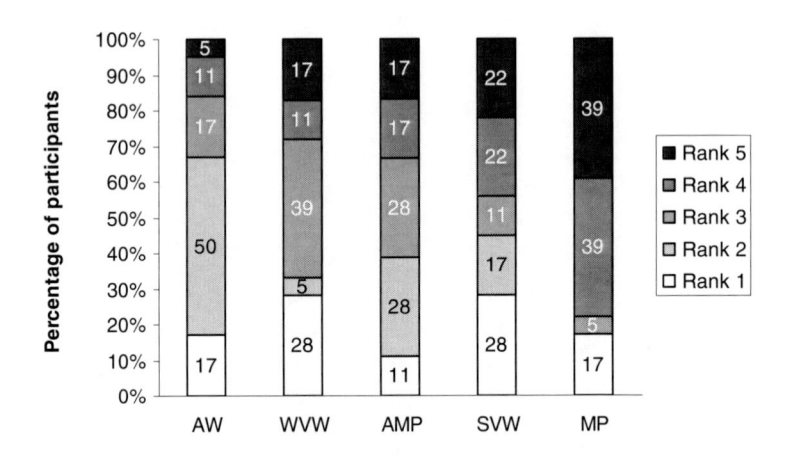

Figure 4. Proportion by rank of the relative classification in order of preference

Discussion

The global effectiveness of mutual control assistance devices was assessed through the observation of the duration of lateral excursion episodes. Only those assistance devices which gave a motor prompt to the drivers yielded a significant performance improvement in that respect. All other assistance devices (warning devices) yielded very few significant improvements compared to the control condition. The benefits associated with the MP approach did not seem to be related to a reduction of steering reaction times. They were rather due to sharper and stronger responses, as evidenced by an increased rate of steering wheel acceleration. These observations are consistent with the result of a previous study (Navarro *et al.*, 2007). The first objective of the current experiment was to refine the understanding of the improved response associated with MP. This was done through a series of comparisons, detailed in the following discussion.

MP can be described as a haptic display that delivers a directional motor prompt to the hands. The main question was to determine whether the motor component of the stimulation is sufficient to explain why MP seems to elicit sharper responses. For this, MP was compared to WVW, which was identical in all points to MP except that it did not deliver a motor incentive. The results showed that both devices with a MP component decreased the duration of lateral excursion and increased the maximum rate of steering wheel acceleration more than WVW. In fact, WVW elicited similar responses to those found with the other warning devices, including SVW. The latter also used the haptic modality to provide directional information but did not stimulate the hands. If the signal was given to the driver specifically through the steering wheel, a small decrease in reaction time might be possible. Indeed, the results showed a slight tendency of WVW, MP, and AMP to reduce the time between the initiation of lane departure and the beginning of the response on the steering wheel. However, this effect was only significant for WVW in bends when compared to the control condition and no significant difference was found between assistance devices. In all cases, this effect did not influence the duration of lateral excursion. Using the haptic modality rather than audition does not appear to have a significant influence on recovery manoeuvres either. AW gave rise to results very similar to those recorded for both vibratory warning devices. This supports previous studies that showed the absence of significant differences between sensory modalities in the domain of lateral control support (Navarro *et al.*, 2007; Suzuki & Jansson, 2003). Thus, neither the fact that the stimulation was delivered to the hands through the steering wheel, nor the use of the haptic modality to convey the signal *per se* appear to be essential in MP. The fundamental mechanism that underlies the improved recovery manoeuvres observed with MP seems to be that the directional cue does not only improve situation diagnosis, as is the case with warning devices. It also acts directly at the motor level and prompts the driver's hands to move.

Assistance devices which deliver motor priming were the only ones found to significantly improve recovery manoeuvres. This result contrasts with other studies where warning devices were also found to be effective (Hoc *et al.* 2006; Navarro *et al.*, 2007; Suzuki & Jansson, 2003; Sayer *et al.*, 2005; Rimini-Doering *et al.*, 2005). A notable difference with Navarro *et al.* (2007) was also found concerning the motor priming modes. In that study, the average duration of lateral excursion observed in the control condition was reduced by 38 percent when MP was used. In the current study, the reduction only amounts to 19 percent. Thus, all assistance devices were globally less efficient. This is related to a large variability in the way critical situations (i.e. lane departure situations) were generated by the reading task. In Navarro *et al.* (2007), the critical situations were provoked by occluding the visual scene. In this study, although the distraction task gave the experiment greater ecological validity, its consequences were much less controllable. Even if the conclusion of the reading task and the triggering of assistance devices informed drivers that they had to look back at the road, drivers could be more or less reactive depending on the degree of attention paid to the reading task. Consequently, the reading task sometimes led to more serious lane departures. Nevertheless, even in these unfavourable conditions, the benefits of MP remained quite significant.

Finally, it was shown that the acceptability of the devices was not related to their efficiency in helping the driver at recovering a safe position in the lane. The way the participants ranked the assistance devices in order of preference and the analysis of post-experimental reports greatly differed across subjects. No significant difference was observed (for more details, see El Jaafari *et al.*, submitted). However, the results tend to confirm the assumption that MP would be less accepted by drivers than a lateralized auditory warning. The combination of motor priming with a lateralized auditory warning was ranked in an intermediate position and may be a reasonable compromise between efficiency and acceptability.

Conclusion

Despite an important variability in the way lane departures occurred, assistance devices based on the motor priming concept clearly remained more effective in improving recovery manoeuvres than warning devices. The results support the hypothesis that MP devices directly intervene at the motor level, in contrast to more traditional warning systems that improve situation diagnosis. The efficiency of MP is essentially due to the incentive nature of the motor signal it delivers. On the other hand, subjective data highlighted the fact that motor priming was not well-accepted by drivers. Combining MP with a well-recognized auditory signal may be a solution to improve acceptability. Future studies where MP will be installed in real cars are now necessary to evaluate the validity of these conclusions in a more complex environment.

Acknowledgments

This study has been supported by the French Program PREVENSOR (ANR/PREDIT). The authors are grateful to Susan Watts for English-language proofreading.

References

Bar, F. & Page, Y. (2002). *Les sorties de voies involontaires* Rueil-Malmaison, F : CEESAR, LAB

El Jaafari, M., Forzy, J.F., Navarro, J., Mars, M., & Hoc, J.M. (submitted). User acceptance and effectiveness of warning and motor priming assistance devices in car driving. *Proceedings of the Humanist conference 2008, Lyon, France.*

Hoc, J.M. (2001). Towards a cognitive approach to human-machine cooperation in dynamic situations. *International Journal of Human-Computer Studies, 54,* 509-540.

Hoc, J.M., & Blosseville, J.M. (2003). Cooperation between drivers and in-car automatic driving assistance. In G.C. van der Veer & J.F. Hoorn (Eds.), Proceedings of CSAPC'03 (pp. 17-22). Rocquencourt, France: EACE.

Hoc, J.M., Mars, F., Milleville-Pennel, I., Jolly, E., Netto, M., & Blosseville, J.M. (2006). Evaluation of human-machine cooperation modes in car driving for safe lateral control in bends: function delegation and mutual control modes. *Le Travail Humain, 69,* 153-182.

Lefeuvre, R., Bordel, S., Guingouain, G., Pichot, N., Somat, A., & Teste, B (2004). La mesure de l'acceptabilité sociale d'un produit technologique : l'exemple des dispositifs d'aide à la conduite : Nouvelles Technologies, Sécurité et Exploitations Routières. *Revue Générales des Routes et des Aérodromes, 832,* 27-32

Navarro, J., Mars, F., Hoc, J.M. (2007). Lateral control assistance for car drivers: a comparison of motor priming and warning systems. *Human Factors, 49,* 950-960.

Rimini-Doering M., Altmueller T., Ladstaetter U., & Rossmeier M. (2005). Effects of lane departure warning on drowsy drivers' performance and state in a simulator. *Proceedings of the International Driving Symposium on Human Factors in Driver Assessment, Training, and Vehicle Design, USA,* 88-95.

Sayer B.T., Sayer J.R., & Devonshire, J.M. (2005). Assessment of a driver interface for lateral drift and curve speed warning systems: mixed results for auditory and haptic warnings. *Proceedings of the International Driving Symposium on Human Factors in Driver Assessment, Training, and Vehicle Design, USA,* 218-224.

Sklar, A.E., & Starter, N.B. (1999). Good Vibrations: Tactile Feedback in Support of Attention Allocation and Human-Automation Coordination in Event-Driven Domains. *Human Factors, 41,* 543-552.

Suzuki K. & Jansson H. (2003). An analysis of driver's steering behaviour during auditory or haptic warnings for the designing of lane departure warning system. *Japan Society of Automotive Engineers Review, 24,* 65-70.

Van Erp J.B.F. & Van Veen H.A.H.C. (2004). Vibrotactile in-vehicle navigation system. *Transportation Research - Part F, 7,* 247-256.

Wickens, C.D., Hollands, J.G., (2000). Introduction to engineering psychology and human performance. In Wickens, C.D., Hollands, J.G. (Eds.) *Engineering Psychology and Human Performance* (pp. 1-14). Prentice-Hall, Upper Saddle River, New Jersey.

Adaptive Automation enhances human supervision of multiple uninhabited vehicles

Ewart de Visser, Don Horvath, & Raja Parasuraman
George Mason University, Fairfax, VA
USA

Abstract

Human operators supervising multiple uninhabited air and ground vehicles (UAVs and UGVs) under high task load must be supported appropriately in context by automation. We examined the efficacy of such *adaptive* automation in a simulated high-workload reconnaissance mission involving four sub-tasks: (1) UAV target identification; (2) UGV route planning; (3) communications, with embedded verbal situation awareness probes; and (4) change detection. Three automation conditions were compared: manual control; static automation, in which an automated target recognition (ATR) system was provided for the UAV task; and adaptive automation, in which individual operator change detection performance was assessed in real time and used to invoke the ATR if and only if change detection accuracy was below a threshold. Change detection accuracy and situation awareness were highest and workload was lowest in the adaptive automation condition compared to the two other conditions. The results show that adaptive automation leads to a levelling of workload and enhances performance both within and across operators under conditions of high task load. The results point to the efficacy of adaptive automation as it is *tailored* to unique human operator needs. We further describe results from a second experiment where the efficacy of adaptive automation was examined in a high-fidelity uninhabited vehicle simulation environment.

Introduction

Uninhabited vehicles (UVs) and other robotic systems are being introduced in rapid fashion into the military to extend manned capabilities, provide tactical flexibility, and act as "force multipliers" (Barnes, Parasuraman, & Cosenzo, 2006; Cummings & Guerlain, 2007). In the US Army's Future Combat Systems (FCS), for example, battlefield force structures will be redesigned to be flexible, reconfigurable components tailored to specific combat missions. The human operators of these systems will be involved in supervisory control of UVs with the possibility of occasional manual intervention. Soldiers will operate multiple systems while on the move and while under enemy fire, as a result of which they will operate under high stress. Because of the consequent increase in the cognitive workload demands on the soldier, automation will be needed to support human-system performance.

An important design issue is what the level and type of automation should be for effective support of the operator in such systems (Parasuraman, Sheridan, & Wickens, 2000). Unfortunately, automated aids have not always enhanced system performance, primarily due to problems in their use by human operators or to unanticipated interactions with other sub-systems. Problems in human-automation interaction have included unbalanced mental workload, reduced situation awareness, decision biases, mistrust, over-reliance, and complacency (Billings, 1997; Parasuraman & Riley, 1997; Sarter, Woods, & Billings, 1997; Sheridan & Parasuraman, 2006; Wiener, 1988).

Adaptive automation (Opperman, 1994)—information or decision support that is not fixed at the design stage but varies appropriately with context in the operational environment—has been proposed as a solution to the problems associated with inflexible automation (Inagaki, 2003; Parasuraman, 2000; Scerbo, 2001). In the context of military operations, adaptive automation is initiated by the system (without explicit operator input) on the basis of critical mission events, performance, or physiological state (Barnes et al., 2006). The adaptive automation concept has also been criticized as potentially increasing system unpredictability (Billings & Woods, 1994), although this is not inevitable, and well-designed systems in which users are given the flexibility to delegate tasks to automation can overcome this limitation (Miller & Parasuraman, 2007).

The efficacy of adaptive automation must not be assumed but must be demonstrated in empirical evaluations of human-robot interaction performance. One method of invocation for adaptive automation includes assessing operator workload in real time (Parasuraman et al., 1992). In addition to measures of workload, assessment of situation awareness might also be useful in adaptive systems (Kaber & Endsley, 2004). Reduced situation awareness has been identified as a major contributor to poor performance in search and rescue missions with autonomous robots (Burke & Murphy, 2004; Murphy, 2004). In particular, transient or dynamic changes in situation awareness might be captured by probing the operator's awareness of changes in the environment. One such measure is *change detection* performance. People often fail to notice changes in visual displays when they occur at the same time as various forms of visual transients (Simons & Ambinder, 2005). Durlach (2004) has shown that this "change blindness" phenomenon is also found with complex visual displays used in various military C^2 environments.

In the present study we conducted two experiments with the goal of assessing the efficacy of adaptive automation, on operator performance in a multi-task scenario involving supervision of multiple UVs in a simulated reconnaissance mission. We developed an in-house simulation capability, the Robotic NCO, designed to isolate some of the cognitive requirements associated with a single operator controlling robotic assets within a larger military environment (Barnes et al., 2006). The goal was to create a microworld with face validity for future military operations involving UAVs and UGVs, while providing for a degree of experimental control. The first experiment used operator change detection performance to drive adaptive automation in the Robotic NCO simulation. In a second experiment, we used task loading to drive adaptive automation in a higher-fidelity simulation environment.

The Robotic NCO simulation required the participant to complete four military-relevant tasks: (1) a UAV target identification task; (2) a UGV route planning task; (3) a communications task with an embedded verbal situation awareness probe task; and (4) an ancillary task designed to assess situation awareness using a probe detection method, a change detection task embedded within a situation map. We conducted one experiment with the Robotic NCO in which we examined the effects of adaptive automation on performance, workload, and situation awareness in the same task under conditions of low and high task load. In a subsequent experiment we used a high-fidelity simulator to study the same effects under slightly different circumstances.

Experiment I

In the first experiment we examined the effects of adaptive automation, based on real-time assessment of operator change detection performance, on performance, situation awareness, and workload in supervising multiple UVs in the Robotic NCO simulation under two levels of task load. We used an adaptive automation invocation method first developed by Parasuraman, Mouloua, and Molloy (1996), known as "performance-based" adaptation. In this method, individual operator performance (i.e., change detection performance) is assessed in real time and used as a basis to invoke automation. In contrast, in static or "model-based" automation, automation is invoked at a particular point in time during the mission based on the model prediction that operator performance is likely to be poor at that time (Parasuraman et al., 1992). This method is by definition not sensitive to individual differences in performance, since it assumes that *all* operators are characterized by the model predictions. In performance-based adaptive automation, on the other hand, automation is invoked *if and only if* the performance of an individual operator is below a specified threshold at a particular point in time during the mission. If a particular operator does not meet the threshold, automation is invoked. However, if the threshold is exceeded in another operator, or in the same operator at a different point in the mission, the automation is not invoked. Thus, performance-based adaptation is by definition context-sensitive to an extent.

To demonstrate the potential benefit of adaptive automation, it is essential to compare its effects not only to performance without automation (manual performance), but also to static automation (Barnes et al., 2006). Accordingly, in the main experiment, we examined performance in the Robotic NCO task under three conditions: (1) Manual; (2) Static Automation, in which participants were supported in the UAV task with an automated target recognition (ATR) system, thereby off-loading them of the responsibility of identifying targets. (3). Adaptive Automation, in which the ATR automation was invoked if change detection performance was below a threshold, but not otherwise. Each of these conditions was combined factorially with two levels of task load, as manipulated by variation in the difficulty of the communications task.

We predicted that change detection performance and situation awareness as assessed using the verbal situation awareness probes, would both be enhanced with adaptive automation, whereas overall mental workload would be reduced, following the logic of Parasuraman et al. (1996, 1999). In turn these benefits would be greater for the adaptive compared to the static automation condition, with both automation conditions being superior to manual performance. Finally, we expected that the selective benefits of adaptive automation, if found, would be greater under high task load than under low task load.

Methods

Participants

Sixteen young adults (8 women, 8 men) aged 18-28 years (mean=21.9) participated. The experiment lasted approximately 2 hours and participants were paid $15.00 per hour.

Procedure

Participants were asked to take the role of a robotic operator in a Mounted Combat System company (MCS). They were asked to conduct a reconnaissance mission for the MCS platoon using their UAV and UGV assets. The paths of the UAV and UGV were pre-planned by the experimenter were not under control of the participant, except for the UGV re-routes. The UAV travelled faster than the UGV and provided surveillance information to the participant as described previously. The UGV followed its routed path and when an obstacle or area of interest appeared, waited for operator input, as described. During supervision of the two robotic assets the operator received either call sign acknowledgments or status queries, and performed the change detection task.

During the scenario the UAV encountered a fixed number of targets (20) while the UGV submitted a fixed number of requests (7) to the operator. These task parameters were combined with either a low (16 own call signs in 20) or high level of uncertainty (4 own call signs in 20) in the communications task to create two conditions of overall task load, low and high. Participants performed 8 simulated missions of 5 minutes each, following familiarization, training, and practice on the Robotic NCO simulation,

As part of the training procedure, participants were informed about the automatic target recognition (ATR) system and were told at the beginning of the experiment this ATR might be invoked during the missions. Each time the automation was about to be engaged, the participant received a message (presented both auditorially through the speaker and visually in the communications window), indicating that the ATR was now available. Even though the ATR off-loaded participants by supporting the UAV task they still had to monitor the results of the ATR, since they were required at the end of the mission to evaluate the best platoon path to take following the reconnaissance mission.

There were three main automation conditions. In the Manual condition, participants performed all four tasks of the Robotic NCO simulation without any automation assistance. In the Static Automation conditions participants were supported in the UAV task with the ATR system, thereby releasing them of the responsibility to identify targets. The automation was engaged in the middle (at ~ 2.5 min) of the 5-min mission (after 4 change detection events) for all participants without regarding their level of performance. A similar procedure was used in the Adaptive Automation condition, except that our software maintained a running count of an individual operator's change detection accuracy. After four change detection events had occurred in the middle (at ~ 2.5 min) of the 5-minute mission, the current change detection accuracy was compared to a threshold. If the threshold was not met, the ATR automation was engaged, but not otherwise. A threshold of 50% accuracy was chosen, based on the results of a baseline experiment as well as pilot work. It was expected that with this threshold most but not all participants would receive adaptive aiding under high task load and many if not all in the low task load condition. After adaptive automation was invoked (or not), the mission proceeded as before, that is the ATR remained on until the end of the mission.

The experiment was a 2 x 3 x 2 within-subjects design. The within- subjects factors were the uncertainty of the communications task (low or high), Automation Condition (Manual, Static and Adaptive Automation), and Block (one and two).

Each participant completed 12 5-minute mission blocks. The order of blocks was counterbalanced. For statistical evaluation of the effects of automation, Mission Phase (pre-automation invocation and post-automation invocation—in the middle of the 5-minute mission) was included as a factor with all other conditions.

Results

Due to space limitations we report only part of the results, including change detection performance and subjective ratings. MANOVAs and ANOVAs were conducted. The within-subjects variables included in the statistical analyses were Communications Task Difficulty, Mission Phase (Pre – Post Automation), Automation Condition, and Block. Because there were no significant interactions involving the Block factor, the data were collapsed across this factor.

Change detection performance

In the low task load/adaptive automation condition, the adaptive ATR was not invoked in three participants when it was first performed (block 1). The ATR was also not invoked in one participant in the second block of this condition. The ATR was invoked in all participants in the high task load/adaptive automation condition. The results of the statistical analyses were the same whether these participants were included or excluded in the data set. Therefore, we first present results with the data from all the participants, regardless of whether the ATR assisted them in the adaptive automation condition.

Results for the change detection accuracy scores showed significant main effects for Automation Condition, $F(2, 30) = 22.8$, $p < .001$, Mission Phase, $F(1, 15) = 151.3$, $p < .001$, and Communications Task difficulty, $F(1, 15) = 37.5$, $p < .001$. In addition, the interaction of Mission Phase and Automation Condition was significant, $F(2,30) = 22.1$, $p < .001$. Separate ANOVAs were computed for both the pre-automation invocation and post-invocation phases to examine the interaction further. In the pre-invocation phase before automation was implemented, there were no significant differences in change detection accuracy between conditions, $p > .10$. However, the effect of Automation Condition was significant for the post-invocation phase, $F(2,30) = 31.3$, $p < .001$. As can be seen in Figure 1, and as verified by pair-wise comparisons, participants detected significantly more situation map changes in the Static and Adaptive Automation conditions compared to the Manual condition, p's $< .001$.

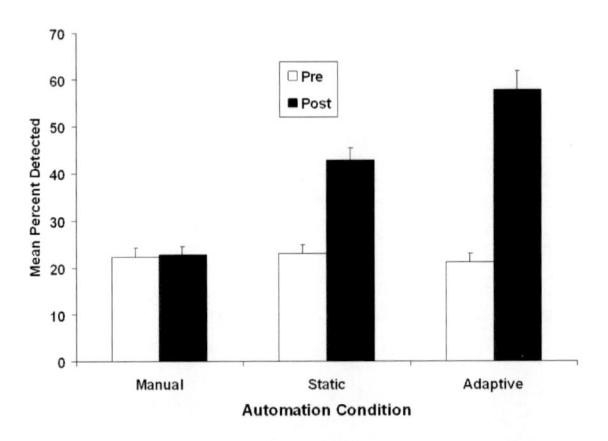

Figure 1. Effects of static and adaptive automation on change detection accuracy, compared to manual performance

Furthermore, change detection accuracy was significantly higher in the Adaptive Automation condition than in the Static Automation condition, $p < .02$. Finally, more changes were detected when the uncertainty of the communications was low ($x=36.0\%$, S.E.=1.22) than high, ($x=27.2\%$, S.E.=.94). However, the Automation Condition x Communications Task difficulty interaction, $F(2, 30) = 1.06$, NS, and the Automation Condition x Mission Phase x Communications Task difficulty interaction, $F(2, 30) = 2.24$, NS, were not significant, although we predicted a greater effect of adaptive automation on change detection accuracy under high than low task load.

Finally, we also compared change detection performance in those participants for whom automation was *not* invoked adaptively with those in whom it was. As described before, the ATR was not invoked in three participants in the low task load condition because they exceeded the threshold (see Figure 2), whereas it was in 13 others. In addition, Figure 2 shows the change detection performance of these two groups in the pre- and post-invocation mission phases.

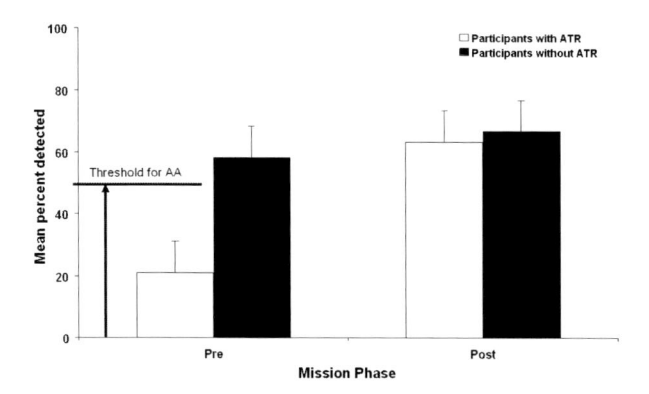

Figure 2. Change detection accuracy in participants for whom automation was not invoked adaptively and in those for whom it was triggered, for pre- and post-invocation periods

By definition, the non-ATR group performed better on the change detection detection task compared to the ATR-group in the pre-invocation phase, as confirmed by a *t*-test adjusted using Levene's test for unequal variances (due to unequal group sample sizes), $t=4.26$, $p <.05$. In the post-invocation phase, however, the two groups did not reveal any significant differences, $t=0.1$, *NS*. Thus adaptive automation had the effect of "pulling up" the performance of the 13 participants up to the level of the three for whom automation assistance was not needed at that point in the mission.

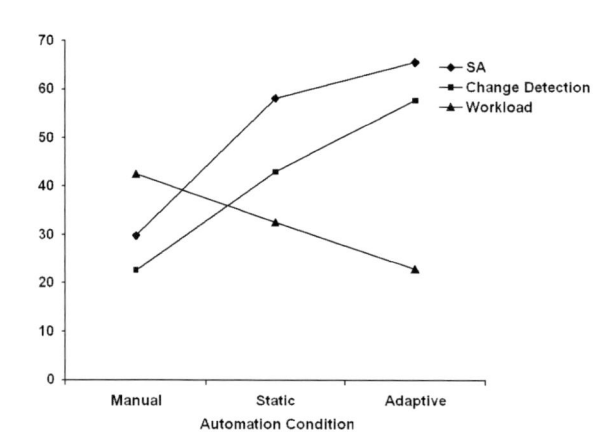

Figure 3. Inter-relationships between effects of static and adaptive automation on change detection, situation awareness, and workload

Relationships between change detection, Situation Awareness, and workload

The inter-relationships between change detection, situation awareness, and workload are shown in Figure 3, which plots mean values for these measures in the post-invocation phase as a function of automation condition. Both automation conditions, increased change detection accuracy and reduced subjective workload, as indicated by the reciprocal relationship between these measures illustrated in Figure 3. For both these measures the adaptive automation showed the most benefits compared to static automation and manual conditions. In addition, automation also enhanced situation awareness compared to the manual conditon, but no significant differences were found between the two automation conditions.

Discussion

We predicted that change detection performance and situation awareness would be enhanced with adaptive automation, whereas overall mental workload would be reduced, and that these benefits would be specifically associated with adaptive (as opposed to static) automation. The results supported this prediction. Compared to manual performance, both static and adaptive automation led to an increase in change detection accuracy and situation awareness, whereas workload was reduced. In addition, in comparison to static automation, there was a further increase in change detection accuracy and concomitant reduction in workload with adaptive automation. This last finding is important, because simply demonstrating performance benefits due to automation is insufficient; rather, the specific benefit, if any, of adaptive automation must be shown, over and above that associated with static automation (Barnes et al., 2006; Parasuraman, 1993). The results thus add to the growing literature pointing to the efficacy of adaptive automation for reducing or balancing operator workload and enhancing performance (Inagaki, 2003; Kaber & Riley, 1999; Parasuraman et al., 1999; Scerbo, 2001), and confirm that these benefits also accrue in the domain of human operator supervision of multiple UVs in a simulated tactical reconnaissance mission.

There was a reciprocal relationship between the different operator performance measures in terms of the effects of adaptive automation. Specifically, static automation led to an increase in both change detection accuracy and situation awareness and a decrease in workload, with a further increase and decrease in these measures with adaptive automation (see Figure 3). However, it should be noted that the additional increase in situation awareness with adaptive automation was not statistically significant. Several studies have documented benefits of adaptive automation for situation awareness (Kaber & Endsley, 2004). It is possible that the non-significant trend we found might simply reflect the relative insensitivity of our verbal probe measure, since that we provided only two such probes in the post-invocation phase of each mission during which automatic target recognition was implemented. Nonetheless, the substantial benefits for change detection and overall workload—a 34% enhancement in the case of the former—argue strongly for the efficacy of adaptive automation. The reduction in overall workload was also reflected in better performance of one of the other three sub-tasks that participants performed, the communications task. Accuracy in responding to communications was higher with static automation than with manual performance, and higher still with adaptive automation. Thus adaptive automation was successful not only in supporting the human operator in an appropriate context—when their change detection performance was low, pointing to low perceptual awareness of the evolving mission elements—but also freed up sufficient attentional resources to benefit performance on other less critical but important sub-tasks.

Our next step was to replicate these results in a higher-fidelity simulation.

Experiment II

In the second experiment we used a simulation environment called the System Integration Lab (SIL). This is a high-fidelity simulator of a planned Army system for mounted soldiers. The simulation required the participant to complete three military-relevant tasks: (1) a UGV target identification task; (2) a UAV target detection task, and (3) an ancillary task designed to assess situation awareness using a probe detection method, a change detection task embedded within the UAV video feed display.

As in the first experiment, we examined performance in the SIL task under three conditions: (1) Manual; (2) Static Automation, in which participants were always supported in the UGV task with an automated target recognition (ATR) system, thereby off-loading them of the responsibility of identifying targets; (3) Adaptive Automation, in which the ATR automation was on only during high workload phases in the mission. Unlike the first experiment, adaptive automation was not invoked based on the operator's performance directly but rather on the context of the mission, which is another of the major invocation strategies identified in previous research (Parasuraman et al., 1992). Each scenario contained low workload and high workload phases. In the low workload condition targets were easier identify because there was less forest density whereas in the high condition targets the forest was much denser making it harder for a participant to detect a soldier's presence. The differences in performance between low and high load conditions were confirmed in a pilot experiment of this study.

We predicted that the performance benefits would be greater for the adaptive compared to the static automation condition, with both automation conditions being superior to manual performance. Finally, we expected that the selective benefits of adaptive automation, if found, would be greater under high task load than under low task load.

Method

Participants

Twelve young adults (2 women, 10 men) aged 18-24 years (mean=21.1) participated. The experiment lasted approximately 3 hours and participants were paid $15.00 per hour.

Procedure

Participants were welcomed into the experimental suite and asked to fill out several demographic questionnaires. They were then trained for about 45 minutes on use of the Tactical Control Unit. Participants were asked to conduct a reconnaissance mission with two unmanned vehicles, a UAV and a UGV. Participants were shown how to place detected targets from the UAV video feed on the touch-screen tactical map. The UGV, meanwhile, traveled to 16 waypoints each time notifying the participants that a picture had been taken at a particular waypoint. They were instructed to switch from the UAV task to the UGV task each time the UGV notification was presented. When participants switched to the UGV task they had to decide whether a soldier was present in the forest displayed on a reconnaissance picture. After they made their decision, they were instructed to switch back to the UAV task until the next UGV notification was presented to them. For the UGV task, participants were either presented with a low density forest image which constituted the low workload condition or with a high density forest image which served as the high workload condition.

At the beginning of the experiment participants were told that an automated target recognition (ATR) system might be invoked during the missions. They were also given training on the ATR prior to the main experimental blocks.

There were three main automation conditions. In the Manual condition, participants performed all three tasks of the SIL simulation without automated support. In the Static Automation conditions participants were always (i.e. at each of the 16 waypoints) supported in the UGV task with an ATR system, thereby off-loading them of the responsibility of identifying targets. In the Adaptive Automation condition, the automation was invoked only for high workload periods in the mission and was off during the low workload periods of the mission.

Participants completed a total of 6 missions lasting around 15-20 minutes each. After each mission they filled out an adapted trust questionnaire (Lee & Moray, 1992), and an electronic version of the NASA-TLX workload measure.

Results

We present preliminary data for the second experiment. ANOVAS were conducted on all the data. The within-subjects variables included in the statistical analyses were Automation Condition, Workload Condition and Block.

UGV detection accuracy

Results for the mean UGV accuracy scores revealed a significant main effect for Workload, $F(1, 13) = 56.8$, $p < .001$. There were no significant interactions. Pairwise comparisons showed that accuracy in the Low Workload condition ($x=94.5\%$, S.E.=1.20) was significantly higher than in the High Workload condition ($x=77.3\%$, S.E.=2.00), p's $< .05$.

UGV detection reaction time

Results for the mean UGV detection time scores revealed significant main effects for Workload, $F(1, 13) = 61.6$, $p < .05$, and Block, $F(1, 13) = 43.3$, $p <.05$. There were no significant interactions. To examine main effects further, post-hoc analyses were performed for each level of Workload and Block. Pair-wise comparisons for Workload showed that reaction time (in seconds) was significantly slower in the High Workload condition (x=10.12, S.E.=0.6) compared to the Low Workload condition (x=6.72, S.E.=0.3), p's < 0.05. Pair-wise comparisons for the second Block showed that reaction time was faster in the second Block (x=8.00, S.E.=0.4) compared to first Block condition (x=8.9, S.E.=0.5), p's < 0.05.

Trust in the automation

Results for the mean trust in automation scores revealed significant main effects for Automation, $F(1, 56) = 5.3$, $p < .05$, and Block, $F(1, 56) = 9.59$, $p <. 05$. There were no significant interactions. To examine the main effect further post-hoc analyses were performed for each level of Automation. Pair-wise comparisons showed that trust in the automation was higher in the Adaptive (x=54%, S.E.=2.8) compared to the Static condition (x=51%, S.E.=3.0), p's < .05 (see Figure 4). In addition, pair-wise comparisons showed that trust in automation was significantly lower in the Second Block (x=49%, S.E.=3.2) compared to the First Block (x=56%, S.E.=2.8).

Self-Confidence ratings

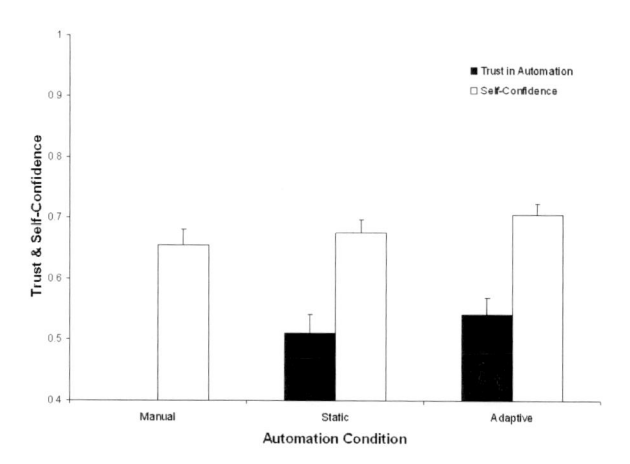

Figure 4. Effects of Automation condition on measures of Trust and Self-confidence

Results for the mean trust in automation scores revealed a significant main effect for Automation, $F(2, 104) = 7.4$, $p < .05$ and Block $F(1, 52) = 16.7$, $p < .01$. In addition, the interaction of Automation Condition and Block was significant, $F(2,104) = 7.6$, $p < .05$. To examine this main effect, further post-hoc analyses were performed for each level of Automation. To examine the interaction further ANOVAs were conducted for each Block. Pair-wise comparisons showed that self-confidence in one's own ability to carry out the task in the automation was higher in the Static (x=66%, S.E.=2.5) compared to the Manual condition (x=61%, S.E.=3.3), p's < .05 (see Figure 4). Self-confidence was higher than both Static and Manual conditions in the Adaptive Automation condition ($x=72\%$, S.E.=1.7), p's < 0.05. However, no such differences were found in the second Block, $F(2, 106) = 0.913$, $p > .05$. Furthermore, self-confidence increased in the second Block ($x=71\%$, S.E.=2.2) compared to the first Block ($x=65\%$, S.E.=2.4), p's < .05.

Workload

Results for the mean workload in automation scores revealed no significant main effects for Automation, $F(2, 28) = 2.1$, *NS*, and Block, $F(1, 14) = 3.5$, *NS*. However, we did find a significant effect in post-hoc analyses. The pair-wise comparison showed that workload in the Static Automation condition ($x=53\%$, S.E.=3.2) was slightly higher compared to the Adaptive condition ($x=50\%$, S.E.=3.6), p's < .05 (see Figure 5).

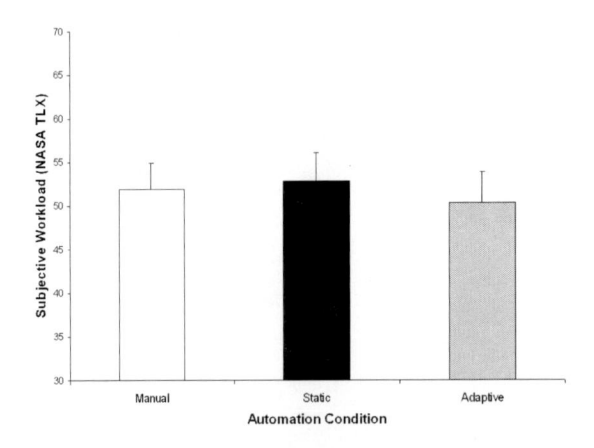

Figure 5. Effects of Automation condition on subjective workload as measured by the NASA TLX

Discussion

High cognitive workload demands on personnel in military systems working with multiple UVs has mandated the use of automation support (Barnes et al., 2006). Because automation does not always achieve the goal of supporting the operator effectively (Billings, 1997; Parasuraman & Riley, 1997; Sarter et al., 1997), providing context-appropriate aiding—adaptive automation—has been proposed (Miller & Parasuraman, 2007; Parasuraman et al., 1992; Scerbo, 2001). In the present study, adaptive aiding was provided to participants supervising multiple UVs, based on real-time assessment of their change detection accuracy. We compared the effects of manual performance, static automation, and adaptive automation on workload, situation awareness, and other aspects of task performance.

The specific advantage that adaptive automation based on assessment of individual performance or physiology brings is its sensitivity to individual differences, whereas other approaches such as model-based adaptive automation are not (Parasuraman et al., 1992). This benefit was apparent when we compared the participants in the first experiment for whom automation was not invoked adaptively in the low task load condition to those for which automation was triggered. We found that while change detection performance was initially higher in the former group (the criterion for *not* invoking automation), both groups had equivalent levels of change detection accuracy in the post-invocation mission phase. Thus in comparison to static automation, adaptive automation, by providing aiding only in certain individuals when they need it, or in the same individual when he or she needs it at different times, has the effect of "levelling" performance between and within individuals, thus providing for more stable system performance. This view is consistent with the findings of a study examining the effects of adaptive aiding on UAV weapons release performance (Wilson & Russell, 2007). Results showed that low performers were particularly challenged in high task load conditions, while high performers showed a large performance decrease when aid was provided randomly. For both types of operators, adaptive automation provided benefits by alleviating task load for the low performers and accommodating the more advanced strategies of the high performers.

In the second experiment we found benefits for adaptive automation compared to static and manual automation levels for workload and trust. Participants experienced less subjective workload in the Adaptive Automation condition compared to the Static Automation condition. Furthermore, self-confidence ratings and trust were higher in the Adaptive Automation compared to the Static Automation condition. The finding of higher trust in adaptive automation contradicts the claim that adaptive automation would be less trustworthy because of its unpredictability (Sarter et al., 1997}. Thus adaptive automation yielded significant benefits over static automation in terms of workload and trust. However, there were no differential benefits of adaptive automation over static automation for UGV detection performance (reaction time and accuracy), in contrast to Experiment 1. This might reflect the different adaptive automation methods used in the two studies. Nevertheless, both experiments confirm the general beneficial effect of adaptive automation.

What are the practical implications of the research reported in this paper? This work is part of a broader Army science and technology program aimed at understanding the performance requirements for human-robot interaction in future battlefields (Barnes et al., 2006). Initial findings from this project indicate that the primary tasks that soldiers are required to perform place severe limits on their ability to monitor and supervise even a *single* UV, let alone multiple UVs. For example, Chen, Durlach, Sloan, and Bowers (in press) examined target detection accuracy in participants given either a single robotic asset (either a UAV, a UGV or a teleoperated UGV) or all three assets. Target detection performance was lower with three than with a single UV, and participants were also less likely to complete their mission in the allotted time. In addition, crew safety may be compromised because soldiers who have to carry out routine tasks such as radio communications and ensuring local security also have to supervise and manage several robotic tasks during high workload mission segments (Chen & Joyner, 2006; Mitchell & Henthorn, 2005). Adaptive automation would therefore be particularly well suited to these situations because of the uneven workload and the requirement to maintain SA for the primary as well as the robotic tasks. Our results point to the efficacy of adaptive automation for supporting the human operator under these conditions.

Acknowledgements

This research was supported by a contract from the Army Research Laboratory under the Army Technology Objective (ATO) on Human-Robot Interaction. Michael Barnes was the technical monitor. The views expressed in this paper are solely those of the authors and do not necessarily reflect official U.S. Army policy. Thanks to Peter Squire for critical discussions of this work.

References

Adams, T.K. (2001). Future warfare and the decline of human decision-making. *Parameters* (2001-2002), 57-71.

Barnes, M., Parasuraman, R., & Cosenzo, K. (2006). Adaptive automation for military robotic systems (chap.7.4). In NATO Technical Report *RTO-TR-HFM-078 Uninhabited Military Vehicles: Human Factors Issues in Augmenting the Force* (pp. 420-440). Neuilly-sur-Seine, France: NATO Research and Technology Organization.

Billings, C. (1997). *Aviation automation: The search for a human-centered approach*. Mahwah, NJ: Erlbaum.

Billings, C.E., & Woods, D.D. (1994). Concerns about adaptive automation in aviation systems. In M. Mouloua and R. Parasuraman (Eds), *Human performance in automated systems: Current research and trends* (pp. 24–29). Hillsdale, NJ: Lawrence Erlbaum.

Burke, J.L., & Murphy, R.R. (2004). Human-robot interaction, situation awareness and search: Two heads are better than one. In *Proceedings of the 2004 IEEE International Workshop on Robot and Human Communication*. Okayama, Japan.

Chen, J.Y.C., Durlach, P.J., Sloan, J.A., & Bowers, L.D. (in press). Human robot interaction in the context of route reconnaissance missions. *Military Psychology*.

Chen, J.Y.C., & Joyner, C.T. (2006). *Concurrent performance of gunner's and robotic operator's tasks in a simulated mounted combat system environment* (ARL Technical Report ARL-TR-3815). Aberdeen Proving Ground, MD: U.S. Army Research Laboratory. (retrieved 01.01.2008 from http://www.arl.army.mil/arlreports/2006/ARL-TR-3815.pdf)

Cummings, M.L., & Guerlain, S. (2007). Devloping operator capacity estimates for supervisory control of autonomous vehicles. *Human Factors, 49,* 1-15.

Durlach, P. (2004). Change blindness and its implications for complex monitoring and control systems design and operator training. *Human-Computer Interaction, 19,* 423-451.

Inagaki, T. (2003). Adaptive automation: Sharing and trading of control. In E. Hollnagel (Ed.), *Handbook of cognitive task design* (pp. 147-169). Mahwah, N.J.: Lawrence Erlbaum Publishers.

Kaber, D.B., & Endsley, M. (2004). The effects of level of automation and adaptive automation on human performance, situation awareness and workload in a dynamic control task. *Theoretical Issues in Ergonomics Science, 5*, 113-153.

Kaber, D., & Riley, J. (1999) Adaptive automation of a dynamic control task based on workload assessment through a secondary monitoring task. In M. Scerbo & M. Mouloua, (Eds.), *Automation technology and human performance: Current research and trends* (pp.129-133). Mahwah, NJ: Lawrence Erlbaum.

Murphy, R.R. (2004). Human-robot interaction in rescue robots. IEEE *Transactions on Systems, Man, and Cybernetics-Part C: Application and Reviews, 32*, 138-153.

Miller, C., & Parasuraman, R. (2007). Designing for flexible interaction between humans and automation: Delegation interfaces for supervisory control. *Human Factors, 49*, 57-75.

Mitchell, D.K., & Henthorn, T. (2005). *Soldier workload analysis of the Mounted Combat System (MCS) platoon's use of unmanned assets.* (ARL-TR-3476), Army Research Laboratory, Aberdeen Proving Ground, MD.

Opperman, R. (1994). *Adaptive user support.* Hillsdale, NJ; Erlbaum.

Parasuraman, R. (1993). Effects of adaptive function allocation on human performance. In D.J. Garland and J.A. Wise (Eds.) *Human Factors and Advanced Aviation Technologies* (pp. 147-157). Daytona Beach: Embry-Riddle Aeronautical University Press.

Parasuraman, R. (2000). Designing automation for human use: Empirical studies and quantitative models. *Ergonomics, 43*, 931-951.

Parasuraman, R., Bahri, T. Deaton, J., Morrison, J., & Barnes, M. (1992). *Theory and design of adaptive automation in aviation systems.* (Progress Report No. NAWCADWAR-92033-60). Warminster, PA: Naval Air Warfare Center.

Parasuraman, R., Mouloua, M., & Hilburn, B. (1999). Adaptive aiding and adaptive task allocation enhance human-machine interaction. In M.W. Scerbo and M. Mouloua (Eds.) *Automation technology and human performance: Current research and trends* (pp. 119-123). Mahwah, NJ: Erlbaum.

Parasuraman, R., & Riley, V.A. (1997). Humans and automation: Use, misuse, disuse, abuse. *Human Factors, 39*, 230-253.

Parasuraman, R., Sheridan, T.B., & Wickens, C.D. (2000). A model of types and levels of human interaction with automation. *IEEE Transactions on Systems, Man, and Cybernetics – Part A: Systems and Humans, 30*, 286-297.

Sarter, N., Woods, D., & Billings, C.E. (1997). Automation surprises. In G. Salvendy (Ed.), *Handbook of human factors and ergonomics* (2nd ed., pp. 1926–1943). New York: Wiley.

Scerbo, M. (2001). Adaptive automation. In W. Karwowski (Ed.), *International encyclopedia of ergonomics and human factors* (pp. 1077-1079). London: Taylor & Francis.

Sheridan, T., & Parasuraman, R. (2006). Human-automation interaction. *Reviews of Human Factors and Ergonomics, 1*, 89-129.

Simons, D.J., & Ambinder, M.S. (2005). Change blindness: Theory and consequences. *Current Directions in Psychological Science, 14*, 44-48.

Simons, D.J. & Rensink, R.A. (2005). Change blindness: Past, present, and future. *Trends in Cognitive Sciences, 9*, 16-20.

Wiener, E.L. (1988). Cockpit automation. In E.L. Wiener and D.C. Nagel (Eds.), *Human factors in aviation* (pp. 433-461). San Diego: Academic Press.

Wilson, G.F., & Russell, C.A. (2007). Performance Enhancement in an Uninhabited Air Vehicle Task Using Psychophysiologically Determined Adaptive Aiding. *Human Factors, 49*, 1005-1018.

A simple minded model for levels of automation

Michela Terenzi & Francesco Di Nocera
University of Rome "La Sapienza"
Italy

Abstract

The lack of a common theoretical and methodological framework for describing Levels of Automation (LOA) makes it difficult to compare results from different studies. Starting from this consideration, previous studies showed that automation could be effectively defined in terms of the amount of information traded by humans and machines. However, the automation aids employed in those studies were always 100% reliable, and it is well know that imperfect automation can affect the operators' reliance, mental workload and performance. Therefore, a critical issue for testing and improving the model is to investigate the differential impact of unreliable automation aids. A new study has been devised for testing human performance and workload in different LOA conditions by changing the reliability of the automation aid during a visual search task. Results showed differential effects of LOA and reliability on performance and mental workload.

Introduction

The pervasive implementation of automated functions into highly complex systems led to the need of examining the nature of the interaction between humans and the automatic tools they use.

The classical definition of automation refers to "the execution by a machine agent (usually a computer) of a function that was previously carried out by a human" (Parasuraman & Riley, 1997 p. 231). Automation may be introduced to reduce costs, to prevent errors, and/or to relieve the human operator of parts of the task. However, the integration of automated functions in the system can change the nature of the interaction between humans and machines, imposing new and unexpected demands on the operators. Despite this, the so-called "technological imperative" impels to automate tasks as full as possible. Of course, this is not always achievable (due to technological limitations), not considering that a vast corpus of studies showed that automation is not always beneficial. Automation can impair performance due to well documented phenomena such as increase of mental workload, decrease of situation awareness, degradation of skills and complacency effects (Parasuraman et al., 1993; Sarter et al.1997). With the aim of minimising the costs and maximising the benefits, automation aids might be conceptualized as different levels (Levels of Automation: LOA) that are flexibly implemented during system operations.

In D. de Waard, F.O. Flemisch, B. Lorenz, H. Oberheid, and K.A. Brookhuis (Eds.) (2008), *Human Factors for assistance and automation* (pp. 301 - 312). Maastricht, the Netherlands: Shaker Publishing.

Several LOA taxonomies have been proposed in the literature. Usually, they describe the automation construct along a continuum ranging from full manual control to full machine autonomy. Between these extremes, several studies have identified different levels of automation (e.g., Billings, 1991; Sheridan & Verkplant, 1978; Parasuraman et al., 2000; Endsley & Kaber 1997, Endsley & Kaber, 1999). In this framework, the LOA is defined as a combination of rules and responsibilities shared by humans and computers for a task to be performed. Therefore, the classic approach to the concept of LOA is qualitative in nature: it simply describes the trading of system control between humans and computers. The main constrain that arise from this approach is that most of the taxonomies and models proposed are strictly dependent on specific types of task to be performed and can be applied only in structured task environments (see Fereidunian et al., 2007).

In the attempt of addressing the LOA issue into a more general design framework, Parasuraman et al. (2000) proposed a model for types and levels of automation based on a four-stage information-processing model (information acquisition, information analysis, decision selection, action implementation). Albeit the focus on cognitive processing allows the generalization to any type of task and operational environment, this model is still limited to the qualitative description of LOA and does not provide the common theoretical and methodological framework that would be needed to describe the LOA concept.

Defining a common platform

Developing a general LOA scale is a somewhat ambitious objective. Of course, the existence of many types of automated systems, each one having its own performance objectives and levels of specificity, makes it difficult to generalise. An opportunity may arise from the straightforward consideration that complex tasks can be fragmented in subtasks units, identifying basic features or low-level processes. According to this consideration, automation could be assessed by simply taking into account the number of these basic features for which automation support has been provided. In other words, a new LOA definition has been proposed (Terenzi et al., 2007) in terms of the *amount* of information traded by humans and machines, thus migrating its definition from qualitative to quantitative.

Consistently with the four-stage model proposed by Parasuraman and colleagues (2000), this research program has firstly taken into account the information acquisition and analysis stage. In this phase, automation may be used to support the operator in the initial analysis of raw data from the external environment and to trigger pattern recognition. An example of information acquisition may be the radar for aircrafts or sonar for ships. These systems may simply collect information and display it on the screen or, when the computer is more involved, certain types of information may be highlighted to grab the operator's attention. In the present study automation has been implemented at different levels during a visual search task and automation aids were provided in order to support the target identification process. According to the quantitative model proposed, LOA may be defined in terms of number of features of an object that the automation has to be identified. Consequently, automation providing reliable information about 1 out of 4 possible features has LOA equal to 0.25, whereas for a system providing specific aid about 2 out of 4 features LOA is equal to 0.50. LOA is equal to 0 when manual control is taken into account. Defining the LOA in this framework is as simple as counting, something that can be done for any type of task. In this study, this quantitative model has been tested under three different degrees of reliability (25%, 50% and 75%). As introduced above, reliability can affect operator–automation interaction in many ways. Previous studies showed that imperfect automation can generate overtrust, undertrust, or calibrated trust (Parasuraman & Riley, 1997), "cognitive laziness" (Skitka et al, 1999), "complacency" (Parasuraman, Molloy, & Singh, 1993), and performance loss (Molloy & Parasuraman, 1996; Rovira & Parasuraman, 2002; Rovira, Zinni, & Parasuraman, 2002; Wickens & Xu, 2002). To date, the most important question refers to how reliability should be represented in a system in order to offer real benefits to the operator. Recently, a vast literature on how automation reliability affects operator performance seems to suggest two main results. Firstly, a 0.7 lower limit reliability should be set in order to grant any beneficial effect of automation. Secondly, more reliable aids tend to result in better operator performance (Wickens & Dixon, 2005). However, the relationship between automation and reliability is quite complex and can be modulated also by other factors such as workload or operator reliance on automated aids. The combined influence of these variables on performance can be relevant for a wide range of applications. However, they have been rarely addressed in a quantitative model. Nevertheless, other attempts have been made in order to develop a general LOA model. For example, Wei et al. (1998) proposed the concept of "degree of automation" (DofA). This approach is quite complex, and requires a very detailed task definition including a workload parameter that is difficult to estimate prior experimentation. Although the model proposed in this study might seem very simple-minded, it is should be noted that more complicated models might be not as functional as a simple one.

Method

Subjects
Thirty-three students (mean age = 22.7, SD = 2.7, 26 females) volunteered in this study. All participants were right-handed, with normal or corrected to normal sight. No cases of Daltonism were reported.

Stimuli
Sixty-four digital rectangular shapes were used as stimuli in the visual search task. Shapes were created by manipulating three features: colour, texture and orientation. Each feature was implemented in four different alternatives. Particularly, colour was implemented as red / blue / black / green; texture was implemented as horizontal / vertical / right-tilted / left-tilted stripes; orientation was implemented as vertical / horizontal / right-tilted / left-tilted shapes. This manipulation allowed obtaining a set of stimuli that presented all combination of the given characteristics.

Automation aids
According to the quantitative model proposed, the automation aids were assessed as follow:

a) Manual control (LOA = 0): a visual search task, with targets presented at each trial;
b) Moderate automation support (LOA = 0.25): same task, but participants received automation support consisting of the highlighting of those shapes sharing at least one out of four characteristics with the target (e.g., all red-coloured shapes were highlighted if the target were red);
c) High automation support (LOA = 0.50): same tasks, but participants received automation support consisting of the highlight of those shapes sharing two out of four characteristics with the target (e.g., all red horizontal shapes were highlighted if the target were red and horizontally oriented).

According to this procedure, in the moderate automation condition participants were required to search the target among 16 shapes highlighted over 64, while in high automation condition the target had to be find among 4 highlighted shapes over 64.

In all conditions the highlighted stimuli were positioned approximately at the same distance from the centre of the screen and no feedback was provided either for the correct or the incorrect response.

System reliability
In the experimental design the reliability level was used as between-subjects factor. Three different levels of system reliability were implemented: 25%, 50%, 75%. The reliability level was assessed by manipulating the accuracy of highlighting in the visual search task. Particularly, automation aids were provided at 25% 50% and 75% of system reliability (respectively 30, 60, 90 correct highlighting over 120 highlighted stimuli per condition).

Procedure

Participants were randomly assigned to one reliability condition (25%, 50%, 75%, 11 participants each). They sat in a dark and sound-attenuated room, received instructions for the execution of the visual search task. The target was always presented at the centre of the screen, followed by a screenshot containing 63 distractors and the target (set size = 64 stimuli). Across trials, targets were evenly scattered among 16 positions within a 4x4 ideal grid dividing the screen space in 16 quadrants (not visible to participants). Stimuli were presented on a 17" LCD monitor (screen resolution = 1024 x 768) with uniform grey background. During the visual search task participants were requested to inspect the target for 1 second, then automatically switched to the search screen and were required to click over the target shape as soon as they found it. The experimental session was composed of 6 blocks of trials and each block consisted of 3 sub-blocks supported by three different LOA (manual, moderate, high) whose order was randomized across blocks. Each sub-blocks consisted of 15 search trials. At the end of each sub-block participants were requested to fill the NASA Task Load Index (NASA-TLX; Hart & Staveland, 1988). Each experimental session lasted for about 30 minutes.

Search times, number of correct responses and NASA-TLX scores were collected for the successive analysis.

Participants were informed that automation was not 100% reliable all of the time. They did not receive any additional information about the reliability levels, and no rationale for failure was provided.

Data analysis and results

Performance data were analyzed by mixed ANOVA designs using LOA (manual vs. moderate vs. high) as repeated factor and System Reliability (25% vs. 50%. vs. 75%) as between-subjects factor. Number of correct responses, target detection time and NASA-TLX scores were used as dependent variables.

Results showed a main effect of LOA on target detection time ($F(2,60)=20.3$, $p<0.0001$). Particularly at LOA = 0.25 subjects spent more time searching target than in Manual and in LOA = 0.50 conditions. Moreover, a significant interaction LOA by System Reliability ($F(4,60)=2.61$, $p<0.05$) was found. Inspecting the interaction separately for level of System Reliability, results showed that when the system is reliable at 25% and 50% the LOA = 0.25 was associated with slower target detection time than the other two LOA (see figure 1). No significant difference was found between Manual and LOA = 0.50 conditions; both were associated with faster target detection times. When the visual search task was highly reliable (75%), target detection times were faster for LOA = 0.50 condition and slower for Manual and LOA = 0.25.

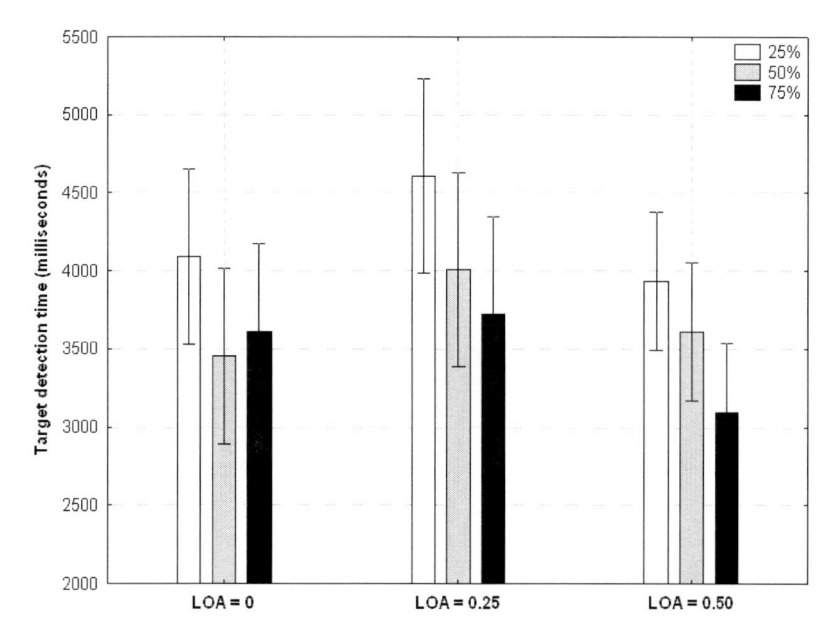

Figure 1. Target detection time by LOA and System Reliability condition. Vertical bars denote 0.95 confidence intervals

Analyses run on number of correct responses showed a similar trend (see figure 2). Results showed a main effect of LOA (F(2,60)=4.95, p<0.01); particularly, higher number of correct responses was found for the Manual and LOA = 0.50 conditions than LOA = 0.25. Moreover, a significant interaction LOA by System Reliability was found (F(4,60)=2,69, p<0.05). Inspecting the interaction separately for level of System Reliability, results showed that when the system was reliable at 25% LOA = 0.25 performance was significantly worse. Contrarily, higher number of correct responses were found for Manual condition and LOA = 0.50. When the system reliability was implemented at 50% no significant differences were found among the three different LOA conditions. At 75% of system reliability LOA= 0.50 showed a tendency towards statistical significance showing higher number of correct responses than the other two conditions. It should be noted that, although reliability did not apply to the Manual condition (because no automation was provided), the analyses included this level of the LOA factor to check for potential widespread effects of the reliability level. Indeed, as proposed by Di Nocera et al. (2006), reliance on automation may possibly transfer to parts of a task that are unsupported by the automated system.

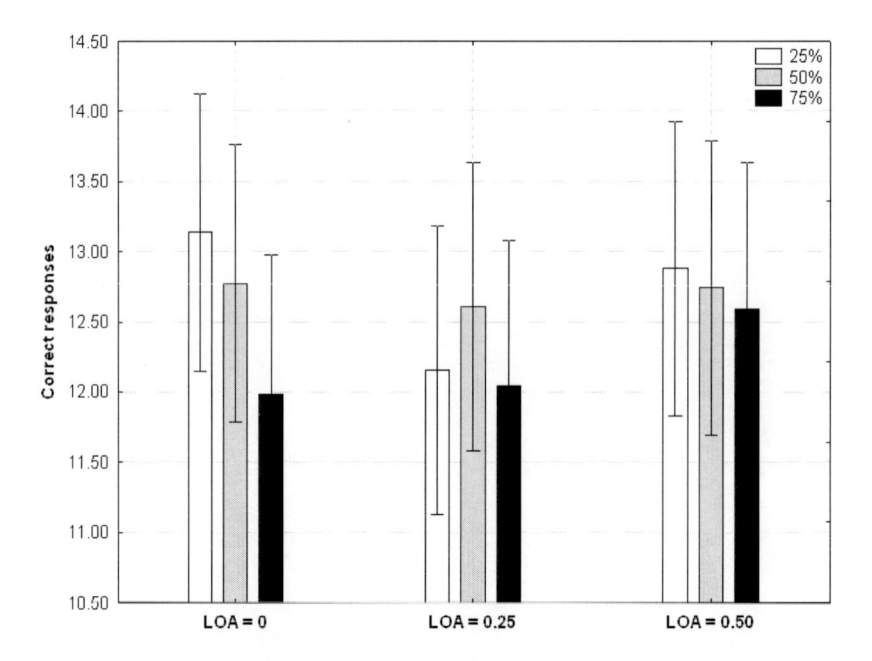

Figure 2. Correct responses by LOA and System Reliability condition. Vertical bars denote 0,95 confidence intervals

The mental workload subjectively assessed by NASA-TLX showed a main effect of LOA (F(2,60)=6.26, p<0.005); particularly LOA= 0.25 was associated to higher perceived mental workload than the other two conditions. A tendency towards statistical significance was found for system reliability (p = 0.12). As showed in figure 3, mental workload experienced at 25% system reliability was greater than that experienced in the other two conditions.

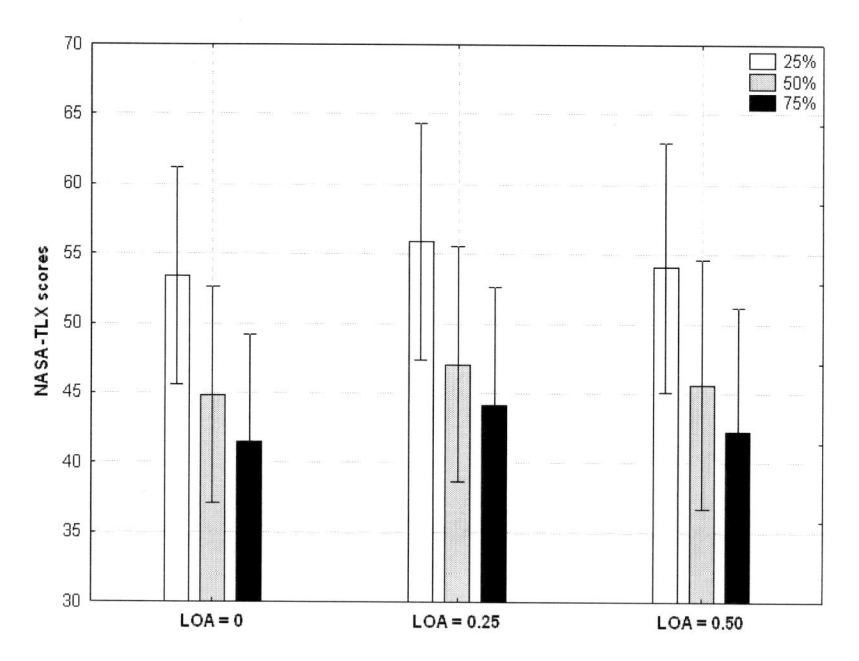

Figure 3. NASA-TLX scores by LOA and System Reliability condition. Vertical bars denote 0.95 confidence intervals

Discussion and conclusions

The present paper confirmed and extended results of previous studies (Terenzi et al., 2007), supporting the proposed definition of LOA in terms of *amount* of information traded by humans and machines. The present study investigated the effects of different levels of system reliability on the proposed quantitative LOA model. Particularly, performance in a visual search task was examined at different levels of automation and at three levels of system reliability (25%, 50% and 75%). Consistently with previous studies, a visual search task was used here. Visual search is a very common task in experimental psychology and much is known about the changes in performance one should expect varying the degree of the difficulty of the task. Visual search activity is also an important aspect of many civilian and military applications such as reconnaissance, tracking, information retrieval, aircraft inspection, baggage screening, medical image screening, and the monitoring of sonar, radar, and other displays. Therefore, knowledge about the behavioural effects of the different levels of automation may have important implications for designing future systems.

The experimental visual search proposed here engages similar cognitive processes that should be triggered for example during aircraft inspection or baggage screening.

Of course, the operative context is quite different. This study mainly represents a methodological investigation, and its primary aim was to add further evidence in the attempt to build a shared platform for investigating the effects of LOA in different operative domains and tasks.

The system reliability levels employed here might be quite different from the real operative contexts (25% system reliability is indeed a too low value to be used in the real world). However, from a methodological point of view, it was mandatory to investigate the model performance across a large range of reliability levels. Stressing downwards the levels of reliability allowed comprehending how the model works in more dynamic situations. Indeed, automation is likely to fail occasionally, causing changes in the way the user interacts with it, or it may fail completely, forcing the operator to perform the task manually. In general, visual search performance should be improved circumventing the limited information processing capabilities of human operators through the use of automation (Mosier & Skitka, 1996). Therefore, during the first two stages of information processing (information acquisition and analysis) participants should be supported to detect the target. In the current study a yellow highlighting around the stimuli was used as direct cue for attention guiding. This pop-out effect, implemented according to the quantitative definition of LOA, is supposed to support (and improve) the performance in the visual search task. Moreover, the different levels of reliability implemented in the system are assumed to modulate the performance.

In general, results of this study showed both expected and unexpected patterns in terms of target detection time and number of correct responses. Level of automation apparently affects performance in the visual search task. As expected, at LOA = 0.25, participants spent more time searching the target and they provided less correct responses than in the high automation condition. However, no significant differences were found between high automation support (LOA = 0.50) and manual condition (LOA = 0), both showing the best performance. Examining results separately for level of system reliability, emerged that -in terms of target detection time- moderate automation led to worse performance when the system was less reliable (25% and 50%). A similar pattern was found in terms of number of correct responses (even if the significance of this effect was limited to 25% of system reliability). How to interpret this unexpected disadvantage of moderate automation? It is clear that automation aids implemented in the visual search task provided a spatial cue that is assumed to assist the participants searching the target. Wickens and colleagues (e.g. Wickens et al., 1999; Yeh et al., 1999; Yeh and Wickens, 2001) have shown how highlighting (or attention cueing) can provide both benefits and costs. Spatial cues should be beneficial because the visual field that needs to be searched is restricted. However, this benefit becomes a cost when the spatial cue is inaccurate. In the present case, the spatial cue provided should be not only inaccurate (depending on level of reliability) but it is also combined with several distracting shapes that are highlighted as well. Consequently, it is possible to hypothesise that the costs found may arise from a combination of these two effects, providing an additional *visual* and *cognitive* load. In the present study, this effect is amplified in the LOA = 0.25 condition, where participants were requested to find the target among 16 highlighted shapes instead of 4 (as for LOA = 0.50). Of course, in the manual condition (where neither the reliability nor the highlighting was implemented) performance was much better, and it was not different from the high automation condition.

Inspection of NASA-TLX scores seems to support this interpretation, showing that participants reported relatively higher perceived mental workload with moderate automation than in the other two experimental conditions. However, it is worth noting that the NASA-TLX values reported were generally quite low, showing a mental workload score ranging from 46 to 49. This may suggest that the visual search task was perceived as quite easy to perform. As a consequence, the lack of significant differences between the manual (LOA = 0) and the high (LOA = 0.50) conditions is not surprising. In fact, it is well-known that the introduction of automation may not always provide an effective increase in human system performance. For example, Rovira et al., (2002) suggested that one factor affecting the automation effectiveness might be related to task complexity. In other words, if the manual task is relatively easy, automating it may not produce an increase in performance. Moreover, there is some evidence attesting that performance benefits became evident only in difficult and high workload task environment (Merlo et al., 1999), or under temporally demanding or uncertain task situation (Dinzolet et al., 2001; Muthard & Wickens, 2001).

The level of reliability implemented also played a relevant role in assessing the quantitative model proposed. In fact, the effects discussed above are largely modulated by the reliability of the system. If, at lower reliability levels (i.e. 25% and 50%), performance seems to be affected by a combination of factors, looking at a more realistic reliability level (i.e. 75%) results showed a more predictable pattern. In fact, as expected, the best performance was found with high automation support both in terms of target detection time and number of correct responses. It is worth noting that participants belonging to the 25% reliability group reported higher mental workload than those belonging to the other two groups. This may indicate that those participants switched to manual control more often that the others. Of course, it was easier to decide whether the computer aid was effective or not when LOA = 0.50, because only 4 highlighted stimuli had to be checked, whereas the switch towards manual control was delayed when LOA = 0.25.

The question that arises from this discussion is what methodological implications could be drawn from these results. Interesting suggestions may come from a comparison with results obtained in a previous set of studies (Terenzi et al., 2007), in which automation support was always 100% reliable. Both in terms of target detection time and number of correct responses, results showed a very similar pattern to that obtained in the current study at 75% of system reliability. Particularly, LOA = 0.50 was associated to shorter detection time and higher number of correct responses than the other two conditions. This result suggests that, when the system performance is quite high, it is possible to detect performance improvements only if almost half of the task is allocated to the system. These results are in agreement with those reported in a review by Wickens & Dixon (2005) who found that a reliability of approximately 0.7 is the lower reliability limit to actually help the operator.

The study reported here represented a further step towards the development of a new quantitative LOA model. More studies will be needed to fully understand the complex relationship between automation and performance. The research agenda includes the generalisation of the same rationale to different stages of information processing (see Parasuraman et al., 2000). Moreover, further studies will be carried out in order to test this model in more operative and ecological environments.

References

Billings, C.E. (1991). Human-centred aircraft automation: A concept and guidelines. *NASA Tech. Memo. No. 103885*. Moffet Field, CA: NASA-Ames Research Center.

Di Nocera, F., Fabrizi, R., Terenzi, M., & Ferlazzo, F. (2006). Air Traffic Control Procedural Errors: Effects of Traffic Density, Expertise and Automation. *Aviation, Space & Environmental Medicine, 77*, 639-643.

Dixon, S.R., & Wickens, C.D. (2004). Reliability in Automated Aids for Unmanned Aerial Vehicle Flight Control: Evaluating a Model of Automation Dependence in High Workload *(AHFD-04-5/MAAD-04-1)*. Savoy, IL: University of Illinois, Aviation Human Factors Division.

Dzindolet, M.T., Pierce, L.G., Beck, H.P., Dawe, L.A., & Anderson, B.W. (2001). Predicting misuse and disuse of combat identification systems. *Military Psychology, 13*(3), 147-164.

Endsley, M. R. & Kaber, D. B. (1997). The use of level of automation as a means of alleviating out-of-the-loop performance problems: a taxonomy and empirical analysis. In P. Seppala, T. Luopajarvi, C. H. Nygard, and M. Mattila (Eds.), *13th Triennial Congress of the International Ergonomics Association*, Vol. 1 (pp. 168–170). Helsinki: Finish Institute of Occupational Health.

Endsley, M.R. & Kaber, D.B. (1999). Level of automation effects on performance, situation awareness and workload in a dynamic control task, *Ergonomics, 42*, 462–492.

Fereidunian,A., Lucas, C., Lesani, H., Lehtonen, M., & Nordman, M.(2007). Challenges in Implementation of Human-Automation Interaction Models. *Proceeding of the 15th Mediterranean Conference of Control & Automation*, Athens-Greece.

Merlo, J.L., Wickens, C.D., & Yeh, M. (1999). Effect of reliability on cue effectiveness and display signaling *(Technical Report ARL-99-4/FED-LAB-99-3)*. Savoy, IL: University of Illinois, Aviation Research Lab.

Molloy, R. & Parasuraman, R. (1996). Monitoring an automated system for a single failure: vigilance and task complexity effects. *Human Factors, 38*, 211-322.

Mosier, K., & Skitka, L.J. (1996). Human decision makers and automated decision aids: Made for each other? In R. Parasuraman & M. Mouloua (Eds.), *Automation and human performance: Theory and applications* (pp. 201-220). Hillsdale, NJ: Lawrence Erlbaum Associates.

Muthard, E.K., & Wickens, C.D. (2001). *Change detection in a flight planning task* of automation and its relation to system performance and mental load. *Human Factors, 40*, 277-295.

Parasuraman, R. & Riley, V. (1997). Humans and Automation: Use, Misuse, Disuse, Abuse. *Human Factors, 39*, 230-253.

Parasuraman, R., Molloy, R., and Singh, I. (1993). Performance Consequences of Automation-Induced "Complacency." *The International Journal of Aviation Psychology, 3*, 1-23.

Parasuraman, R., Sheridan, T.B., & Wickens, C.D. (2000). A model for types and levels of human interaction with automation. *IEEE Transactions on Systems, Man, and Cybernetics. Part A: Systems and Humans, 30*, 286-297.

Rovira, E., & Parasuraman, R. (2002). Sensor to shooter: Task development and empirical evaluation of the effects of automation unreliability. Paper presented at the *Annual Midyear Symposium of the American Psychological Association*, Division 10 (Military Psychology) and 21 (Engineering Psychology). Ft. Belvoir, VA.

Rovira, E., McGarry, K., & Parasuraman, R. (2002). Effects of unreliable automation on decision making in command and control. *Proceedings of the Annual Meeting of the Human Factors Society*, (pp. 428-432). Santa Monica, CA, USA: HFES.

Rovira, E., Zinni, M., & Parasuraman, R. (2002). Effects of information and decision automation on multi-task performance. *In Proceedings of the 26th Annual Meeting of the Human Factors and Ergonomics Society.* (pp. 327-331). Santa Monica, CA: HFES.

Sarter, N. B., Woods, D. D., & Billings, C. (1997). Automation surprises. In G. Salvendy (Eds.), *Handbook of Human Factors Ergonomics* (2nd ed.). New York: Wiley.

Sheridan, T.B., & Verplank, W. (1978). *Human and Computer Control of Undersea Teleoperators.* Cambridge, MA: Man-Machine Systems Laboratory, Department of Mechanical Engineering, MIT.

Skitka, L., Mossier, K., & Burdick, M. (1999). Does Automation Bias Decision-making? *International Journal of Human-Computer Studies, 51*, 991-1006.

Terenzi, M., Camilli, M., & Di Nocera F. (2007). Making it quantitative: early phases of development of a new taxonomy for levels of automation. In D. de Waard, G.R.J. Hockey, P. Nickel, and K.A. Brookhuis (Eds.) (2007), *Human Factors Issues in Complex System Performance* (pp. 313 - 323). Maastricht, the Netherlands: Shaker Publishing.

Wei, Z., Macwan, A.P., & Wieringa, P.A. (1998). A quantitative measure for degree of automation and its relation to system performance and mental load. *Human Factors, 40*, 277-295.

Wickens, C.D. & Xu, X. (2002). *Automation trust, reliability and attention.* (AHFD-02-14/MAAD-02-2). Savoy, IL: University of Illinois, Aviation Research Lab.

Wickens, C.D., & Dixon, S.R. (2005). Is there a magic number 7 (to the minus 1)? The benefits of imperfect diagnostic automation: A synthesis of the literature *(Tech. Rep. AHFD-05-01/MAAD-05-1).* Savoy, IL: University of Illinois, Aviation Human Factors Division.

Wickens, C.D., Canejo, R., & Gempler, K. (1999). Unreliable automated attention cueing for air-ground targeting and traffic maneuvering. *Proceedings of the Human Factors and Ergonomics Society 43rd Annual Meeting* (pp. 21-25). Santa Monica, CA: HFES.

Yeh, M., & Wickens, C.D. (2001). Display signaling in augmented reality: Effects of cue reliability and image realism on attention allocation and trust calibration. *Human Factors, 43*, 355-365.

Yeh, M., Wickens, C.D., & Seagull, F.J. (1999). Target cuing in visual search: The effects of conformality and display location on the allocation of visual attention. *Human Factors, 41*, 524-542.

Inter-individual differences in executive control activity during simulated process control: comparison of performance and physiological patterns

Peter Nickel, Adam C. Roberts, Michael H. Roberts, & G. Robert J. Hockey
The University of Sheffield
United Kingdom

Abstract

Task duration and environmental stressors have the potential to threaten maintenance of sustained attention, especially for tasks in process control environments that impose high demands on working memory and executive control processes. Experimental sessions of three hours were conducted to investigate effects of noise and prolonged work on operator performance. Ten highly trained operators performed monitoring and control tasks, including fault management, under different environmental conditions. Although integrity of central aspects of task performance was maintained at a high level, inter-individual differences in primary task performance were identified and could be attributed to differences in control and information sampling behaviour. Further comparisons of relatively high and low level performers showed differential effects for effort investment as indicated by heart rate variability analysis. They also revealed differences in frontal midline theta activity consistent with the view that high performers were better able to maintain or adjust level of executive control activities, such as applying different monitoring and control strategies over the session. These findings suggest that individual differences in monitoring and control strategies may be masked by overt system performance, but have a significant impact on detection and selection of appropriate support for operators in adaptive automation scenarios.

Introduction

In automation-enhanced human-machine systems the operator is required to continually adapt to new and unforeseen changes in the dynamic process under control, and to determine whether and what actions are required to prevent or correct for drifts or faults of the technical subsystem (Meshkati, 2003). In continuous process control, monitoring and control operations not only impose high demands on working memory and executive control processes but also require the operator to maintain sustained attention over prolonged work periods and under variations of environmental stressors that add to the threats to system safety and performance. Consequently, the human operator is exposed to dynamic task demands, with executive control functions becoming a major determinant in operator's attempts to control the task (Royall et al., 2002; Shallice, 2005).

In D. de Waard, F.O. Flemisch, B. Lorenz, H. Oberheid, and K.A. Brookhuis (Eds.) (2008), *Human Factors for assistance and automation* (pp. 313 - 323). Maastricht, the Netherlands: Shaker Publishing.

In complex task environments human cognition and action are more likely to be triggered by the pursuit of top-level mission goals (e.g., system safety, productivity, operational availability) with the prospect of keeping operator motivation at high levels. Because of the complex nature of the task and the greater degrees of freedom for performance, regulatory aspects of operational behaviour depend on characteristics of operator, task and also the level of workload, which in turn is moderated by the operator's adopted strategy (Sperandio, 1971, 1978). The cognitive energetical model of operator state regulation (Hockey, 1997) argues that operators have various compensatory strategies at their disposal, even if maintaining overt task performance under high demands. However, the use of such compensatory strategies also attracts costs or 'latent decrements': the adoption of less demanding (riskier) strategies, secondary task decrements, and increases in both subjective strain and physiological activation. Furthermore, prolonged periods of sustained effort in the performance protection mode are likely to generate fatigue after-effects (Hockey, 1997). Because of this, the analysis of effects of task demands on executive control processes in the regulation of operator behaviour in complex task environments is not limited to changes in overt primary task performance; it must also consider mismatches between current and target operator states indicated by latent decrements. In addition to the effects of task demands, task duration and environmental stressors may also reduce the effectiveness of the performance regulation (Hitchcock et al., 2003; Hockey et al., 2007; Molloy & Parasuraman, 1996). Hockey and Hamilton's (1983) analysis of stressor effects shows that, noise is one of the class of stressors that increase anxiety and distraction (reduced attention to task-relevant events). Hockey et al. (2007) found no effect on primary control activity under noise, but impairment of lower priority tasks that made demands on executive control.

Micro-worlds, as scaled-down models of real world situations, can provide suitable environments to study executive control strategies in complex task situations, since they represent a compromise between experimental control and realism, facilitate experimental research within dynamic, complex decision-making situations (Gonzalez et al., 2005), and produce multiple behavioural variables of human-machine interactions potentially relevant for the investigation of specific concepts of interest (Howie & Vicente, 1998; Sauer et al., 2000b). The compensatory control model (Hockey, 1997) calls for a multi-measures approach for the investigation of executive control strategies in a complex task environment. This paper explores individual differences in the use of executive control under variation of task and environmental stressors. Psychophysiological measures have specific advantages as markers of effort and strain, since they are relatively unobtrusive and provide continuous data even in the absence of apparent behaviour (Kramer, 1993). A number of measures of both autonomic and central nervous system activity have been associated with changes in mental effort and executive activity. For this study the strongest markers identified by previous work were included: the 0.10 Hz component of HRV (Mulder et al., 2004; Tattersall & Hockey, 1995), and the two EEG components of the Task Load Index (TLI): frontal midline theta and parietal alpha activity (Gevins & Smith, 2003; Manzey, 1998; Onton et al., 2005).

Method

Participants

Ten male postgraduate student operators from engineering and science departments of The University of Sheffield received extensive training on the process control environment autoCAMS. Participants were financially compensated for taking part in the study.

Testing environment and tasks

The experimental sessions were scheduled in the air-conditioned Human Factors Laboratory furnished with desks for the participant and the experimenter and movable walls for visual isolation. The participant computer provided the Cabin Air Management System (CAMS; Hockey et al., 1998; Sauer et al., 2000b; Wastell, 1996) in its automation-enhanced version (autoCAMS; Lorenz et al., 2002), developed to study problem solving and effects of environmental stressors in simulated complex task settings. The monitoring and control task consisted of cognitive processes relevant to information acquisition and analysis, decision making and action selection. It required participants to interact with the dynamic semi-automatic system that simulates control of interactive parameters of the atmospheric environment (e.g. nitrogen, oxygen, carbon dioxide, humidity and temperature) of a closed system such as a space capsule or submersible. Monitoring by the operator is mainly related to the autoCAMS interface and refers to continuous assessments of the dynamic changes of flows, states of the subsystems and processes under automatic and/or manual control. Based on these assessments the operator should always be able to deduce information about whether the system is in a normal, transitional (uncertain) or off-normal (fault) state. Control operations were mainly relevant during fault management, i.e. when the operator has reasons to assume displays and controls to malfunction or parameters of the atmospheric environment tending to deviate from normal range. Using systematic reasoning, sources of malfunctions and faults had to be identified and repairs authorised if necessary. During ongoing maintenance of autoCAMS the operator was also in charge of ensuring operations within safe limits using manual control.

Performance and psychophysiological recording and data reduction

Levels of cabin air parameters of autoCAMS (nitrogen, oxygen, carbon dioxide, humidity, temperature) were sampled at 1 Hz. Logged data was classified as system parameters within or out of normal operational range and calculated as percentage of time any of the cabin air parameters of the interacting subsystems were in or out of normal range during monitoring and control operations, respectively. Events of flow view enquiries and control actions were logged in order to gain information about the operators' monitoring and control behaviour. The Active Two Base System (Biosemi, The Netherlands) was used for continuously acquiring psychophysiological data, for ECG (electrocardiogram), EOG (vertical, horizontal electrooculogram), mastoids for post hoc referencing (in the BioSemi design the ground electrode during acquisition was formed by the Common Mode Sense active electrode and the Driven Right Leg passive electrode), and EEG (electroencephalographic signals from Fz and Pz sites extracted from 10-20 system). Data acquisition for all signals sampled at 2048 Hz was controlled by the ActiView 6 interface (BioSemi, The Netherlands).

With Brain Vision Analyzer Professional (version 1.05; Brain Products, Germany) EEG activity was high and low pass filtered (1.6 Hz and 40 Hz) before segmentation and ocular and baseline corrections were applied. The epoched EEG signals were analysed via Fast Fourier Transform and resulting power in individual frequencies were collapsed into theta (4-7.5 Hz) and alpha (8-12.5 Hz) band activity. The ECG signal was composed of signals recorded from Nehb's triangle (Nehb, 1938). R-peaks were detected with the 'EKG Markers' procedure provided by Brain Vision Analyzer Professional and artefacts were corrected with functions provided by the Vision Analyzer interface, if necessary. Application of LabVIEW (National Instruments, USA) virtual instruments and the procedure described in Nickel and Nachreiner (2003) resulted in the measure for the 0.1 Hz component of heart rate variability (HRV).

Procedure

The experimental session lasted 3 hours, divided into consecutive 1 hour periods of quiet, noise and quiet. All sessions were carried out at the same time of day, to control for circadian effects. After fixing electrodes and acquiring psychophysiological data participants performed simulated process control operations with autoCAMS. During the two quiet periods operators received broadband noise at 55 dB(A) through headphones. The intervening noise period presented superimposed a variety of industrial noises (machinery, etc) at 80dB(A). Each task period contained alternating task segments of fault management and monitoring, consisting of six fault scenarios and subsequent monitoring scenarios. Fault scenarios were designed to include comparable sets of faults across the different periods. Psychophysiological and performance data were continuously recorded on different computers during simulated process control operations. Neither breaks nor system resets interrupted the sense of continuity of data acquisition.

Data analysis

The primary goal of the analysis was to assess the impact of effects of noise and prolonged work on operator performance. Operators received a considerable amount of training and gained expertise from participation in previous experimental studies and therefore quality of primary task performance was very high. Nevertheless, even at this high level of performance inter-individual differences could be identified. Variations in total amount of time the system parameters were out of normal operating range allowed to divide participants into two groups of 5 having either very high (H) or normal (N) control performance. Repeated measures analyses of variance were computed using SPSS for a mixed 2 x 3 x 2 design with the two performance Groups as between subject factor and the within subject factors Periods (3; quiet/noise/quiet), each containing two Tasks (2; fault management and monitoring operations). Analyses were based on untransformed pre-processed data integrated across task segments. (The present study is part of a research project on adaptive automation, with the requirement to acquire and analyse data on-line and in real-time. Since we will not include log transformation in our procedures for on-line data analysis we adopted the most similar procedure for the off-line analysis presented here.) Huyhn-Feldt corrections for degrees of freedom were applied when Mauchley's test showed significant departures from sphericity.

Results

Performance

Control error
By definition, as the basis for our classification of participants, control error (the percentage of time the system parameters went out of normal operating range) was lower for group H than for group N, $F(1,8) = 41.54; p < .001$ (Fig. 1). Also unsurprisingly, there was a main effect of Task, $F(1,8) = 350.20; p < .001$, with control error occurring largely under the more demanding fault management phases of the task. More interesting for the present study there was a large Group x Task interaction, $F(1,8) = 19.16; p < .01$. As Figure 1 clearly illustrates, the difference between ability groups occurred only when there was a requirement to carry out fault management. There was no main effect of periods or any interactions with other factors (all $p > .10$).

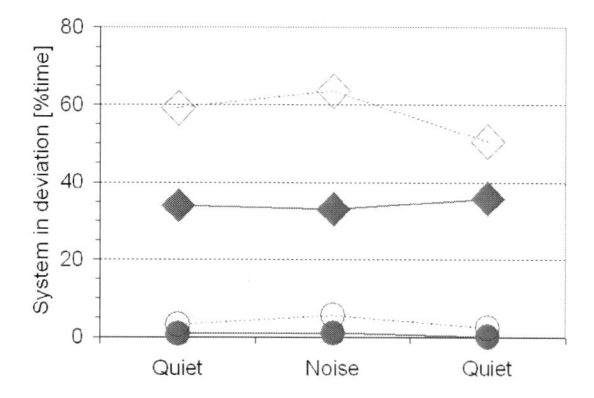

Figure 1. Percentage time total system is out of normal range during fault management (diamonds) and monitoring (circles) for subgroups H (filled) and N (open)

Operator interactions
In addition, two kinds of operator interactions with the system were examined. Feed flow sampling is an essential component of effective monitoring behaviour, while operator control actions are required primarily when fault states are present and need to be managed. The pattern of the data in Figure 2 generally supports this distinction. Feed flow sampling there was higher during monitoring, though the effect of Task was only marginally significant, $F(1,8) = 4.81; p < .10$. There were no effects of time periods/noise, $F(2,16) = 1.71; p > .10$, or Task x Period, $F(2,16) < 1$. Despite the appearance of increased sampling by the Group H participants, there was also no

difference between sub-groups, $F(1,8) = 1.18; p > .10$, or interactions with Group (all F values < 1). In contrast, control actions were significantly higher for fault management, $F(1,8) = 177.76; p < .001$, with very few interventions for monitoring phases of the task. Because of this the main effect of Group was not significant, $F(1,8) = 3.18; p > .10$. However, the difference between tasks was more pronounced for group H than group N, as shown by the marginally significant Group x Task interaction, $F(1,8) = 4.90; p < .10$. There was also a significant monotonic decrease in control actions over periods, $F(2,16) = 3.82; p < .05$. No other effects were significant (all $p > .05$).

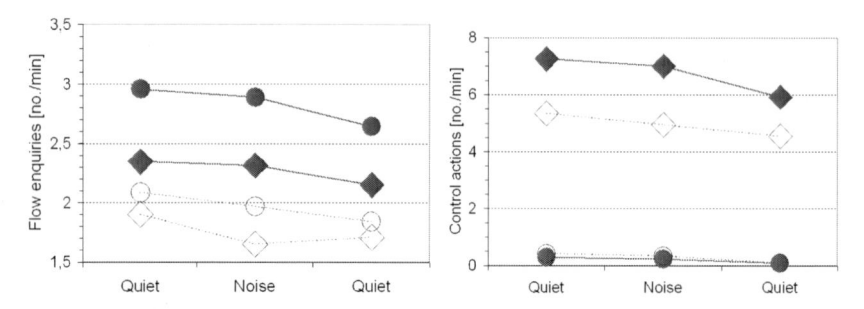

Figure 2. Rate (number per minute) of Feed Flow enquiries (left) and Control Actions (right) during fault management (diamonds) and monitoring (circles) for subgroups H (filled) and N (open)

Psychophysiology

Heart rate variability

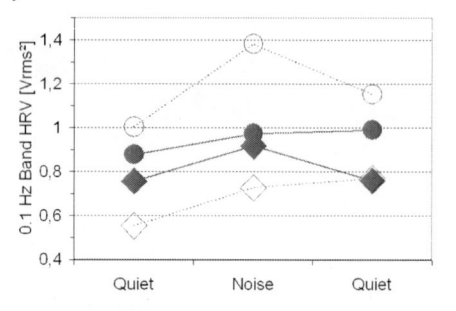

Figure 3. Energy in 0.1 Hz component of HRV during fault management task (diamonds) and monitoring tasks (circles) for subgroups H (filled) and N (open)

Figure 3 shows that suppression of heart rate variability in the mid frequency (0.1 Hz) band was driven primarily by differences between tasks, $F(1,8) = 5.56; p < .05$; indicative of higher effort investment during fault management. The data suggest an overall effect of increased effort over time, with monitoring for group H and fault management for group N attracting more effort under noise. There was a marginally significant effect for Period, $F(2,16) = 2.65; p < .10; \varepsilon_{HF} = 0.98$. It appears that HRV suppression is reduced under noise, but a separate test of noise (period 2) vs quiet (periods 1+3) did not show a significant difference ($F(1,8) = 4.02; p > .05$). Figure 3 also suggests that the HRV response for group H participants is more consistent across the two tasks. However, the Group x Task interaction was not significant, $F(1,8) = 1.80; p > .10$.

EEG activity
Data for EEG theta and alpha activity are shown in Figure 4. Theta activity measured at Fz was marginally sensitive to task manipulations, $F(1,8) = 3.60; p < .10$, suggesting that level of activity (and thereby investment of executive control resources) was, as expected, higher during fault management phases. Despite the apparent higher levels of theta for group H the difference between sub-groups was not significant, $F(1,8) < 1$. The marginally significant Group x Period interaction, $F(2,16) = 3.82; p < 0.10; \varepsilon_{HF} = 0.81$, was caused by an increase in theta activity for the better (group H) performers and a decrease for group N. Again, there were no effects of periods/noise or other interactions (all $p > .10$). Alpha activity at Pz showed a significant main effect for Task, $F(1,8) = 7.76; p < 0.05$, with the level of relative activity higher (and, therefore, general activation lower) for monitoring. Again, there was no overall difference between sub-groups, $F(1,8) < 1$, though there was a marginal Group x Period interaction, $F(2,16) = 2.87; p < 0.10; \varepsilon_{HF} = 0.92$. Group H maintained a higher level of general activation (lower alpha power) across the first two periods before it decreased in period 3. In contrast, group N showed a decrease in general activation during period 2 (noise).

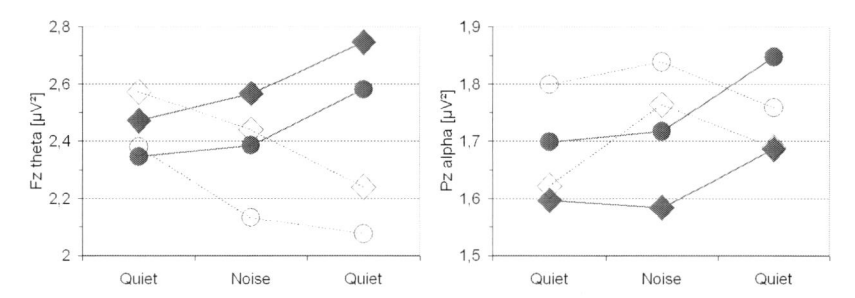

Figure 4. EEG activity at Fz in theta band (left) and at Pz in alpha band (right) during fault management task (diamonds) and monitoring tasks (circles) for subgroups H (filled) and N (open)

Discussion

During three hours of continuous process control operations with variations in noise level, operators managed to maintain overt task performance at a constant and relatively high level. Despite extensive training in small groups and practice gained from previous experimental studies of a long-term research programme (Hockey et al., 2006); inter-individual differences occurred at a high level of task performance, attributable mainly to applications of different strategies in executive control activity.

Hockey's compensatory control model (Hockey, 1997) suggests that the major source of individual differences at high level overt task performance is that of variation in the choice of regulatory strategies. Because of the very small sample sizes available for the analyses, some of the apparent differences in the means were not significant. However, the frequency of system intervention and information sampling for the two groups of very high (H) and normal (N) levels of control performance suggest that the group N adopted a less demanding and riskier strategy, resulting in longer deviations out of normal range for the main system parameters of the atmospheric environment. Similar conclusions can be drawn from research in automation supported task environments by Lorenz et al. (2002), who found information sampling behaviour (at higher frequencies) an important cue for shifts in attention allocation and maintenance of system awareness during reliable automation. Other studies using the CAMS simulation (Hockey et al., 2007; Sauer et al., 2000a) were able to establish close links between system interventions or information sampling behaviour and different training approaches. Both studies found that operators who received system-based training made more frequent control and information sampling actions than operators who received procedure-based training. They assumed the more frequent information sampling especially during monitoring operations to be indicative of more active involvement with process characteristics and the more frequent control actions especially during control operations to be rather indicative of a more reactive mode (Hockey et al., 2007), i.e. driven by system necessity rather than by anticipatory planning.

Consistencies of such results with those of the present study are not surprising, at least from an instructional point of view. In the long-term research programme the training approach was guided by Rasumussen's (1981) classification of skill-based, rule-based and knowledge-based behaviour, intending to enhance operators' competencies for a broad range of process control scenarios while at the same time provide learning cues that meet operators' needs at different levels of learning development and over time. Training was aimed at promoting flexible reasoning in operators, allowing them to solve emerging problems and sustain high level mission goals, rather than a normative approach with routine procedures available for well-known scenarios. This approach resulted in operators being trained with a system-based emphasis. The differences in system interventions or information sampling behaviour for H and N groups are likely to be related to the higher level of system-based competencies attained by the very high performers in the course of training and practice.

Evidence for differential effects on effort investment, as another compensatory control strategy (Hockey, 1997), can be gained from psychophysiological data. Results for the 0.1 Hz component of HRV for the better performers (group H) indicate a stable pattern of effort investment across tasks and periods. However, task differences in effort investment for the normal performers (group N) suggest that monitoring task periods were used as recovery from the more effortful control tasks. A similar effect was found by Röttger et al. (2007) in their study of workload effects associated with different levels of operator assistance. In the present study there is a suggestion that reduced flow sampling is associated with a more relaxed (reactive) monitoring effort, with reduced opportunities for anticipatory control activities, and resulting in less stable and more error-prone control performance. The hypothesis of a low effort strategy for group N is further supported by the falling investment of executive control resources over the 3-hr session, as indicated by theta activity at Fz (Gevins & Smith, 2003; Onton et al., 2005). In contrast, better performers seemed to increase their investment of executive control resources (increase in theta activity) to match task demands, allowing maintenance of performance over the session at very high level; in addition, they were able to maintain a high level of general activation (lower alpha power) even under noise before signs of deactivation become noticeable towards the end of the experimental session.

In conclusion, the present study provides evidence for individual differences in executive control activities in the regulation of human performance while maintaining overt system performance at a relatively high level. Such differences may be due to two major influences. First, the specific training approach chosen and operator responses to it may have resulted in different training stages being reached for subgroups of operators, consistent with the findings in Hockey et al. (2007) and Sauer et al. (2000a). Second, such differences may be due to stable between person characteristics, such as general mental abilities and cognitive style (Burkholter et al., 2007). As suggested by Gopher (1996) executive control functions are likely to be affected by individual differences (e.g., training, personality) and moderated by task requirements. Additionally, the results presented highlight the importance of investigating strategies in task performance to enable effective and efficient individual support of operators in training and automation.

Acknowledgements

This study is part of the project funded by the UK EPSRC grant GR\S66985\01. We thank the anonymous reviewers for helpful comments on an earlier version of the manuscript.

References

Burkholter, D., Kluge, A., Schüler, K., Sauer, J., & Ritzmann, S. (2007). Cognitive requirements analysis to derive training models for controlling complex systems. In D. de Waard, G.R.J. Hockey, P. Nickel, and K.A. Brookhuis (Eds.), *Human factors in complex system performance* (pp. 475-484). Maastricht, The Netherlands: Shaker Publishing.

Gevins, A., & Smith, M.E. (2003). Neurophysiological measures of cognitive workload during human-computer interaction. *Theoretical Issues in Ergonomics Science, 4*, 113-131.

Gonzalez, C., Vanyukov, P., & Martin, M.K. (2005). The use of microworlds to study dynamic decision making. *Computers in Human Behaviour, 21*, 273-286.

Gopher, D. (1996). Attention control: Explorations of the work of an executive controller. *Cognitive Brain Research, 5*, 23-38.

Hitchcock, E.M., Warm, J.S., Matthews, G., Dember, W.N., Shear, P.K., Tripp, L.D., Mayleben, D.W., & Parasuraman, R. (2003). Automation cueing modulates cerebral blood flow and vigilance in a simulated air traffic control task. *Theoretical Issues in Ergonomics Science, 4*, 89-112.

Hockey, G.R.J. (1997). Compensatory control in the regulation of human performance under stress and high workload: A cognitive energetical framework. *Biological Psychology, 45*, 73-93.

Hockey, G.R.J., & Hamilton, P. (1983). The cognitive patterning of stress states. In G.R.J. Hockey (Ed.), *Stress and fatigue in human performance* (pp. 331-361). Chichester, UK: Wiley.

Hockey, G.R.J., Wastell, D.G., & Sauer, J. (1998). Effects of sleep deprivation and user-interface on complex performance: A multilevel analysis of compensatory control. *Human Factors, 40*, 233-253.

Hockey, G.R.J., Nickel, P., Roberts, A.C., Mahfouf, M., & Linkens, D.A. (2006). Implementing adaptive automation using on-line detection of high risk operator functional state. In ISSA Chemistry and Machine and System Safety Sections (Eds.), *Design process and human factors integration: Optimising company performance* (pp. 1/004:1-13). Paris, France: INRS.

Hockey, G.R.J., Sauer, J., & Wastell, D.G. (2007). Adaptability of training in simulated process control: Knowledge- versus rule-based guidance under task changes and environmental stress. *Human Factors, 49*, 158-174.

Howie, D.E., & Vicente, K.J. (1998). Measures of operator performance in complex, dynamic microworlds: advancing the state of the art. *Ergonomics, 41*, 485-500.

Kramer, A.F. (1993). Physiological metrics of mental workload: A review of recent progress. In P. Ullsperger (Ed.), *Psychophysiology of mental workload* (pp. 2-34). Heidelberg, Germany: Haefner.

Lorenz, B., Di Nocera, F., Röttger, S., & Parasuraman, R. (2002). Automated fault-management in a simulated spaceflight micro-world. *Aviation, Space, and Environmental Medicine, 73*, 886-897.

Manzey, D. (1998). Psychophysiologie mentaler Beanspruchung. In F. Rösler (Ed.), *Ergebnisse und Anwendungen der Psychophysiologie* (Enzyklopädie der Psychologie, C/I/7) (pp. 799-864). Göttingen, Germany: Verlag für Psychologie, Hogrefe.

Meshkati, N. (2003). Control rooms' design in industrial facilities. *Human Factors and Ergonomics Manufacturing, 13*, 269-277.

Molloy, R. & Parasuraman, R. (1996). Monitoring an automated system for a single failure: Vigilance and task complexity effects. *Human Factors, 38*, 311-322.

Mulder, L.J.M., Kruizinga, A., Stuiver, A., Vernema, I., & Hoogeboom, P. (2004). Monitoring cardiovascular state changes in a simulated ambulance dispatch task for use in adaptive automation. In D. de Waard, K.A. Brookhuis, and C.M. Weikert (Eds.), *Human factors in design* (pp. 161-175). Maastricht, The Netherlands: Shaker Publishing.

Nehb, W. (1938). Zur Standardisierung der Brustwandableitungen des Elektrokardiogramms. *Klinische Wochenschrift, 17*, 1807-1811.

Nickel, P., & Nachreiner, F. (2003). Sensitivity and diagnosticity of the 0.1 Hz component of heart rate variability as an indicator of mental workload. *Human Factors, 45*, 575-590.

Onton, J., Delorme, A., & Makeig, S. (2005). Frontal midline EEG dynamics during working memory. *NeuroImage, 27*, 341-356.

Rasmussen, J. (1981). Models of mental strategies in process plant diagnosis. In J. Rasmussen, and W.B. Rouse (Eds.), *Human detection and diagnosis of system failures* (pp. 241-258). New York, USA: Plenum.

Röttger, S., Bali, K., & Manzey, D. (2007). Do cognitive assistance systems reduce operator workload? In D. de Waard, G.R.J. Hockey, P. Nickel, and K.A. Brookhuis (Eds.), *Human factors in complex system performance* (pp. 351-360). Maastricht, The Netherlands: Shaker Publishing.

Royall, D.R., Lauterbach, E.C., Cummings, J.L., Reeve, A., Rummans, T.A., Kaufer, D.I., LaFrance, W.C., & Coffey, C.E. (2002). Executive control function: A review of its promise and challenges for clinical research. *Journal of Neuropsychiatry and Clinical Neuroscience, 14*, 377-405.

Sauer, J., Hockey, G.R.J., & Wastell, D.G. (2000a). Effects of training on short- and long-term skill retention in a complex multiple-task environment. *Ergonomics, 43*, 2043-2064.

Sauer, J., Wastell, D.G., & Hockey, G.R.J. (2000b). A conceptual framework for designing micro-worlds for complex work domains: A case study on the Cabin Air Management System. *Computers in Human Behavior, 16*, 45-58.

Shallice, T. (2005). The fractionation of supervisory control. In M.S. Gazzaniga (Ed.), *The cognitive neurosciences III* (pp. 943-956). Cambridge, USA: MIT Press.

Sperandio, J.-C. (1971). Variation of operator's strategies and regulating effects on workload. *Ergonomics, 14*, 571-577.

Sperandio, J.-C. (1987). The regulation of working methods as a function of workload among air traffic controllers. *Ergonomics, 21*, 195-202.

Tattersall, A.J., & Hockey, G.R.J. (1995). Level of operator control and changes in heart rate variability during simulated flight maintenance. *Human Factors, 37*, 682-698.

Wastell, D.G. (1996). Human-machine dynamics in complex information systems: the 'microworld' paradigm as a heuristic tool for developing theory and exploring design issues. *Information Systems Journal, 6*, 245-260.

Aviation Human Factors

Augmented reality environments for airport control tower

In D. de Waard, F.O. Flemisch, B. Lorenz, H. Oberheid, and K.A. Brookhuis (Eds.) (2008), *Human Factors for assistance and automation* (p. 325). Maastricht, the Netherlands: Shaker Publishing.

Modelling the allocation of visual attention using a Hierarchical Segmentation Model in the augmented reality environments for airport control tower

Ella Pinska[1] & Charles Tijus[2]
[1]Ecole Pratique des Hautes Études
[2]Université Paris 8
Paris
France

Abstract

Augmented Reality technology allows the incorporation of computer generated graphical and textual information into a real visual environment. Augmented Reality (AR) applied to the Air Traffic Control domain is expected to enhance the performance of tower controllers. However, the introduction of additional information to the visual scene must be conducted with care in order not to increase the clutter, saturation or complexity of the scene. The Hierarchical Segmentation Model (HSM), based on contextual categorisation, and expressed by the mathematical formalism Galois Lattice, provides a visual complexity index derived from the distribution of the features over the objects. The model can be used to calculate the efficiency indicator for particular objects in the scene that represents the strength of the object in the given context to attract the visual attention of the observer. This paper presents the predictions derived from HSM compared to the experimental result of the eyes gaze data of sixteen participants scanning the controller's devices and the view out of the tower's windows. The results, expressed as frequency and dwell time of visiting the areas of interests (AOI), indicated that augmenting the objects with additional superimposed information attracted the visual attention of the observer to those objects.

Augmented reality for airport control tower

Controlling the aircraft at the airport surface requires complex operational knowledge exercised under time pressure. Introducing new technology to the safety critical domain like Air Traffic Control (ATC) should be carried out with special care.

The analysis of the activity of tower controllers showed that the major occupation of tower controllers is to monitor airport surfaces by looking out of the window (Hilburn, 2004a, 2004b; Pinska, 2007). Other activities are scanning the strips, synthesised flight plan and searching for various information provided through the

In D. de Waard, F.O. Flemisch, B. Lorenz, H. Oberheid, and K.A. Brookhuis (Eds.) (2008), *Human Factors for assistance and automation* (pp. 327 - 342). Maastricht, the Netherlands: Shaker Publishing.

displays, such as two-dimensional ground movement radar presenting airport layout and traffic.

Merging the information, included on the strips to a head-up display, is expected to bring benefit for the controllers while working in dense traffic. Direct observation through the tower's windows including augmented identification data, together with the information provided by ground movement radar can be sufficient to build a mental picture of the traffic situation. Presenting the information on a head-up display can also reduce head-down time and eliminate the constant attention switching between short and long distances that were reported as generator of risk in the tower controller work (Hilburn, 2004a).

Augmented reality technology allows the overlaying of graphical and textual information on the real environment The information that today is spread among different head-down sources, can be displayed using a panoramic transparent display or head-mounted display (HMD) at the view out of the tower's window. Augmenting the information directly in to the field of view of the observer can improve the performance of a visually-dependent task. However introducing additional information might increase the complexity of the visual scene and the mental cost of processing the scene. The design of the interface for augmented reality technology should highlight the information that is significant for the controllers without increasing the complexity of the visual scene nor the cognitive complexity of processing the scene.

To satisfy these requirements we use a Hierarchical Segmentation Model (HSM) that predicts the segmentation of the visual inputs into the hierarchy of categories and subcategories depending on the features of the objects. The higher number of categories indicates the higher visual complexity and higher cost of cognitive processing of the visual scene. The augmented reality allows to manipulate the distribution of the external features over existing categories in a simple way. The design process should aim to introduce the external features that will generalise existing categories to simplify the segmentation. The model provides a relative efficiency index that represents the cognitive cost of the processing of the objects.

Model of hierarchical segmentation of visual scene

Processing the visual scene by humans is based on the segmentation of the inputs. Transforming the visual stimuli into the internal representation is interpreted variously according to different approaches.

The basic approach, *Feature Integration Theory* (Treisman & Gelade, 1980), a model of *Guided Search* (Wolfe, 1994) and *Feature Localisation model* (Asby et al., 1996) states that the visual scene is decomposed by humans in two stages: preattentive and attentive. In the preattentive stage, the features such as colour and shape are processed independently, whereas, in the attentive stage, the features are combined into the objects based on stored knowledge (long-term memory).

The second approach, *synthetic point of view*, states the coexistence of two processes, the grouping of the features into the objects and segmentation of visual information at the same time (Di Lollo et al., 2000).

The third approach concerns theories stating that visual attention is distributed sequentially to the visual scene (Tijus & Reeves, 2004) and (Prinzmetal et al., 2002). The model of Hierarchical Segmentation integrates previous approaches and goes down to deeper comprehension of processing the bottom up inputs. According to the model, processing of the visual scene depends on the distribution of the features and perceptual groupings in this scene. The extraction and segmentation of visual dimensions is dependent on the dimension's position in the hierarchy. When processing the features, the colour is extracted before the shape whereas the spatial localisation is processed at every extraction level of other perceptual dimensions. After features have been processed, the hierarchy of processing the dimensions of the objects provides better feature's integration of dependent dimensions such as colour and shape, than for independent dimensions such as two different shapes (Figure 1). The localisation is present at every level of processing.

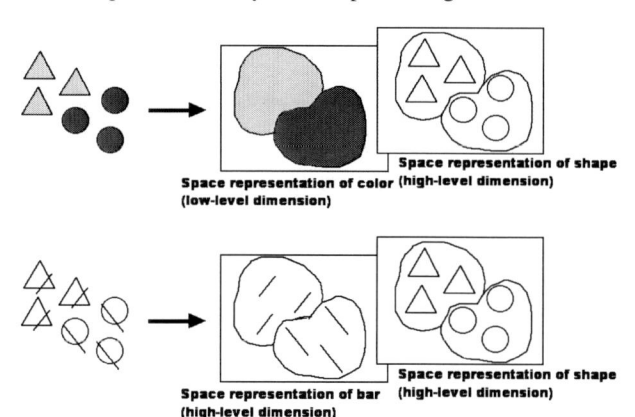

Figure 1. The hierarchy of processing of dimensions (colour before shape: bars and geometrical figures) provides a better features integration of dependent dimensions (colour and shape) than for independent dimensions (bars and shape)

The features are categorised based on context (Tijus, 2001) and (Poitrenaud et al. 2005). Categorisation is a basic mechanism of human cognition, and is based on the long-term memory, whereas contextual categorisation is based on the current situation, when the objects are grouped according to context.

One category describes a set of features (descriptors) that distinguish a set of the objects. The co-occurrence of the object with its features is called an association. The contextual categorisation, according to the descriptors, applieds to the objects different types of association.

- independence (the presence of one descriptor does not allow to infer the presence of the other),

- equivalence (the presence of one descriptor implies the presence of the other and vice-versa),
- exclusion (the presence of one precludes the presence of others),
- implication (the presence of one descriptor implies the presence of the other, the reverse being not necessary)

The final organisation of the values of dimension is derived from a Galois lattice organisation.

A *Galois lattice* called also *Concept lattice* is a mathematical theory of concepts and concept hierarchies (Wille et al., 2005). The Galois lattice is a set of formal concepts. Each formal concept is defined as a triplet of an object O_i, a descriptor D_j, and a binary relation between the object and descriptor. Galois lattice is expressed by the formula On x Dm that indicates if each of n objects has an m descriptor - feature. Galois relations and sets of formal concepts can be visualised by a line diagram. The Figure 2 represents the pair of original (a) and augmented (b) photos and their corresponding lattices (c and d). On the lattice diagram, letter f indicates the feature (descriptor) and letter o - an object. To read the diagram one should follow the nodes bottom up to define all the objects and descriptors of the concept. A given object has a given descriptor only if there is an upwards leading path connecting them. The bottom up descriptors represent the features that are not shared, whereas the features shared among the objects are placed higher up in the hierarchy of lattice.

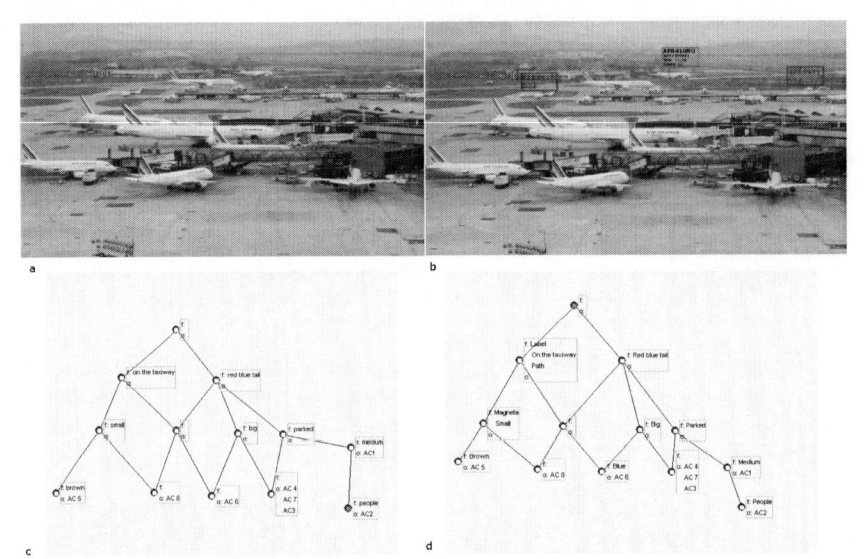

Figure 2. Original (a) and augmented (b) view out of the tower's window and their corresponding lattices

The objects in the original picture (Figure 2a) are aircraft (e.g. AC 1, AC 2...), whereas descriptors referring to those objects are: *small, medium, big, brown, red-*

blue-tail, parked or moving or on the taxiway. The descriptors were chosen to indicate important characteristics from the traffic control point of view. For example, the colour of the tail indicates the airline that is needed to identify the aircraft. Also, the aircraft that is parked requires less attention than the one on the taxiway as it is not moving.

The lattice diagram (c) of photo (a) represents the objects divided into two main categories: *on the taxiway* and *red blue tail*, that indicates the aircraft of one company. The subcategories were *small, big, medium* and *parked*.

By analysing the way in which the descriptors are distributed over the objects according to the categories of relations, one can determine the visual complexity of the scene. The more features are shared among the object, the more difficult it is to access those features. The model provides a value of the discriminability of the objects in the scene represented as an efficiency index. The value of efficiency of an object reflects its cognitive complexity: the higher the efficiency value, the lower the cognitive cost for processing. The objects that are more efficient require less effort to be extracted from the scene, thus they will attract the visual attention of the observer first. To obtain the efficiency it is necessary to calculate complexity of the scene and the power of the descriptors.

Complexity

The object is more complex when its representing lattice is composed of many descriptors and categories or when the descriptors are distributed variously over categories. Unit of complexity is the one couple category-descriptor. Complexity is defined by the product of the number of categories (Nc) by the number of descriptors (Nd):

$$COMP.obj = Nc \times Nd$$

Descriptive power

Some descriptors are more useful than others because they apply to several categories. We can define the usefulness of descriptors by their applicability to the total number of categories. The applicability of a descriptor to categories is defined by its scope or extension that is by the number of categories to which it applies:

$$EXT.descrip = Nc.descrip$$

The power of a descriptor reflects the relationship between its extension and the total number of categories to which it could be possibly applied:

$$POW.descrip = EXT.descrip / Nc$$

Finally, the power of a description corresponds to the sum of each description power:

$$POW.obj = \sum POW.descrip.$$

Efficiency

A description is all the more efficient, that its power is compensated by its complexity. We express this efficiency by the relationship between power and complexity. The efficiency of a description of an object:

$$EFF.obj = POW.obj / COMP.obj$$

Calculating the efficiency values of the objects provides the major complexity contributors in the scene.

The indexes can be used during the design process of a new interface to control the complexity that could be affected when we add new elements. The hierarchical segmentation model offers the hierarchy of the processing of the scene by extracting the general features that are shared among many objects e.g. *red-blue* tail in the Figure 2. The single object having a not-shared feature that applies only to this object, will pop-out in the scene due to saliency. According to the Figure 2, the example of salient feature is *people* as it is only applicable to the object AC 2.

In addition, the model allows us to identify what are the less discriminative components of the scene. Such information can be employed in the design of the new interfaces by introducing the visual indicators that attract the visual attention of the user to the components that might be neglected during high mental load.

The HSM provides the objective analysis of the complexity of the visual scene. It allows to decompose the scene into the elements and to verify, which elements are major complexity contributors. Also, it provides the information about the discriminability of the elements in the given scene.

This method is especially valuable for augmented reality applications, where the change of the distribution of the features is easy to manipulate. The hierarchical segmentation theory advocates for careful introduction of a new object to a real world by generalising the features that correspond to the descriptors. To evaluate possible solutions we can compute the complexity metrics and compare them in order to choose the simplest solution.

Hierarchical segmentation model applied to augmented reality interfaces

Introducing new HMI solutions to a safety critical domain should be carried out with precaution. Various studies carefully analyse the effect of colour coding in Air Traffic Control devices (Xing, 2006). According to Yuditsky (2002), there was benefit gained when three different methods of colour coding were individually applied to ATC interfaces. However, the benefit was lost when all of them were merged into one display. Similar results were reported by Cummings and Tsonis (2006) showing the effectiveness of colour coding involving six colours which was lost when nine colours were used. Those results indicated that the new methods should be evaluated integrated within the whole environment. Using the complexity metrics, we might be able to predict whether the introduction of the additional colours to the system would increase the complexity of the whole system.

Augmented reality interfaces can bring benefits for the performance of visually dependent tasks (Pinska, 2007). However, presenting the information on head-up displays faces the new problems of obscuring the real view and perceptual load (McCann & Foyle, 1994). Those issues can be alleviated by careful application of new descriptors (colour, shapes, organization of the textual information) over the existing objects and by generating the prediction of cognitive load using appropriate metrics.

Another problem concerning the head-up displays is attentional tunnelling (Prinzel & Risser, 2004). The phenomenon occurs when the main attention of the user is directed towards displayed digital information, risking omitting less discriminative components of the view. Prediction of the processing of the visual scene based on the HS model is expected to provide the means to mitigate this problem.

In the current study, the HSM predictions are used to design the augmented reality interface for the tower controllers. The new elements of the interface were defined by operational requirements derived from existing tower environments. The HSM predictions were used to evaluate the efficiency of the new elements in the interface to guide the visual attention of the controllers. The aim of the design was to merge the additional elements into the visual scene in a way to facilitate the objects dicriminability. The additional information should direct the visual attention of the controllers to the particular objects that were identified as important according to operational criteria. Also, the information should attract the controllers as they were crucial for the task performance.

Method

Interface design

The analysis of the traffic situations occurring at the airport provided the indications of the information necessary for the controllers to perform the tasks. Various interface augmentation solutions were proposed depending on the information sources available at the tower: view out of the window, ground movement radar and ATIS (Automatic Terminal Information Service) display. Table 1 presents the augmentation scheme for those devices.

Stimuli

Five series of photos were taken at Orly airport in Paris. The photos presented the tower's devices and view out of the window. Each series of photos was taken within a short time (maximum 15 s), reflecting the state of traffic at the airport.

The pictures presented:

- Ground movement radar – 2D airport layout

- View out of the tower's window, capturing the view behind, aside and in the front of the control position
- ATIS display – presenting meteorological information and general information about the status of the airport.

The original pictures were augmented by adding graphical information according to the rules described in the section before. The applied modifications were consistent with the currently existing interface, respecting the colour coding and information representation layout. The scheme of modifications assumed that the additional information concerned the objects of high importance regarding operational aspects and included the information necessary to perform the tasks. The hypothesis was that the objects that were modified in augmented photos were important and therefore the visual attention should be directed to them.

The frequency of visit of modified objects (Areas Of Interest) was expected to be higher in augmented photos than in original photos. The information included in the augmentation were crucial for the task performance therefore it was expected that the duration time (dwell) of visit the modified objects (modified area of interests) should be longer in augmented photos that in original photos. Those hypotheses concerned all series of photos including the view out of the window, ground movement radar and ATIS display.

The detailed process of the application of the modification and the HSM predictions is presented here for one sample photo presenting the view out of the window (Figure 2b). The modifications consist in adding the labels coloured depending on the status of the aircraft and the paths indicating direction of the movement.

After applying the modification, the new lattice was created (Figure 2d). The new lattice includes new descriptors of the aircraft such as *label, path, blue and magenta*. The new features were applied to the scene in order to maintain the categorisation created by the lattice representing original photo. The additional feature *label* divides the objects into categories *blue label* and *magenta label*, without creating new categories.

In addition, we calculated the efficiency metrics for both pictures (Figure 2a and 2b). The values are presented in Table 2.

According to the picture, the aircraft are numbered:

AC 1 – middle size aircraft parked in the front,
AC 2 – middle size aircraft parked on the left of the picture,
AC 3 – big aircraft parked in the centre of the picture,
AC 4 – parked aircraft partly hidden by AC 3,
AC 5 – small aircraft at the crossing of the taxiways,
AC 6 – big aircraft, taxiing in the second plane,
AC 7 – big aircraft parked in the second plane,
AC 8 – small aircraft, partially hidden on the taxiway.

Table 1. Modification scheme for augmented photos. (The example of the augmentation applied to view out of the window, ATIS display or ground movement radar images can be found at http://extras.hfes-europe.org *; Photo 1, 2, 3, 4, 5)*

The view out of the window	
Identification of the aircraft	
	If the aircraft is parked: non-active status – no label displayed the AC is departing from attached to the aircraft including information:– AC
	If callsign – Transponder code number – Type of aircraft – Runway assigned for take off – Time o the time presentation: in case of CTOT bold font in the frame o in case of the flight plan time – regular font – The letter indicating the ATIS update version – Instant wind
	If the AC is arrival traffic –magenta label with the parking position indication
	If the AC is towed – white label indicating the aircraft type
Identification of the vehicles.	
	If the vehicle is parked – no label displayed
	If the vehicles are moving on the taxiways – white frame around the vehicle
Detection of a possible conflict	
	In case of conflict on the taxiway between:
	Aircraft and aircraft - highlighting the path of both aircraft in orange
	Aircraft and vehicle - highlighting the path of the aircraft and the frame of the vehicle in orange
Representation of the direction of the movement of the aircraft	
	If the AC is moving – highlighting the yellow line in the front of the aircraft predicting aircraft movement, the length of the line should indicate the speed of the aircraft
Runway occupancy	
	In case of only one aircraft taking off from the runway – no indication displayed
	When there is one aircraft passing the holding point and another aircraft or vehicle entering the runway from the taxiway, the runway will be highlighted in red
	In case of take off and landing at the same runway, both aircraft should be visible by highlighting the labels in yellow
Unidentified objects on the runway	
	In case of unidentified objects on the runway, the runway will be highlighted in orange
Ground Movement Radar	
	If an aircraft is parked, non-active status – the label is not displayed
	If an aircraft is departing from the airport – blue label attached to the aircraft including following information: – Aircraft callsign – Type of aircraft
	If aircraft is arrival traffic – magenta label with the parking position indication
	If the aircraft is towed – white label indicating the aircraft type
Representation of the direction of the movement of the aircraft	
	If the aircraft is moving – highlighting the yellow line in front of the aircraft predicting aircraft position. The length of the line should indicate the speed of the aircraft
ATIS display	
	Displaying the callsign of departure traffic

Our main interest is given to the aircraft AC 5, AC 6 and AC 8 as these are the ones that are placed on the taxiways, which indicates that they are moving. Those aircraft on the augmented photo have labels and paths. According to Table 2, the order of those aircraft and the efficiency values were changed. The HSM predicts that the distribution of the attention of the air traffic controller should change between original and augmented photos in following direction:

In both photos AC 5 is the most efficient object, however in the augmented photo it should be visited less frequent than in the original. The AC 8 and AC 6, in the augmented photo, should be visited more frequent than in the original photo. The AC 1, AC 2, AC 3, AC 4 and AC 7 are not expected to change the frequency of visits.

Table 2. Efficiency values of various aircraft represented by Figure 2a and 2b

Original Photo		Augmented Photo	
AC 5	0,75	AC 5	0,71
AC 2	0,53	**AC8**	0,68
AC 8	0,5	**AC 6**	0,56
AC 6	0,42	**AC 2**	0,53
AC 1	0,42	AC 1	0,42
AC 4	0,33	AC 4	0,33
AC 7	0,33	AC 7	0,33
AC 3	0,33	AC 3	0,33

According to the assumption, the modifications included the significant information. Therefore, we expected that duration time of visit of the modified aircraft would be longer. In sample photo the modified objects were AC 5, AC 6 and AC 8, thus it is expected that the duration time of visits should increase.

Data collection

The eye gaze data of participants scanning the original and augmented photos were collected using the Facelab eye tracker (Seeing Machines, 2008). The measurements of the gaze position were taken every 20 ms. The photos were presented at the 21 inches display, 1600x1200-screen resolution, placed 90cm from the participants. The participants were seated and had a support to fix the head in order to increase the precision of the measurements. Before the experiment and in between there was a calibration performed to assure the correct measurements.

Procedure

The participants first read a written instruction of the experiment and completed a personal questionnaire.

They were introduced to the sample pictures representing ground movement radar, three different window views from the tower and an ATIS display. They received extended explanation about the devices and the information that they present, including the colour coding scheme, the airport layout and the traffic organisation at this particular airport. The task given to participants was:

- to understand the traffic situation,
- to identify the aircraft,
- to follow the traffic progress.

There was no objective verification of the performance after the experiment. However, participants were asked to provide the feedback regarding the experiment in a form of comments using the post - exercise questionnaire.

The experiment started with one trial run whose results were not considered in the analysis. The measurements were taken while presenting the original and augmented photos. The experimental run consisted of five photos presented twelve seconds each, between each photo there was a black buffer slide. There were five runs of original and five runs of augmented photos presented to the subject. The presentation of the series and the conditions was counterbalanced to avoid learning effect.

Participants

Twenty-six novice participants took part in the experiment. The data of ten subjects were excluded from the analysis due to the low quality of gaze tracking. Only the data of subjects that reached 70% of gaze tracking time were taken into account in the analysis. Therefore the following analysis came from the results of sixteen novice subjects, ten men and six women. They were between 22 and 45 years old and they were staff of the research centre where the experiment was conducted. All participants reported correct vision.

Results

The analysis of the results started by defining areas of interest (AOI) for each photo. One AOI corresponds to one object e.g. an aircraft or a vehicle. The AOI were the same size for a corresponding pair of original and augmented pictures. The AOI were divided into two groups: modified and unmodified. The modified AOI indicated the objects that were highlighted in the augmented photos and therefore objects that are important for the controllers such as moving aircraft or vehicles on the runway. The unmodified AOI indicated the objects that were not changed in the augmented photos such as parked aircraft, which are considered as less important for the controllers.

The data describing one AOI consist of the count of visits (frequency) and mean duration time (dwell) for one visit. The following analysis is based on the pooled data of modified and unmodified AOI in original and augmented conditions for twenty-five photos.

Frequency

The frequency of visits of modified and unmodified AOI was compared between the original and augmented photos. The mean values for visiting unmodified AOI was at comparable level (3.7 and 3.9) whereas the mean value for visiting the modified AOI increased from 3.3 to 4.4 (F(1,16) = 23.3, p<0.0001). Those results confirmed that subjects scanning the augmented photos were visiting the modified AOI more frequent. However, the frequency of visiting the unmodified AOI between original and augmented photo was not significant. These results confirmed that the objects that were unseen in the original photos attracted the attention of the subject when the visual information was added.

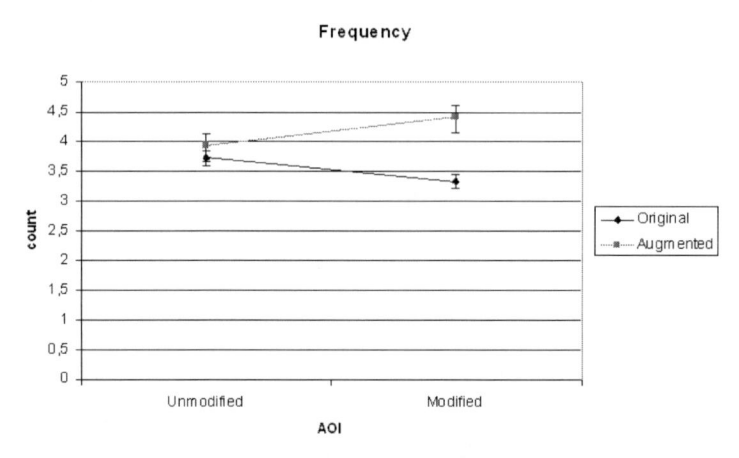

Figure 3. Frequency of visits for modified and unmodified AOI for original and augmented photos

Duration

We calculated the mean duration of one visit to modified and unmodified AOI for original and augmented photos. The results presented in Figure 4 show the decrease of the time spent in unmodified AOI from mean value 0.23 s for original photos to 0.17 s for augmented photos. The time spent in the modified AOI increased from 0.16 s to 0.19 s (F(1,16) = 21.5, p< 0.001).

The results indicated that the distribution of the attention between the original and augmented photos had changed. The objects that in the original photos did not attracted attention, popped up in the augmented photos, gaining the attention of the observers. Additionally the distribution of attention between modified and unmodified AOI is more balanced in augmented photos than in original ones. The difference for time spent in the important (more critical) and less important AOI in the original photos is higher compared to the augmented photos. In the augmented photos, the subjects return their attention to the modified - important objects, however, still they give attention to the unmodified - less important objects.

modelling the allocation of visual attention 339

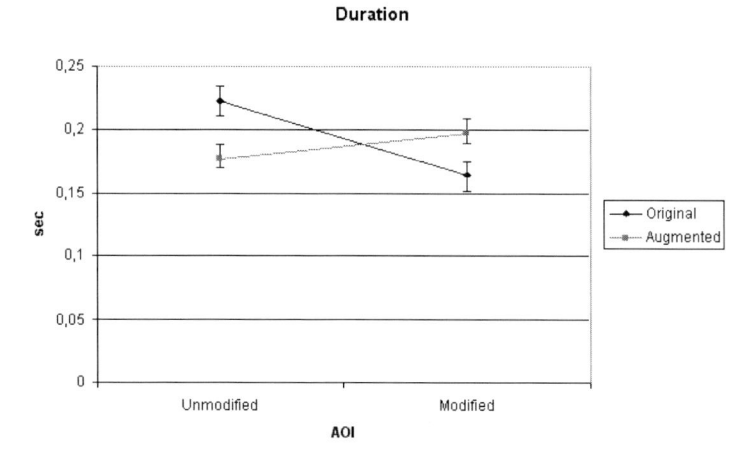

Figure 4. Mean duration of one visit to modified and unmodified AOI in original and augmented photos

Comparing the AOI within one sample picture

To analyse the prediction of the hierarchical segmentation model we compared the frequency of visits to various AOI in the sample Figure 2a and 2b. According to Figure 5, the frequency of visits for various AOI representing the aircraft has changed between original and augmented photos. According to Table 2 the efficiency of AC 6 and AC 8 should increase, thus, the objects in the augmented photo should be more discriminative and the subject's visual attention should be directed more frequently to those objects.

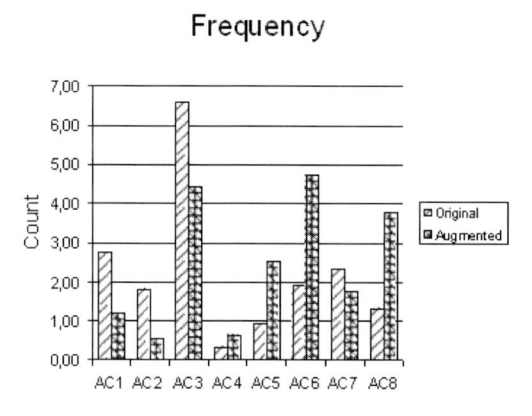

Figure 5. Frequency of visits for AOI in original and augmented sample photo

The results showed an increase of the values for the AC 6 and AC 8, which reflects the prediction derived from the hierarchical segmentation model. However, the frequency of visiting AC 5 increased as well what was not expected according to the model. The frequency of visit of AC 1, AC 2, AC 3, AC 4 and AC 7 was not expected to change whereas the data showed the decrease.

Mean duration

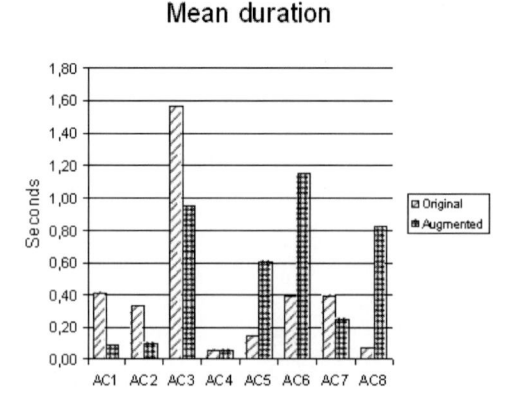

Figure 6. Mean duration for visiting AOI in original and augmented sample photo

The duration of visit was expected to increase for the AOI corresponding to AC 5, AC 6 and AC 8. The results confirmed these expectations, showing that observers focus their attention on indicated objects and their augmentations.

Discussion

The results of the experiment confirmed that augmenting some visual information in the scene influenced the distribution of the attention given to all the objects in the scene. The visual indices, such as labels and paths can successfully guide the subject's attention to the objects, which are important from an operational point of view but which might be easily overlooked. The participants were able to return their attention to less critical zones, but not so frequently and for a shorter time than in the original scene.

The presented study should be considered as a preparation step verifying the hierarchical segmentation model predictions about processing the visual scene and introducing visual indicators to the scene in the air traffic control domain. The experimental results confirmed the model's prediction regarding the objects that can be considered as the object of prime importance, but not the entire scene. However, the participants of the experiment were novices that were unfamiliar with the operational information like callsign, aircraft types, as real air traffic controllers would be, what affected the results. Also, the air traffic controllers are trained to monitor the information using specific scanning pattern, what was missing in the presented experiment. In this experiment, the comparison was performed within the original photos and the augmented photos that included additional information. Further investigation should concern the analysis of the HSM regarding the presentation of the same information in two different ways. As well, the introduction of a task that requires more engagement, under a time pressure can bring interesting results regarding top - down impact. Using the air traffic control experts as participants would provide the information about the specific scanning pattern of the scene.

Conclusions

The paper introduces the hierarchical segmentation model, which aims to reflect how humans process the visual scene, applied to the design of the new interface that incorporates computer-generated information into the real view of the observer.

The model predicts the efficiency of the objects to attract the visual attention of the observer, based on the analysis of the distribution of the objects' features and perceptual groupings in the visual scene.

The augmented reality application provides the means to manipulate the object's features in a simple way, and can easily be adjusted to real objects in order to reduce visual complexity or guide the visual attention of the observer.

Results from the experimentation confirmed that the hierarchical segmentation model provides a promising means to evaluate the complexity of the visual scene. Further investigation of visual attention guidance using augmented reality should be based on a comparison of different variants of augmenting the information and should be conducted in dynamic scenarios. As well, a task should increase the participants' engagement to an exercise, what would bring the top - down effect of processing the scene.

Acknowledgement

Ella Pinska was supported by a PhD scholarship from the Innovative Research Programme at the Eurocontrol Experimental Centre, Bretigny sur Orge, France.

References

Ashby, F.G., Prinzmetal, W., Ivry, R., & Maddox W.T. (1996). A formal theory of feature binding in object perception. *Psychological Review*, *103*, 165-192.

Cummings, L. & Tsonis, C. (2006). Partitioning Complexity in Air Traffic Management Tasks. *International Journal of Aviation Psychology*, *16*, 277-295.

Di Lollo, J., Enns, T., & Rensink, R.A. (2000). Competition for consciousness among visual events: the psychophysics of reentrant visual processes. *Journal of Experimental Psychology*, *129*, 481-507.

Hilburn, B. (2004). *Cognitive Complexity in Air Traffic Control - A Literature Review*. (EUROCONTROL EEC Note No. 04/04). Bretigny sur Orge, France: EUROCONTROL Experimental Centre.

Hilburn, B. (2004a). *Head-down Time in Aerodrome Operations: A Scope Study*. The Hague, the Netherlands: Center for Human Performance Research.

Hilburn, B. (2004b). *Head-Down Time in ATC Tower Operations: Real Time Simulations Results,* The Hague, the Netherlands: Center for Human Performance Research.

McCann, R.S., Foyle, D.C. (1994). Superimposed symbology: Attentional Problems and Design Solutions. *SAE transitions: Journal of Aerospace, 103,* 2009-2016.

Pinska, E. (2007). *Warsaw Tower Observations,* (EUROCONTROL EEC Note no. 02/07). Bretigny sur Orge, France: EUROCONTROL Experimental Centre.

Poitrenaud, S., Richard, J-F., & Tijus, C. (2005). Properties, categories and categorization. *Thinking and Reasoning, 11,* 151-208.

Prinzel, L.J., Risser, M. (2004.) *Head-Up display and Attention Capture* (NASA/TM-2004-213000) Hampton, Virginia 23681-2199: Langley Research Center.

Prinzmetal, W., Ivry, R.B., Beck, D. & Shimizu, N. (2002). A measurement theory of illusory conjunctions. *Journal of Experimental Psychology: Human Perception and Performance, 28,* 251-269.

Seeing Machines, (2008). http://www.seeingmachines.com/facelab.htm

Tijus, C. & Reeves, A. (2004). Rapid Iconic erasure without masking. *Spatial Vision, 17,* 483-495.

Tijus, C. (2001). Contextual Categorization and Cognitive Phenomena. In V. Akman, P. Bouquet, R., Thomason, and R.A. Young (Eds), *Modelling and Using Context.* (pp 316-329). Berlin: Springer-Verlag.

Treisman, A., & Gelade, G. (1980). A feature-integration theory of attention. *Cognitive Psychology, 12,* 97-136.

Wille, R., Stumme, G., Ganter, B. (2005). *Formal Concept Analysis, Foundation and Application,* LNAI 3626, Berlin Heidelberg: Springer – Verlag.

Wolfe, J.M. (1994). Guided Search 2.0: A Revised Model of Visual Search. *Psychonomic Bulletin & Review, 1,* 202-238.

Xing, J. (2006). *Color and Visual Factors in ATC Displays,* (DOT/FAA/AM-06/15). Oklahoma City, OK 73125: Civil Aerospace Medical Institute, Federal Aviation Administration.

Yuditsky, T., Sollenberger, R., Della Rocco, P., Friedman-Berg, F., Manning, C. (2002). *Application of Color to reduce Complexity in Air traffic Control.* (DOT/CT-TN03/01). Atlantic City International Airport, NJ08405: Wiliam J. Hughes Technical Center.

Empowerment of the planning controller in the use of Controller-Pilot Data-Link Communication (CPDLC)

Renée Schuen-Medwed, Bernd Lorenz, & Stefan Oze
Eurocontrol-CRDS
Budapest, Hungary

Abstract

It is common practice in air traffic control that two operators on the ground, the executive controller (EC) and the planning controller (PC) provide the air traffic control service for their sector. The working method used is based on rather rigid task-sharing between the EC and the PC. Voice communications between the ground and the air crew, i.e. issuing instructions via radio to the pilot and listening for requests and read backs from the pilot, represent a large proportion of the EC`s task. Controller-Pilot Data-Link Communication (CPDLC) provides a second parallel and independent communication channel that can be used by either controller.

This paper presents the results of a simulation conducted at the CEATS Research, Development and Simulation Centre (CRDS) in Budapest between 27 November and 8 December 2006. The main aims of this study were to develop a flexible task-sharing method between the EC and the PC and to explore the potential of this method to ease EC workload. The task-sharing between the EC and the PC is flexible in the sense that the EC can delegate CPDLC tasks to the PC and resume them. This method was evaluated against a use of CPDLC with no delegation. Both methods were compared under various levels of traffic density and with different proportions of aircraft in the airspace being equipped with CPDLC. Indicators of workload, situation awareness, controller communication performance, and assessments of subjective team performance were used for these comparisons. Results indicate that the expected performance benefits could not be fully exploited in higher traffic because of an increased demand on intra-sector communication between EC and PC for which the system did not provide enough procedural and HMI support.

Introduction

Today, air-traffic control operation relies on a division of the airspace into sectors. Basically, sectors are three-dimensional volumes of airspace stacked at various altitudes. Commonly, two operators on ground, the executive controller (EC) and the planning controller (PC) possess control authority for one sector. It is the primary role of the EC to ensure, that all aircraft she/he is responsible for are separated. The aircraft can be separated either vertically or horizontally according to standards of the sector mandated by the International Civil Aviation Organization (ICAO). Subordinate to this primary goal, the EC has to ensure an orderly and expeditious sector transit of all aircraft. Both goals are achieved by giving instructions to pilots and by listening to their requests. The role of the PC is to ensure all aircraft have optimal entry and exit into and out of the sector. The PC has therefore to solve possible conflicts in the planned trajectories of aircraft via telephone coordination with the upstream sectors before the new aircraft enters his airspace. Similarly the exit conditions are coordinated downstream.

Normally, the EC uses radio telephony for communication with pilots. Therefore, each EC has a dedicated radio frequency to which each aircraft transiting his sector has to be tuned. In view of the steady increase in air traffic, the congestion of such a single frequency is becoming a serious problem. Pilots may have difficulties accessing the frequency, or simultaneous transmissions interfere with each other and require their repetition. To some degree these difficulties can be alleviated by splitting the sector into two smaller sectors. This sector split involves a further controller team supplied with an additional frequency channel for air-ground communication. However, sector splitting creates a workload overhead due to the necessary between-sector coordination. That may offset the benefit of this strategy.

Controller/Pilot Data-Link Communication (CPDLC) allows pilots and air traffic controllers to communicate with each other via text messages. These messages can be pre-formatted or may be submitted as free text messages similar to text messaging capabilities via the internet or mobile phone networks. Unlike communication by voice, CPDLC occurs in a point-to-point rather than in a broadcast mode, a CPDLC message is transmitted only to one aircraft and not to all aircraft. This removes the so-called "party-line", i.e. pilots can no longer monitor the air-ground communication of other aircraft using the same frequency channel anymore (Midkiff & Hansman, 1993). However, voice communication is instantaneous and its tone can convey urgency, whereas CPDLC, as it is implemented now, involves transmission delays and is textual and silent. Therefore, CPDLC is currently not useful for clearances to aircraft, where an immediate pilot response may be necessary. Presently, CPDLC is only recommended in combination with radio telephony (Eurocontrol, 2006b, Sollenberger et al., 2005). The benefits of such a bi-modal communication environment are expected to be fivefold:

1. Increasing the availability of the voice radio frequency for the delivery of time-critical clearances by using CPDLC for non time-critical communication.
2. Reducing the controllers' workload by the automation of certain communication tasks.
3. Improving the workload balance within the sector team by optimizing the sharing of communication tasks.
4. Down-linking of various aircraft preferences and parameters. This includes details of an equipped aircraft's preferred flight level or top of descent, meteorological conditions, such as wind direction and speed, turbulence, air temperature and pressure etc. This down-linked data will contribute to improved ATC awareness of an aircraft's operational preferences and also to better

aeronautical meteorological reports and forecasts and better ATCO and aircrew environmental awareness.
5. Improving safety by reducing the vulnerability of voice radio frequency for miscommunications.

Numerous research activities were carried out in the US and Europe, ranging from fast-time and real-time simulations, to pre-operational field trials, in order to investigate the feasibility, acceptability, cost-benefits and safety aspects of initial data link concepts (FAA, 1995; Eurocontrol, 2000; 2002. Prinzo, 2001; Shingledecker et al., 2005).

A central theme in these studies is the potential of CPDLC to reduce the workload of the executive controller. As there will be a transition period during the CPDLC implementation when not all airlines will be equipped with CPDLC it was investigated at what equipage rate workload savings for the EC can be achieved. Eurocontrol (1999) derived a linear relationship between the percentage of aircraft equipped with CPDLC and the workload reductions. This relationship was inferred from a model-based simulation without real-time controller intervention, by modelling various communication message triggering events and estimates of message durations to predict communication load. Other controller tasks, flight data handling, monitoring, conflict detection, conflict resolution, were modelled using estimates of task execution times to derive controller "workload". Using this rather "mechanical" view of controller workload -which would be better referred to as task-load- it is not surprising that increasing CPDLC equipage communication load and hence overall workload decreases. The calculated workload reductions were then translated into capacity gains. From this an 11.2% capacity is expected to be gained at a 75% level of CPDLC equipage.

Real-time simulations could not always validate the expected workload reductions using controller subjective ratings (Eurocontrol, 2004; 2005, 2006a; Prinzo; 1998; Ruitenberg, 1998). It appears that the workload contribution of the CPDLC task in an environment supporting both voice radio and CPDLC has been underestimated up to now. It must be pointed out that in a mixed (CPDLC and voice) environment the controller has to switch between both communications modes. Voice communication is used for aircraft that are not equipped with the CPDLC features and is also used for aircraft that are equipped with CPDLC in time-critical situations. Therefore, the EC first has to differentiate between equipped and non-equipped aircraft and, second between time-critical and non-time-critical situations when she/he has to decide which communication mode to use.

Current research and development activities in support of CPDLC implementation aim at a further exploitation of the workload saving potential of CPDLC. Its delegation to automation (e.g. automated sector transfer) and/or to the PC was identified as promising in this regard (Eurocontrol 2004, 2005, 2006a, 2007b). How should the distribution of work be optimally designed?

- Should there be a fixed partition of aircraft designated to PC and EC?
- Should there be a fixed allocation of all CPDLC tasks to the PC?

- Should there be a division of labour by the type of clearances (e.g. EC gives all level clearances)?
- Should there be a flexible allocation which allows both controllers, the EC and the PC to give CPDLC clearances according the situation.

A flexible task allocation would not only allow for adaptation to the actual operational situation but would also require an assessment of the situation. Subsequently the decision is made as to which CPDLC task should remain with the EC and which CPDLC task should be delegated to the PC. Based on the fact that the ultimate responsibility for aircraft separation lies with the EC, the task of delegation should be part of the role of the EC.

In 2006 the CEATS Research, Development and Simulation Centre (CRDS) was tasked by Eurocontrol's LINK2000+ Programme (Eurocontrol, 2000) to develop and validate in a real-time simulation a working method involving to the planning controller, the aim being to make best use of CPDLC in an en-route environment.

Controller working method

Both the rigid working method, represented by a fixed task allocation and the flexible working method including delegation are based on the general principle that CPDLC shall only be used in the context of non-time-critical communications. Time-criticality is determined by the ATC traffic situation and the end-to-end performance (systems and flight crew/controller response time). The CPDLC capability was implemented in the CRDS strip-less environment, meaning the controller has access to all relevant data of a certain aircraft via the interactive data label on the radar screen. By using the interactive label the controller updates the system after voice clearances, i.e. the controller enters the changed values for heading, speed, level, etc. using a pop-up menu interactive label (see figure 1). When opting for CPDLC the controller simply has to select the "D/L"-button to submit the clearance. Aircraft equipped with CPDLC capability are indicated by a filled triangle used for the aircraft track symbol as compared to filled circles used for unequipped aircraft. Additionally, the data block for equipped aircraft is framed whereas it is unframed for unequipped aircraft. The controller working positions for EC and PC are identical with respect to all HMI functions. Therefore, both controllers have equal access to the CPDLC capability. Moreover, each controller can monitor his colleague's CPDLC communication actions on her/his own radar screen.

In the rigid working method the PC is only allowed to issue Transfer of Control (TOC) messages via CPDLC. The flexible working method is characterized by the possibility to shift any CPDLC task from the EC to the PC. During the simulation there were no mandatory procedures laid down for using the flexible method by the EC and the PC. Each controller team was encouraged to exploit the possibility of sharing the CPDLC task as much as possible for the benefit of improving the workload balance and for an optimal adaptation to the operational situation. This task shift, however, had to occur at the discretion of the EC either by EC consent upon PC request or by explicit EC delegation.

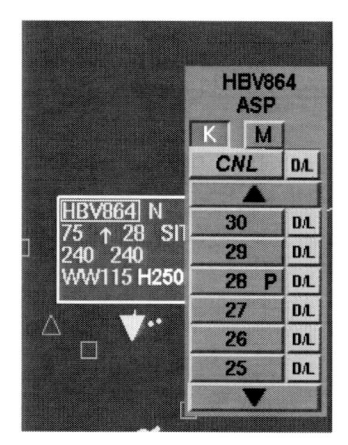

Figure 1. Controller HMI for the transmission of CPDLC messages. The controller can open pop up menus for speed, heading, level, and waypoints via the interactive flight data block with a left mouse click. In the illustrated example the pop up menu for speed has been activated for the CPDLC equipped aircraft HBV864. By clicking on the D/L button right to the speed value a new cleared speed is up-linked. CPDLC equipped aircraft are indicated on the radar screen by, first a framed data block and second by a filled triangle as the aircraft track symbol.

Hypotheses

It was assumed that the new working method reduces EC workload due to the transfer of CPDLC communication load from the EC to the PC. Moreover, it was expected that the new working method also empowers the proactive role of the PC with the result that less tactical intervention, thus less voice communication by the EC would be necessary and that this task shift from the EC to the PC facilitates shared situation awareness and team work. These assumptions lead to the following hypotheses:

H1: The new working method reduces EC workload due to the transfer of CPDLC communication load from the EC to the PC.
H2: The new working method reduces voice communication as a result of the proactive role of the PC.
H3: The new working method results in better team work and increased shared situational awareness.

These hypotheses were investigated in a two-week real-time simulation experiment during which the new working method was compared with the traditional working method. Additionally, traffic load and data link equipage rates were systematically varied. Measurements were derived from communication events (voice, CPDLC) and taken from the controller working positions of three sectors. Furthermore, subjective ratings on workload, situational awareness and perceived team work were collected after each simulation exercise. Ratings of working method acceptability and operability were obtained in a semi-structured interview at the end of the simulation experiment.

Method

The CRDS real-time air traffic control simulator comprises 26 controller working positions (13 EC and 13 PC), all located in one operational room and up to 20 "pseudo pilot" positions. "Pseudo pilots" are responsible for the manoeuvres of the simulated aircraft. They enter the controllers' clearances into the simulation computer to enact them at the controllers' radar screen and confirm the instructions via a simulated radio line in the same way as real pilots do.

Experimental design

The experimental plan for this study involved four independent variables (ANOVA factors):

- Traffic Load (TL): Low versus high, with low representing the traffic forecast for the year 2009 (~80 aircraft per hour) and high representing the traffic forecast for the year 2014 (~95aircraft per hour)
- Working Method (WM): rigid working method versus flexible working method
- Data Link Equipage (DLE): 20% versus 50% of aircraft equipped with CPDLC
- Sector (SEC): 3 different sectors were simulated. The simulated measured area consisted of parts of the Austrian and Hungarian airspace. (D5, D6E and D6W)

This represents a 2x2x2x3 Analyses Of Variance (ANOVAs) design with repeated measures on the first two factors. The factors DLE and SEC were varied between subjects. Data collected from the EC and the PC position were analysed in separate ANOVAs. Statistical analyses were conducted with the GLM (General Linear Model) repeated measures procedure of the Statistical Package of the Social Science (SPSS^{TM}) (SPSS, 2006).

During the two weeks of the simulation every controller worked on at least eight runs (4 times as an EC and 4 times as a PC) each lasting one and a half hours. After each exercise the ATCOs had to fill in the post exercise questionnaires. Each day three or four exercises were simulated. On days with three exercises additional debriefings were conducted. A rotation plan ensured the balanced assignment of the controllers to the conditions of the experimental variables. Generally working methods were only changed between days and not between exercises of the same day to avoid any misunderstandings as to what working method was used.

Measurements

Measurements were derived from objective and subjective data. First, the communication load was measured on the basis of controller input automatically recorded by the system. It was recorded how often in an exercise the controller used Voice and CPDLC for communication with the flight deck.

In addition to these objective measurements, the following subjective data were collected:

Subjective workload ratings. The Instantaneous Self Assessment (ISA) method (Eurocontrol 1996) was used to collect subjective data workload as perceived by the controller. The participants were asked to respond every two minutes by pressing one of five buttons appropriate to their perceived workload at that moment. Buttons were labelled Very Low (under-used), Low (relaxed), Fair (comfortable), High and Very High (excessive).

Situational Awareness (SA). SA was measured on the basis of self-rating scales developed within the Eurocontrol project called Solutions for Human-Automation Partnerships in European ATM (SHAPE) (Eurocontrol, unknown). The self-rating scale used in the simulation was SASHA-Q (Situational Awareness for SHAPE Questionnaire). The scale consisted of a set of statements on different aspects of Situational Awareness (e.g. "I was ahead of the traffic") that needed to be answered on a 7-point rating scale ranging from "never" to "always".

Team Work. The SHAPE Teamwork Questionnaire (STQ) (Eurocontrol, unknown) offers to assess the quality of teamwork by means of post-exercise 7-point rating scales. The long version used here consisted of six subtests with four items each resulting in a total of 24 items. The six sub-scales were: Team Situational Awareness, Team Roles & Responsibilities, Team Cooperation, Team Climate, Team Error Management and Team Communication. For example participants were expected to give ratings to statements like "it was clear to me which tasks were my responsibility" using a 7-point scale ranging from "never" to "always". A psychometric analyses completed in a study involving 24 air traffic controllers of AUSTROCONTROL (Eurocontrol, unknown) revealed sufficiently high overall internal consistencies (Cronbach's alpha = 0.88) as well as satisfactory internal consistencies for the four sub-scales (Cronbach's alpha range from 0.68 to 0.87). These results justify the calculation of a global score as well as scores for the six sub-scales.

Subjective usability and acceptability ratings. Usability can be defined as "the extent to which a product can be used by specified users to achieve specific goals with effectiveness, efficiency and satisfaction in a specific context of use" (ISO 9241-11; 1998(E)). Acceptability refers to the opinion of the user of the concept. Both the usability and the acceptability of the working methods were assessed simply by asking controllers in interviews to give a rating ranging from 1 not acceptable (not usable) to 5 very acceptable (very usable).

Participants

Seventeen controllers from ten different air traffic control centres (ATCCs) in Europe participated in the study. One female and 16 male controllers with experience as licensed controllers for between 5 up to 30 years. The age of the 17 controllers ranged from 33-54 years.

Results

Voice communication

The frequency of the R/T communication was recorded automatically by the system. The ANOVA revealed a significant main effect of traffic load (F(1,11) = 212.53; p < 0.001) and of sector (F(2,11) = 65.14; p < 0.001). Both effects are trivial in expressing the fact, first that a higher traffic load causes more air-ground communication and second that in one of the sectors less communication was needed because of less dense traffic patterns. The impact of the two working methods on the voice communication did not differ significantly (F < 1). It was expected that voice communication would decrease with a higher rate of Data Link equipage. In fact, such an effect was observed in the expected direction, however, the difference was too small to be confirmed with statistical confidence (F (1, 11) = 4.19; p = 0.065). Instead, the ANOVA revealed a significant Data Link Equipage (DLE) x Working Method (WM) interaction effect (F (1, 11) = 5.47; p = 0.039). The pattern causing this effect is illustrated in figure 2. The graph presented in figure 2 depicts the average number of radio calls plotted as a function of the two working methods separately for the two levels of CPDLC equipage. The resulting pattern suggests that the level of CPDLC equipage induced the expected decrease in the use of voice communication only when the rigid working method is used, whereas voice communication levels were very similar for 20% and 50% equipage when the flexible working method was used.

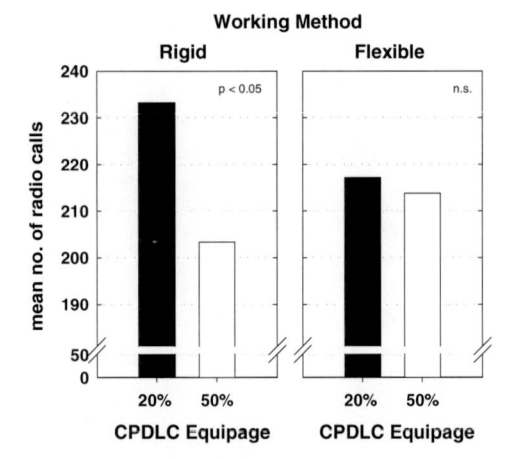

Figure 2. Mean number of Radio Calls plotted as a function of CPDLC equipage and separated for the two different working methods

CPDLC – communication

It was expected that the EC would delegate CPDLC communication to the PC when adopting the flexible working method. This, in fact, occurred in the simulation as per figure 3 and was confirmed by the outcome of the two ANOVAs performed separately for the EC and the PC data, respectively (EC: F (1, 11) = 8.06; p = 0.016;

PC: $F(1, 11) = 20.56$; $p = 0.001$). Thus, the EC issued fewer CPDLC messages when using the flexible as compared to the rigid working method and the reverse was true for the PC.

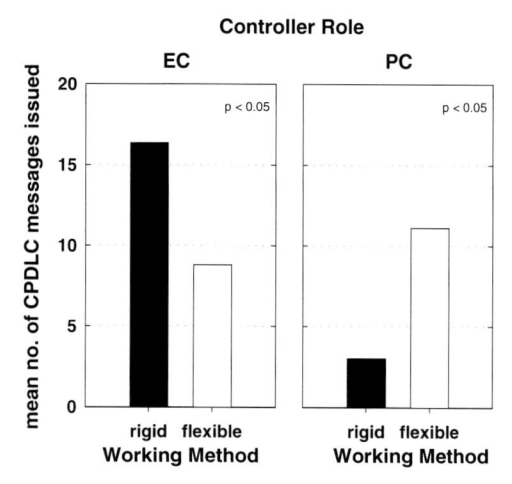

Figure 1. Mean number of CPDLC messages plotted as a function of working method and separated for controller role (EC vs. PC). (Note, that the number of CPDLC message is greater than zero for the PC using the rigid working method, because the working method allowed the PC to issue transfer of control messages)

Workload

One of the central objectives of the study was to show the extent of the shift of communication task load from the EC to the PC when using the flexible working method and how this task load shift translates into reductions in workload perceived by the EC. The analysis of the data does not confirm a straightforward workload reduction for the EC. The average ISA rating given by the EC remained between 2 and 3, thus between "low" and "fair" with both working methods (the ANOVA resulted in an F-value < 1). As expected simulation runs with the higher traffic density caused the EC to give higher average subjective workload ratings. This is confirmed by a significant main effect of traffic load ($F(1, 11) = 8.08$; $p = 0.016$). No further effects became statistically significant.

Situational Awareness

Neither the ANOVA computed on the data collected at the EC working positions nor the one on the data collected at the PC working positions revealed a significant effect on situational awareness due to any of the experimental factors. No significant effect due to any interaction between the experimental factors was detected.

Team Work

No significant effects were found for the PC data. For the EC data only the Working Method (WM) x Task Load (TL) interaction effect became significant (F (1, 11) = 6.519; p=0.027). The pattern that caused this interaction was as follows: At the lower level of traffic load team work ratings favoured the flexible rather than the rigid working method whereas the reverse was true at the higher traffic load. In the post-exercise interviews, participants mentioned that the flexible working method requires an extra amount of communication in the sector team. Apparently, this extra communication occurred for the benefit of improved teamwork as long as the level of traffic allowed it. This suggests that at a higher traffic load participants cannot afford the extra communication and teamwork deteriorates.

Acceptability and usability

The average ratings were analysed by paired t-tests. This revealed that the flexible working method was rated significantly more acceptable than the rigid working method (t (16) =-2.487; p= 0.024). Comments made during the interview confirmed that the idea of shifting CPDLC tasks from the EC to the PC was highly appreciated by the controllers. However, controllers felt that some work outlining explicit delegation rules is needed to avoid ambiguity in the resulting new roles of the sector team. This drawback of the flexible working method offers an explanation for the results obtained with the analysis of the controllers' usability ratings. Here, the rigid working method was rated significantly more usable than the flexible working method (t(16)=2.496; p=0.024).

Preference of the working methods

Finally, asking controllers which working method they prefer revealed that the majority (N=13) prefer the flexible working method, only a minority (N=4) prefer the rigid working method. This statement agrees with the higher acceptability of the flexible working method. The higher acceptability and preference of the flexible working method contrasts its lower usability.

Discussion

Reductions in the usage of radio communication represent the key element in the mechanism by which workload benefits are expected to evolve in the use of CPDLC. There is undoubtedly a close relationship between workload and the amount of voice communication required to maintain control over air traffic. Voice communication is time consuming because the frequency channel is available only in one direction and each controller message needs to be read back by the pilot. As usage of the radio frequency is increased, high workload is induced by a proportional reduction in the time available to carry out required air-ground communications. Thus, much of the workload associated with high traffic is related to this time limitation, rather than to an inherent cognitive limitation in the ability of the controller to handle the increased air traffic of the airspace. Therefore, it is necessary to search for opportunities to exploit the CPDLC capability to reduce controller workload and create space to expand the current capacity limits.

Delegating CPDLC communication to automation or to the PC is one promising approach. The latter is explored in this paper. Exploring the benefits of an empowered PC started with the development of a new more flexible working method and was complemented by validation in a real-time simulation (Eurocontrol, 2005; 2007a). The advantage of the flexible working method is that EC and PC can simultaneously communicate via CPDLC and can so resolve more conflicts at a pre time-critical stage, thus leaving fewer aircraft to be handled via voice. Here, data is reported obtained in the Eurocontrol (2007a) real-time simulation, which allowed a systematic comparison between the new flexible and the traditional (rigid) working method under different levels of traffic load and CPDLC equipage rates.

First of all, when using the flexible working method a shift in the CPDLC task from the EC to the PC was observed suggesting that the sector team generally used the flexible working method in the intended way. The hypothesized benefit that the flexible working method also promotes a reduction in voice communication was also observed in the simulation experiment. However, this benefit of the flexible working method was only detected using a CPDLC equipage level of 20%. The benefit disappeared at the equipage level of 50%. Controllers reported that the flexible working method required more communication within the sector team. Thus, in an environment supporting a flexible distribution of tasks within the sector team new communication demands emerge which add to the controllers' cognitive workload. This workload can be assumed to increase at higher CPDLC equipage levels since there were more aircraft for which the CPDLC task needed to be distributed among the sector team. This leads to the interpretation that the flexible working method supported a more proactive working mode at a CPDLC equipage level of 20% but at a 50% of data link equipage the additional demand on communication interfered with the proactive working mode, resulting in more time-critical EC interventions to be issued via voice.

Despite the fact that in traffic involving 20% CPDLC equipage the flexible working method reduced the air-ground communication load for the EC (fewer CPDLC clearances and fewer voice radio calls) this task-load reduction did not translate into a perceived workload reduction. This was indicated by an absence of the measured impact on subjective controller workload ratings. This can be explained by the assumption that the reduction in air-ground communication load was offset by an increase in the load due to increased internal communication within the sector team.

Increased flexibility may have a downside if roles and responsibilities within the sector team become ambiguous. Workload associated with increased internal communication and collaboration need to be properly understood and hence planned for. However, controllers rated the flexible working method as more acceptable and, overall, preferred it over the rigid working method. Thus, the controllers support the concept of an empowered PC. Controllers identified the need for procedural support of the shift of the tasks, like, e.g. introducing the word "checked" to inform the team member that the information was received and understood. However not only rules for certain tasks but for certain traffic pattern might be identified. Controllers mentioned in the interviews that the use of the flexible working method enables the EC to delegate whole traffic flows as soon as they are conflict free. Furthermore, a need to enhance the system to provide support for the delegation of tasks was recognized.

It becomes evident that the role of the EC changes with the emergence of the delegation of tasks. It has been shown that the flexible working method requires additional communication and collaboration. When using the flexible working method, communication is needed to reduce unpredictability for both team members. This unpredictability can be directly related to the workload required to maintain awareness of actions taken by the team partner.

This creates a trade-off to be solved in order to balance the desirable workload-reducing effect of a more flexible EC-PC task sharing against its undesirable workload-enhancing effect caused by the emerging element of delegation. On the one hand, the CPDLC communication task is a good candidate to be delegated to the PC because of the non-time-criticality, thus the more strategic characteristics of CPDLC instructions. This enables the EC to better focus on time-critical clearances. On the other hand, placing the EC in charge of task delegation creates a twofold burden in requiring the EC to think strategically as to how best to share non time-critical interventions with the PC while simultaneously finding immediate solutions for time-critical problems.

It is interesting to note that the problems with the delegation task discussed so far have also been found to be an issue in the implementation of dynamic function allocation in the use of automatic controller assistance. Dynamic function allocation, which is also referred to as adaptive automation (Hilburn et al., 1997) has been described "as a form of automation that allows for dynamic changes in control function allocations between a machine and human operator based on states of the collective human–machine system" (Kaber et al., 2001, p.37). Vanderhagen et al. (1994) compared different implementation modes for dynamic task allocation between the human controller and a conflict resolution tool to support en-route air traffic control. This means that sometimes the controller resolved the conflict and sometimes the tool did. For that purpose the controller was provided with a special dialog interface. In an explicit mode, the human air traffic controller managed the task allocator through this dialog interface. So the controller decided to allocate the task either to himself or to the tool. In the implicit mode, task allocation was managed automatically based on some rules that ensured that easy tasks were allocated to the tool when the overall task demand of the controller was high. Controller performance was best in the implicit mode, although the controllers subjectively preferred the explicit mode. The problem with the explicit mode was apparently the increased workload imposed on the controller by accomplishing the additional task of making a task allocation decision. A subsequent study (Lemoine et al., 1996; Hoc & Lemoine, 1998) revealed that the problem with the explicit mode was not only workload but also a conflicting mismatch in role assignment. This time, the explicit mode was compared with an assisted explicit mode in which the planning controller was assigned the role of the task allocator who was supported by automatic advisories. Performance evaluation was in favour of this latter mode, apparently because task allocation management was a strategic task for which the planning controller was better suited than the more tactically engaged executive radar controller. These results suggest that the flexible allocation of the CPDLC task would be better assigned primarily to the PC rather than the EC. One solution could be that the PC suggests aircraft to be delegated and asks for EC consent. Such a procedure could be supported by an appropriate HMI (e.g. a dialog interface sensu Vanderhaegen et al., 1994).

In conclusion, given the high acceptability and the preference of the controllers towards an empowered role of the PC future improvement of the flexible working method should focus on roles and responsibilities and forms of procedural and HMI support. The aim should be, first to improve the usability of the delegation task by minimising its negative impact on EC workload and, second to facilitate internal communication and collaboration.

References

Eurocontrol Experimental Centre (EEC) (1996). *ERGO (Version 2) for Instantaneous self assessment of workload in a real-time ATC simulation environment.* EEC Note No. 10/96. Bretigny, France: Eurocontrol Experimental Centre. Retrieved 26.03.2008 from
http://www.eurocontrol.int/eec/gallery/content/public/documents/EEC_notes/1996/EEC_note_1996_10.pdf.

Eurocontrol (1999). *LINK 2000+ fast time simulation. To assess the impact of DATA LINK on sector capacity.* Brussels: Eurocontrol Headquarters. (Retrieved 26.03.2008 from
http://www.eurocontrol.int/link2000/gallery/content/public/files/documents/CAPANSimLINK2k.pdf.

Eurocontrol (2000). *LINK 2000+ business case development simulation*, final report, EEC Note No. **/00, Project L2KBC Version 1, Bretigny, France: Eurocontrol Experimental Centre. Retrieved 26.03.2008 from
http://www.eurocontrol.int/link2000/gallery/content/public/files/documents/EECSimLINK2k.pdf.

Eurocontrol (2002). *PETAL-II transition and final report. Data Link as an addition to R/T*. Vol. 2. Document No. OPR.ET1.ST05/2000/P2/19. Brussels: Eurocontrol Headquarters. Retrieved 26.03.2008 from
http://www.eurocontrol.int/agdl/gallery/content/public/docs/P2FINAL.pdf

Eurocontrol (2004). *LINK2000+ France real-time simulation project*. EEC Report No. 395. Bretigny, France: Eurocontrol Experimental Centre. (Retrieved 26.03.2008 from
http://www.eurocontrol.int/link2000/public/standard_page/specific_docs.html

Eurocontrol (2005). *CEATS SSRTS 5 data link*. CRDS Note no. 13. Budapest, Hungary: Eurocontrol CEATS Research, Development and Simulation Centre. Retrieved 26.03.2008 from http://www.eurocontrol.int/crds/gallery/content/public/reports/4535-CRDSSIMRTS-OZE.pdf

Eurocontrol (2006a). *CASCADE STREAM 1 real–time simulation.* EEC Report No.404. Bretigny, France: Eurocontrol Experimental Centre. Retrieved 26.03.2008 from http://www.eurocontrol.int/cascade/gallery/content/public/documents/report.pdf

Eurocontrol (2006b). *ATC data link manual for Link 2000+ services V 4.0.* Brussels: Eurocontrol, Headquarters.

Eurocontrol (2007a). *LINK2000+ small scale real-time simulation No.2 (LINK2000+SSRTS2)*, Final Report, CRDS Note No.26. Budapest, Hungary: Eurocontrol CEATS Research, Development and Simulation Centre. Retrieved 26.03.2008 from http://www.eurocontrol.int/crds/gallery/content/public/reports/6701-CRDSSIMRTS-OZE-V01.00.pdf

Eurocontrol (2007b). *CASCADE real time simulation No. 2.* Simulation report, CRDS Note No. 27, Budapest, Hungary: Eurocontrol CEATS Research, Development and Simulation Centre. Retrieved from http://www.eurocontrol.int/crds/gallery/content/public/reports/7192-CRDSSIMRTS-PET-01.30.pdf

Eurocontrol (unknown). The new SHAPE questionnaire: a user guide (Edition No. 0.1). Retrieved 26.03.2008 from http://www.eurocontrol.int/humanfactors/gallery/content/public/docs/SHAPE_questionnaires/SHAPE%20User%20Guide%20v0.1.pdf

Federal Aviation Administration (FAA) (1995). *User benefits of two-way data link ATC communications: Aircraft delay and flight efficiency in congested en route airspace*, FAA Report FAA/CT-95-4. Washington, DC: FAA. Retrieved 26.03.2008 from http://stinet.dtic.mil/oai/oai?verb=getRecord&metadataPrefix=html&identifier=ADA292927

Hilburn, B. J., Byrne, E.,&Parasuraman, R. (1997). The effect of adaptive air traffic control (ATC) decision aiding on controller mental workload. In M. Mouloua & J.M. Koonce (Eds.), *Human–automation interaction: Research and practice*, (pp. 84–91). Mahwah, NJ: Lawrence Erlbaum Associates.

Hoc, J.M., & Lemoine, M.P. (1998). Cognitive evaluation of human-human and human-machine cooperation modes in air traffic control. *International Journal of Aviation Psychology, 8*, 1-32.

ISO 9241-11(1998). *Ergonomic requirements for office work with visual display terminals (VDTs) --Part 11: Guidance on usability.* Geneva, Switzerland: International Organization for Standardization.

Kaber, D.B., Riley, J.M., Tan, K.-W., & Endsley, M.R. (2001). On the design of adaptive automation for complex systems. *International Journal of Cognitive Ergonomics, 5*, 37–57.

Lemoine, M-P., Debernard, S., Crévits, I., & Millot, P. (1996). Cooperation between humans and machines: first results of an experiment with a multi-level cooperative organisation in air traffic control. *Computer Supported Cooperative Work. Journal of Collaborative Computing, 5*, 299-321.

Midkiff, A.H. & Hansman, R.J. (1993). *Identification of important "Party Line" information elements and the implications for situation awareness in the data link environment.* (SAE 922023). Warrendale, PA: SAE International.

Prinzo, O.V. (1998). *How controller-to-pilot data link communications might affect feeder controller workload in a terminal option.* Paper presented at the 17th Digital Avionics Systems Conference sponsored by IEEE/AIAA, Bellevue, WA.

Prinzo, O.V, (2001). *Data-linked pilot reply time on controller workload and communication in a simulated terminal option,* Final Report Civil Aeromedical Institute, Federal Aviation Administration (FAA), DOT/FAA/AM-01/8, Oklahoma City, Oklahoma: FAA.

Ruitenberg, B. (1998). The human factors of CNS/ ATN. *The Controller, 7,* 25-7.

Shingledecker, C., Giles, S., Darby, E.R., Jr. Pino, J., Hancock, T.R. (2005). *Projecting the effect of CPDLC on NAS capacity.* Proceedings of the Digital Avionics Systems Conference, 2005 (Vol. 1). DASC 2005.

Sollenberger, R. L, Willems, B., Della Rocco, P. S, Koros, A., Truitt, T. (2005). *Human-in-the-Loop Simulation the Collocation of the User Request Tool, Traffic Management Advisor, and Controller Pilot Data Link Communications: Experiment 1 – Tool Combinations,* DOT/FAA/CT-TN04/28. Atlantic City International Airport, NJ: FAA William J. Hughes Technical Center.

SPSS (2006). *SPSS Advanced Models™ 15.0 Manual.* Chicago, IL: SPSS.

Vanderhaegen, F., Crévits, I., Debernard, S., & Millot, P. (1994). Human-machine cooperation: Toward activity regulation assistance for different air traffic control levels. *International Journal on Human-Computer Interaction, 6,* 65-104.

Designing scenarios: the challenge of a multi-agent context for the investigation of authority distribution in aviation

Sonja Straussberger & Guy Boy
European Institute of Cognitive Sciences and Engineering (EURISCO International)
Toulouse
France

Abstract

To better understand the issue of authority distribution between airborne (cockpit) and ground (air traffic control) side in the aeronautical domain, the design of scenarios and related a priori-evaluations is essential before implementing resource-intense human-in-the-loop simulations. Currently used methods are considered more as "a bunch of guys sitting around the table" (BOGSAT). Describing the aeronautical domain as a complex multi-agent system with distinctive needs, the development of scenarios requires a structured multi-faceted procedure to meet defined success criteria as well as the variety of demands from different scenario users. We present a multi-stage process used in defining scenarios for the specific situations of separation and collision avoidance within the future air traffic management. Part of this process was the application of two Group Elicitation Method (GEM) sessions with a total of 12 domain experts to obtain the essential scenario elements based on experience. A categorization allowed a systematic assignment of collected viewpoints to structural categories, which supported the description of both declarative and procedural perspectives. The methodological relevance of such an integrated approach through combining the advantages of qualitative and quantitative methods will be discussed.

Introduction

Scenarios for developing the future ATM

Within the air traffic management (ATM) domain, the demand has been noted to improve the technologies and systems currently in use through implementing European-wide initiatives in the Single European Sky Aviation Research (SESAR) program. The essential question is how we can adapt today's ATM services to deal with future air traffic increases, which according to EUROCONROL STATFOR (2006) are estimated to multiply by a factor between 1.7 and 2.1 until 2025. Due to profound technological changes, an in-depth change of the role of human actors can be expected, which will consequently impact the ways of sharing or distributing authority. There is, however, a clear statement that the human operator will remain

In D. de Waard, F.O. Flemisch, B. Lorenz, H. Oberheid, and K.A. Brookhuis (Eds.) (2008), *Human Factors for assistance and automation* (pp. 359 - 372). Maastricht, the Netherlands: Shaker Publishing.

central within this system (SESAR Consortium, 2007). This is also in line with arguments towards human-centred automation, which have been stated more frequently, since Billings (1997) pointed out the need of a human-centred approach in aviation system developments. Despite such premises to keep the human in the loop when managing air traffic, such an approach has hardly been systematically tackled when proposing scenarios.

The design of scenarios is one of the methods that enable us to assess possible complex future human-machine systems. Thus, to anticipate the behaviour of the actors of the future ATM, the selection of appropriate scenarios is essential. But already today, they play a central role in aviation. Numerous formats to represent scenarios exist, and conditions range from investigating the specific impact of tools or technologies to running operational simulations in nominal and non-nominal situations. However, in the past, most of the aviation research projects have focused on one segment, be it the air or ground, to describe scenarios. One exception is represented by the work of Oberheid, Lorenz, and Werther (2006), who used scenarios to investigate function allocation solutions between human and technologies in an integrated air-ground system through representing the delegation of separation from ground to air. However, such an approach may be of generic importance for evaluating the effects of implementing new technologies in a complex system, as is aviation. Within this setting, a human-centred approach applied to scenarios means integrating user experiences through an iterative process throughout the development. For the purpose of integrating both air and ground perspectives in a generic scenario definition approach the definition of scenarios is reviewed in more detail.

Designing scenarios beyond BOGSAT

According to Carroll (2000), scenarios can be seen as stories that comprise characteristic settings. Designated elements include agents or actors, goals or objectives, and actions and events. Plihon, Ralyté, Benjamen, et al. (1998) describe scenarios as a "possible behaviour limited to a set of purposeful interactions taking place among several agents". To better classify the different types of scenarios, Rolland, Achour, Cauvet, et al. (1998) proposed a framework that differentiated four different views. Textual, graphical and image aspects can be distinguished as regards their form. The content view categorizes knowledge as far as it addresses the internal world of a system, the external world or the interaction between both. This classification contains both behavioural information (e.g. actions, events) and object-related information (e.g., entities, data, attributes). Finally, the purpose view captures the role of a scenario (descriptive or explorative), and the lifecycle view addresses if scenarios evolve and are thus transient or persistent. It is noted that in this conception, the frequently used notion of use-cases represents a particular type of scenario. According to Bittner and Spence (2002), a use case describes a specific way of using the system by performing some parts of its functionality. It specifies all the existing ways and interaction between actors and actions. Similarly, the procedure proposed by Dearden, Harrison, and Wright (2000) pointed out the relevance of the definition of characterized agents, the situation or context, the scenario baseline, objectives, and actions. There is, however, no indication on how a structured approach may support such a scenario definition process.

To carry out this definition process, few authors proposed a structured method for developing these elements. As Ericsson and Ritchey (2002) described, the dominating approaches can be characterized as a "Bunch of Guys sitting around the Table" (BOGSAT). To tackle this issue within the field of aviation, within the MAEVA project guidelines were proposed (Isdefe, 2001) on the variables to be considered when developing scenarios. Still, these indicators mainly address factors related to the air traffic structure. A more systematic approach is provided by Gordon et al. (2004), who selected safety scenarios for simulations based on results of a human error analysis.

To tackle these restrictions, Ericsson and Ritchey suggested the use of a morphological analysis for scenario development. Through a non-quantified representation of a problem with a number of categorized parameters a certain number of different conditions can be described. Applied to military missions, they demonstrated how this analysis allowed the consideration of complex cases through a structured distinction of relevant variables such as possible destinations or involved weapon systems. A similar procedure was proposed with the Modular Scenario Composition Method (Sato, 2004). It supports the integration of conditions and the represented context from separate modules within a unified information platform. Such modules include actions, time, users, objects, intentions, tools, and different types of contexts. In a next step, this method links such descriptive field information with formal analytical models. Still, these techniques remain focused on global structural elements of scenarios. Hickey, Dean, and Nunamaker (1999) argued for a collaborative iterative approach on scenario elicitation to overcome limitations of different scenario formats. Using this approach in the space domain, improvements of scenario quality and efficiency were expected.

Scenarios for demonstrating authority issues in a multi-agent context

A human-centred approach for scenario development requires identifying specified needs of aviation air and ground actors. A multi-agent approach is proposed to reflect the variety of interactions between a number of actors trying to achieve their mission goals in this complex aeronautical system. In addition, future operational concepts lead us to expect a relocation of assigned authority in separation and collision-avoidance functions (e.g. Ruigrok & Hoekstra, 2007). To reveal critical issues within such a context requires thus developing initiatives in this multi-agent context. Whereas in the past mainly one-agent systems - human or machines – were addressed, even though they dealt with the interaction between board and ground, a multi-agent approach is valuable when implementing operational concepts such as Airborne Separation Assurance System (ASAS) or establishing a contract between air and ground based on a 4-dimensional (4D) representation of the aircraft trajectory (Wichman, Lindberg, Kilchert & Bleeker, 2004) taken up in the SESAR operational concept (SESAR Consortium, 2007).

Multi-agent systems (MAS) are envisaged as an environment, a set of agents, and their interactions with the environment. An agent is defined as an entity that acts locally and autonomously to collectively solve a problem in the environment. Such systems involve two levels of description: one for individual behaviour on the local agent level and one to express the collective phenomenon. For example, individual behaviour in ATM may cover the efficient operations of an air traffic controller, whereas capacity measures may capture the global phenomena. In this context, a transfer of tasks or functions between segments and increased levels of automation will lead to different allocations of authority between human and machine agents. Authority can be defined as the power to act directly on a process or indirectly through assigning or refusing goals. There are, however, different variants of how authority is assigned to the actors in the system. Depending on the objectives, authority can be allocated to one specific actor through a concrete assignment and consequently delegated. Sharing authority describes the behaviour of using a resource jointly with others, thus achieving single functions through sharing control (Sheridan, 1992). Trading (Inagaki, 2003) can be defined as a voluntary exchange of services, while distributing is associated with transferring services from one location to another. Even though various examples of these different forms and combination of authority transfer can be imagined in aviation, currently only few have been empirically investigated (e.g. Parasuraman, Mouloua & Molloy, 1996). To analyze the effects of different forms of authority distribution on the interaction between human and machine agents on both the air and the ground side, the choice of the appropriate scenario is important.

Requirements for a structured approach of scenario development

According to Carroll (1999), several good reasons underline the need of scenarios, such as the integration of end-user experiences, enabled reflection and discussion on multiple solutions, and multiple forms of presentation. Following these considerations, it is preferable to obtain elements involved in this interaction through a human-centred and experience-based approach. As in an integrated air-ground context the operators of each segment are frequently not aware of the impact of local activities on the global system, this analysis of interactions is useful for demonstrating relevant effects on human performance within this context. Through putting structural scenario elements from both segments in interaction, human factors issues are made explicit, that otherwise would remain covered until a late stage of the concept validation process.

Reviewing existing methods for scenario design showed the difficulty of systematically collecting structural and functional elements to build scenarios in a single approach. A further requirement is to execute the definition work in a multi-disciplinary environment with a certain number of representatives of mixed operational backgrounds and various experience levels. In addition, these representatives may have different, even contradictory, objectives. Beyond that, scenarios need to enable an analysis and evaluation of "soft" human-factor based performance criteria from different agent perspectives. Therefore, the question is how qualitative and quantitative methods can be combined to arrive at consensus-based scenarios. At the same time, risk factors need to be tackled, such as incompleteness, the failure to provide critical information, or inefficiency.

Thus, the following criteria are defined for scenarios describing the authority issue with regard to the future ATM development. Scenarios need to:

- Enable the comparison of today's system with the future situations in a different technological environment due to traffic increases
- Comprise normal and abnormal situations
- Tackle the issues of separation and/or collision avoidance in enroute/cruise and approach
- Comprise conditions in which both the air and the ground side are present as well as human and machine agents
- Represent a realistic mapping of the reality and associated high-level functions
- Integrate different stakeholder perspectives of scenario description.

At the same time, risk factors were identified to

- Construct scenarios that are not relevant enough or neglect important components
- Consider that consequences may affect differently the stakeholders of the ATM system
- Choose the appropriate level of abstraction between high-level tasks and moment-to-moment views
- Correctly integrate different views of different users.

Proposing a method for an integrated iterative approach for developing ATM scenarios

Defining the global objectives for integrated scenarios in an aeronautical multi-agent context

An approach to systematically develop scenarios from the perspective of an integrated air-ground system has been applied within the context of a French national project, PAUSA*. The specific demand is to integrate users' experience from an early stage in the ATM improvement process. Scenarios are one of the key components of the first project phase, as it allows the early integration of the viewpoints from operational practitioners (pilots, air traffic controllers), engineers, and researchers. The goal of these scenarios is to demonstrate the impact of authority distribution on the interaction between human and machine agents and involved human factors issues within the future ATM. Examples for considered human factors issues are workload, situation awareness, responsibility, and collaboration. The process of developing scenarios is embedded within different initiatives and executed as an iterative cycle.

---

* PAUSA is a French acronym for Authority Distribution in the Aeronautical System (Partage d'AUtorité dans le Système Aéronautique); see http://www.eurisco.org/pausa/

Within this scope, the high level objectives of scenarios are to integrate user-experience from separate ground and airborne segments in an early stage to propose scenarios that demonstrate human factors issues in relation to alternative forms of authority assignment within these segments and human and machine agent types. Because of the far-reaching consequences of any outcomes, and despite expressed limitations or constraints, consensus-based work is essential to develop generally accepted scenarios. Beyond that, scenarios serve as an input for modelling the global and local ATM system with the objective of understanding the variety in behaviours and emerging cognitive functions. These functions consist of cognitive processes resulting from an adaptation of the technical system requiring new forms of operator activities. Subsequently, related observations will be made in human-in-the-loop simulations when introducing new functionalities. Hence, these requirements for scenarios necessitate an explicit consideration of different structural and functional elements to make clear at which moment of a scenario new cognitive functions may be expected and which are the evoking conditions for such functions.

A multi-step procedure for obtaining scenarios

To achieve consensus- and experience-based scenarios under consideration of previously defined constraints and limitations, the following steps are proposed and exemplified with the process of developing PAUSA scenarios:

Step 1: Definition of basic scenario structure. The first step summarizes which form of characterization is required with regard to the overall objectives and selects the appropriate form of scenario views and elements.

To start the description of authority sharing scenarios with domain experts from different professional backgrounds, a separate description of structural and functional elements was envisaged to approach the complex setting through focusing on single elements. The notions of declarative and procedural expressed these static and dynamic elements. Declarative scenarios describe the necessary elements involved to achieve the mission's goals. The essential components are agents, tasks or functions, procedures, and context. Such descriptions necessarily lead to the way objects and agents interact with each other, and consequently to application use cases. Procedural scenarios describe chronologies of events and interactions among objects and agents and thus add the temporal component to this interaction. The implication in stories and episodes demonstrates how agents are instantiated.

Step 2: Elicitation of scenario concepts. This process applies a consensus-based method with multi-disciplinary user and stakeholder groups.

To obtain declarative elements, the group elicitation method (GEM; Boy, 1996) was used. The advantage of this method is that it allows participants reaching a consensus by considering individual statements and emerging discussion points at the same time. This method is composed of several phases where participants are asked to address an open question. In a first round, viewpoints are collected individually in a computer-based setting and alternately completed by all other participant's agreements and disagreements. The subsequent open discussion leads to an integration of single viewpoints into elaborated concepts. Finally, these concepts are exposed to a rating and re-evaluated in a final critical discussion. This technique of defining concepts helps to complete the list of necessary scenario elements based on the experience of involved users. It shall be mentioned that the representation of the links between the elicited issues in the form of a domain ontology (e.g. Reiss, Barnard, Moal, et al., 2006) supported the characterization process. Defining basic terms and relations using the vocabulary of a specific area, this type of ontology describes physical and functional resources and allows the expression of semantic complexity.

Step 3: Categorization of concepts. This process summarizes relevant concepts with regard to required characteristics and prioritizes them in a hierarchy.

Within the current process, these elements were assigned to categories characterizing manipulated and measurable experimental variables such as air space, agents and roles, normal and abnormal conditions, technologies, human factors issues, etc.

Step 4: Integration of declarative scenario concepts with procedural component. This process links the elements by adding the temporal components and sets them in action.

In the next step 4, the elements were incrementally merged towards synthesized generic scenarios integrating declarative and procedural components. For this purpose, the procedural part was developed in guided focus group meetings through adding the time component to the different elements developed in the specific context. To ease the process of interdisciplinary work, textual tabular and graphical process representations were used applying a domain-oriented language for describing the content.

Step 5: Iteration on scenarios. A first iteration reviews the scenarios with the objective to complete or correct the representations with regard to the original objectives.

Step 6: External Validation and Re-iteration of scenarios. This step conducts an evaluation of scenarios by assessing their relevance by domains experts external to the project.

The final iteration and external evaluation is an ongoing process. External experts are independently evaluating the relevance of the scenarios with regard to a French and a European application. For this purpose, the choice of a representative sample and the construction of appropriate instruments for quantifiable results were required. This instrument considers criteria with regard to the representations of authority problems as well as the domain's characteristics. A kind of criteria checklist is attained for applying scenario validation instruments that guarantee the integration of core authority sharing/distribution issues in each scenario throughout the evaluation process.

An example of applying the scenario development approach: Eliciting scenario concepts on issues relevant for authority sharing

The following section describes how the introduced process was applied to elicit and categorize scenario elements and presents selected results from two GEM sessions.

Participants

Twelve (one participant joined two sessions) participants of various backgrounds representing air and groundside segments joined two group elicitation sessions with an average duration of 6 hours without breaks. The participants of the first session contained one test pilot, two former air traffic controllers, one human factors engineer board, one human factors engineer ground, one interface engineer board, and one interface engineer ground. In the second GEM session one airline pilot, an active and a former air traffic controller, one business-jet test pilot and a business-jet operational pilot participated. All participants can be characterized as highly experienced in the domain as an operator, researcher or engineer.

Procedure

The participants were exposed to the question, which were the most important cases to consider and develop in situations of separation and collision avoidance in the enroute/cruise as well as the approach phase. They were also supposed to reflect on possible solutions to deal with these cases and factors to be considered. The viewpoints with agreements and disagreements as well as the synthesized concepts were stored as raw data. Subsequently, in a first step a categorization of the concepts was accomplished by the authors with reference to the categorisation enabled by the AUTOS framework (Boy, 1998), which organises concepts related to Artefact, User, Task, Organisation, and Situation. This framework allows for the organization of different elements with reference to the relevant category. In a second step, a more detailed analysis adapting Grounded Theory techniques (Strauss and Glaser, 1990) was conducted with reference to the elements required for defining characteristics of agents or contexts as well as their interaction in situations relevant for authority sharing and distribution.

Results

Overall, 17 concepts were identified that addressed technologies, user involvement, the task, the situation, or the organisation. Two of them where explicitly related to authority sharing. Table 1 presents these concepts that were consequently used to make emerge elements to define the structure of scenarios as described in Step 3. Equally integrated in this step are categorized issues related to authority, as the central objective of the current research is to understand the issue of authority assignment in the future ATM. Any items referring to this concept are characterized in detail to make sure they are contained in different types of scenarios. Out of 117 synthesised viewpoints, 23 items explicitly addressed authority related issues. These items are summarized in Table 2 according to the interaction segments they address.

Table 1. The list of synthesized scenario concepts

Dimension	Concept
Artifact	Visualization of traffic environment
	Common expression of 4D contract
	TCAS linked with automatic pilot
	Technologies for collision avoidance in converging parallel approaches
	Other technological solutions
User	Pilot experience
Task	Change of control task
	Management of collision avoidance in advance
	Function allocation
	Management of fully automated piloting and navigation functions
	Definition of basic piloting modes
	Authority sharing and distribution between systems
	Definition of separation criteria
	Definition of scenario evaluation criteria
Organization	Changing type of profession
	Other organisational solutions
	Juridical and technical frame of work
	Articulations in collaborative work setting
	Creation of an agency of evaluation and mediation
Situation	Authority sharing regarding weather conditions
	Consideration of environmental factors
	Consideration of structuring variables (e.g. airspace)
	Consideration of special events or incidents
	Consideration of cancelling 4D contract in abnormal or emergency situation
	Consideration of additional situational variables

Discussion and conclusion

The presented research is embedded in the first phase of a project, which is part of a bigger initiative, the redesign of the future European ATM. In this phase, scenarios are an essential element, as they allow the integration of diverging user needs. Scenarios include the work place as well as the social situation, embody information about resource constraints, explain why users do what they do, take users' goals and context explicitly, and imagine what-if situations. For this reason, a multi-step procedure was developed to design scenarios through integrating multi-disciplinary user experience in the aviation domain. The challenge of such a procedure is to profit from the experience of different user classes that are not necessarily used to interact with each other outside the operational context. A consensus-based approach relating authority issues between the air and ground segment in not yet existing situations requires using today's experience to anticipate the impact of new technologies. The group elicitation method appeared as an appropriate method to uncover relevant issues and obtain essential elements as an input for scenarios. At the same time, this approach does not lack the specificities of each local segment.

Table 2. Specific authority issues to be considered in future systems

Segment	Agent Type	Issue
On board	Human-Machine	To rely on trustworthy automats for automatic piloting and navigation
		Increased responsibility of pilot, but no total delegation of task to machine
		Automatic piloting, but with the option to switch back to manual piloting
		Strategy and coordination of equipment for the converging parallel approach
Board-Ground	Human	Transfer of tasks and authority from ground to the board
		Responsibility for recovery after failure
		Common understanding, e.g. 4D contract (n=2)
		What is transferred to board needs to be anticipated by the ground (predictable behaviour)
		Delegation of separation with responsibility of pilot
Board-Ground	Human-Machine	Manage situations where an information conflict is possible
		Delegation concept to be better defined
		The pilot remains on board in case of total system failure
		Different philosophies in the development of systems for board (eg., EGPWS et TCAS) and ground
		Sharing intention in common contract
		Processing and managing of contract between board and ground in new technological configurations
		Responsibility for diagnostics on board delegated to pilot or ground
		Manage weather conditions
		Task presentation may be different on board and ground if tasks are shared
		Take back a situation from a decision-maker after an incident or automation failure
		Development of collaboration with increased automation in ATM
		Automatic information in case of cancelling 4D contract to everyone concerned before entering the controlled sector

Notably, a high number of reported concepts take up the interaction between air and ground segment as well as human and machine agents at the same time. Specifically, interaction with any new technologies in the operational context evokes concerns. Thus, authorities related issues are central for operators in the future ATM, and were confirmed to be specifically addressed in the evaluation of scenarios. This supports the argument for a continuous multi-agent approach when advancing the future ATM system.

Additional questions came up at different stages concerning the quality of emerging scenarios. The level of detail to describe scenarios evoked increasing difficulties the farther the temporal horizon imagined. Also, the type of evaluation criteria to be suggested for verifying the appropriate scenario is reviewed. This point appears to be uncomplicated, but becomes more complex if referred to the different nature of agents' goals. Various objectives were proposed to characterize successful future ATM, such as increased safety, efficiency, capacity, etc. On the other hand, the represented human operators have different problems to tackle and a performing system would thus decrease these problems. As a result, different types of criteria need to be developed in such a complex integrated system. Finally, the combination of user experience with the requirement of a certain level of global acceptability requires a systematic approach combining a purely qualitative development approach expressed in the steps 1 to 5 with a more quantitative approach applied in step 6.

For this reason, the group elicitation method was just one part of the overall process. It can be characterized as a purely qualitative approach that helps to understand problems when human and machine agents are interacting. Its advantage is credibility through active involvement and a continuous integration of experience to progressively advance the scenario description from the known to explore the unknown. It does not, however, allow quantifiable conclusions. Regarding the overall project objectives, it is important to have a general acceptance of proposed scenarios. For a future evaluation and validation of scenarios beyond the selected French context and to check for the transfer of conclusions obtained in oncoming work, the final scenarios will need to be validated in an extended context.

Even though such a process appears to be resource-intense, the consideration of divergent perspectives from an early stage consensus may contribute to avoid time-consuming discussions at a later stage. Through integrating what users have experienced in the past, the anticipation of the potential future evolvement of the ATM system is favoured. Different forms of representation enable the use of adequate levels of granularity during the different phases. To cover all relevant situations an interdisciplinary setting is used to achieve a certain level of result quality.

In summary, the described process demonstrates how an iterative approach in designing scenarios includes user experience and general acceptance through a shift from the application of qualitative elicitation and categorization towards quantitative validation methods. One essential objective is to obtain consensus between different agents, which will help to better understand the domain and anticipate potential problems due to human factors. This process is however continued beyond this phase of scenario definition. Future implementations of scenarios are instantiated in descriptive models of organizational and local agent's levels. Dynamically represented scenarios will serve to pre-evaluate scenarios with reference to human performance criteria and emerging functions before launching human-in-the-loop simulations. Thus, scenarios are an essential part of developing the future European ATM system throughout the process as a means to favour comprehension between stakeholders.

Acknowledgements

This work is part of the research project PAUSA funded by the French Civil Aviation Authority Department. The authors wish to thank the partners and participants involved in the process of developing this method.

References

Billings, C. E. (1997). *Aviation Automation: The Search for a Human-Centred Approach*. Mahwah, NJ: Lawrence Erlbaum Associates.
Bittner, K. & Spence, I. (2002). *Use case modeling*. Boston, Ma: Addison Wesley Professional.
Boy, G.A. (1996). The Group Elicitation Method for Participatory Design and Usability Testing. *Proceedings of CHI96*, the ACM Conference on Human Factors in Computing Systems (pp. 87-88). ACM Press, New York.
Boy, G. (1998). *Cognitive function analysis*. Westport, CT: Ablex, Greenwood Publishing Group.
Carroll, J.M (1999). Five reasons for Scenario-Based Design. *Proceedings of the 32nd Hawaii International Conference on System Sciences*, Volume 3, 3051-3057.
Carroll, J.M. (2000). *Making Use: Scenario-Based Design of Human-Computer Interactions*. Cambridge: MA MIT Press.
Dearden, A., Harrison, M., & Wright, P. (2000). Allocation of function: scenarios, context and the economics of effort. *International Journal of Human-Computer Studies, 52*, 289-318.
Eriksson, T. & Ritchey, T. (2002). *Scenario Development using Computerized Morphological Analysis*. Adapted from Papers Presented at the Cornwallis and Winchester International OR Conferences, England. Available at www.swemorph.com.
Eurocontrol STATFOR (2006). *EUROCONTROL Long-Term Forecast: IFR Flight Movements 2006-2025* (DAP/DIA/STATFOR Doc216). Brussels : Eurocontrol.
Gordon, R., Shorrock, S.T., Pozzi, S., & Boschiero, A. (2004). *Using human error analysis to help to focus safety analysis in ATM simulations: ASAS Separation*. Paper presented at the Human Factors and Ergonomics Society 2004 Conference, Cairns, Australia, 22nd - 25th August, 2004.
Hickey, A., Dean, D. & Nunamaker, Jr., J. (1999). Establishing a Foundation for Collaborative Scenario Elicitation. *The DATABASE for Advances in Information Systems*, 30 (3&4), 92-110.
Isdefe (2001). MAEVA. *A Master ATM European VAlidation Plan*. D 2.3: Scenario Definition for Validation Exercises. MVA/AEN/WP2 23DA_20.
Inagaki, T. (2003). Adaptive automation: Sharing and trading of control. In E. Hollnagel (Ed.), *Handbook of cognitive task design* (pp. 147-169). Mahwah, NJ: Lawrence Erlbaum Associates.

Oberheid, H., Lorenz, B., & Werther, B. (2007). Supporting Human-Centered Function Allocation Processes Through Animated Formal Models. In D. de Waard, G.R.J. Hockey, P. Nickel, and K. Brookhuis (Eds). *Human Factors Issues in Complex System Performance* (pp. 361-372). Maastricht: Shaker Publishing.

Parasuraman, R., Mouloua, M. & Molloy, R. (1996). Effects of adaptive task allocation on monitoring of automated systems. *Human Factors,* 28, 665-679.

Plihon, V., Ralyté, J., Benjamen, A., Maiden, N.A.M., Sutcliffe, A., Dubois, E., & Heymans, P. (1998). *A reuse-oriented approach for the construction of scenario based methods.* Proceedings of the International Software Process Association's 5th International Conference on Software Process (ICSP'98), Chicago, Illinois, USA, 14-17 June 1998.

Reiss, M., Moal, M., Barnard, Y., Ramu, J.-Ph., Froger, A.(2006). Using Ontologies to Conceptualize the Aeronautical Domain. In F. Reuzeau, K. Corker, & G. Boy, *Proceedings of the International Conference on Human-Computer Interaction in Aeronautics* (pp. 56-63). Cépaduès-Editions, Toulouse, France

Rolland, C., Ben Achour, C., Cauvet, C., Ralyte, J., Sutcliffe, A., Maiden, N., Jarke, M., Haumer, P., Pohl, K., Dubois, E. & Heymas, P. (1998). *A Proposal for Scenario Classification Framework.* Requirements Engineering Journal, 3(1), Also available as CREWS Report Series No. 96-01.
http://citeseer.ist.psu.edu/rolland96proposal.html.

Ruigrok, R.C.J & Hoekstra, J.M. (2007). Human factors evaluations of Free Flight Issues solved and issues remaining. *Applied Ergonomics,* 38(4), 437-455.

Sato, K. (2004). Context-sensitive Approach for Interactive Systems Design: Modular Scenario-based Methods for Context Representation. *Journal of Physiological Anthropology Applied and Human Science,* 23(6), 277–281.

SESAR Consortium (2007). *Milestone Deliverable D3 – The ATM Target Concept* (DLM-0612-001-02-00). Toulouse. Retrieved 03.10.2007 from http://www.sesar-consortium.aero

Sheridan, T. (1992). *Telerobotics, automation, and supervisory control.* Cambridge: MIT Press.

Strauss, A., & Corbin, J. (1990). *Basics of Qualitative Research: Grounded Theory Procedures and Techniques.* London: Sage.

Wichman, K., & Lindberg, L., Kilchert L., & Bleeker, O. Four-Dimensional Trajectory Based Air Traffic Management. *AIAA Guidance, Navigation, and Control Conference and Exhibit,* Providence, Rhode Island, Aug. 16-19, 2004 (AIAA-2004-5413).

Head-Mounted Display
– evaluation in simulation and flight trials

Sven Schmerwitz, Helmut Többen, Bernd Lorenz, & Bernd Korn
DLR, Institute of Flight Guidance
Braunschweig, Germany

Abstract

Pathway-in-the-sky displays may drive pilots' attention to the aircraft guidance task at the expense of other tasks particularly when the pathway display is located head-down. A pathway HUD may overcome this disadvantage. Moreover, the pathway may mitigate the perceptual segregation between the static near domain and the dynamic far domain and hence, may improve attention switching between both sources. In order to more comprehensively overcome the perceptual near-to-far domain disconnect alphanumeric symbols could be attached to the pathway leading to a HUD design concept called 'scene-linking'. Experimental studies were completed by conducting two different flight tests. The first mainly focused at usability issues. Visual and instrument tasks were evaluated comparing HMD navigation with standard instrument or terrestrial navigation. The study revealed limitations of the HMD regarding its see-through capability, field of view, weight and wearing comfort that showed to have a strong influence on pilot acceptance rather than rebutting the approach of the display concept as such. Additionally it was found that pilots had difficulties during segment transitions while using the HMD. In a second flight test a redesigned pathway-predictor-director concept was implemented. The trials were designed as a high workload task. The results exposed more or less the same difficulties regarding the usability of the display but showed a much better pathway following especially during segment transitions.

Introduction

Curved approach profiles to large airports in low visibility pose a particular challenge. Pathway displays increase flight path awareness of pilots enabling them to fly difficult, e.g. curved trajectories with high accuracy (Haskell, 1993; Grünwald, 1996; Wickens, 2004). Hence, by using pathway guidance systems air travel capacity improving the efficiency of airspace use could be improved as well as accident rates (e.g. CFIT reduction – controlled flight into terrain) could be reduced. However, these displays may drive the pilot's attention head-down at the expense of monitoring the outside scene referred to as attention fixation or attention capture (Flemisch, 2000; Thomas, 2004; Wickens, 2005). One way to overcome the disadvantage of increased head-down times associated with head-down pathway displays is to present pathway guidance symbology on a head-up display (HUD) or head- or helmet-mounted display (HMD). But it has been found that alike head-down pathway guidance, attention fixation problems with the use of HUD technology also occur (Fadden, 2001; Fischer, 1980; Wickens, 1995; Lorenz, 2005). McCann et al. (1993) identified differential motion between the HUD imagery (static domain) and structures of the outside environment (dynamic domain) as the crucial element that promotes both information sources to be visually segregated as separate object domains during processing. The concept of scene-linking is based upon this claim. Central to this concept is the attempt to mitigate HUD-induced attention fixation by the creation of common motion between objects of the far-domain and the near-domain HUD imagery. In fact, Sheldon, (1997) demonstrated a performance advantage of scene-linking in low fidelity flight performance. Fadden et al. (2000) also implemented scene-linked HUD symbology. They attached altitude and airspeed readings to the moving symbology elements of the flight path. Although the pathway has no physical counterpart in the real world upon which it can be overlaid in the way this can be done with the desired track on ground, it can be argued that the pathway represents a virtual referent to the outside scene providing a suitable means to create common motion.

A pathway guidance display concept that uses scene-linking has been developed for a monocular HMD (Többen, 2005). Its feasibility was tested during a series of high-fidelity simulated approach-and-landing scenarios where 18 pilots completed 8 curved approaches also involving unexpected runway obstacles to be detected upon landing (Lorenz, 2005). The implementation of scene-linking was done in a similar way as Fadden et al. (2000) and a significantly delayed runway incursion detection using the HMD was observed compared to using standard ILS guidance located head-down. Pilots flying during low visibility conditions with head-down instruments detected the runway incursion approximately 2 seconds earlier than the pilots using the head-mounted display. Nevertheless, the expected benefit in regard of flight path tracking was high. Using the HMD the horizontal cross track accuracy increased up-to 10 times, the vertical cross track accuracy was 3 times better and speed-tracking also accomplished 30% better accuracy compared to head-down guidance.

However, the delayed incursion detection was not as severe as to reject the idea of using scene-linking in airborne pathway guidance. During the high-fidelity experiments it was not directly compared if the scene-linked symbology dominates compared with symbology having a fixed location on the HUD screen. In order to do so a low-fidelity PC-based simulation task was developed (Schmerwitz, 2006). The participating pilots completed a series of simulated low-altitude flights through mountainous terrain supported by pathway guidance and were instructed to detect friendly from hostile surface-to-air missile (SAM) stations hidden in the outside terrain while reacting to information changes displayed either at a fixed or scene-linked location. The hypothesis that scene-linked HUD imagery facilitates the division of attention between near-domain (primary flight information - PFI) and far-domain (outside scene) monitoring could only been partially proven. When information was displayed scene-linked it led to faster detection and discrimination of targets hidden in the outside terrain. No performance trade-off between display and scene event detection could be found. However, scene-linking clearly deteriorated pathway following performance. Moreover, it was found that with the more difficult flight path trajectory there was also a detrimental effect of scene-linking on performance in the display detection and adjustment task. Besides the difficulty to explain the performance loss in flight path tracking, the observed benefit of scene-linking in the division of attention between the two event detection tasks is regarded as supporting evidence for this display concept.

Simultaneous to the laboratory experiment a first real flight test focusing mainly at the usability of a HMD onboard a Do228 began. The tests consisted of two different sets. One was a local pattern flight scenario using the HMD or terrestrial navigation, the other was an area instrumental flight using the HMD vs. VOR/NDB/ILS (conventional navigation aids) navigation (Schmerwitz, 2006). With the findings of the first flight test the display was modified and tested in a second flight test further exploring the feasibility of the monocular HMD. The first flight test revealed difficulties of the pilots especially during segment transitions. Therefore, implementation changes of the pathway symbology were combined with a predictor-director guidance. The enhanced display used less of the scene-linking concept, but intended to improve segment transitions, track re-intercepting and readability of primary flight information. The scenario was a closed virtual departure, cruise and approach task. It was designed to generate a high workload with several different climbs, descents, curved segments and setting changes. The navigational aids were either RNAV-GPS (satellite based area navigation) or the pathway HMD. After a brief description of the two concepts details of the findings during the flight tests will be presented in this paper.

Pathway-predictor guidance concept

Pathway

The pathway-in-the-sky symbol consists of two different representations shown in figure 1 and 2. For better printability all images in this paper are inverted. Therefore, white regions in the figures are transparent and the grey levels represent red levels on the NOMAD HMD display. For the area in the vicinity of the aircraft, the pathway consists of horizontal bars alike crossties that are bend up at each end providing a good horizontal guidance but less vertical error margins. The aim of the design was to achieve a good balance between precision requirement and demand on visual attention resources rather than to provide the highest possible precision cue. The crossties have a width of 267 feet (ft) and an average distance of 1500 ft. The width derives from the primary flight information along the pathway. It is 200 ft tall what leads to a width of 267 ft (81 meter) when using the displays aspect ratio of 4:3. At distances larger than 2.5 nautical miles (nm) the trajectory is represented by tunnel symbols. When these pathway symbols are viewed from the side they appear similar to an arrow pointing into the flight direction. This is done to provide early cues for pilot about upcoming curves. At distances larger than 5 nm the pathway symbols are darkened to reduce clutter. The trajectories represented by the pathway were generated using the DLR 4D flight management system (Czerlitzki, 1994).

Predictor

A basic element of the display is the predictor. A stylized aircraft symbol is presenting the expected position of the aircraft within 10 seconds into the future in case of HMD use and 6 seconds in case of HUD use. The calculation of this prediction is based on a simple formula for horizontal curves additionally the present climb angle and the wind vector were included. The predicted flight trajectory is represented by a thin line between the aircraft nose and the predictor. This line allows easier detection of the predictor symbol and visualizes the current aircraft's velocity vector and turn rate. In case of flying a curve the appearance of the predictor smoothly changes into a 3D symbol.

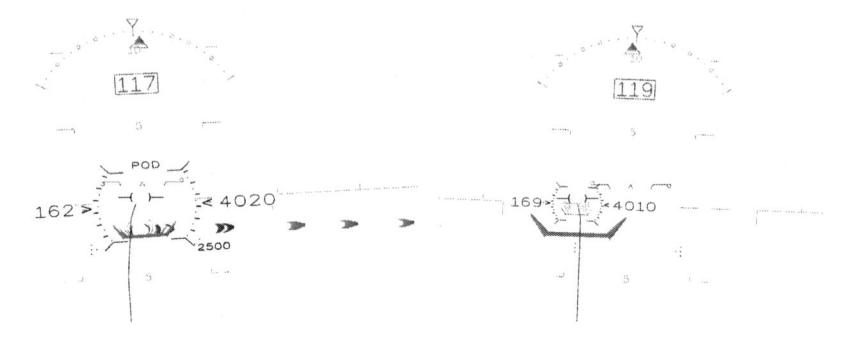

Figure 1 (left). Pathway-predictor HMD symbology before descent and curve,
Figure 2 (right). Pathway-predictor HMD symbology at straight horizontal flight

Primary Flight Information

In an attempt to apply the technique of scene linking developed by Foyle et al. (1995) the presentation of primary flight information was modified. Attaching numerical values of the aircraft's speed and altitude to a pathway winglet made this information appear as virtual elements of the outside environment. Figure 1 and 2 show the pathway with attached airspeed and barometric height values. Thin bars outside the tunnel provide information about acceleration and climb rate. As soon as the aircraft passes through a pathway symbol with the attached primary flight information, this information is presented anew in a larger distance, typically at a position three pathway symbols ahead, dependent on the distance to the aircraft. The new pathway symbol is faded in softly and over a time of 1 second when two PFI-pathway linked symbols are visible. The pilots needed some training to get used to this dynamically changing display, but after a short training they were able to synchronize their scanning pattern with the display.

First flight test

Method

Design and participants
The experiment consisted of two different sets. The first was a local pattern flight using either visual terrestrial navigation or HMD guidance. The second was an area instrumental flight with either HMD or VOR/NDB/ILS guidance. All trials contained a departure, cruise, arrival and low approach segment. Every single task was repeated resulting in a total of 2 (instrument and visual navigation) x 2 (with and without HMD) x 2 (first and second attempt) = 8 trials. Five male test pilots from DLR's flight operations department in Braunschweig participated in the experiment. Their age averaged to 40 years and they had an average flight time of 4000 hours. The test took place in the Braunschweig (EDVE) area which all pilots were well familiar and trained regarding the area's navigational procedures.

Apparatus and symbology
The HMD used was a red monochrome monocular retinal scanning HMD. In combination with an optical headtracker the right seat of a Dornier 228 was modified for HMD testing. The onboard experimental system delivers a different data set than the fixed base simulator, so that the original display had to be modified in order to work with the available data. The navigational aid was generated from DGPS and INS. Unfortunately neither torque nor indicated airspeed could be presented. As supplement ground speed was presented. Minimum but adequate configuration change information together with position and transition information were presented in discrete (alpha-) numeric callouts. The trajectories of the scenarios were generated in advance with a laboratory flight management system (FMS). Due to a missing adequate FMS onboard the Do 228 only a fixed pathway could be presented not being able to adapt to the actual wind situation during trials. Because the laboratory flight management system generates autopilot flyable courses it was obviously flyable manually in most wind situations. The pathway therefore ensured in all trials with HMD to lead along exactly the same track. However, altitudes flown after a true track (HMD) are harder to compare with altitudes from a barometric track (without HMD).

Flight test scenario and procedure
In advance, all pilots were trained using the HMD and got briefed on the experimental task. At the beginning there was a presentation of HMD symbology followed by a relatively short session of about 20 minutes length in the fixed base simulator to get used to the HMD technology, mounting the display and the display symbology. The eight trials were flown in one airborne session leading to approximately 2 hours block time. The vertical path was comparatively easy. After the initial climb phase to 4000 ft instrumental or 1500 ft visual, respectively, a more or less short cruise phase led to the initial descent. In case of the instrumental task the descent transitioned into an ILS approach procedure. After the experimental session pilots were given a debriefing questionnaire. Besides NASA-TLX (Hart, 1988) and SART (Taylor, 1990) they were asked to rate and comment on aspects

such as handling, usability and workload, which was supplemented by an interactive debriefing session with the experimenters.

Results

Flight path following
In visual trials, the precision analyses did not show as significant differences between flight path following with and without HMD. Pilots reported about difficulties following the HMD path due to the need of much more frequent configuration changes during local pattern flights. They had to use head-down instruments in addition to gather all necessary information. At the present state of the HMD system not all of the necessary information could be presented head-up. This drawback almost negates the major HMD advantage of reducing head-down scanning. Moreover, in some occasions the downwind had to be stretched in order to follow air traffic control (ATC) advices adding problems to the precision analyses. Still the flight data from HMD tracks appear to be more precise. However, further evaluation of the data was discarded because of the drawback mentioned above. For the instrumental trials the evaluation showed the expected significant differences in favor of HMD navigation. Figure 3 shows an overview of all instrumental session trials. During the approach segment before or at the initial approach fix ATC gave several traffic advises leading abeam track or into holding.

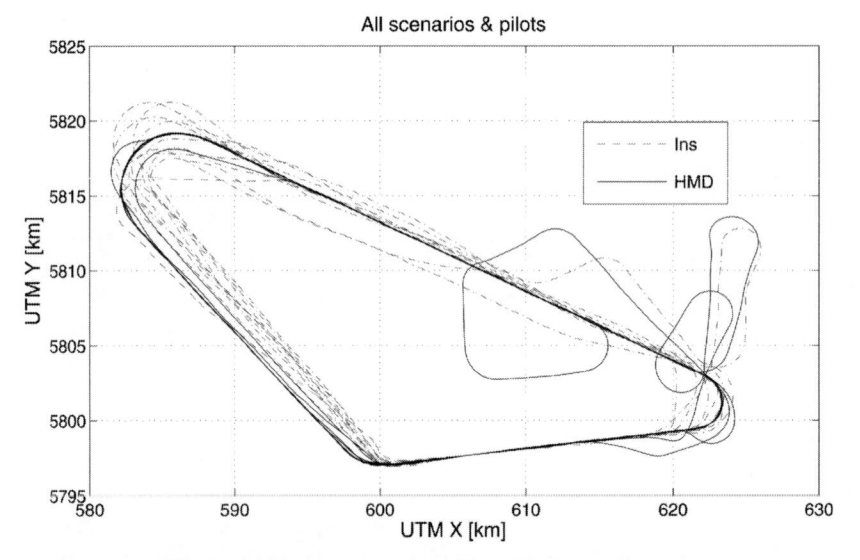

Figure 3. First flight test, flight track overview of all instrumental flight session

Compared with prior results from fixed base simulation (Korn, 2005), the advantage of HMD navigation in real flights was smaller. One reason lessening the dominance of the pathway HMD may be that in the simulation the display had more functionality. In fixed base simulation, indicated airspeed and power setting could be presented leading most likely to superior usability and, in particular, more precise flight path following. In addition to the precision analyses of straight segments the handling qualities during the trial were investigated. The pilot's behavior in altitude tracking as well as their timing pattern of turning into and out of curves was investigated. By intention, the pathway-in-the-sky lacks a definitive altitude constraint above the pathway. Therefore, pilots could choose different altitudes above the pathway. The performance of altitude tracking was in most cases adequate. With this scene-linked approach it was not fully surprising that pilots had difficulties to exactly time the turning into the curves as demanded. The same happened with climbs and descents. However, most pilots more or less steered adequately along the pathway, some missed the beginning of curves, climbs or descents and left the pathway boundaries (see figure 3 – Universal Transverse Mercator (UTM) is a Cartesian coordinate system). It became obvious during trials that the symbology did not provide proper guidance cues at certain segment transitions. Even when the pilots stayed on the pathway their commands into curves for example were delayed or too early and with an aggressive gain at times. The steering commands observed at the end of turns likewise occurred either too early or too late. In some rare cases the pilots did not turn with a constant but constantly changing bank angle. One possible reason might be a lack of training with the display. Carry-over effects caused considerable problems for the scoring of flight path tracking performance. In the case of a lost track situation during curves it was obvious that pilots began the next segment with a great offset before getting back on track.

Table 1. HMD usability questionnaire

	agree	slightly agree	neutral	slightly disagree	disagree
A. Total Pilot Skill					
1. aircraft control is easy	1	1	1	1	1
2. following pathway is easy		3			2
3. planning and decision making is easy	2				3
4. attitude control is easy	1	2		2	
5. communication is easy	3			1	1
6. traffic awareness is easy		1	1	2	1
7. Scanning (attention switch) is not difficult			1	1	3
B. Image quality of HMD					
1. color of symbols are good	3		2		
2. display symbol is clear	1	1		1	2
3. image is smooth for head moving			1		4
4. the sense of distance is good		1		1	3
C. HMD Symbols					
1. predictor gives good guidance information	1	3	1		
2. pathway gives good guidance information	1	1		2	1
3. PFI HMD helps control	1	2	1		1
4. situation awareness is improved by HMD	1	1		1	2
D. HMD installation feeling					
1. I feel comfortable with HMD					5
2. I don't feel fatigue in my eyes			3		2
E. Total evaluation					
1. I don't have stress with HMD	2			2	1
2. Control is easier with HMD than with ILS		1	2		2
3. Control is easier with HMD than visually	1		1		3
4. I want to use this HMD	1	1			3
5. I don't think the HMD should be improved					5

Questionnaires and comments in the debriefing session
Table 1 shows most items used in the debriefing questionnaire to collect pilot ratings on their subjective experience made with the HMD during the experimental flights. The table also contains the frequency counts across the five rating categories obtained by the five participants. There was a rather consistent feeling of discomfort with wearing the HMD (table 1, D1) and a concern regarding the smoothness of the HMD image during head movements (B3). Thus, pilots found latency in imagery in case of head movement to be a major drawback. Overall, these concerns resulted in the consistent statement that the HMD needs improvements (E5). The answers in the other groups of items are mostly scattered across the entire range of the scale that does not allow to draw further conclusions, e.g. in group C addressing the symbol quality. Items for which no ratings in the mid range but only at both extremes were obtained pose particular difficulties in this regard. For example, three pilots slightly agreed that following the pathway was easy whereas two pilots completely disagreed (A2). This was even more so with the item A3 addressing planning and decision making. Comments regarding this item in the debriefing discussion revealed that the unfamiliarity with the HMD and the effort needed to follow the pathway interfered with long-term planning and decision making tasks. Pilots also commented that they had difficulties with overlapping symbology and with the sensitive predictor at times. The most critical comments made in the final debriefing session concern the hardware layout. The wearing comfort of the head-mounted device was felt improper for flight tests. In addition, the headset does not fit well over the head mount of the HMD allowing too much external noise that pilots compensated by increasing the COM volume. All pilots described difficulties in simultaneously perceiving input from the HMD and the outside scene pointing to binocular rivalry problems of the monocular HMD. Perspective interpretation was further exacerbated by the small field of view. Moreover, the construction of the combiner reduced peripheral vision due to its hard framing. The luminance and contrast in bright ambient light condition doesn't secure readability, or even worse making the HMD-eye drain with tears making the display unreadable for up to a minute. Changing ambient light condition forced the pilot adjust the luminance manually because of a missing ambient light sensor.

Pathway-predictor-director guidance

The presented results of the first flight test led to a redesign of the pathway-HMD. In a second flight test the focus was set to improve segment transitions as well as introducing a trajectory re-intercept ability. During the first flight test ATC had the need to redirect the aircraft several times. In those cases while flying with the original HMD the pilot had to fallback to standard navigation. At the present state of integration the test bed doesn't have a flight management system and the display is driven by pre-planned trajectories. For off-track scenarios the need of a director leading back to the track was obvious. At the same time a director could be a solution for better guidance during segment transitions. Integrating a predictor-director guidance made it impossible to use world referenced gates with PFI without generating further clutter problems. The scene linking concept was reduced to display the PFI at the director plain. The director symbol provides horizontal and vertical guidance. Not being able to display the original scene-linked gate at the pathway plates and the pilot's rating of lacking virtual depth perception the pathway-in-the-sky was changed to tunnel-in-the-sky symbology. The first concept, by intention, delivered only little cues for altitude tracking. By introducing a director the possible accuracy to follow the trajectory should be higher perhaps generating a higher workload. Figure 4 and 5 show the improved display for a curved and a straight segment (see figure 1 and 2 for comparison).

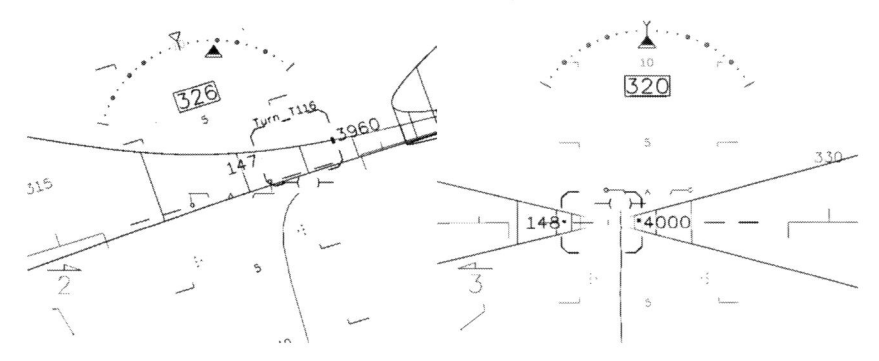

Figure 4 (left). Pathway-predictor-director HMD during curved flight. Figure 5 (right). Pathway-predictor-director HMD at straight horizontal flight

During an on-track situation the director appears as a gate centered in the middle of the reference track with the same normal to its plain. To steer along the track the predictor needs to be inbound of the director that is displayed at the same distance. If an off-track situation occurs where maximum bank or pitch would be needed the director leaves the reference track giving guidance back to the track with the maximum allowed bank or pitch. Because pilots reported having trouble finding the predictor again after an off-boresight scanning (off the center reference line), e.g. head down scanning or looking out for traffic, further symbols were integrated. To ease the pilot's effort to realign his view with the director or the predictor two symbols at the edge of the display pointed the pilot the direction where to turn his head, one to the director and one to the predictor. The dimension of the tunnel was chosen alike the 95% Required Navigation Performance Area Navigation (RNAV-RNP) proposal (Korn, 2005). The dimension of the director was chosen similar with 100% of tunnel height and 33% tunnel width. During approach the tunnel dimension is smoothly reduced down to a selectable precision. For this experiment CAT II (weather dependent ILS precision approach decision height category) was picked as the final approach type which is set at a distance of 2.75 nm (5 km) to the runway. The CAT I dimension is reached earlier at a distance of 5.5 nm (10 km). Therefore, the tunnel dimension is smoothly reducing while getting closer to the runway until the chosen category is reached.

Second flight test

Method

The second flight test was planned as a high workload task in order to evaluate the symbology of the predictor-director guidance at segment transitions. It was planned to be flyable after GPS navigation system onboard the Do-228 and with the HMD. All segments are shorter than 30 seconds. The bank angles of turns rank from 10° up to 35° bank. Climb rates were planned from 900 feet per minute (fpm) up to 1500 fpm, descents from 750 fpm up to 1500 fpm and several speed and configuration changes occurred during the task. To get used to the improved display and the complex scenario 4 participants were trained to fly this task via HMD in a simulator session. During the flight test two rounds of the task were flown with HMD and two rounds with GPS. Unfortunately a malfunction of the head tracker occurred during 2 HMD trials reducing the collected data. Therefore 8 GPS trials and 6 HMD trials could be evaluated. The trajectory was planned for zero winds, but the wind during the experiments changed heavily. The participants needed a high effort to adapt to the wind situation, because the pre-planned bank angles for some turns needed to be adapted during the flight to keep on track. During and after the flight trials questionnaires were taken. After the trials of each navigational method a NASA-TLX was gathered and on ground an interactive debriefing questionnaire was collected.

Results

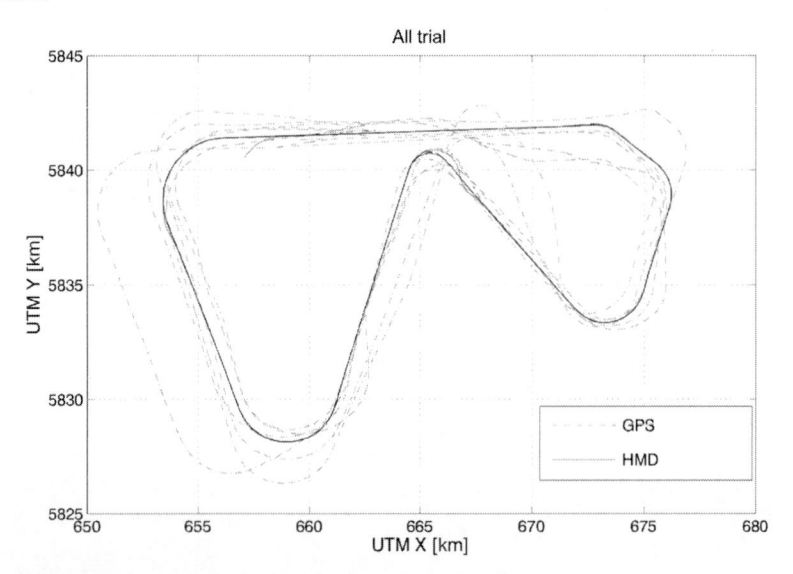

Figure 6. Second flight test, flight track overview of all trials

To be pointed out at first the small number of participants did not preserve enough data to bring up significant findings. The questionnaire could not deliver fundamental results for the same reason. In comparison with prior results they can be assumed to be equal. The overall hardware related problems during the first flight test were expected to be reproduced and did show alike the first flight test. Even so the task was very hard to follow there was only one noticeable off-track situation while using the HMD. It can be stated that using the HMD the course could be followed well to the expense of a high workload. In case of GPS navigation it wasn't as easy to follow the reference track. But for both methods the pilot flying had to relay on the safety pilot for certain assistance. For HMD flight the pilot none flying had to set throttle and keep the reference speed. For GPS navigation the pilot none flying watched and controlled the GPS system and commanded the pilot flying when to commence each segment transition by counting down from three. The weather was a hazard as well. Flying partially in clouds the pilot using the HMD had to adapt to constantly changing ambient lighting. The track was partially underneath overcast (low light, moderate contrast), partially in clouds (moderate light, no contrast) and partially above cloud (bright light, high contrast). To be able to read the display at all times the participants needed to cover the head-mount with dark foil and reduce the transparency alike wearing sunglasses but with the HMD underneath. Figure 6 illustrates the expected guidance superiority of a pathway HMD. Table 2 shows the performance of each participant and the mean RMSE.

Table 2. Cross track RMSE

	GPS	HMD	GPS/HMD
Pilot A	566 m	59 m	9.6
Pilot B	973 m	47 m	20.7
Pilot C	1037 m	25 m	41.5
Mean	859 m	44 m	19.5

The overall pathway following performance is 20 times better than with GPS navigation. During the first real flight test pilots didn't have problems timing the segment transitions using standard navigation aids but while using the HMD they did. This experiment showed the exact opposite. The task was much more complex and very hard to follow with GPS navigation. Sometimes the beginning of a turn was missed but never while using the HMD (see figure 6). Several times the reason was the GPS system. The needed navigational performance to follow the trajectory was slightly above the performance limit of the GPS system. Contrary to this it can be stated that the pathway HMD did allow track following at all times. In comparison with the prior results where the track was lost several times using the HMD the tested predictor-director concept performed as intended. Accordingly the NASA-TLX showed a higher subjective workload during GPS trials for two of three participants. The complete rating is shown in table 3.

Using the pathway-predictor HMD pilot's inputs at transitions were more frequent and oscillating in both angles roll and pitch with a sometimes high gain. The second flight test didn't show as much over corrective maneuvers. Oscillating corrections appeared more seldom and at most times similar to those measured during GPS trials. Corrections still occurred more frequent compared with the GPS trials. Pilot's inputs showed to be different among the participants while using the HMD. On the one hand pilot A showed oscillating behavior in both axis pitch and roll and a much smother control during GPS trials. Pilot B on the other hand does show little oscillating inputs and tends to control equally during the GPS trials. The reason might be insufficient training of pilot A with the HMD or might be an installation problem therefore a readability issue. Further experiments should carefully investigate this topic. The introduced predictor-director guidance also needs to be investigated for a remaining benefit related to the scene-linking concept.

Table 3. NASA-TLX Score

	(G)PS/ (H)MD	Very low	Low	Slightly low	Slightly high	High	Very high
MD	G			1		1	1
	H		1		1	1	
PD	G		1		2		
	H			1			2
TP	G		1	1		1	
	H		3				
OP	G		2				1
	H		2	1			
EF	G	1		1	1		
	H			1	2		
FR	G			1	2		
	H	1		1			1

Conclusion

The presented set of experiments intended to find a viable solution for a pathway HMD to be used in civil fixed wing aviation. The experiments covered the full span of complexity: fixed-base simulation, laboratory experiments and real flight testing. It was intended to proof the reliability of a experimental setup using a low cost HMD for flight tests. The overall goal was to reduce attention fixation due to 3d-compellingness of pathway head-down displays and as well mitigate the perceptual segregation of the displays near domain and the out of the window far domain with the use of head-up devices. The benefit of scene-linking was investigated for airborne tasks. Where laboratory experiments found evidence in favor for scene-linked pathway head-up or head-mounted devices the real flight tests could not fully support this display concept. Several system related difficulties were found that led to a latency which then generated frequent misalignments of conformal symbology with their referents. This problem poses a particular challenge for the implementation of the scene-linking concept. At the same time it was difficult to ensure readability of the displays content due to changing ambient lighting. An improvement of the flight test equipment might ensure a more reliable system for augmented reality flight tests, but it seems that today's low cost HMDs are still insufficient for airborne testing. In spite of the allocated adversities the performance benefits regarding pathway following were once more proven. How far this performance benefit relays on the display concept or might only be a trade-off with a higher workload could not be revealed fully. Further investigations need to further prove this approach. The reported disturbance of the long term planning abilities and difficulties of traffic awareness as well as self separation reduce the usability of the tested display concept as expected. Even so, in all studies evidence has been found that the head-up pathway concept could be superior to present head-up solutions. To fully unfold the potential of this concept it is needed to integrate a planning display that is coupled to a FMS providing long term planning abilities. Today's HUD's combined with enhanced vision systems provide aircraft credits to approach and land manually at adverse weather conditions. Performance and therefore the safety margins, e.g. aircraft separation during bad weather, remain unchanged. If a pathway head-up or head-mounted display, like the one presented, could prove superior performance under those conditions it might be possible to even further improve an airports capacity by using EVS together with pathway head-up guidance: A challenging topic for further research.

Acknowledgement

The paper presented results that were gathered by the help of former co-workers Dr. Bernd Lorenz, who was a DLR employee and now works for EUROCONTROL in Budapest, Hungary and Dr. Tomoko Iijima, who was formerly working for DLR in a scientific exchange program and is now back at JAXA, Japan.

References

Czerlitzki B. (1994). The experimental flight management system: Advanced functionality to comply with ATC constraints. *Air Traffic Control Quarterly, 2*, 159-188.

Fadden, S., Ververs, P.M., & Wickens, C.D. (2001). Pathway HUDs: Are they viable? *Human Factors, 43*, 173-193.

Fadden, S., Wickens, C.D., & Ververs, P. (2000) Costs and benefits of head up displays: An attention perspective and a meta analysis. Paper presented at 2000 World Aviation Congress. Warrendale, PA: Society of Automotive Engineers.

Fischer, E., Haines, R.F., & Price, T.A. (1980). *Cognitive issues in head-up display* (NASA Paper 1711). Moffett Field, CA: NASA Ames Research Center.

Flemisch, F.O. & Onken, R. (2000). Detecting usability problems with eye tracking in airborne battle management support. Paper presented at the RTO HFM Symposium on "*Usability of information in battle management Operation*", Oslo, Norway, April 2000. Published in RTO MP-57, 23-(1-14).

Foyle, D.C., McCann, R.S., & Sheldon, S.G. (1995). Attentional issues with superimposed symbology: Formats for scene-linked displays. In R.S. Jensen and L.A. Rakovan (Eds.), *Proceedings of the Eights International Symposium on Aviation Psychology* (pp. 98-103). Columbus, OH: Ohio State University.

Grünwald, A.J. (1996). Improved tunnel display for curved trajectory following: experimental evaluation. *Journal of Guidance, Control, and Dynamics, 19*, 378-384.

Hart, S.G., & Staveland, L.E. (1988): Development of NASA-TLX (Task Load Index): Results of empirical and theoretical research. In P.A. Hancock and N. Meshkati. (Eds). *Human mental workload* (pp. 139-183). Amsterdam: North Holland.

Haskell, I.D. & Wickens, C.D. (1993). Two- and three-dimensional displays for aviation: A theoretical and empirical comparison. *The International Journal of Aviation Psychology, 3*, 87-109.

Korn, B. (2005). OPTIMAL-Deliverable D1.1 "State-of-the-art", European FP6 project OPTIMAL, www.optimal.isdefe.es

Lorenz, B., Többen, H., & Schmerwitz, S. (2005). Human performance evaluation of a pathway HMD. *Proceedings of the SPIE Defense & Security Symposium.* Bellingham, WA: SPIE.

McCann, R.S., Foyle, D.C., & Johnston, J.C. (1993). Attention limitations with head-up displays. In R.S. Jensen (Ed.), *Proceedings of the Seventh International Symposium on Aviation Psychology* (pp. 70-75). Colombus, OH: Ohio State University.

Schmerwitz, S., Lorenz, B., & Többen, H. (2006). Investigating the benefits of 'scene-linking' for a pathway HMD: From laboratory flight experiments to flight tests. *Proceedings of the SPIE Defense & Security Symposium. Bellingham*, WA: SPIE.

Sheldon, S.G., Foyle, D.C. & McCann, R.S. (1997). Effects of scene-linked symobology on flight performance. *Proceedings of the 41st Annual Meeting of the Human Factors and Ergonomic Society* (pp. 294-298). Santa Monica, CA: HFES.

Taylor, R.M. (1990). Situational Awareness Rating Technique (SART): The development of a toll for aircrew systems design. In *AGARD Conference Proceedings* No 478, Situational Awareness in Aerospace Operations. Aerospace Medical Panel Symposium, Copenhagen, October, 2-6, 1989.

Többen, H., Lorenz, B., & Schmerwitz, S. (2005). Design of a pathway display for a retinal scanning HMD. *Proceedings of the SPIE Defense & Security Symposium.* Bellingham, WA: SPIE.

Thomas, L.C. & Wickens, C.D. (2004). Eye-tracking and individual differences in off-normal event detection when flying with a synthetic vision system display. *Proceedings of the Human Factors and Ergonomics Society 48th Annual Meeting.* Santa Monica: Human Factors and Ergonomics Society, 2004.

Wickens, C.D., Alexander, A.L., Thomas, L.C., Horrey, W.J., Nunes, A., Hardey, T.J. & Zheng, X.S. (2004). Traffic and flight guidance depiction on a synthetic vision system display: the effects of clutter on performance and visual attention allocation. Technical Report, AHFD-04-10/NASA(HPM)-04-1. Urbana, IL: University of Illinois, Institute of Aviation, Aviation Human Factors Devision.

Wickens, C.D. (2005). Attentional tunnelling and task management. *Proceedings of the 13th International Symposium on Aviation Psychology* (pp. 620-625), Oklahoma City, OK, April 18-21.

Wickens, C.D. & Long, J. (1995). Object- vs. space-based models of visual attention. Implications for the design of head-up displays. *Journal of Experimental Psychology: Applied, 1*, 179-193.

Modelling and Simulation

Everyday mistakes: confidence or cognition?

In D. de Waard, F.O. Flemisch, B. Lorenz, H. Oberheid, and K.A. Brookhuis (Eds.) (2008), *Human Factors for assistance and automation* (p. 387). Maastricht, the Netherlands: Shaker Publishing.

Everyday mistakes: confidence or cognition?

Robert R.A. van Doorn & Fred R.H. Zijlstra
Maastricht University
The Netherlands

Abstract

This study focuses on why some individuals report having a higher propensity for making everyday mistakes than others. These lapses in otherwise skilled behaviour have been related to less efficient basic attention mechanisms in laboratory tasks. Three hypotheses were tested by comparing objective task performance level with self rating of performance and invested effort. First, the identified attention deficit may be too small in grain size to have an impact on complex task performance. Second, the attention deficit may be successfully compensated at the expense of more invested effort. Third, making more everyday mistakes may be related to a lower confidence and self image. This should become visible as a consistently lower self assessment of performance level. To test these hypotheses, participants were divided in two groups on the basis of their score on the cognitive failures questionnaire (CFQ), a frequently applied instrument to measure the tendency of making everyday mistakes. The results show that individuals with high CFQ scores systematically rate their performance level on two tasks as lower, while they do not differ regarding the actual performance level and the invested effort. These outcomes show that the effect of a stronger tendency toward making mistakes on behaviour in everyday task situations is related to confidence and self image.

Introduction

Making mistakes is an individual's everyday experience. Everyone knows daily examples of mistakes such as forgetting appointments, inadvertently dropping things, or overlooking traffic signs. When people are questioned regarding these everyday failures, they typically differ as to the frequency of these types of everyday mistakes. It is unresolved how this self perception of making more everyday mistakes relates to an individual's behaviour at home and at work. For example, individuals who report relatively more everyday errors may also show generally lower performance levels. In addition, these individuals may rate their performance as lower and possibly experience that they have to invest more effort to successfully complete the task at hand. The present study appraises how the self report of making more everyday mistakes relates to the way tasks are performed and experienced.

Everyday mistakes are frequently viewed as lapses in cognitive mechanisms including perception, memory and handling. An often used instrument to determine the perceived frequency of everyday mistakes is the cognitive failures questionnaire

In D. de Waard, F.O. Flemisch, B. Lorenz, H. Oberheid, and K.A. Brookhuis (Eds.) (2008), *Human Factors for assistance and automation* (pp. 389 - 400). Maastricht, the Netherlands: Shaker Publishing.

(CFQ) (Broadbent, Cooper, FitzGerald, & Parkes, 1982). Its usefulness has been shown in a number of settings. CFQ has been applied in the clinical field and was able to predict personal indices such as neuroticism and fantasy proneness (Merckelbach, Muris, Nijman, & de Jong, 1996; Merckelbach, Muris, & Rassin, 1999). In a work setting, the inventory has also been related to vulnerability to stress and burnout (Matthews, Coyle, & Craig, 1990; van der Linden, Keijsers, Eling, & van Schaijk, 2005). In addition, large scale studies indicate that individuals with high CFQ scores are more frequently involved in actual mishaps in work environments, such as in the military, industry, and in traffic (Larson & Merritt, 1991; Larson, Alderton, Neideffer, & Underhill, 1997; Wallace, Kass, & Stanny, 2001; Wallace & Vodanovich, 2003a, 2003b; Wallace, Vodanovich, & Restino, 2003). These outcomes are often used to support the general notion that high CFQ individuals may experience more workload (Merckelbach et al., 1996), or may be less able to cope with stress (van der Linden et al., 2005), possibly instigated by failures of concentration (Matthews et al., 1990).

Based on these ideas it is conceivable that everyday mistakes are related to possible deficits of attention mechanisms. Indeed, basic research on selective attention revealed that individuals scoring high on the CFQ are more easily distracted (Broadbent et al., 1982; Broadbent, Broadbent, & Jones, 1986; Tipper & Baylis, 1987; Wallace et al., 2001). More specifically, high CFQ individuals often take longer to identify targets embedded in a distractive visual environment, while the number of selection errors remains unaltered. Individuals who are more vulnerable to making errors seem to be less able to inhibit intrusions of automatic cognitive processes (Merckelbach et al., 1999; Tipper & Baylis, 1987; van der Linden et al., 2005).

The present study investigated whether an individual's inclination to everyday mistakes is related to actual and perceived everyday task performance. One issue is whether the fine grained cognitive hampering (Broadbent et al., 1982; Broadbent et al., 1986; Tipper & Baylis, 1987) that appears to distinguish high from low CFQ individuals will also appear in everyday task performance. At first sight, it seems fair to expect that the difference between CFQ scores is not related to the level of everyday performance. This makes sense as CFQ is believed to appraise the frequency of events that depart from the normal smooth flow of function and ability (Broadbent et al., 1982). This may imply that the identified cognitive component does not affect the level of complex everyday task performance (Broadbent et al., 1982; Broadbent et al., 1986). It is possible that the grain size of the identified attention component may be too small to have an impact on performance of more complex tasks that involve many successive decisions and manual responses to a dynamically changing stimulus environment. One may also assume that additional cognitive mechanisms are required to perform such a task. These mechanisms may flood the relatively small difference in selection time between low and high CFQ individuals. On the basis of these arguments the first hypothesis is: The tendency to make more everyday mistakes is not related to the task performance level.

The reported tendency to make more everyday mistakes may also be related to how workers perceive their performance. Indeed, the finding that high CFQ individuals are more susceptible to stress and burnout (van der Linden et al., 2005) may indicate that these individuals experience more effort during task performance (Merckelbach et al., 1999; Muris & Merckelbach, 1995). One reason why individuals that score high on the CFQ also perceive more effort may be related to the earlier described slower attention processes (Matthews & Wells, 1988; Tipper & Baylis, 1987). Even though the hampering of such a process may not affect the level of task performance (hypothesis 1), it may still be perceived as effortful. Another reason why more effort may be experienced is that the attention component has a potential influence on task performance but that individuals with a high CFQ score successfully compensate the slower selective attention processes (Merckelbach et al., 1999; Tipper & Baylis, 1987; van der Linden et al., 2005). Such a compensation may then prevent that the actual performance level drops at the expense of increased effort investment (Vidulich & Wickens, 1986; Zijlstra, 1993). The second hypothesis therefore is: The reported tendency to make everyday mistakes is related to the perception of higher mental effort.

Finally, individuals that score high on CFQ may also perceive their own performance level differently from individuals with low CFQ scores. Aware of their tendency to make more errors, these individuals may perceive the level of their performance as consistently lower. Such an outcome may imply that CFQ scores are also indicative of a person's confidence and self-image (Broadbent et al., 1986). The third hypothesis is: The reported tendency to make more everyday mistakes is related to a lower self-image and confidence. In the present study the three hypotheses were tested by having low and high CFQ-groups of participants perform two tasks that were varied in performance difficulty.

Method

Participants

Sixty-four undergraduate students from the Maastricht University participated for course credits. Participants' average age was 23 year (SD = 4.1 year), and they all had normal or corrected-to-normal vision. Every participant gave informed consent before commencing the test session.

Material

Cognitive Failure Questionnaire (CFQ)
We employed a direct translation of the English CFQ (Broadbent et al., 1982) into a Dutch version (Muris & Merckelbach, 1995). The questionnaire consists of 25 items such as "Do you drop things?" or "Do you forget appointments?" or "Do you fail to notice signposts on the road?" A participant rates the personal frequency during the past months of such everyday errors from 'Very often' (4) to 'Never' (0), resulting in a total score from 0 to 100. CFQ has been used in a number of practical settings to measure a single dimension. Close examination of the inventory has revealed separate dimensions of cognitive failures (Larson et al., 1997; Matthews et al., 1990;

Wallace et al., 2001), but has also shown the highest internal consistency when employed as a single dimension (Broadbent et al., 1982; Houston, 1989; Merckelbach et al., 1996; Merckelbach et al., 1999; Tipper & Baylis, 1987). Internal consistencies (Cronbach α) of CFQ as a single dimension ranged from .75 to .81, and it showed a significant test-retest correlation of .8 (Broadbent et al., 1982; Merckelbach et al., 1996).

Two tasks were performed, namely a search and a tracking task. The tasks were selected because they represent everyday complex tasks in that they involve multiple decisions and responses to an altering stimulus environment. Another reason for selecting these tasks is their difference with respect to the need for mental matching and visual motor matching. For every task, a participant sat on an office chair that was adjustable in height. All information necessary to perform a task was provided on a 17 in. computer monitor positioned approximately 50 cm in front of the participant. The reader is referred to figure 1 for screen examples of both tasks.

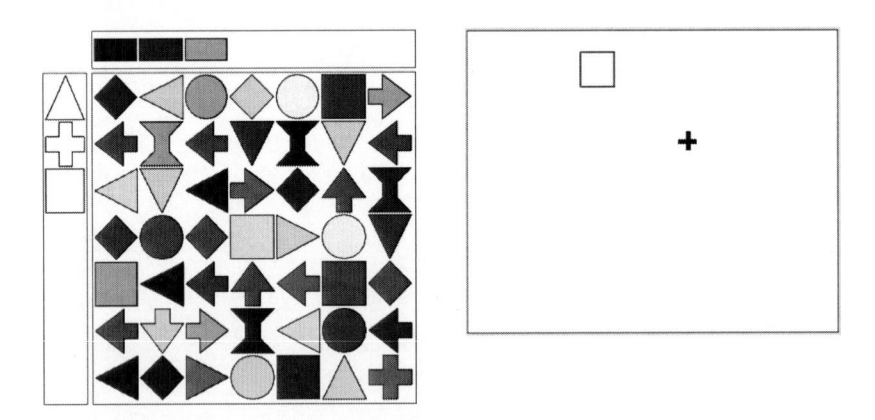

Figure 1. Representations of the search and tracking tasks

Search task
The task entailed locating targets in an environment of distractors on a computer screen where all stimuli were presented in a matrix. Targets and distractors differed as to form and color. The task involved a so-called conjunction or serial search, as distractors shared none, one or two features (color and shape) with the targets. Targets and distractors were randomly assigned from 12 possible shapes (including, circle, rectangle, diamond, hour glass, four types of arrows and four types of triangles), and 9 possible colors. The randomization process ensured that targets were not presented at adjacent positions in the matrix. On top of the matrix were presented one or more colored rectangles. On the left of the matrix one or more white shapes were presented. In order to find a target in the stimulus matrix, participants mentally combined the colors displayed above the matrix with shapes presented on the left to form a mental representation of a target. Identification of a target was confirmed by a mouse click on the target surface in the matrix. When clicked the target stimulus changed into a neutral distractor without a target feature.

A search was completed by a final click on a designated button. Participants were instructed to locate the two to four presented targets as quickly as possible.

Task difficulty was manipulated by the total number of stimuli (size of the matrix) and the number of color-shape combinations required to identify all targets. In the low difficulty condition the stimulus matrix comprised 25 stimuli (5 x 5), while a target had one specific form and shape. In the medium difficulty condition, the matrix covered 6 by 6 objects and two colors and two shapes had to be combined individually resulting in four possible target identities. In the high difficulty condition the matrix was 7 by 7 and three colors had to be individually combined with three shapes providing nine possible target identities. Since the matrix's screen surface remained unaltered, all stimuli became smaller as difficulty increased.

Tracking task
In this task, participants spatially matched the mouse pointer (cross hair) to the surface of a moving target consisting of an open rectangle (zero order tracking). The target followed a predictable path and bounced back (with equal angles of incidence and reflection) from the inner sides of the 10 by 8 cm rectangular display area in the center of the computer screen. The target started to move in the center of the display area and was colored black. The participant had 2 seconds to enter the moving rectangle before error registration commenced for a period of 20 s. During the recording period, the target was colored green when the cross hair was inside and red when it was outside the target area. Errors were sampled with a frequency of 57 Hz. The total number of errors per trial was measured. Task difficulty was manipulated by target surface size which differed about 600 pixels between difficulty conditions, 1600 pixels (high difficulty), 1024 pixels (medium difficulty), 400 pixels (high difficulty).

For every task, participants first performed 10 trials in the low difficulty condition to familiarize to the task. Ten experimental trials were subsequently carried out in each difficulty condition. These 30 trials per participant entered subsequent analyses. Task order (tracking versus search) was counterbalanced across participants. The order of the difficulty conditions was fixed for all participants. This was done to provide participants a firm basis for rating both their performance and invested mental effort. The reader will note that both tasks fulfilled the special purpose of creating distinct difficulty levels. Subsequent statistical analyses on performance data will confirm that the difficulty manipulation affected all data.

Upon completing the ten trials of a difficulty condition, participants rated their performance on a visual analog scale (VAS) ranging from 0 (minimum) to 10 (maximum), presented on screen. They then rated the invested effort via the rating scale mental effort (RSME). This scale has been developed to rate mental effort on one dimension (Zijlstra, 1993). Previous study established the reliable positions of verbal labels that indicate separate intervals on this VAS ranging from 0 (minimum) to 150 (maximum). The rating scale showed good reliability and validity (Zijlstra, 1993).

Analysis

For the search task, the time to complete a trial was measured, while for the tracking task the number of tracking errors was determined. Per task, a participant's performance measure (search time, tracking errors, performance rating and RSME) was averaged across trials per difficulty condition. Participants were split at the median CFQ score resulting in 31 low CFQ and 33 high CFQ individuals. For performance measures two-way mixed design ANOVAs were performed per task, across difficulty conditions (3) and CFQ (2). For performance rating and RSME a two-way mixed design was used: task (2), difficulty condition (3) and CFQ (2). Violation of sphericity was corrected by Greenhouse-Geisser epsilon, when it affected the significance level of a test.

Results

Tables 1a and b provide an overview of all measures for search (a) and tracking (b) as a function of difficulty condition and CFQ-group.

Table 1 (a) Search task: Average and standard deviations of the dependent measures as a function of search difficulty (low, medium high) and CFQ group (low, high). Search task performance was measured in seconds. (b) Tracking task: Average and standard deviations of the dependent measures as a function of tracking difficulty (low, medium high) and CFQ group (low, high). Tracking performance was measured as the average number of errors per trial.

(a)

Search Difficulty	CFQ	Performance Time	SD	Performance Rating Score	SD	RSME Score	SD
Low	Low	4.2	0.6	7.6	1.6	30.4	19.6
	High	4.5	0.9	7.5	1.4	33.0	21.3
	Mean	4.3	0.8	7.6	1.5	31.8	20.4
Medium	Low	14.9	3.4	5.9	1.3	54.1	21.9
	High	16.3	5.0	5.3	1.6	64.9	21.6
	Mean	15.6	4.3	5.6	1.5	59.7	22.2
High	Low	24.7	6.1	5.4	1.5	67.2	25.3
	High	26.9	7.0	4.3	1.6	75.5	24.3
	Mean	25.8	6.6	4.8	1.6	71.4	25.0

(b)

Tracking Difficulty	CFQ	Performance Errors	SD	Performance rating Score	SD	RSME Score	SD
Low	Low	343	120	6.0	1.3	67.3	28.6
	High	378	110	4.9	1.6	77.2	24.1
	Mean	361	115	5.4	1.6	72.2	26.7
Medium	Low	474	123	5.6	1.4	73.9	27.0
	High	510	118	4.1	1.4	85.5	22.7
	Mean	492	120	4.8	1.6	79.9	25.3
High	Low	771	108	4.4	1.7	84.8	28.9
	High	797	106	2.5	1.6	94.6	21.8
	Mean	784	107	3.5	1.9	89.9	25.7

everyday mistakes 395

CFQ

CFQ scores ranged from 11 to 90 with a median value of 41, yielding a Cronbach α of 0.89 across the 25 items for the used sample of 64 individuals.

Task performance

Figure 2 shows that the difficulty manipulation had strong effects on the performance of both tasks.

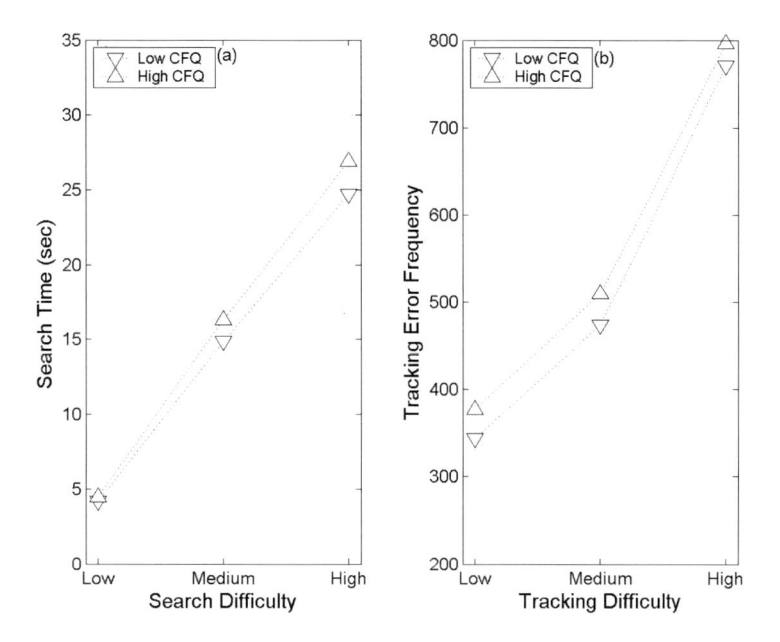

Figure 2. (a) Search time (s) as a function of search difficulty (horizontal axis) and CFQ group (markers). (b) Tracking error frequency (number of average tracking errors per trial) as a function of tracking difficulty (horizontal axis) and CFQ group (markers)

Search time
There was a strong main significant effect of task difficulty on search time, $F(2, 124) = 587.42$, $p < 0.001$. Pair wise comparisons revealed a significant 11.26 s increase from low to medium difficulty ($p < 0.001$), and the time difference between medium and high was 10.20 s. ($p < 0.001$). The 1.2 s difference between low and high CFQ groups was not significant ($p > 0.1$). There was no significant interaction.

Tracking error
There was a main significant effect of tracking difficulty on tracking error, $F(2, 124) = 279.18$, $p < 0.001$. Pair-wise comparisons revealed an increase of 131 errors from low to medium tracking difficulty ($p < 0.001$), and an error increase between medium and high of 291 ($p < 0.001$). There was no significant difference between

CFQ groups (p > 0.2). The interaction between tracking difficulty and CFQ groups was not significant (p > 0.6).

These results show that the difficulty manipulation resulted in separate levels of performance, but that CFQ score had no additional effect on performance level (hypothesis 1).

Effort rating (RSME)

RSME was measured to investigate the relation between CFQ groups and the perceived effort (hypothesis 2). CFQ groups did not significantly differ with respect to their effort rating, leading to the rejection of the second hypothesis. Comparison of tables 1 and 2 reveals that the tracking task caused participants to perceive more effort investment than the search task. There was indeed a main effect of task, $F(1, 62) = 95.04$, $p < 0.0001$. Difficulty manipulations had a main effect on RSME, $F(1, 124) = 182.82$, $p < 0.001$, confirming the validity of the effort scale. The difficulty manipulation had a stronger effect on effort rating of the search task than on the tracking task, as indicated by a significant interaction between task and task difficulty, $F(2, 124) = 43.18$, $p < 0.001$. The difference for search and tracking between low and medium difficulty was 27.9 and 7.7, and the difference between medium and high was 11.7 for the search task and 10.0 for the tracking task. This means that participants found the increase from medium to high about equally difficult in both tasks.

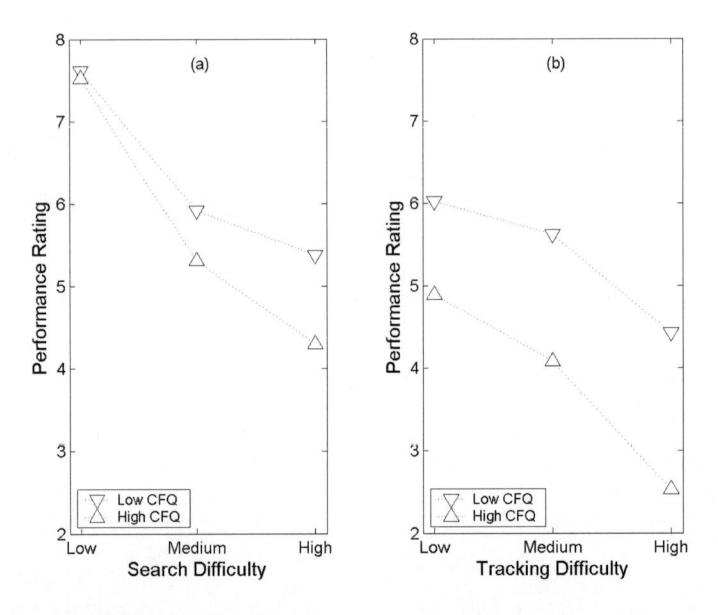

Figure 3. (a) Performance rating as a function of search difficulty (horizontal axis) and CFQ group (markers). (b) Performance rating as a function of tracking difficulty (horizontal axis) and CFQ group (markers). Performance rating was measured on a VAS ranging from 0 (low) to 10 (high)

Performance rating

Performance rating was measured to appraise the relation between CFQ group and confidence/self image (hypothesis 3). Figure 3 displays performance rating as a function of task and CFQ group. The results suggest that both task difficulty and CFQ group had effects on performance rating. The search task was rated as generally 1.4 higher than the tracking task, $F(1, 62) = 50.43$, $p < 0.0001$. CFQ groups significantly differed across tasks, $F(1, 62) = 16.20$, $p < 0.0001$. The difference in performance rating between CFQ groups was 1.1 larger for the tracking task than for the search task, as statistically confirmed by the significant interaction between task and CFQ group, $F(1, 62) = 5.45$, $p < 0.05$. The difficulty manipulation had a strong effect on performance rating across tasks, $F(2, 124) = 163.50$, $p < 0.001$. This effect was more outspoken for the tracking task than for the search task, as confirmed by the significant interaction between task and difficulty, $F(2, 124) = 25.29$, $p < 0.001$. The difference between CFQ groups increased with task difficulty in both tasks, as apparent from the significant interaction between task difficulty and CFQ group, $F(1.3, 80.9) = 5.73$, $p < 0.05$. The three-way interaction involving the factors task type, task difficulty and CFQ group was not significant.

Discussion

Individuals differ regarding their reported tendency to making everyday mistakes. The present study examined whether this difference relates to the level of actual performance or associates with the perception of task performance (Broadbent et al., 1986; Merckelbach et al., 1996; Muris & Merckelbach, 1995). A reliable instrument to verify this perceived tendency is the so-called cognitive failures questionnaire (CFQ). Participants in the present study who scored low or high on CFQ performed a search and a tracking task on three difficulty levels. Dependent variables were task performance, as well as self rating of performance and the effort invested. Three hypotheses were tested.

The results provide support for the hypothesis that the inclination to making more everyday mistakes is related to self image and confidence. On both tasks, high CFQ individuals rated their performance as systematically lower than the low CFQ individuals. Moreover, the difference in self ratings between CFQ groups grew when the tasks became more difficult. The CFQ scores may therefore represent self-image and confidence regarding personal performance. This concurs with findings from clinically oriented studies where self-image plays an important role (Houston, 1989). Indeed, high CFQ scores have been reported to correlate with personality indices such as anxiety and neuroticism, and have been found to relate to depressive symptoms (Matthews & Wells, 1988; Matthews et al., 1990; Merckelbach et al., 1996; Merckelbach et al., 1999; Muris & Merckelbach, 1995).

The present results also provide support for the notion (hypothesis 1) that the tendency to make more everyday mistakes does not relate to the level of performance. This makes sense as a higher frequency of everyday mistakes implies relatively infrequent slips that form a departure from the normal smooth flow of function and ability of skilled behaviour (Broadbent et al., 1982). The present results indicate that the attention component, identified to affect reaction time tasks in laboratory, does not relate to the performance of complex tasks (Broadbent et al., 1982; Broadbent et al., 1986). Such a task involves successive perception-decision-response chains in a dynamically changing stimulus environment. It is fair to assume that such a complex setup requires additional cognitive mechanisms than the one identified. It is possible that the natural variation of these mechanisms across all individuals, floods the relatively small difference in selection time between low and high CFQ individuals.

It has often been suggested that high CFQ individuals are more vulnerable to external influences such as increased strain (Matthews & Wells, 1988; Matthews et al., 1990; van der Linden et al., 2005), and thus perceive more mental effort during task performance (Merckelbach et al., 1996). Moreover, high CFQ individuals may successfully compensate the deficient attention mechanism to sustain their level of performance at the expense of additional mental effort investment. These notions formed the basis of the hypothesis stating that high CFQ individuals perceive proportionally more effort and even more so with increasing task difficulty. This hypothesis was not confirmed by the present data on mental effort. There were no significant differences between CFQ groups regarding RSME-ratings. Perceived mental effort monotonously increased with task difficulty, confirming the validity of the RMSE (Zijlstra, 1993).

It is feasible that the differences between CFQ groups with respect to perceived effort may have been too weak to surface in the present setup. The results indeed show that the average RSME values differ between CFQ groups, but that these differences remain below statistical significance. We may speculate that RSME scores are also affected by other personal differences. This variation may have flooded the possible distinction between CFQ groups, and created a large unexplained variation. In future research, RSME scores may be also employed to find these possible additional personal indices that are currently treated as statistical error.

The tasks used in this study were selected to elicit relatively complex processing in order to represent everyday performance. Both tasks required successive manual responses to dynamic stimuli presentations. They were also selected on their difference with respect to the need for mental matching versus visual motor matching. The search task was thought to involve more mental matching as it required the mental combination of color and shape in order to locate and identify targets. The tracking task required prolonged perceptual motor matching (van Doorn & Unema, 2004; van Doorn, Unema, & Hendriks, 2005; van Doorn & Unema, 2006) to manually track the movement of a target. Despite these differences, both tasks yielded similar results regarding the effect of difficulty on performance and self assessment measures, and the specific effect of CFQ groups on performance rating. Indeed, the factor CFQ group had a strong main effect on performance rating of both tasks and this effect increased with higher task difficulty. Since these outcomes remained robust across tasks, they give additional substance to the conclusion that CFQ groups differ regarding confidence and self-image.

Even though the used tasks led to similar main results, there were a number of apparent differences. The data showed larger effort investments for the tracking task. This means that this particular task was perceived as relatively more difficult. Also the performance self-rating was considerable lower for the tracking task. We believe that this may be related to the task's strong motor component in combination with the error measurement. This may have created a situation in which the participants perceived that the increased effort only had a minor influence on the prevention of the visibly not abating errors. It is conceivable that high CFQ individuals are more vulnerable when the task requires preventing errors, while making errors is unavoidable (Matthews & Wells, 1988; Matthews et al., 1990; van der Linden et al., 2005). Indeed, the tracking task showed a stronger difference between CFQ groups regarding performance rating than the search task. This outcome strengthens the conclusion that the perceived inclination to making more everyday mistakes is related to self image and confidence.

References

Broadbent, D.E., Cooper, P.F., FitzGerald, P., & Parkes, K.R. (1982). The Cognitive Failures Questionnaire (CFQ) and its correlates. *British Journal of Clinical Psychology, 21*, 1-16.

Broadbent, D.E., Broadbent, M.H., & Jones, J.L. (1986). Performance correlates of self-reported cognitive failure and of obsessionality. *British Journal of Clinical Psychology, 25*, 285-299.

Houston, D.M. (1989). The relationship between cognitive failure and self-focused attention. *British Journal of Clinical Psychology, 28*, 85-86.

Larson, G.E., & Merritt, C.R. (1991). Can accidents be predicted? An empirical test of the Cognitive Failures Questionnaire. *Applied Psychology: An International Review, 40*, 37-45.

Larson, G.E., Alderton, D.L., Neideffer, M., & Underhill, E. (1997). Further evidence on dimensionality and correlates of the Cognitive Failures Questionnaire. *British Journal of Psychology, 88*, 29-38.

Matthews, G., & Wells, A. (1988). Relationships between anxiety, self-consciousness, and cognitive failure. *Cognition and Emotion, 2*, 123-132.

Matthews, G., Coyle, K., & Craig, A. (1990). Multiple factors of cognitive failure and their relationships with stress vulnerability. *Journal of Psychopathology and Behavioral Assessment, 12*, 49-65.

Merckelbach, H., Muris, P., Nijman, H., & De Jong, P.J. (1996). Self-reported cognitive failures and neurotic symptomalogy. *Personality and Individual Differences, 20*, 715-724.

Merckelbach, H., Muris, P., & Rassin, E. (1999). Fantasy proneness and cognitive failures as correlates of dissociative experiences. *Personality and Individual Differences, 26*, 961-967.

Muris, P., & Merckelbach, H. (1995). De 'Cognitive Failures Questionnaire' (CFQ). *Gedragstherapie, 28*, 123-128.

Tipper, S.P., & Baylis, G.C. (1987). Individual differences in selective attention: The relation of priming and interference to cognitive failure. *Personality and Individual Differences, 8*, 667-675.

Van der Linden, D., Keijsers, G.P.J., Eling, P., & Van Schaijk, R. (2005). Work stress and attentional difficulties: An initial study on burnout and cognitive failures. *Work and Stress, 19*, 23-36.

Van Doorn, R.R.A., & Unema, P.J.A. (2004). Influence of different modes of real time visual information on single aimed movements. *Acta Psychologica, 116*, 309-326.

Van Doorn, R.R.A., Unema, P.J.A., & Hendriks, E. (2005). The locus of adaptation to altered gain in aimed movements. *Human Movement Science, 24*, 31-53.

Van Doorn, R.R.A., & Unema, P.J.A. (2006). Impact of the visual environment on the execution of aimed movements. *Motor Control, 10*, 55-68.

Vidulich, M.A., & Wickens, C.D. (1986). Causes of dissociation between subjective workload measures and performance: Caveats for the use of subjective assessments. *Applied Ergonomics, 17*, 291-296.

Wallace, J.C., Kass, S.J., & Stanny, C. (2001). Predicting performance in 'Go' situations: A new use for the Cognitive Failures Questionnaire? *North American Journal of Psychology, 3*, 481-490.

Wallace, J.C., & Vodanovich, S.J. (2003a). Workplace safety performance: Conscientiousness, cognitive failure, and their interaction. *Journal of Occupational Health Psychology, 8*, 316-327.

Wallace, J.C., & Vodanovich, S.J. (2003b). Can accidents and industrial mishaps be predicted? Further investigation into the relationship between cognitive failure and reports of accidents. *Journal of Business and Psychology, 17*, 503-514.

Wallace, J.C., Vodanovich, S.J., & Restino, B.M. (2003). Predicting cognitive failures from boredom proneness and daytime sleepiness scores: An investigation within military and undergraduate samples. *Personality and Individual Differences, 34*, 635-644.

Zijlstra, F.R.H. (1993). *Efficiency in Work Behaviour: A Design Approach for Modern Tools*. Delft University, Delft.

A model based approach to Cognitive Work Analysis and work process design in air traffic control

Christoph Möhlenbrink, Hendrik Oberheid, & Bernd Werther
German Aerospace Center (DLR)
Braunschweig, Germany

Abstract

In the DLR (Deutsches Zentrum für Luft und Raumfahrt) project RApTOr (Remote Airport Tower Operation research) new working processes for air traffic controllers will be designed. As a basis for this development a cognitive work analysis (CWA) was conducted of the presently existing work environment and decision processes at Leipzig airport. Understanding decision making in such highly complex and dynamic environments with multiple competing goals is generally a difficult endeavour. A cognitive work analysis is a possible approach to investigate the environmental and cognitive constraints of the operator. However, in the work domain of air traffic control timing plays a crucial role and cannot be neglected. In fact, the operators' task can only be understood when time-critical and causal interdependencies are analyzed. From the authors' point of view it is not the task of a CWA itself to look at time-critical and dynamic dependencies, but model based approaches have to address this issue. With executable models specific scenarios can be simulated, analyzed and discussed with respect to temporal and causal dependencies. A human-machine-model is needed to be able to analyze the reciprocal interdependencies of an air traffic controller's task and the dynamic processes on the airport.

In line with Cacciabue's approach the model consists of a human-model, a machine model (process-model) and an interaction model. The paper introduces how the results of the CWA are transferred into the executable human-machine-model FAirControl, realized with coloured petri nets. Such models allow formal verification by graphical state space analysis.

The controller model (PG-Model) and the airport process model (AP-Model) are represented as resource limited systems. More assumptions for both models are specified. The tool FAirControl contains a visualisation of the airport processes and a user-interface to be able to investigate work process design by human-in-the-loop-experiments.

The authors suggest a formal state space analysis to identify multiple task situations. These situations for the controller are determined by the timing constraints implemented in the model. An example is given, that contrasts a controller-model with limited resources (timed) to a controller-model with unlimited resources

In D. de Waard, F.O. Flemisch, B. Lorenz, H. Oberheid, and K.A. Brookhuis (Eds.) (2008), *Human Factors for assistance and automation* (pp. 401 - 414). Maastricht, the Netherlands: Shaker Publishing.

(untimed). For one simple air traffic scenario the multiple task situations are analyzed.

Although the example is simple, it underlines that conclusions for the controllers work process design should not only consider qualitative approaches, but also the timing of processes within the human mind and on the airport.

Introduction

Within the DLR project RApTOr (Remote Airport Tower Operation research) an operational concept and prototype technical systems are developed for the remote control of aircraft movements on small airports (Schmidt et al., 2006). While a technical objective of the project lies in building a high-resolution video panorama system with augmented vision support for the remote controller, the future work processes of the controller within the novel operational setting and related human factors issues are also investigated in RApTOr. In order to design adequate work processes and automation assistance to support them, the potential control strategies and working constraints of the domain have to be considered carefully.

Cognitive Work Analysis (CWA) is a structured approach for gaining an understanding of complex work environments of human operators. The framework has been discussed in the literature (Vicente, 1999) and intends to support new system design in a formative way. CWA starts by analysing the work domain constraints imposed by relevant physical processes (ecological perspective) and ends up with investigating the cognitive constraints of the operator. However in time-critical domains such a qualitative approach is not satisfying. Cummings (2006) points out this limitation of CWA not ending up in executable models and therefore not allowing the prediction of dependent process variables. Especially for intentional domains such as air traffic control (ATC) the analysis of temporal and causal interdependencies of the work environment and the evaluation of their influence on potential operator strategies remains ill supported by traditional CWA techniques and tools.

Degani et al. (2002) showed in their paper, how a formal model can be applied for system design, by their analysis of a pilot's interaction with an autopilot onboard a modern commercial aircraft. While this approach is referred as indeed general by the authors themselves, they claim that its adaptation to timed and hybrid systems remains on open challenge for the future.

This paper proposes a transfer of CWA results into a formal executable model. It includes the effort to come up with a formal analysis for a timed complex system. The aim is to achieve a better tractability of dynamical and temporal effects in the interaction of the operator with the work system. To develop this approach an integrated human-machine model is presented consisting of human-model, process-model, and interaction-model which should subsequently inform work process design. The entire model is built using a uniform Coloured Petri Net (CPN) (Jensen, 1997), a modelling technique which facilitates formal analysis. To avoid discussing specific model parts with technical implementation a technical description of the petri net model is given in Werther (2006). A special feature of the petri net based approach is that petri nets support the calculation of state spaces or reachability graphs, thereby allowing the systematic investigation of all possible paths along which a traffic scenario can develop.

In the first part of the paper the architecture of an integrated human-machine model is presented, named FAirControl. Such an integrative approach is meant to support the analysis of temporal effects in complex work processes. It is indicated where and how the CWA results from the RApTOr project were fed into the different FAirControl model parts. Then an overview is given into the model's basic operating mode by mapping the most important model processes to Rasmussen's Decision Step Ladder (Rasmussen, 1986). The FAirControl microworld is briefly introduced which can be used for experimental model validation and comparison of subjects' behaviour with human-model behaviour. In the second part of the paper a small case study is conducted where the effects of two potential controller strategies in handling a traffic scenario are compared under different assumptions on cognitive processing speed of the controller. It is shown that timing plays a crucial role in the process and how this can be analysed on the presented model.

The FAirControl-Tool

Model architecture and experimental system

The objective of the CPN model FAirControl is to incorporate the essential functional relations and constraints of a physical airport process and the cognitive decision processes of the human controller resulting from CWA within one integrated dynamic human-machine-model. The model has a tripartite structure using an architecture proposed by Cacciabue (1998) consisting of a human-model, a machine-model (process-model) and an interaction model. The function of the interaction model is to decouple the human- and process model so they can evolve independently for certain time periods. For example the mental model of the controller is not necessarily updated to the newest process state of the process model at all times, which adds further indeterminism to the process.

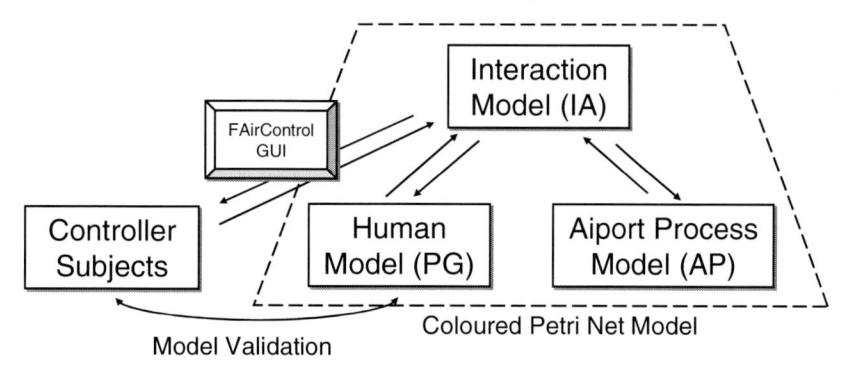

Figure 1. FAirControl model architecture

The CPN model allows for scenario-based simulation and automated formal analysis using state space techniques. However, the FAirControl-tool is not only a CPN model, but is extended by a graphical user interface (FAirControl GUI). The automated-controller can thus be replaced by human subjects to interact with the same modelled airport process for experimental validation and comparison. The FAirControl model architecture is depicted in Figure 1.

Transfer of CWA results into executable models

The development of the FAirControl CPN model is based on a cognitive work analysis (CWA) conducted for a tower controller's working position at Leipzig airport (Werther & Uhlmann, 2005). According to Vicente the five phases of a CWA approach are summarized in Figure 2. The figure indicates how the different phases of the CWA relate to the three modules of the FAirControl model (black labels). Thus, results of the Work Domain Analysis (1) concerning aircraft movements and traffic limitations enter in the airport process model (AP-Model), while the characteristics of available information sources are fed into the Interaction Model (IA-Model). The results of Control Task Analysis (2) looking at pre- and post condition of individual tasks, and the Strategy Analysis (3) with the focus on handling multiple goals under limited cognitive resources end up in FAirControl's PG-Model. The Organisational and Coordination Analysis (4) is predefined for FAirControl and encompasses a single tower controller interacting with the airport traffic by radio and far view. The Competence Analysis of the controller is not considered here.

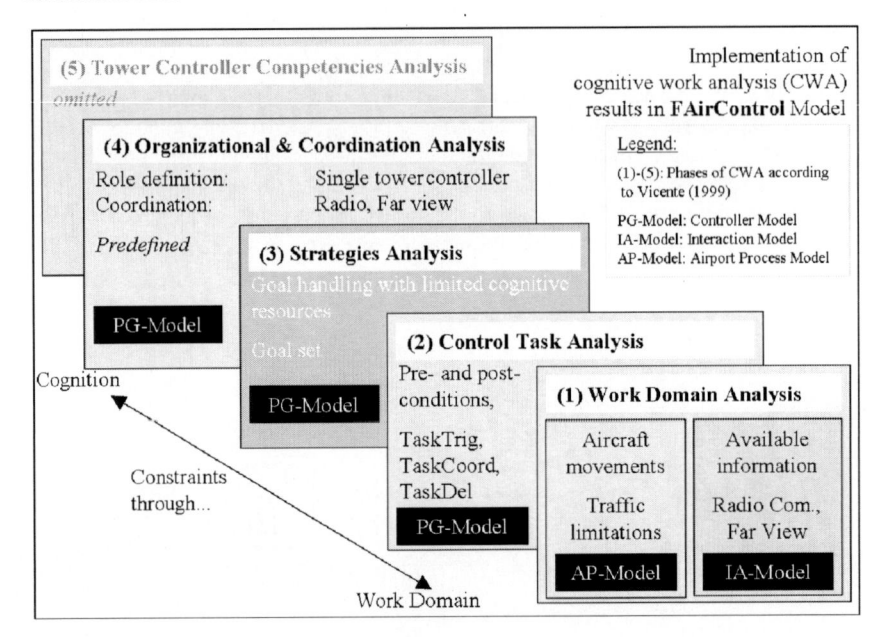

Figure 2. Relation between CWA and FAirControl model

FAirControl human-machine model

In this section the basic assumptions and operating mode of the executable FAirControl Model are introduced and differences and similarities to the Man-Machine Integration Design and Analysis System (MIDAS) are pinpointed (Corker et al., 1993).

With MIDAS a whole cognitive architecture for mental modelling based on production rules is worked out. The human model in FAirControl differs from the operator models in MIDAS in some respect. While MIDAS aims for modelling all aspects of the operator, including perception, attention and memory, FAirControl is looking for a minimalist approach. The four analysis steps (activation, observation, identification, evaluation) and the four planning steps (Interpretation, Task definition, Planning, Execution) of human information processing according to Rasmussen (1986) are modelled. (See Figure 3) Further, the tower controller in FAirControl is modelled as a resource model with limited cognitive processing capacity (Simon, 1957; Werther, 2006). Consequently, the controller can only observe one area of the airport at a time and can only transmit or receive one radio message at once. A further assumption of the model is that while the controller may identify various goals simultaneously, he can only actively pursue a single goal at a time.

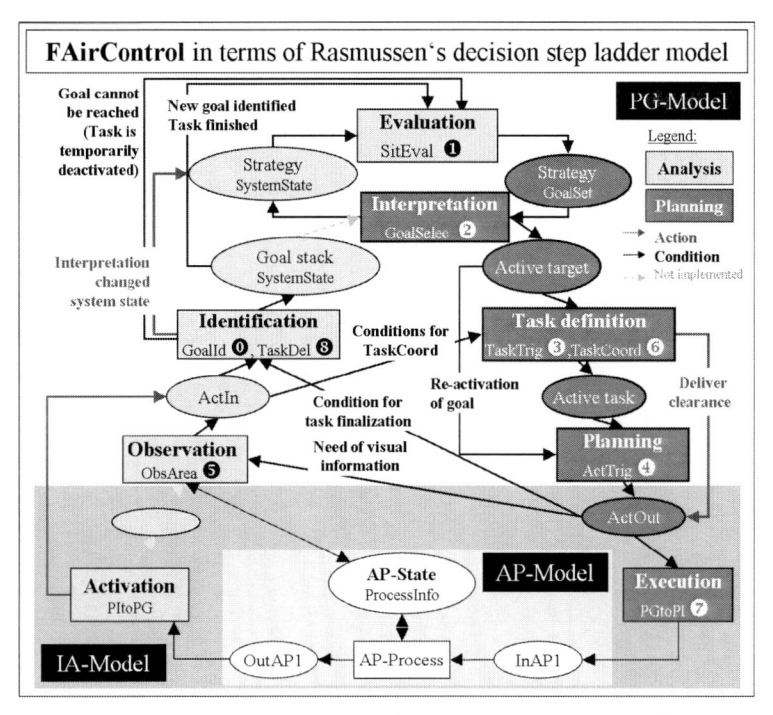

Figure 3. FAirControl in terms of Rasmussens decision step ladder model (abbreviations from petri net model)

A motivational character is assumed for the work of the controller and is also implemented in the model: If a pilot requests a certain clearance the controller will always adopt the issuance of this clearance as a conscious goal and the request will be processed.

According to Hacker (1986) a goal is understood to be a presumption which includes the anticipated final result of the associated action. In the next part, a concrete example should make clear to the reader on which level the tower controller is represented in the controller model.

The processing cycle starts with the *identification (0)* of a new goal which the controller has to pursue, e.g. triggered through the request for a certain ATC clearance by a pilot. It is then possible (see above) that the controller holds only a single goal or several goals at the same time. In the latter case, an *evaluation of the working situation (1)* will bring up several goals competing for the cognitive resources of the controller. Through the cognitive process *interpretation (2)* one single goal will be activated as the active target (e.g. "issue landing clearance for aircraft X"). In terms of Heckhausen's rubicon model (Heckhausen, 1987), this is equivalent to leaving the motivational state of awareness, weighting up different goals against each other and taking a decision for the realization of a particular goal. Specific *control tasks are now defined (3)* and *planned (4)* to reach the activated target. Subsequently the necessary conditions for the execution of these control tasks are checked. This includes the *observation (5)* of relevant process information and comparison of this information with *nominal system states (6)*. If the conditions for delivering a certain clearance or control task are fulfilled, then the task will be *executed (7)* and *finalized (8)*. A further *interpretation (2)* and *evaluation (1)* of the remaining goals on the goal stack follows. If the active target could not be realized, due to non-fulfilment of a necessary pre-condition a new *evaluation (1)* is triggered. The previously active target is deactivated and a new target is chosen.

Even though there are differences in the scope of the human model in FAirControl compared to the operator model of MIDAS, both approaches exceed simpler static forms of a task analysis and account for the context sensitive demands in a dynamic procedural simulation.

Experimental environment

For validation of the controller-model (PG-Model) an experimental environment was created allowing subjects to fulfil manually and in real time the same tasks which have been modelled by the PG-Model. The AP-Model (petri net based) simulates the airport process which is visualised in a schematic top view of the airport surface. The different areas of this surface are initially hidden by a coloured overlay und thus invisible. In order to collect information about the state of the airport process the subject has to uncover different areas of the airport in a serial manner.

The subject controls the process over a touchpad user interface. Aircraft are selected on the basis of flight strips at the bottom of the screen and interactive buttons enable the subject to give clearances by a touch pen.

The advantage of an extension of the petri net model for an experimental environment allows analyses of decision strategies of subjects in a dynamic system. The assumptions made within the PG-Model about information processing and decision strategies can be tested.

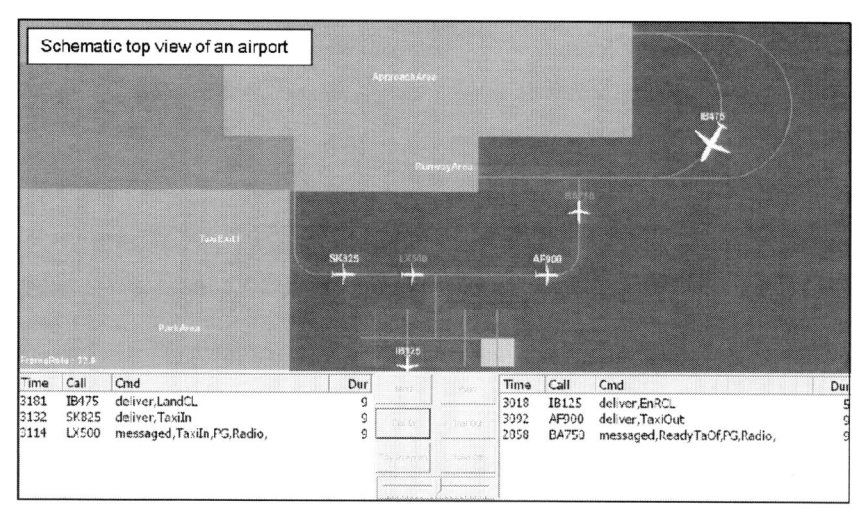

Figure 4. FAirControl experimental environment

Case Study: strategies of traffic evaluation for goal activation

The following case study investigates two different tower controller strategies for goal activation under the assumption of limited cognitive and limited airport resources. Both controller strategies are represented as a different heuristic in the controller model (PG-model) and guide the selection of the next active goal, based on an evaluation of the current airport traffic situation. Each of the strategies can be tested under different assumptions about the controller's cognitive processing capacity, i.e. the speed by which the controller can make observations and check the necessary preconditions for issuing a specific clearance. The section starts with the description of a small example traffic scenario and a problem analysis of the controller's task in handling traffic on the modelled airport. On the basis of the scenario, some characteristics and effects of the two considered strategies and the influence of cognitive constraints in multiple task situations are then discussed.

Scenario description

The simulated traffic scenario includes five aircraft (callsigns IB148...IB152) which are handled by the PG-Model. All of these aircraft are initialized airborne and subsequently pass a complete inbound/outbound cycle (that is, they are guided through a complete approach procedure, land on the runway, taxi to the apron where they park and after some delay turn around, taxi back to the runway and start from the runway). The strategic aim for the tower controller in handling the scenario is to enable an efficient and safe flow of traffic over the airport and to permit each aircraft to complete the procedure consuming as little time as possible. According to this aim, a simple performance metric to measure overall process efficiency is formulated as being the sum of all processing times which each individual aircraft consumes when passing through the process.

Problem analysis

Conceptually, the aircraft movements on and around the modelled airport can be assigned to six different task categories according to the type of clearance an aircraft is expecting next in order to proceed with its flight. These six categories are:

C1) Landing: Aircraft expecting a landing clearance
C2) TaxiIn: Aircraft expecting a taxi clearance from runway to apron
C3) TaxiInApron: Aircraft expecting a clearance into the apron to the parking stand
C4) PushBack: Aircraft expecting a Pushback clearance from the stand onto the apron
C5) TaxiOut: Aircraft expecting a clearance to taxi from apron to runway
C6) TakeOff: Aircraft expecting a take off clearance

The categories themselves can easily be distinguished and aircraft can easily be assigned to one category for each situation in the scenario. Recording the number of aircraft in each individual category in a vector of clearance request presents a meaningful characterization of the current traffic situation and distribution of traffic load at the airport. The result is therefore expected to be a significant factor in a controller's decision which task should be prioritized and which clearance should be issued next. The clearance request vector also builds the basis for the controller strategies discussed below.

Despite the simple categorisation of tasks however, the processing of tasks in the separate categories to control and balance the traffic load is highly interdependent from a controller's point of view. Some of the clearances and associated aircraft movements share the same physical airport resources and can thus not be executed concurrently. The resource usage by the different task categories is summarized in Table 1.

Table 1. Usage of Airport resources for different task categories

Airport resources	C1 Landing	C2 TaxiIn	C3 TaxiInApron	C4 Pusback	C5 TaxiOut	C6 TakeOff
Runway area	x					x
TaxiIn		x				
TaxiOut					x	
Apron Area (part 0)			x	x		
Apron Area (part 1 or 2)			x	x		
Stand Area (part 1 or 2)			x			

The shared use of resources by tasks from different categories easily leads to bottleneck situations in the airport process. In Table 1 the two potential bottlenecks can be identified: 1.) runway bottleneck 2.) apron bottleneck. The Runway bottleneck describes a situation where arrival traffic (expecting a landing clearance,

C1) requires the same physical resource as departing traffic (expecting a take off clearance, C6) that is, the *Runway area* resource. The Apron bottleneck describes a situation where traffic entering the apron (expecting a TaxiInApron clearance, C3) competes with traffic intending to leave the apron (expecting TaxiOut clearance, C5) for different *Apron Area* resources. The two bottlenecks and corresponding conflicts are shaded in grey in Table 1.

Implemented Strategies

Within this case study two different strategies have been implemented in the PG-Model. The basic idea of the two strategies follows the principle of fast and frugal heuristics introduced by Gigerenzer and Todd (1999). The strategies are fast in that controller decisions do not rely on time consuming projections of airport processes into the future. They are considered frugal because an exact acquisition of all available information about the airport process (exact aircraft positions etc.) is not performed and instead the decision is based on an evaluation of the coarse clearance request vector introduced above. A fast and frugal approach is supported by the fact that in practice the tower controller's decision process also requires adaptive choices with a minimum of time, knowledge and computation effort.

The two strategies can be explained in terms of the three important building blocks of simple heuristics as follows:

A single-stage evaluation strategy:
1.) The search rule defines the order in which the controller looks for clearances. In this strategy, the controller searches through the list of current requests (clearance request vector) in the following order: TaxiIn, Land, TakeOff, TaxiOut, TaxiInApron, Pushback-Clearance.
2.) The stopping rule stops the search for requested clearances after the first clearance has been encountered.
3.) The decision rule defines the first clearance found as the next goal the controller has to follow.

After that the controller will check the preconditions that have to be fulfilled for the clearance to be given. If the Clearance cannot be given the controller will go back to the search rule (stage 1).

The second strategy in comparison is more extensive, as it asks for more information and uses quantitatively weights for clearances in different airport areas before deciding on a specific clearance in a second step.

A two-stage evaluation strategy:
1.) The search rule sums up separately the number of clearances requiring the runway-resource (Land, TaxiIn, TakeOff) and number of clearances requiring apron-resources (TaxiInApron, TaxiOut, Pushback).
2.) This strategy does not possess a stopping-rule on the dimension of clearances but searches all requested clearances and then stops.

3.) The decision rule determines if there are more clearances requiring the runway-resource (case 1) or Apron-resource (case2). In case 1, it chooses the first clearance requiring the runway resource in order of priority *TaxiIn, Land, TakeOff* as the next goal the controller has to follow. In case 2 to it chooses the first clearance requiring the Apron-resource in order of priority *TaxiOut, TaxiInApron, Pushback*

Again, the controller will check the preconditions that have to be fulfilled before the clearance can be given. If the Clearance cannot be given he will go back to the search rule (stage 1).

State space analysis for the case study

The analysis of the petri net for the case study focuses on the investigation of temporal effects within the human-machine model and the importance of time critical situations for controller strategies. To study these effects, the state space has to be generated of the model for the traffic scenario introduced above. The state space is a directed graph containing all different system states the model could ever adopt through simulation starting from a defined initial state plus the connections (arcs) between these states. As state spaces quickly get very large for systems with many degrees of freedom as the one considered here, a special graph reduction technique was introduced in (Möhlenbrink et al., 2007). This branching graph technique allows to search within the entire graph the states where critical actions (operator decisions or physical actions) are possible that irrevocably affect the final process results (a critical branch between paths to different terminal states). The technique excludes the states and arcs of the graph where an actor or process inevitably performs a number of sequential steps with no definite impact on the final result of the process. The branching graph technique helps to identify in large state spaces the critical situations where different controller strategies (or heuristics) actually have different effects and produce different observable actions and results. In the following the technique is applied for the above traffic scenario.

In a first experimental condition and to serve as an optimal performance baseline, the automated controller (human-model) is assumed to have unbound cognitive processing speed. This means there is no time required by the controller for information acquisition with regard to the preconditions of a certain clearance (airport observation). There is also no time required for the controller to submit a clearance to an aircraft.

The generation of the state space for this extreme condition leads to a state space graph with 558 nodes (intermediate states) but just one terminal node. The analysis reveals that due to the assumed unbounded speed of the controller there is always just one goal (one clearance) for the controller to process at a time. Multiple task situations do not occur, and thus the implemented strategies for goal prioritization do not even come into play. For this condition, the scenario has degraded to a deterministic process, with just one possible process trajectory and outcome. The throughput time summed up for all aircraft is 7280 time units (IB148=1056, IB149=1256, IB150=1456, IB151=1656, IB152=1856).

CWA and Work Process Design in ATC 411

In a second experimental condition a more realistic perspective of the controller's cognitive performance is taken. Information acquisition and communication in practice consumes time, thus it will now be assumed the modelled controller needs three time steps to observe a certain area of the airport and three time steps for the communication of a clearance via radio.

The generation of the state space graph for this condition yields 939 nodes with three possible terminal states. The process has become a non-deterministic one and three different process outcomes are possible for the same initial condition and scenario definition. The branching graph technique identifies two branching situations where multiple task situations exist for the controller, and where the two strategies introduced above would in fact lead to different goal activations and actions by the controller. The identified branching situations are depicted in Figure 5.

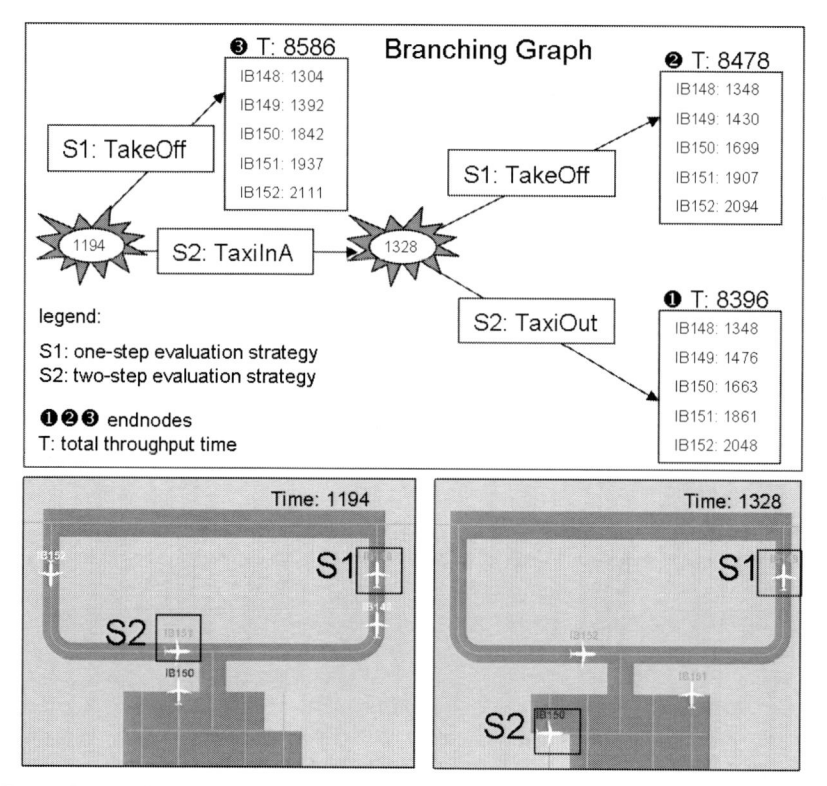

Figure 5. Branching graph for evaluation of the two strategies for handling of multiple goal situations

The schematic branching graph in the upper part of Figure 4 shows two situations in which the two controllers' strategies differ in their goal-activation. The corresponding traffic situations in the experimental environment are graphically illustrated in the lower part of the figure. At time point 1194 strategy one directs the controllers' goal-activation towards a 'takeoff-clearance for IB148', while strategy two directs to activate the 'taxi-in-apron clearance for IB151'. Which goal is activated at time point 1194 is crucial for later traffic situations of the airport and aircraft processing times. The reduced state space explains that if the automated-controller-model acts inline with strategy one, no further situations will occur in which strategy one enables a decision different from strategy two. The traffic scenario will result in terminal state 3 with a total processing time of 8586 time units. However, to activate the goal 'taxi-in apron clearance for IB151' at time point 1194 according to strategy two leads to a further situation in which the two controller strategies yield different decisions. According to strategy one at time point 1328 the 'take-off-clearance for IB149' is activated, in contrast to the 'taxiout-clearance for IB 150' for strategy two. The traffic scenario will result in terminal states 1 or 2 with processing times of 8396 and 8478 time units respectively.

Comparing the results of the two experimental conditions with unbounded and bounded cognitive processing speed of the controller shows that it is as important to consider the limitations of the human mind and the interaction with the environment as the initial process state of the system itself to understand the situations in which the controller has to make his decision.

The strength of this kind of analysis for an experimental system lies in the consideration of all possible time-critical interdependencies of the human-model and the process-model. Different behaviour of the automated-controller can result in complex decision situations or not. Different goal-activations at one time point cause different future situations in which a decision has to be made. In dynamic task environments like that of the generic airport it is difficult to understand the evolution of the scenario over time. However, a model based approach supports understanding the time-depended evolution of the scenario.

Conclusion and outlook

A human-machine-model FAirControl was developed based on a CWA. To go one step further and transfer the results of a descriptive CWA into an executable model has one main advantage: executable models allow the identification of the dependencies between the airport process and the controller's working process under specific constraints with respect to time and resources. The presented formal model is not aiming at a verification of inconsistent system design, but exceeds the approach of Degani et al. as it includes this timing aspect. In dynamic systems like the air traffic management domain it is important to understand the consequences of an action at a point in time for the system state at a later point in time. To understand what kind of multiple goal situations can occur cannot be answered by qualitative approaches.

The detailed analysis of the Case study was presented to explain the methodological approach. Understanding a multiple task situation of one controller, who has to decide, which task to prioritize seems not to be an interesting question. But imagine the controller is asked to fulfill two tasks at a time, but both tasks require his full attention. Indeed, such a multiple task situation is of interest for work process design.

Simpler static forms of task analysis do not offer support, to evaluate the situations a controller has to solve, if time is neglected.

The FAirControl user interface is valuable for experimental trials. While in MIDAS a whole cognitive architecture is suggested to evaluate a holistic human-machine model, the experimental environment allows studying the value of the simplified controller model within FAirControl. It is an open question, if the simple controller model presented here is sufficient to analyze multiple task situations, or if model refinement has to take place.

References

Cacciabue, P.C. (1998). *Modelling and simulation of human behaviour in System control*. London: Springer.

Cummings, M.L. (2006). Can CWA Inform the Design of Networked Intelligent Systems? In *Moving Autonomy Forward* Conference, Grantham, UK, 2006.

Corker, K.M. & Smith, B.R. (1993). An Architecture and Model for Cognitive Engineering Simulation Analysis: Application to Advanced Aviation Automation. Presented at AIAA Confernce on Computing in Aerospace, San Diego, CA.

Degani, A., & Heymann, M. (2002). Formal verification of human -automation interaction. *Human Factors, 44*, 28-43.

Gigerenzer, G., Todd, P. M., & the ABC Research Group (1999). *Simple Heuristics that make us smart*. New York: Oxford University Press

Hacker, W. (1986). Arbeitspsychologie. *Psychische Regulation von Arbeitstätigkeiten*. Bern, Switzerland: Huber Verlag.

Heckhausen, H. (1987). Intentionsgeleitetes Handeln und seine Fehler. In H. Heckhausen, P.M. Gollwitzer, and F.E. Weinert (Eds.), *Jenseits des Rubikon: Der Wille in den Humanwissenschaften* (pp.143-175). Berlin: Springer Verlag.

Jensen, K. (1997). Coloured Petri Nets. Berlin: Springer Verlag.

Möhlenbrink, C., Oberheid H., & Werther B. (2007) Model based work process design in air traffic control. In Onera DLR Aerospace Symposium ODAS 2007, Göttingen, Germany.

Rasmussen, J. (1986). A framework for cognitive task analysis in systems design. In E. Hollnagel, G. Mancini, and D.D. Woods (Eds.), *Intelligent decision support in process environments* (pp. 175-196). Berlin: Springer Verlag.

Schmidt, M., Rudolph, M., Werther, B., & Fürstenau, N. (2006). Remote Airport Tower Operation with Augmented Vision Video Panorama HMI. In Proceedings of the 2nd International Conference on Research in Air Transportation ICRAT, Belgrade (pp. 221 – 230).

Simon, H.A. (1957). *Models of man*. New York: Wiley.

Vicente, K.J. (1999). *Cognitive Work Analysis*. Mahwah, NJ: Lawrence Erlbaum Associates.

Werther, B. (2006). *Kognitive Modellierung mit farbigen Petrinetzen zur Analyse menschlichen Verhaltens*. PhD Thesis, DLR-Institute of Flight Guidance, Braunschweig.

Werther, B., & Uhlmann, H. (2005). Ansatz zur modellbasierten Entwicklung eines Lotsenarbeitsplatzes. In Zustandserkennung und Systemgestaltung, Fortschritt Berichte VDI, 22, 291 -294.

Werther, B. (2006). Colored Petri net based modeling of airport control processes. In Proceedings International Conference Computational Intelligence for Modelling, Control & Automation (CIMCA) Sydney.

An integrated model for working environments and rail human factors

Malte Hammerl, Bärbel Jäger, & Karsten Lemmer
German Aerospace Center DLR
Braunschweig
Germany

Abstract

Railway human factors represent a subject that has not been given high priority over a considerable period of time. However, a number of reasons have led to a special interest in human factors in the railway domain since about the turn of the century: European research has emphasised consideration of several characteristics of the subject. In order to enable an overview on ergonomics science in railway applications this paper provides an integrated model that covers a range of rail human factors topics. The scheme is based on a work science related approach for work environments in railway operation. Besides, the model encompasses the psychological aspects of work and the interrelation between variables such as work conditions and desired outputs, i.e. a safe and high performing railway operation. Design of human-machine interfaces, workload analysis and possible attempts for assistance systems are presented as topics currently covered by research of the Institute for Transportation Systems in the German Aerospace Center. An outlook on the domain of rail human factors as well as on further research concludes the paper.

Introduction

Human factors had traditionally not been considered to a relevant extent in the railway domain. Developments and changes had been driven by technical achievements or basic examinations for the train driver's task. Some standards have been established to support ergonomics at the work places and some research on human reliability has been carried out for railway environments. Publications concerning the "railway human factors" have been few over a considerable period of time, until recently.

Rail automation systems, computational developments and new technology changed work places for both train drivers and signallers remarkably. The will of the European Union to transform national railways and to open markets for commercial railway enterprises led to particular changes in many countries. All adaptations created modified human factors issues in the railway domain. Additionally, the introduction of ERTMS (European Rail Traffic Management System) brings along new topics of international railway ergonomics consideration. Unfortunately, some severe railway accidents in Europe underlined the impact of human factors on safety, too.

Human factors in aviation have a certain lead to railways due to its complexity and the obvious safety link of human factors. The railway domain is ready to learn from research results already achieved for pilots and air-traffic controllers.

Since the late 1990s European publications concerning railway human factors have become more frequent and more sophisticated. Efforts particularly in Great Britain contributed to the denotation and the foundation of the subject. Notably, the 1st and 2nd European Conferences on Rail Human Factors, 2003 in York (Wilson et al. 2005) and 2005 in London (Wilson et al. 2007b), have brought together experts of the new domain.

However, it is to state that a large amount of the publications concentrates on a certain topic without integrating research into the field of human factors of the railway system as a whole. For example, the domain is sometimes only approached from an ergonomists point of view focusing on physical ergonomics. On the other hand, rail human factors encompass much more than just human error.

With the background of the recent evolution of the subject this paper provides an integrated approach with the focus on human factors of railway staff. Therefore, the following chapter gives basic considerations of the subject, a very brief literature review of latest publications and integration into familiar sciences. When the focus is set on railway operations staff, the approach sketches a pattern on the basis of ergonomics structure. Later, a model of working environments is developed, functional chains are explained and safety related approaches are presented.

Railway Human Factors

Comprehensive reviews of ergonomics in the railway domain have recently been published (Wilson & Norris, 2006; Wilson et al. 2007a). The monographs comprising the proceedings of the conferences (Wilson et al., 2005, Wilson et al. 2007b) represent the most worthy sources for the subject. The review work shall not be replicated in this paper; instead it is briefly supplemented by a viewpoint from German publications. Due to cultural differences in railway background, a summary of the German state of the art and technology is essential for the further application of human factors on German railway working environments.

Because of a great belief in technical reliability, the human impact on safety of railway systems has always been tried to cover by technical safety measures. A substantial study of human error in railways was presented by Hinzen (Hinzen, 1993). In Germany, research on human factors had been limited to punctual considerations of workload at the train driver's and signaller's work place for many years. The small amount of publications is certainly also due to the fact that the German railway was the public monopolist: there was minor need to disseminate research results. Three early reports concerning workload are available (Huber, 1985; Myrtek et al., 1991; Leutner, 1995). Later, workload studies were motivated by major technical modifications at the work places (Mussgnug, 2000; Rentzsch & Liesemeier, 2000). After the turn of the century, the focus was set on ergonomic design of work environments (Oberländer, 2002) – partly supported by international projects (Rentzsch et al., 2002). Two publications are available concerning human factors and border crossing in the framework of the introduction of ERTMS (Metzger, 2004; Schmidt, 2004). Training of train drivers continuously represented a topic of research (e.g., Schmitz & Maag, 2007).

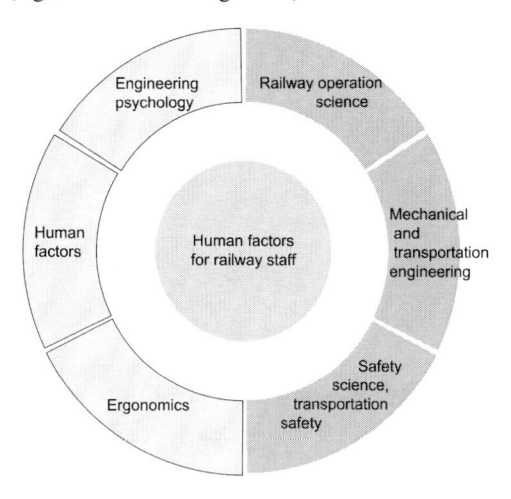

Figure 1. Interaction of sciences concerning human factors for railway staff

Summing up the brief review, basic considerations for ergonomics and workload were conducted. However, the majority of reports approached the workplaces from an ergonomic point of view. In German publications, the involvement of people in the performance and safety of the railway system – linked by human factors and human error – was not highlighted explicitly. This relationship has been established by the intensification of international railway human factors research in the last decade.

In spite of the recent development of the subject, a concrete description has not yet been explicitly worded in literature. A definition for *railway* human factors was derived from the well-known general outline of human factors by the Human Factors and Ergonomics Society (HFES, 2007). Here, the following definition is proposed:

> "The subject of rail human factors represents the scientific branch under the discipline of human factors concerned with the understanding of interactions among humans in the railway system and between humans and the railway system, aiming at the optimization of human well-being as well as performance and safety of the railway system."

While the proper functioning of the entire system conventionally was the focus of railway engineers, human well-being and physical design of working environments represented the aim of ergonomists. Both disciplines are included in the definition above. Note that due to its importance in transportation, safety is mentioned as a particular goal. The integration of all these issues characterizes the subject of rail human factors. A long list of stakeholders indicates diversity as another quality of human factors in the railway domain: for example, workload and situation awareness have a significant impact on the job of a signaller, train drivers have to cope with problems of vigilance, workload and reliability. Different aspects of ergonomics are to be considered when regarding passenger comfort and safety. Level crossings represent interfaces where people of the general public interfere with the railway system: to discourage road users from crossing the railway line when a train is approaching represents an important rail human factors research topic, too.

Understanding rail human factors as an umbrella for research on interaction between human beings and the railway system enables an integrated overview and provides a sophisticated basis for further detailed investigations. The definition of the research field given above covers a considerable list of stakeholders, similarly presented in (Wilson et al., 2007a).

- staff
    - signallers
    - controllers
    - drivers
    - station and on-board personnel
    - managers
    - planners and engineers
    - maintenance personnel
- passengers
- general public

In the later sections of this paper, consideration of human factors will be limited to staff in the railway system. Performance and safety are directly affected by this party, employees are continuously present in the system and last but not least interactions between employees play an important role. Some online descriptions of human factors in the railway domain are even limited to ergonomics of work environments (RSSB, 2007; ORR, 2007).

When focussing on human factors of railway staff, research in the subject is linked to a set of interdisciplinary neighbour sciences. See figure 1 for an illustration of relations to human-centred and rather engineering-focused sciences. A collaboration of psychologists, designers, engineers and railway experts is preferably suitable for research and development within human factors in the railway domain.

Human Factors of railway operation staff

The focus in this chapter is not only on railway operations staff, but is limited to train driver, signaller and maintenance personnel as well. This concentration is justified by the high importance their work has on safety and performance of the railway system *during operation*. Due to differences in railway architecture, international research defines signallers and controllers diversely. Here, it is not referred to British task scopes of signallers and controllers, instead to the word signaller as the major controller of railway operation in general.

A structure for ergonomics has been suggested by the International Ergonomics Association; the breakdown separates physical, cognitive and social-organisational ergonomics. In the following, the way is shown how the structure is applied on work places in the railway domain. Figure 2 shows the latest research areas in human factors of railway operations personnel. The table represents the major topics and is not to be understood as a complete list.

	Train driver	Signaller	Maintenance personnel
Physical ergonomics	Cab design	Control center interfaces design	Design of work and equipment
	Signals		
Cognitive ergonomics	Vigilance devices and fatigue	Vigilance and fatigue	
	Driver models and performance	Signaller models and performance	
	Loss of skills, Situation awareness		
	workload		
	human error and reliability		
Organisational and social ergonomics	Safety culture and climate, rules and violation		
	Roster planning and shifts		
	Education, training, competences		

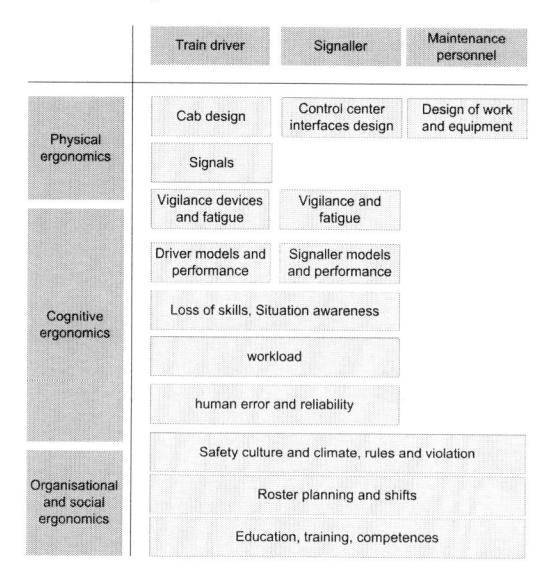

Figure 2. Human factors of railway staff

Applying physical ergonomics on the working environment of a train driver, cab design certainly plays the key role. In addition, the settings of information outside the vehicle have an impact on his work, i.e. the design and positioning of signals. Vigilance devices are somewhat a part of cab design but are mentioned particularly because of their high importance for railway operation performance and safety. The location of vigilance in the table in figure 2, i.e. on the border between physical and cognitive ergonomics, has been chosen, due to the psychological source and the physical counter measures of the problem.

The physical working environment of a modern signaller workplace is characterized by control centre interface. The work at a computer screen is to a great extent covered by the research fields of human-computer interaction and usability. Maintenance personnel encompasses employees maintaining and repairing railway vehicles and track. Design of their work and equipment does not represent a core research field of rail human factors as the scope of their tasks has only a low mental component. Note that this subject is not regarded less important but is covered by different ergonomics principles.

Human performance, human reliability, workload and automation represent some of the most important keywords of human factors for train drivers and signallers. Despite the wider range of aspects, review publications revealed the concentration of research on the latter subjects. Modelling and understanding train drivers and signallers work supports the comprehension between task, workload and performance. Although the two tasks are linked very closely, cognitive modelling has to consider a set of particularities of the work systems, the boxes in figure 2 are therefore split up.

Human failures can be divided into errors and violations. Both have a close link to workload and to safety culture, but violation of a certain rule means a contravention of safety culture. As safety culture and climate represent issues of organisational ergonomics and mental workload belongs to the cognitive point of view, the box of safety culture and violations has been set between the two fields. Certainly, roster planning, shift lengths and human resources management as well as education and frequent training are matters of organisational ergonomics.

Finally, it is to say that the chart in figure 2 enables a quick overview on the human factors subject for railway operation personnel. However, subjects that cannot be easily shown by the diagram exist; especially, the communication between staff is relevant and involves safety pertinence. The integration of human factors subjects into the product development process has also been a major concern in latest research in Great Britain (RSSB, 2006). Furthermore, the introduction of new technologies such as ERTMS has asked for the examination of international human factors experts in Europe.

Integrated model

In this chapter, a model covering a wide range of ergonomics and human factors issues is presented. The approach is based on a common model that was introduced by the German REFA organisation (Association for Work Design/Work Structure, Industrial Organization and Corporate Development) in 1983 (cited in Luczak, 1998).

In this model, the work system is characterized by the interaction of a human with his task, his work subjects and instruments. An input set of material, information and energy supplies the work process. The work system core is reasonably governed by the job to be performed. Social and physical impact factors have an influence on the work system core and its cooperation of human being, task, instruments and work matter. Classically, the input of raw material was processed to a product as the output. The approach of the present paper derives and supplements the general model and applies the principles on the work of train drivers and signallers.

In order to present the model for railway personnel, some particularities are to be considered. Work material is not applicable in either of the two railway jobs. The job and the task may be interpreted as a joint item of a common train driver – for example – it is constant during the work period. Thus, the work system core can be defined by an interaction of human, task and instruments. The impact factors of physical and organisational-social ergonomics remain in the model as well as the input and output of work, see figure 3. So, the horizontal axis represents the work system in the moment of operation while the impact factors at the top and at the bottom are less dynamic and continuously influence the core.

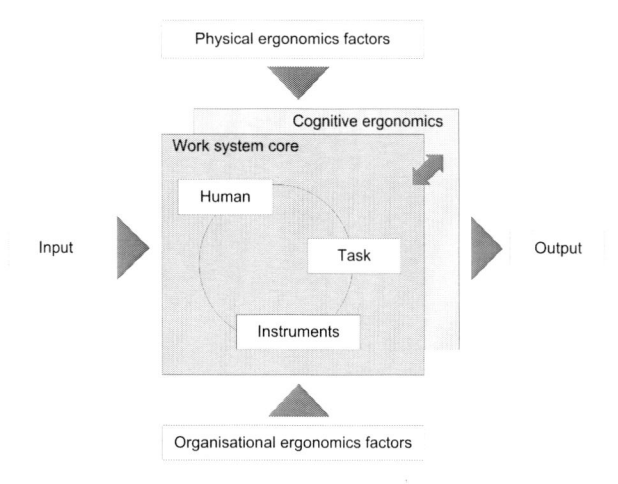

Figure 3. Model of working environments (1): first layer

Due to the important mental part of the train driver's and the signaller's work, a second layer including cognitive ergonomics has been added to the model. The background layer comprises the psychological components of work such as mental workload, vigilance, situation awareness etc. As these subjects have a strong link especially to the human being and the task, the arrow represents a very tight interaction between these layers.

The following paragraph supports the approach by applying the model on the train driver's work. Information from instruments and displays on the train, wayside equipment and signals represent the major input for the train driver (depicted on the left hand side in figure 3). Actuating and controlling of the train movement, door opening and closing, passenger information and train supervision can be interpreted as the main outputs. The impact factors of physical ergonomics comprise design of the train cab, the interfaces and the issues of "classic" ergonomics, like anthropometry, noise, vibrations etc. Note that the general design of the human-machine-interface belongs to physical ergonomics while the information displayed in a certain situation represents an input. Finally, roster planning, training, rules and safety culture and personnel management are meant by social and organisational ergonomics impact factors.

The advance of the model representation is the differentiation between set variables and dependent variables. The boxes at the top and at the bottom shall be interpreted as variable while the core is determined by the surroundings. The functioning of the work system core depends on the influence factors of physical and organisational ergonomics. The task has been included in the work system core because of its high relevance in the interaction with human beings and instruments. For the train driver, the task is certainly fixed, however the signaller's task scope is a little more designable – so it may also be considered as a set variable although being placed in the centre. If the task is separated, the centre changes to a dual system: the human-machine interaction.

In the centre of figure 4, the second layer including the cognitive ergonomics aspects is set in the foreground. The layer still belongs to the work system core. As stated before, workload, vigilance, fatigue, human reliability, stress and situation awareness represent important psychological subjects when driving a train or controlling its movement. The understanding is based on a close interaction between the two layers. The layer of cognitive ergonomics is to be regarded rather as a set of dependent variables – influenced by organisational and physical factors. For example, the human reliability and the workload can be changed by emphasised training (organisational ergonomics) or a modified display device (physical ergonomics).

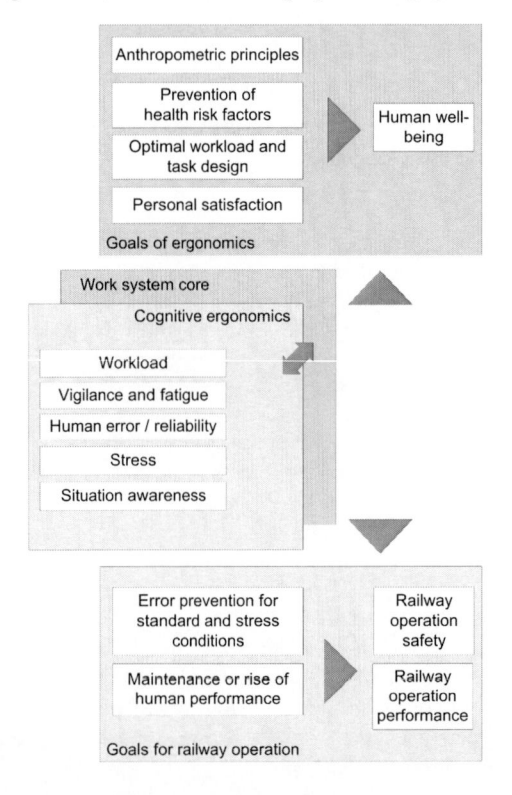

Figure 4. Model of working environments (2): second layer and targets

Following the definition given in the second chapter, the discipline of rail human factors pursues the objectives of human well-being and a safe and well-performing railway operation. The goals are depicted by the upper and lower boxes in figure 4.

The objectives of this model of a working environment are equally linked to the work system core and cognitive ergonomics.

The four aims of ergonomics were derived from an approach for work assessment published by Hacker, 1980 (cited in Luczak, 1998). Fulfilment of anthropometric principles and the prevention of health risk factors represent basic postulations for workplaces. A comfortable task design and consideration of personal satisfaction are further basics for optimization of working environments, supporting the overall goal of human well-being. While ergonomics deal with person-centred goals, the focus of railway engineering is performance-oriented. From the operational point of view, the major objectives of research on human factors are safety and performance of the railway system. An improvement shall be obtained by the prevention of human error and possibly the rise of human performance. Here, human failures play the key role between ergonomics, engineering and cognitive psychology and railway operation engineering.

Summing up this chapter, a model for working environments was presented and applied on the railway domain. The convenience of this scheme allows a separation of independent and dependent variables. The objectives of rail human factors were incorporated in the approach. In the future, the model will be complemented by a deeper analysis of train drivers and signallers.

Lines of action

Human well-being, safe and well-performing railway operation can be supported by a set of research approaches. This chapter structures the attempts and integrates them into the model presented in the last section.

Physical and organisational ergonomics were identified as set variables for the working environment. Defining an optimum and setting these impact factors influence the work system core. A better interaction in the centre results in an improved output. Targets of an augmented human well-being or a functioning railway operation are thus fulfilled in a better way.

This functional chain is applicable for a reasonable number of approaches and lines of actions. They can be interpreted as being straightforward when taking the direction from left to right in the model. Some examples of combinations are listed below. The steps generally represent the chain "action" > "reaction in the work system core layer or the cognitive ergonomics layer" > "effect" > "target" whereas effect and target join in some cases.

- improved training > enhanced situation awareness > reduction of human error > higher performance and safety
- reformulation of rules and guidelines > reduction of human error > higher performance and safety
- rise of safety culture > avoidance of violations (reduction of human error) > higher safety

- optimization of roster planning, reorganisation of shifts > vigilance > higher performance and safety, optimal task design, personal satisfaction
- introduction of (driving) assistance systems > workload > maintenance or rise of human performance
- modification of human-machine-interface > easier interaction of the human with the instrument > prevention of health risk factors, optimal workload and task design
- fulfilment of anthropometric guidelines > easier interaction of the human being with the instrument > anthropometric principles, prevention of health risk factors

Another way to address human factors problems are investigations of existent problems and analysis of outputs or core items in order to learn about the relations of work components and ergonomics. With this knowledge, the optimal setting of variables is easier to determine. These approaches are regarded as backward: they start the examination at the result (on the right hand side in the model), identify problems and provide theory how to optimise the system in the next step. Because of the opposite direction, the wording differs from the forward approaches.

- Accident and incident analysis (Output > Error identification > Recommendations for the avoidance of errors > Redesign of physical or organisational ergonomics)
- Human error analysis (Human error > Investigation on the correlation with workload > Recommendations for task design)
- Workload analysis (Workload > Recommendations for task design)

This section has shown that the model is well suitable for the illustration of lines of action, too. Thereby, a wide range of approaches – forward and backward – can be illustrated by the model. So, the model and corresponding lines of actions contribute to a structured overview on railway human factors research.

Approaches

The Institute of Transportation Systems traditionally examines railway operation and railway safety. Human factors have to be integrated in the development process, in the consideration of railway performance as well as into safety cases. So, the focus is not set on anthropometric and organisational ergonomics but on a major functional chain dealing with workload and human error:

- Design of human-machine interfaces > Workload > Reduction of human error > Railway system safety and performance

However, the mentioned chain can be considered as one of the most complex. Workload represents a term that is hard to describe and not easy to measure reliably. In contrast to technical errors, human errors are subtle to predict and are influenced by a set of human factors.

The chain begins with the design of human-machine interfaces. One focus of the approach is the integration of intelligent assistance systems. Degraded situations play an important role in railway operations. In order to re-establish a certain level of performance, the attempt is to overcome the out-of-the-loop-syndrome. Thereby, the integration of intelligent assistance systems into the workplaces of signaller and train driver represents an approach to guide the human being in cases he needs it most. In order to identify the situations of stress, the assessment of mental workload is necessary. The institute disposes of a railway laboratory with a driver desk (Jaschke et al., 2005) and is therefore able to conduct subject studies and to simulate high-demanding task situations outside of real railway operation.

Human performance is to a high extent influenced by workload, stress, fatigue etc. Research tries to establish a link between the domains of human factors and human reliability. As stated before, the subject human error is of high interest for safety assessment in the development process, too. Safety cases currently lack a sophisticated comprehension of human error (Vanderhaegen, 1999). A differentiation of human failure into intentional and unintentional errors is needed (a sophisticated structure in RIAC, 2007). As stated before, intentional violations of rules represent an opponent to the safety culture of the railway organisation. Both types of errors have to be assessed and tried to be set in relation to the input workload, situations of stress and actual fatigue.

Another safety related approach by the Institute of Transportation Systems is to depict railway safety as a combination of safety layers. In the model of safety layers (MoSiS, Pelz et al. 2007) – based on the Swiss cheese model (Reason, 2000) – safety is educible and scaleable. The attempt identifies safety layers by dint of a Why-Because Analysis (Ladkin, 2007). Thereafter, the understanding contributes to redesign of tasks and reassignment of safety impact of railway staff.

Conclusion and outlook

This paper established a definition and an outline for the new research field of rail human factors. The model for working environments for train driver and signaller provided an integrated overview on the different field of ergonomics and the targets of human factors in the railway domain. The approach tried to distinguish subjects into set and dependent variables and depicted functional chains. Finally, lines of action were illustrated and two safety related approaches were presented. Thus, the approach contributes to the fundamentals of rail human factors and supports the comprehension of the domain as an integrated field of research. Further automation and an augmented safety and security concern will certainly request deeper consideration of rail human factors in the future.

References

HFES (2007). Online resources, retrieved on 10.09.2007 from
    http://www.hfes.org/Web/EducationalResources/HFEdefinitionsmain.html
Hinzen, A. (1993). *Der Einfluss des menschlichen Fehlers auf die Sicherheit der Eisenbahn*, PhD thesis. Dortmund, Germany: Verlag für neue Wissenschaften

Huber, K. (1985). *Beanspruchung bei Belastung durch Informationsverarbeitung von DB-Triebfahrzeugführern in unterschiedlichen Einsatzbereichen*, Munich, PhD thesis.

Jaschke, K.P., Hartwig, K., Meyer zu Hörste, M., & Lemmer, K. (2005). A facility for testing ERTMS/ETCS conformity and human factors. In J. Wilson, B. Norris, T. Clarke, and A. Mills: *People and Rail Systems: Human Factors at the Heart of the Railway* (pp. 167-174). Alershot: Ashgate publishing

Ladkin, P. (2007). The Why-Because Analysis Homepage, http://www.rvs.uni-bielefeld.de/research/WBA/, retrieved 12.09.2007

Leutner, D. (1995). Psychologische Aspekte der Belastung von Schienenfahrzeugführern im öffentlichen Personennahverkehr und Entwicklung eines simulatorgestützten Belastungs-Reduktions-Trainings, In VDI 1219 (Ed.) *Simulation und Simulatoren für den Schienenverkehr* (pp. 49-59). Düsseldorf, Germany: VDI-Verlag.

Luczak, H. (1998). *Arbeitswissenschaft* (2nd edition). Berlin: Springer.

Metzger, U. (2004). Human factors und Ergonomie im ERTMS/ETCS - Einheitliche Anzeigen und Bedienung, *Signal und Draht*, 96 (12), 35-40.

Mussgnug, J. (2000). Quantifying Mental Workload of Operators in Future Control-Centre of the Deutsche Bahn AG. Proceedings of the 14th IEA and the 44th HFES meeting. HFES: Santa Monica, CA.

Myrtek, M., Deutschmann-Janicke, E., Strohmaier, H., Zimmermann, W., Lawerenz, S., Brügner, G., & Müller, W. (1991). *Psychophysiologische Untersuchungen zur Belastung und Beanspruchung bei Fahrdienstleitern, Lokomotivführern und Busfahrern*. Research report, Freiburg, Germany

Oberländer, W. (2002). Fahrdienstleiter-Arbeitsplätze in Betriebszentralen der Deutschen Bahn. *Der Eisenbahningenieur*, 53 (7), 40-45.

Office of Rail Regulation (2007). Retrieved 30.09.2007 from http://www.rail-reg.gov.uk/server/show/nav.1117

Pelz, M., Schwartz, S., & Meyer zu Hörste, M. (2007). Model of Safety Layers in the Railway System - MoSiS. In University Žilina (Ed.): *EURNEX - Zel 2007* (pp. 85-92). Žilina, Slovac Republic: EDIS Žilina.

Reason, J. (2000). Human error: models and management. *British Medical Journal*, 320, 768-770

Rentzsch, M. & Liesemeier, B. (2000). Belastung und Beanspruchung von Triebfahrzeugführern im regionalen Eisenbahnverkehr, *Komplexe Arbeitssysteme – Herausforderungen für Analyse und Gestaltung*. Dortmund, Germany: GfA Press.

Rentzsch, M., Rothbauer, M., & Sorin, J. (2002). Modular design of a driver's desk for cross-border railway traffic, In *Second International Conference on Occupational Risk Prevention*, Gran Canaria.

RIAC (Railway Industry Advisory Council), Human Factors Working Group (2007). *Human reliability*. Dossier. Retrieved on 12.09.2007 from http://www.rail-reg.gov.uk/upload/pdf/hf-integration.pdf

RSSB (2006). *Understanding human factors* – A guide for the rail industry, Rail Safety and Standards Board.

RSSB (2007). Fact sheet Human Factors, accessed 19.07.2007 http://www.rssb.co.uk/ humanf.asp, "[Rail] Human Factors is the discipline of optimising human performance in the workplace. It considers the working environment from a human-centred viewpoint looking at the whole system and its influence on the way people behave and interact with the railway."

Schmidt, A.L. & Miller, R. (2004). Faktor Mensch und die Sicherheit des grenzüberschreitenden Schienenverkehrs, *Eisenbahntechnische Rundschau*, 53, 832-836.

Schmitz, M. & Maag, C. (2007). 2TRAIN - TRAINing of TRAIN drivers in safety relevant issues with validated and integrated computer-based technology. In *Proceedings of the 15th International Symposium EURNEX - Žel 2007* (vol. 2). Žilina, Slovac Republic: EDIS Žilina.

Vanderhaegen, F. (1999) APRECIH: a human unreliability analysis method – application to railway system. *Control Engineering Practice*, 7, 1395-1403

Wickens, C.D., Lee, J.D., Liu, Y., & Gordon-Becker, S. (2003). *An introduction to human factors engineering*. Upper Saddle River, NJ: Prentice Hall.

Wilson, J.R., Norris, B., Clarke, T., & Mills, A. (2005). *Rail Human Factors: Supporting the Integrated Railway*. Aldershot, UK: Ashgate Publishing

Wilson, J. & Norris, B. (2006). Human factors in support of a successful railway: a review. *Cognition Technology, & Work*, 8, 4-14

Wilson, J.R., Farrington-Darby T., Cox, G., Bye, R., & Hockey, G.R.J. (2007a). The railway as a socio-technical system: human factors at the heart of successful rail engineering, *Proceedings of the I MECH E Part F Journal of Rail and Rapid Transit*, Vol. 221, No. 1 (pp. 101-115). London: Professional Engineering Publishing.

Wilson, J.R., Norris, B., Clarke, T., & Mills, A. (2007b). *People and Rail Systems: Human Factors at the Heart of the Railway*. Aldershot, UK: Ashgate publishing

Human Machine Interface

Travel information: quick, safe and innovative

In D. de Waard, F.O. Flemisch, B. Lorenz, H. Oberheid, and K.A. Brookhuis (Eds.) (2008), *Human Factors for assistance and automation* (p. 429). Maastricht, the Netherlands: Shaker Publishing.

The effects of travel information presentation on driver behaviour

Karel A. Brookhuis, Matthijs Dicke, Rosanne L. Rademaker, Sander R.F. Grunnekemeijer, Jorien N.J van Duijn, Maarten Klein Nijenhuis, Nynke A. Benedictus, & Elmas Eser
University of Groningen
The Netherlands

Abstract

The aim of this study was to investigate the effects of providing travel information to drivers about a traffic jam ahead and a potential detour or short-cut. Two groups of participants, native and non-native Dutch speakers were requested to drive in a driving simulator under both calm and dense traffic conditions. Travel-information was presented by means of three systems; nomadic systems such as SMS and PDA, and via the simulator mock-up vehicle's audio system. Conclusions with regard to usability are that SMS was evaluated worse than the other two systems, and native participants believed any information-providing system to be less useful than non-native participants did. With regard to cognitive processing, SMS caused more subjective (i.e. experienced) workload than the other two systems. Participants remembered more of the information when traffic was dense and natives remembered more than non-natives. With regard to performance and safety, driving performance was better when traffic was calm, as compared to dense traffic; however, compensation was shown by lowering driving speed in the latter condition. After participants were provided with travel information, their driving performance with respect to the consequences of distraction differed between systems. The auditory information provision system generated the best driving performance; the other two systems required the participants to look away from the road (too) long compromising safety, while reading an SMS took longer than scanning a PDA.

Introduction

In modern society, 24 hours a day, 7 days a week, a host of information is available. Numerous messages of all kinds are distributed through telephones, computers, televisions, radios, for instance about travel opportunities. Transport companies and related business are starting to realize the feasibility and the financial benefits of providing travelers with information to make adequate choices while traveling. With the right resources and an adequate synchrony, it is technically possible to create a dynamic travel information provision system for travellers (De Jong & Brookhuis, 2005).

There are many situations in which it would be beneficial to know what to expect while traveling. Whether there is a fuel station, a traffic jam or a delayed train; all of these situations require reliable information and flexible means to reroute ourselves to solve the problem at hand.

Commuters, professional travellers and even recreational travellers predominantly use a motor vehicle to travel (Dicke, 2008); therefore the interest of this study was specifically how to provide car drivers with travel-information. In order to closely resemble an actual driving task while avoiding hazardous situations controlling the environment, the study was conducted in a driving simulator.

Travel information can be presented to drivers in a variety of ways, through a number of available systems, depending on type of information, situation and recipient. When travelling in the country of origin, for instance, the information content may be familiar irrespective of presentation, but travelling in a foreign country, local information may lead to misunderstanding and errors. The question is what is the best feasible travel information system (off the shelf) when considering usability, safety, perception, comprehension and memory, and task demands? To this end, travel information was presented to two different groups of participants (natives and non-natives) in three different ways, through popular nomadic systems SMS and PDA, and a car audio system, in two kinds of traffic conditions (calm and dense).

Theory and previous research

When designing a system that aims to provide people with information while they are driving, it is essential to find a good match between the human information processing capabilities and information provision in the task environment. Mismatches between human being and task environment may cause misperceptions that might very well lead to accidents involving injury or death (Casey, 1993; Wickens & Hollands, 2000).

Information processing, i.e. mental activities such as recognising, rehearsing, planning, understanding, decision-making and problem solving occur in working memory. Working memory is the temporary, effort demanding store where travel-information is kept active while or until we use it (Wickens et al., 2004). Travel-information can be provided by both visual and auditory means; it is important to know the limitations and possibilities of the information processing systems that have to do with the modalities of vision and audition. The limited processing capacity of working memory may cause a disruption in performance when two tasks are executed concurrently (Baddeley & Hitch, 1974). However, there might be little interference when two concurrent tasks entail information from different modalities (Wickens et al., 2004). Examples of this type of time-sharing are driving a car while trying to listen to the news on the radio at the same time or perceiving and understanding travel information.

The level of automation has an influence on resource demands as well. In the present study, travel-information needs to be attended to during driving, the latter being a highly automated task for experienced drivers. The visual channel is most important for the driver and the concurrent execution of a visual task, like in-vehicle looking at a cell phone, competes for visual attention with the primary driving task (Brookhuis et al., 1991). But auditory (or motor or cognitive) tasks potentially create some conflict with monitoring and processing visual information as well. Because of this competition for limited resources, driving performance may decline. Alternatively, compensation may be sought by the driver in lowering speed. Therefore, several ways of providing information were investigated in order to find out if, and how, driving behaviour, performance and safety are affected. Additionally, the study aims to find out whether there are differences in information processing or in preferences between native Dutch speakers and non-native Dutch speakers.

Method

Participants

For the present study a total of twenty-four paid volunteers were recruited. In order to compare native and non-native speakers of the Dutch language, two separate groups were selected. The first group of participants consisted of twelve people of non-Dutch nationality (8 male and 4 female) with an average age of approximately 33 years. Participants in this group had been living in the Netherlands for 4.5 years on average. They were recruited at a School in Groningen with an adult education program. The participants were taking Dutch lessons (2nd grade), i.e. Dutch for Foreigners, third year students who graduated from level 2. The level of proficiency in the Dutch language of these participants was such that they were able to understand all instructions, as well as provide reasonably adequate answers in Dutch. The group, referred to as 'non-natives' from now on, was selected on driving proficiency as well.

The second group of participants consisted of twelve Dutch volunteers (8 male and 4 female) with an average age of 26.3 years. All native participants were in possession of a Dutch driver's license and had more than one year of driving experience. They were recruited at the University of Groningen.

Apparatus

To understand how different sources of information affect driver behaviour, the 'Driving Simulator' of the former Traffic Research Centre was used. The simulator was located at the academic hospital of Groningen, the "University Medical Centre Groningen", in the laboratory of the Neuropsychology department.

The simulator consists of a driver seat with steering wheel, three pedals, clutch and seatbelt, partly surrounded by three large screens, each two meters wide with a diameter of 4.5 meters. The driver seat is situated in front of the three screens, directly facing the middle screen; the two other screens are attached on the left and right of the middle screen in two 60-degree angles. Three beamers project the simulated world onto the screen, displaying the car's dashboard and rear-view mirrors, as well as the virtual reality outside the simulator. The position of the driver seat in front of the three large screens gives participants a 210-degree horizontal, and 60-degree vertical view on the virtual world projected before them.

Both the driver seat and beamers are connected to a central computer system, consisting of five interconnected personal computers (PCs); one central control PC with a Graphical User Interface, one PC that controls the traffic flow in the virtual world, and three PCs that render the virtual world on the left, middle and right screen. The PCs are interconnected through a Local Area Network (LAN). The software[*] used in the driving simulator contains several modules, one of which is a module for designing the layout of the simulator road environment. Another module is a script design tool to create conditions on the virtual route. The traffic module is a real-time simulation program, the render module renders the virtual world, and together they are the actual runtime modules.

A special feature of this particular simulator and its software is the use of the so-called 'autonomous agents' technique; all of the traffic in the virtual world exhibits autonomous, interactive behaviour. This means that the virtual vehicles in the simulation are not only capable of responding to each others' behaviour, they are also able to anticipate and react to the behaviour shown by the simulator-car. The specifics of the desired behaviour of the surrounding traffic can be set by programming through the scenario script design tool.

A customized virtual-reality driving environment was created, modelled after a standard motorway, to fit the explicit requirements of this study. This motorway was several kilometres long and had eight exits. Road signs indicating the appropriate destination marked each exit. Four of these exits led to make-believe cities with names such as 'Westdorp' or 'Noorddorp'. By taking such an exit, the participants were able to bypass potential congestion problems on the highway and reach the final destination, by secondary roads or 'Park and Ride' (P+R) alternative; the alternative means that participants were offered the option to park their car and continue their journey by means of public transportation.

Procedure and design

The present study is composed of one between-subjects factor and two within-subjects factors, respectively language proficiency, information provision system and traffic density. Language proficiency stands for comparing natives versus non-natives. Providing information compares the three different ways participants in this study were receiving information, i.e. by mobile phone SMS (using short text messages on a *Nokia* 3310 mobile phone), by PDA/palmtop (*Hewlett-Packard* iPAQ Pocket PC), and by auditory messages (using recorded voices over the car audio system). The density of the surrounding traffic had two levels, busy and calm.

---

[*] *The software is developed by StSoftware* ®

To avoid learning and sequence effects, six different scenarios were developed in which different problems asked for different solutions, the order in which scenarios were presented to the individual participants incorporated in a Latin square. The scenarios involved three types of problems (traffic jam, road constructions, accident) with two general types of solutions (take a secondary road or go to the P+R and take the bus). Although the delays caused by problems in different scenarios varied (20-30 minutes), the alternative offered by the information system always shortened the delay by 15 minutes.

Prior to the onset of each experiment, every participant received a brief introduction to the study, an informed consent, and all the questionnaires used for practice. After participants read the introduction and completed an initial questionnaire (asking them about their age, nationality, gender, etc.) they received instructions about how to read an SMS on the mobile phone, how to read the information on the PDA, and they were introduced to the voice messages giving them the route information the auditory way. To familiarise with the virtual reality of the simulator they were thereupon first asked to complete a practice session in the simulator for about ten minutes. Allowing the participants to get acquainted with driving the simulator gave the researcher a chance to check for possible simulator sickness. The participants were instructed to drive as they would normally do, following the motorway until the message occurred after which they could decide to continue or take the exit.

Including the ten-minute practice trial, the total amount of time a participant spent inside the simulator was about 40 minutes. After each trial, the participant was asked to complete a couple of questionnaires that inquired about usability and comfort of the way the information was presented. Participants were also interviewed about their own driving-performance and their perceived workload during the previous trial.

Information provision

In each trail, 1500 meters before the exit that led to the alternative route, the computer emitted a sound over the loudspeakers that alerted the participant to attend to the incoming information. In both SMS and PDA conditions, participants were then supposed to press a button to display the information. In the auditory condition the information followed automatically after the tone.

SMS
The first system used in this study was a mobile phone; short text messages (SMS) were used to display the travel information to the participant. The mobile phone was located in the simulator near the participant by means of a 'car kit'. Previous research has shown that the best way to present route information on a mobile phone is to divide the information into small, meaningful 'chunks', displaying a specific part of the entire message (Dicke & De Groot, 2005). An example of what this looked like is shown in Figure 1.

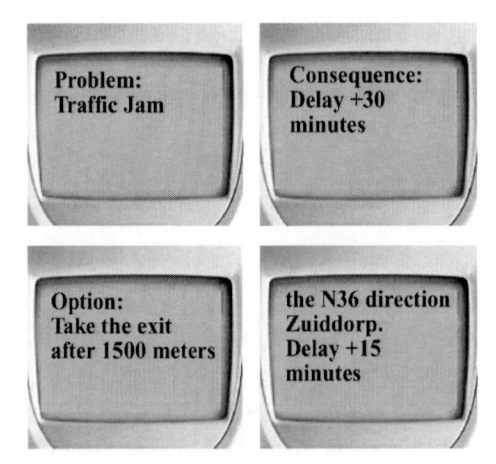

[Figure showing four mobile phone screens displaying:
- "Problem: Traffic Jam"
- "Consequence: Delay +30 minutes"
- "Option: Take the exit after 1500 meters"
- "the N36 direction Zuiddorp. Delay +15 minutes"]

Figure 1. Example of scenario 1 displayed on a mobile phone

PDA

A second system used for presenting the travel information in this study was a palmtop, or PDA. As with the mobile phone, information was divided into small 'chunks', consisting of three general themes central to each message (problem, consequence and option). Pictograms (such as a P+R sign and a bus) were incorporated in the messages on the PDA, making the message shorter, clearer, and more universal.

Auditory

The third system provided route information by auditory messages. Once again, the information was divided into three small and meaningful units, similar to the way this was done for the SMS and the PDA screens. The spoken texts were pre-recorded, and uploaded into the 'digital script' of the simulator. To enhance the messages and distinguish between the three chunks of information, two different voices were recorded; a male voice and a female voice. The male voice stated the title of each chunk of information ('problem', 'consequence', 'option') and the female voice gave the relevant information.

Dependent variables

Dependent variables were collected from three different sources. Participants were requested to complete a number of questionnaires, they were filmed during the experiment with the aim to record eye movements, and additionally a number of driving performance measures were collected and saved on the simulator computer.

Self-report measures were used to gain insight into the (experienced) workload, the (self-evaluation of) driving performance, and the (perceived) usability and comfort of the three systems by which travel-information was provided. To check whether the participants understood the route information and remembered the message, they were asked what they remembered from its content.

Self-reports of workload, driving performance and acceptance

Mental workload
To measure cognitive workload under the different conditions of this study, all participants were required to fill out the rating scale of mental effort (RSME) after each trail, a Dutch self report scale to measure at an interval level the mental effort (Zijlstra, 1993), from 0=no effort, to 100=extreme effort.

Driving performance
Participants were asked to rate their self-perceived driving performance after each trial. This measure gives an indication on how different manipulations affect the way participants believed they were driving (Brookhuis et al., 1985), from -5=very poor to +5=very well.

Acceptance
Acceptance of the systems was measured with a subjective nine-item scale to measure specific acceptance in two dimensions; a usability dimension and a comfort dimension (Van der Laan et al., 1997), while after the participants completed the experiment, a second, general usability scale measure was acquired.

Memory
To test how much of the information provided was remembered, participants were asked after each trial what they remembered from the messages that contained six items each, hence, the minimum score on this variable was 0 and the maximum was 6. The score derived is considered an indication of how well the drivers actually perceived, processed and understood the travel information while they engaged in driving.

Objective measures of safety
Perhaps the most important factor that needs to be considered when using a system designed to provide information while driving is safety. Several studies have demonstrated that when using support systems at the manoeuvre level, such as collision warning, driving performance improves, while support systems at the strategic level, such as information provision about detours, might compromise safety (see Carsten & Brookhuis, 2005). In this study, safety related driving performance was assessed by using specific objective measures. Two captures of driving behaviour were used as an indicator for driving performance and safety around the critical period (i.e. shortly before and after the message providing information); speed and time-headway to cars-in-front. The two simulator variables were measured for a total duration of 45 seconds before and 45 seconds after introduction of the auditory signal warning the participants of incoming travel-information.

Driving a car at 120 km/h, 33.3 meters of highway is covered during each and every second that the driver's gaze drifts off the road. To assess the potential danger of mobile phone or PDA usage in this sense (cf. Mourant & Rockwell, 1972), both the number of times and the total amount of time each participant looked away from the road (while reading the information on the screen) was measured through video-analysis.

Results

Self-reports

Workload

The experienced workload is depicted in Figure 2. The difference in experienced workload between the two groups of participants shows a small but non-significant effect. Natives appear to experience less workload while driving the simulator than do non-natives. A post-hoc analysis indicates that only the native participants show a significant effect for the way information is provided ($F(1,11)=4.2$, $p<0.028$). Providing information by SMS differs significantly from the other two systems in the sense that participants experienced more workload in the SMS condition compared to the other two systems.

Reported Effect (RSME)

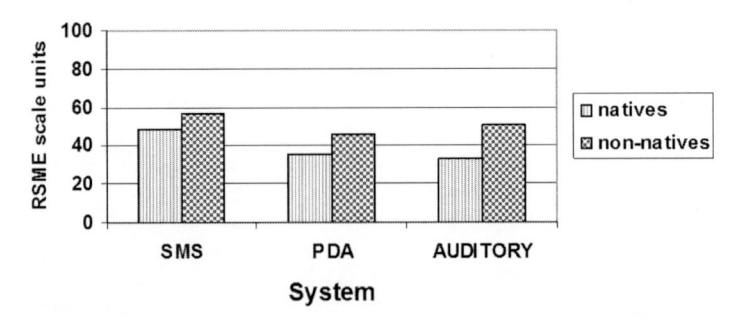

Figure 2. Comparing the two groups on self-reported effort (0 = no effort, 110 = extreme effort)

Driving performance

The system used for providing information had a small effect on how people in this study evaluated their own driving performance ($F(1,22)= 2.8$, $p<0.074$). Participants evaluated their own driving as worst when information was provided to them by SMS, and they believed their performance was best when information was provided by sound.

Usability and comfort factors in the acceptance scale

Acceptance of each system was differentiated in two dimensions, usability and comfort. Prior to experience, i.e. before driving in the simulator, and after each trial in the simulator participants were given the acceptance scale through which they were asked to indicate how useful and comfortable they thought an information-provision system (as described or experienced) would be to them.

Usability dimension

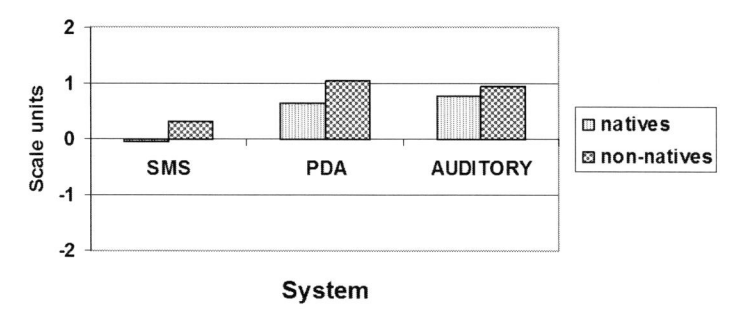

Figure 3. Comparing groups and systems on usability (scale from -2 to 2)

Comfort dimension

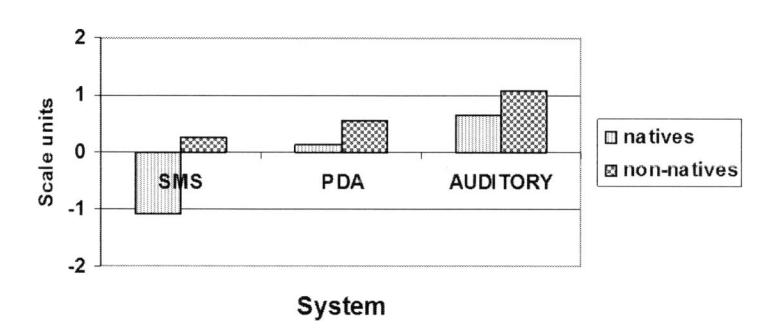

Figure 4. Comparing groups and systems on comfort (scale from -2 to 2)

The small difference that was found indicates that the native participants evaluated the information-providing systems as less useful than did the non-natives (Figure 3). Worth mentioning is that the usability scale measured prior to the research trails (the anticipated usability) already showed a similar difference between the two groups. For the dependent variable 'usability', a small interaction (F(1,22)=2.8, p<0.071) was found between the independent variables language proficiency, system and traffic density.

When looking at the graph in Figure 4, it is clear that the two groups of participants differ in the degree of comfort they experienced while information was provided to them. Non-natives experienced higher levels of comfort than natives (F(1,22)=6.1, p<0.022), with a small interaction (F(1,22)=2.7, p<0.081) with the system used. The two groups seem to differ more in their opinion of the degree in which a system is comfortable in the case where information is received by SMS. The systems used to provide information were evaluated significantly different with respect to how comfortable they were (F(2,21)=14.9, p<0.001).

Memory

Traffic density seems to have only a small effect on the number of items remembered (F(1,22)=3.5, p<0.073), contrary to expectations. However, the difference in remembered items between the two groups of participants shows a significant effect (F(1,22)=8.7, p<0.007). Natives remembered more items from the information provided to them than did non-natives, while no difference was found between the three systems (Figure 5).

Memory of information

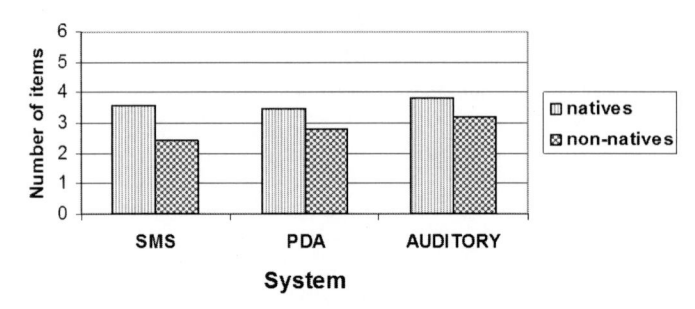

Figure 5. Average number of items remembered for each group (from 0 to 6)

Objective measures, performance and safety

Speed

The two participant groups differed slightly but not significantly in the speed they kept. After the warning signal was presented and participants were required to attend to the provided travel-information, they reduced their speed by about 7 km/h (F(1,22)=39.6, p<0.001), however, quite differently for the various conditions, see Table 1.

Table 1. Mean speed (km/h) before and after the signal indicating incoming information

	SMS		PDA		AUDITORY	
	pre	post	pre	post	pre	post
Natives (Busy)	95.8	96.9	93.6	90.1	96.9	93.3
Natives (Calm)	112.7	111.9	108.2	98.1	107.2	92.7
Non-Natives (Busy)	88.5	73.2	90.7	86.3	92.9	88.8
Non-Natives (Calm)	102.8	89.2	94.9	86.1	97.2	92.7

A rather small, non-significant effect was found for the system by which information was provided, whereas a large effect was found for traffic density on mean driving speed (F(1,22)=12.2, p<0.002). Participants tended to drive significantly faster when traffic density was low, compared to the situation in which traffic density was high, as is illustrated in Figure 6.

Speed

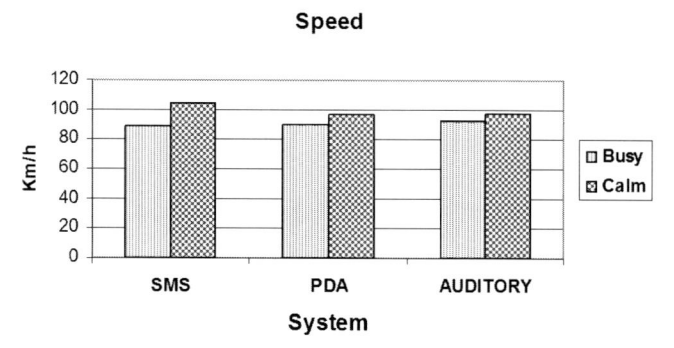

Figure 6. Average mean speed in different traffic density conditions

Time headway

Mean time headway varied from 1.49 up to 4.77 seconds throughout the different conditions. One of the (non-native) participants brought the simulator car to a complete standstill after the signal in the busy condition where information was provided by SMS, a dangerous outlier. It turned out that time headway only slightly increased after the signal but not significantly. Thus the distance participants kept did not change after the signal indicative of incoming information.

The mean percentage time headway under 5 seconds gives the amount of time the participants drove within a 5 second distance from a leading car. A higher percentage of time headway under 5 seconds indicates less safe driving behaviour. With respect to the three different systems, no significant effect was found but traffic density did have a significant effect ($F(2,21)=7.3$, $p<0.013$) as visible in Figure 7. In the busy condition the mean percentage of time headway under 5 seconds is longer, compared to the calm condition, indicating that in busy traffic, participants drive relatively close to the car in front of them for a longer amount of time than they would under calm traffic conditions.

Percentage time headway under 5 seconds

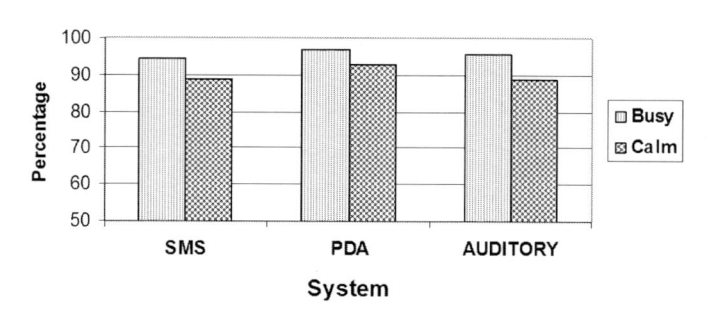

Figure 7. Percentage of time headway under 5 seconds for the systems in different traffic density conditions

Duration of eye fixations

To investigate how long it takes for a participant to read and understand the message provided to them, the total amount of time they shifted their gaze off the road to the information providing device was analysed. Figure 8 indicates that when participants received the information by SMS, they looked away from the road and at the system for a longer total amount of time than when they received the information on the PDA ($F(1,22)=4.2$, $p<0.054$). Density of the surrounding traffic did not have an effect at all, while the two groups differed slightly in the total amount of time their gaze was directed away from the road ($F(1,22)=3.9$, $p<0.061$). From Figure 8 it can be concluded that, as might be expected, non-natives take more time to inspect the message presented to them.

Total time of eye fixations

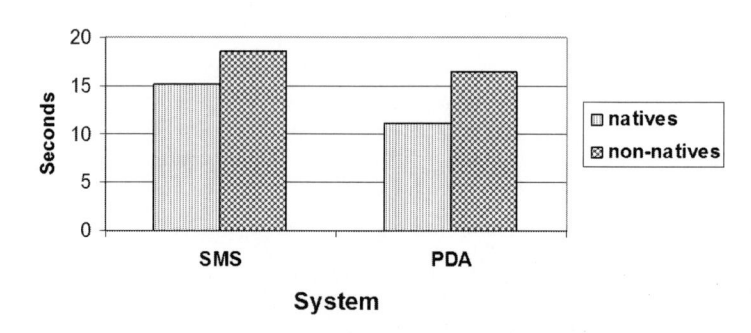

Figure 8. Duration of time the two groups attended to the information providing system

The average duration of each eye fixation differed considerably when the information is provided by SMS, 1.2 s, as compared with the PDA, 1.0 s ($F(1,22)=15.6$ $p<0.001$).

Discussion

When comparing the systems used for providing information, SMS is consistently evaluated as the least usable and comfortable. The other two systems generally do not differ much from one another, but when they do, auditory information provision is evaluated as being most comfortable. Non-native speakers consistently view the information-providing systems as more useful and comfortable than natives do. This might be attributed to a genuine difference between the two groups, perhaps because the non-native participants would benefit more from such a system. Non-natives also tend to evaluate the SMS better than natives do, which might be caused by the fact that with an SMS it is possible to read the message more than once, which is beneficial when you are not fluent in the language at hand. As was expected, the degree of traffic density didn't have an effect on usability.

It was hypothesized that visual systems would compete more strongly for attention with the driving task than the auditory system. But only SMS was found to cause higher experienced workload. From this it can be concluded that pictograms (PDA) and auditory information work better than SMS does. Memory performance wasn't affected at all by the type of system used. Natives did remember more of the Dutch travel-information as was expected in SMS and auditory conditions, but they also remembered more when symbols were used (PDA). A confounding factor may be the level of education, which was almost consistently lower for non-native participants.

When participants were asked about their own driving performance, they believed they drove best when an auditory system presented the information, slightly worse when a PDA was used, and worst when an SMS was used. Although this is consistent with the results found for usability, the objective driving performance measures did not show that the type of system used had an effect on driving performance. Driving performance, and thus safety, does not seem to be affected much by the type of system used. However, when comparing SMS and PDA with respect to the amount of time participants looked away from the road, there is a difference between the two. As expected, drivers took longer to look at an SMS than they did with a PDA, implying that a PDA is safer than an SMS, however, both systems might still compromise safety. Previous research has found that drivers feel safe when glances are shorter than 0.8 seconds (Green, 1999). In this study, looking at the SMS takes about 1.2 seconds, and looking at the PDA occupies participants close to a full second. Moreover, specifically SMS violated the 15 second-rule (Green, 1999), i.e. the total time needed for acquiring information from a device should not exceed 15 seconds, be it that the rule holds for non-moving vehicles. A possible solution to this problem might be to focus on creating a more user-friendly design for the displays, or by making use of auditory displays. However, even verbal tasks still lead to some degree of interference (Lee et al., 2001) and they do not eliminate distraction altogether (Strayer et al., 2003).

Conclusion

The present study shows that reading an SMS on a mobile phone while driving, even if it is placed in a car-kit, is not safe. Despite this finding, the mobile phone is used predominantly by the current traveler population. Nevertheless, new technological advances are a big step forward compared to the old paper roadmap. In this study, PDA and auditory provision of information were found to be relatively safe ways to provide drivers with the information they need. The other side of the coin is that many advances in technology may also cause substantial distraction to the driver. Cohen and Graham (2003) have estimated that eliminating the use of mobile phones would save 2,600 lives and prevent 330,000 injuries annually in the U.S. alone. Information is power, but at what price...

Acknowledgement

This study was carried out in the framework of the Dutch Transumo Program (www.transumo.nl)

References

Baddeley, A.D., & Hitch, G.J. (1974). Working Memory. In D. Groome (Ed.) *An introduction to cognitive psychology* (vol. 5). East Sussex, UK: Psychology Press Ltd, Publishers

Brookhuis, K.A., De Vries, G., & De Waard, D., 1991. The effects of mobile telephoning on driving performance. *Accident Analysis and Prevention, 23*, 309–316.

Brookhuis, K.A., De Vries, G., Prins van Wijngaarden, P. (1985). *The effects of increasing doses of Meptazinol (100, 200, 400 mg) and Glafenine (200 mg) on driving performance (Report VK 85-16)*. Haren, The Netherlands: Traffic Research Centre, University of Groningen.

Carsten, & Brookhuis, K.A. (2005). The relationship between distraction and driving perfomance: towards a test regime for in-vehicle information systems. *Transportation Research Part F, 8*, 191-196.

Casey, S. (1993). *Set phasers on stun and other true tales of design, technology and human error.* Santa Barbara, C.A., USA: Aegean.

Cohen, J.T. & Graham, J.D. (2003). A revised economic analysis of restrictions on the use of cell phones while driving. *Risk Analysis, 23*, 1-14.

De Jong, C., & Brookhuis, K.A. (2005). *De haalbaarheid van betrouwbare reisinformatie*. (The practibaility of reliable travel information). Report to ProRail, Groningen: University of Groningen.

Dicke, M., & De Groot, J.I.M. (2005). *Passende oplossing voor reisinformatie op mobiele telefoons* [Fitting solution for travel information on mobile phones]. In R. Clement (Ed.), (pp. 1687-1706). Rotterdam: Colloquium Vervoersplanologisch Speurwerk.

Dicke, M. (2008). PhD Thesis, University of Groningen.

Green, P.A. (1999). *Visual and task demands of driver information systems* (UMTRI 98-16). Ann Arbor: University of Michigan Transportation Research Institute.

Lee, J.D., Caven, B., Haake, S., & Brown, T.L. (2001). Speech-based interaction with in-vehicle computers: The effect of speech-based e-mail on drivers' attention to the roadway. *Human Factors, 43*, 631-640.

Mourant, R.R., & Rockwell, T.H. (1972). Strategies of visual search by novice and experienced drivers. *Human Factors, 14*, 325-335.

Strayer, D.L., Drews, F.A., & Johnston, W.A. (2003). Cell phone-induced failures of visual attention during simulated driving. *Journal of Experimental Psychology: Applied, 9*, 23-32.

Van der Laan. J.D., Heino, A., & De Waard, D. (1997). A simple procedure for the assessment of acceptance of advanced transport telematics. *Transportation Research—Part C: Emerging Technologies, 5*, 1–10.

Wickens, C.D., & Hollands, J. (2000). Engineering psychology and human performance (3rd ed.). Upper Saddle River, N.J., USA: Prentice Hall.

Wickens, C.D., Lee, J.D., Liu, Y., & Becker, S.E.G. (2004). *An introduction to human factors engineering* (2nd ed.). Upper Saddle River, N.J.: Prentice Hall

Zijlstra, F.R.H., (1993). *Efficiency in work behavior. A design approach for modern tools.* PhD Thesis, Delft University of Technology. Delft, the Netherlands: Delft University Press.

Supporting the localisation of task-relevant information

Martin Groen & Jan Noyes
University of Bristol
Bristol, UK

Abstract

There is ample evidence that humans find it difficult to locate information in a variety of information sources. Therefore, providing support for humans to locate task-objective relevant information could be beneficial. Providing generic search support implies that task and context assumptions need to be considerably relaxed as they may vary a great deal between and within humans. Relinquishing identification of task and task context, the question concerns which elements of an information set (called 'relevancy markers') are used in human search to locate task-relevant information in a variety of presentation formats. Results from a series of experiments that examined these markers as they are used in dialogues are reported. Participants read, or listened to, 3 dialogues and indicated where goal-relevant information was exchanged. The results show that across domains, languages and media formats people appear to orient on the same markers to locate task-relevant information. Due to the communicative nature of information supply, these results can be used to inform the design of search support interfaces.

Introduction

When carrying out tasks, people often need to consult external information sources in order to gather missing or additional task-relevant information. It is well documented, however, that humans have difficulty locating relevant information in a variety of information sources in different media, such as the Internet (Jenkins et al., 2003; Schacter et al., 1998; Shneiderman, 1997; Smith et al., 1997); academic libraries (Haynes & Wilczynski, 2004; Waldman, 2003); and workflow systems (Dustdar & Gall, 2003). In addition, it is not just the characteristics of the information source that makes the search task difficult, but also the characteristics of the search task itself appear to be problematic (Brand-Gruwel et al., 2005; Lazonder, 2000). Therefore, providing computational support for humans to locate task-objective relevant information could be beneficial, leading to shorter search time and a swifter task accomplishment. To design computational search support, a low-level bottom-up approach could be adopted in order to identify task-relevant features of an utilised information source and then to highlight them in some way in order to alert the information searcher to their usefulness. This approach is adopted in this study.

There are some issues with providing search support. One difficult aspect is the determination of the task objectives of the user. This relates to the notorious *frame problem*. With the frame problem, philosophers (e.g., Dennett, 1978; Fodor, 1983) raised the issue of how a rational agent decides which entertained beliefs need to be changed and which can be ignored when an action is executed? In terms of providing support, the frame problem translates to the challenge, posed to providers of support, of defining what constitutes an adequate identification of the perceived task objectives of the intended support recipient(s). This vexing problem is compounded by a particular characteristic of search problems. That is, determining the search task objectives is also difficult for the searcher (Brand-Gruwel et al., 2005; Van Merriënboer, 1997). Often, the objectives only become clear incrementally (e.g., Langley et al., 2005; Lave & Wenger, 1991; Wieth & Burns, 2000). Given the problems with determining the search task and its context and the incremental nature of the identification of search goals, is it possible to provide generic support without having information of the task or its context? This study sets out to examine this issue.

Another problem with providing support is the extent into which the task goals have been clarified. If the goal is not known, the task search space is, in principle, unconstrained. Unconstrained search is generally assumed to be too costly to be justified (e.g., Colburn, 1995; Fu & Gray, 2006; Newell, 1990). The problem of unconstrained search is that with every step, all possible successive actions need to be considered before deciding which step is going to be next. After just a few steps, this will lead to a combinatorial explosion of possibilities which will be impossible to consider exhaustively in a reasonable time frame. Colburn (1995) illustrated this problem with a robot that has to make a decision whether it will turn north-, east-, south- or west. If the robot is instructed to do an unconstrained search and therefore has to weigh all the alternatives, a maximum of $4^3 = 64$ different possibilities will need to be considered. With less straightforward problems, and even with problems of modest difficulty, the search space of possible successive steps quickly becomes too large. Therefore, in these situations, it is likely that knowledge of constraints, in particular those that do not require an understanding of the task or task context in hand, is available to aid the searcher in locating potential useful task-relevant information. That is, if there was no knowledge of constraints, then the search could, in principle, go on indefinitely. These structural constraints will be made available by the originator of the information in order to aid the searcher to localise task-relevant information. Paragraphs, for example, will be created by the originator to ease the perusal of the provided text, or an arrow will be drawn to highlight the region of interest in a diagram. But this raises the question of why would the originator want to aid the searcher in locating the task-relevant information.

A communicative approach to solving and supporting search problems

It is taken as a given that the originator of the information source decided at some point that it would be useful for other people to know of the information too. S/he created therefore an information source in some format in order to communicate it to an intended audience (that is, a communicative intention, e.g., Bangerter & Clark, 2003; Grice, 1989; Hommel et al., 2001). To improve the chances that the task-relevant information is recognised, and acted upon, the originator includes structural constraints in order to guide the searcher towards the location of the information deemed relevant for completing a task. It is in his or her interest to ensure that the task-relevant information is recognised as such, because communicating the task-relevant information was the overriding motivation to create the information source in the first place. But how can the originator be assured that the structural constraints are recognised as signals aimed at suggesting to the addressee the location of potential task-relevant information?

It is suggested, that the originator benefits from the conventional status of these structural constraints. Conventions (Lewis, 1969) are defined as "a community's solution to a recurrent coordination problem" (Clark, 1996, p. 70). To enable their operation, conventions need to be recognised as such in a community and members expect that they are conformed to (see Lewis, 1969 for an extensive definition and elaboration of the social machinery of conventions). Highlighting task-relevant information is a recurrent problem in a community, as people have to, intend to or need to, exchange task-relevant information regularly in order to realise their goals or assist in realising other people's goals. Using conventional signals to do the highlighting task improves the chance that this information is recognised as intended by the originator. So, it is in the originator's interest to use conventional signals to highlight relevant information. That is, if the originator decided to use non-conventional signals, the chances are that the relevant information might go unnoticed. By using conventional markers this possibility is less likely to happen. In addition, it communicates to the addressees that the originator intends the highlighted information to be recognised as relevant.

The question of interest is then, which conventional markers, or *relevancy markers*, are potentially used by originators to highlight task-relevant information? To produce an initial candidate list of relevancy markers, a straightforward choice considering the communicative approach to information search is to consult the field of linguistics.

In linguistics, *discourse markers* seem to play this accentuating role as ascribed to relevancy markers. Discourse markers (Fraser, 1999; Schiffrin, 1987; Schourup, 1999) are word-like forms that are used at the beginning of a contribution to a discourse to signal the contribution's relationship to the current status of the discourse. For example, they can be used to mark changes in the global discourse structure, as exemplified by the terms 'by the way' to mark the start of a digression and 'anyway' to mark the return from one. Take the following excerpt from the British National Corpus (*British National Corpus, Version 2 (BNC World)*, 2000). In it two people are talking about lodgers. Suddenly, one of the dialogue partners remembers that someone told her that he wanted to borrow something which is introduced by the discourse marker 'by the way': "...I suppose Carol rented that place with you. mm, so do I, I would have said. Oh by the way, Dave wants to borrow that. Keith come back here and say have you bought out a video and I'll say oh no I couldn't, Carol wouldn't let me bring it..." (*British National Corpus, Version 2 (BNC World)*, 2000).

In the context of this research, a contribution to a dialogue is the item said when the speaker starts talking until the addressee takes his or her turn to speak (Sacks, 1995). Schiffrin's (1987) classic work identifies 11 discourse markers which are adopted as a first candidate set of word-like forms that can be tested whether they could function as relevancy markers. There are two important aspects of the definition of discourse markers that make them suitable for accepting them as a candidate set of relevancy markers. First, discourse markers are used to highlight a diversion in the supplied information, indicating to the addressee that the information presented afterwards should be treated as such. This could therefore be indicative of the upcoming presentation of relevant information. Second, they are uttered at the beginning of a contribution. This limits the potential problem of the multitude of grammatical roles particular words could play. For example, the word 'and' is considered to be part of the set of discourse markers. However, 'and' could also be used as a conjunction of two sentences. By limiting the presence of discourse markers to the beginning of a contribution a potential misattribution as discourse markers could be prevented. A precise definition of what 'beginning of a contribution' precisely entails could not be found in the literature. Therefore, for this research this was defined as in the first *four* words of a contribution.

As indicated, the presence of relevancy markers in an information source is one of the effects of the communicative act of providing information. The results of the activities of these communicative acts can probably be most directly studied in dialogues. The reason for this is that this medium demonstrates almost instantaneously the effects of the presentation of relevancy markers by studying the response of the addressee(s). It is suggested that a similar mechanism is also utilised in other media formats, as they are produced to serve communicative means too. There is a problem, however, in that dialogues are mostly ephemeral phenomena, making it difficult to study particular patterns of information localisation in them. Therefore, transcripts of dialogues were used in order to enable the precise determination of the particular stretches of dialogues that are highlighted as relevant by the participants. In face-to-face dialogues, a multitude of relevancy markers is available for selection, but in situations where speaker and addressee cannot see each other, the choice of signals is constrained. For example, in telephone conversations, speakers can only highlight goal information by using verbal relevancy markers or a subset of non-verbal relevancy markers. As telephone conversation transcripts were used in the present work, we decided to focus on verbal relevancy markers at this stage. In addition, whether the identified relevancy markers are also used in actual, real-time dialogue was also examined.

In summary, the research reported here examines which relevancy markers provide a benefit for originators and searchers when providing or gathering new or additional task-relevant information. As said, the originators will adhere to conventions in their choice of particular markers and searchers expect the originators to conform to them. Note, that to ensure that the addressee recognises the highlighted relevant information as such, the originator is expected to use conventional relevancy markers to highlight the segment of information that is deemed relevant. To investigate this, the consistency of their use across different contexts, domain of work, topic domain and language community, is considered.

Experiments

The first objective is to determine a candidate set of relevancy markers. This was done in a pilot study which is described in the next section. Following this, three experiments are reported that examine whether the candidate relevancy markers are of benefit to participants in their localisation of relevancy markers and whether these markers are consistently present in the responses of the participants across topics, work domains and language communities. These experiments are also discussed in Groen and Noyes (2007) and are included here to help establish the argumentation. Additionally, a further experiment is reported that was conducted to study the localisation of relevant information in recordings of spoken dialogue. This last experiment has not been reported yet.

Pilot work

Schiffrin's (1987) classic work identified 11 discourse markers which are adopted as a first candidate set of word-like forms that can be tested to check their function as relevancy markers. In the current study, a data set of 15000 email messages in Dutch was used to search for discourse markers in the first four words of every utterance in all messages. These email messages concerned questions relating to financial services. It should be noted that this set of emails was only used to assess the usage frequency of discourse markers. The experiments were conducted with transcripts of telephone conversations. The following list was used to this end: 'well', 'now', 'so', 'but', 'oh', 'because', 'or', 'I mean', 'and', 'Y'know' and 'then' (adopted from Schiffrin, 1987). With the assistance of a native English speaker these 11 terms were translated to Dutch and the frequency of occurrence was gauged in the collection of email messages. After establishing the frequency, a random sample from 20 recorded telephone conversations, collected by one of the participating organisations for evaluation and training purposes, was drawn to inspect the usage in this target type of dialogues. These conversations were transcribed and analysed. The telephone conversation transcripts were used to see which of the high frequent discourse markers were turn-initial in these transcripts. The discourse markers that had the highest frequency of usage *and* that were turn-initial in the transcripts were tagged as candidate relevancy markers. These were selected to be tested empirically in the experiments. The words 'well', 'so' and 'but' had the highest frequency in a turn-initial position. Although the words 'and', 'or' and 'then' had higher frequencies, they were not considered because they were not turn-initial in any of the contributions in the random sample from transcripts of telephone conversations.

The experimental task

The task consisted of a relevant information search task in transcripts of three dialogues. The dialogues were naturalistic telephone dialogues, between customers and representatives of a financial services organisation and a telephone helpdesk for logistics problems. The dialogues were matched for time duration. Participants were requested to read the first transcript and then answer four questions about the dialogues. Three of these questions were aimed to elicit subjective judgements about the success of the dialogue partners in establishing their goals with the dialogue. The crucial question for present purposes, however, was the fourth question. This item requested the participants to mark on the transcripts which segments of dialogue they used to answer the first three items. With the responses to this fourth question, the localisation of relevant information could be examined with regard to whether this was influenced by the presence of a candidate relevancy marker. With this selection of dialogue transcripts, relevancy markers could be studied across work (i.e., participants were recruited from two different commercial organisations) and topic domains (i.e., format consistent dialogues concerning financial services and logistics).

Experimental findings

Text experiments

The experiments were conducted first with Dutch language speakers (n = 31), and later, with English (n = 66) and Mandarin-Chinese (n = 37), since these latter languages have a more widespread use. Participants received a booklet with transcripts of three actual telephone conversations. In the Dutch version of the experiment, two dialogues had a topic from the financial-services domain and one was about a logistics problem. In the English and Mandarin-Chinese experiment sessions all three dialogues were about logistics problems. The order of the dialogues was counterbalanced. After reading each transcript, the participants answered four questions relating to the transcripts. The first three questions requested the participants to assess the extent to which the dialogue partners, from each of the transcripts they had just read, succeeded in establishing their own or mutual goals. The fourth question requested participants to indicate on the transcript which stretches of dialogue were used to answer the first three questions. The distance in words (or characters in Mandarin-Chinese) was calculated between the participants' indicators and the candidate relevancy markers. See Figure 1 for an example of a text with relevant information highlighted by a participant. The results indicate that in Dutch, English and Mandarin-Chinese the participants consistently oriented on, or were oriented by, the presence of the relevancy marker ('so', 'well' or 'but') to locate task-relevant information. This outcome was not influenced by the participants being members of a particular language community (i.e., Dutch, English and Mandarin-Chinese), topic of the dialogue or domain of work. More details of these experiments and the pilot can be found in Groen, Noyes and Verstraten (submitted).

Audio experiment

In the audio experiment the stimuli were recordings of three spontaneous conversations. Whereas in the text experiments, the transcripts were from dialogues with representatives from organisations, the recordings are between interlocutors where the relationship is less formal. If the candidate relevancy markers aid the information searcher in this type of material too, then there is additional credence to their functioning as conventional relevancy markers. That is not only across languages, topics and domains, but also across presentation formats. Eighty-four participants listened to the recordings via headphones. Order of playback of the recordings was counterbalanced. Participants were seated at a computer and pressed the Enter key whenever they thought that goal-relevant information was presented. Key strokes were recorded and the reaction times were analysed. The results showed that in 80% of the cases participants indicated the presence of goal-relevant information within a 10 second time frame after the candidate relevancy markers were presented. Details of this experiment and the text experiments can be found in Groen et al. (submitted).

```
utt61 :     and then <silence> back down to Corning <silence> to load <silence> the
            <silence>
            oranges
utt62 : A:  okay so it'll take three hours to get to Dansville
utt63 :     and uh we'll <silence> pick up <silence> the <silence> boxcars <silence>
            and go back to Corning
utt64 :     and then we gotta load oranges we'll take another hour
utt65 :     um and then two hours to get to Bath so that will total <breathe>
            will take <silence> uh seven hours to do that
utt66 :     + <click> okay +
utt67 : B:  + is there <silence> + there's no other quicker way right
utt68 : A:  uh I <silence> don't <silence>
            think so let's see what so the one that takes the longest is to make
            <silence> the orange juice
utt69 :     and there doesn't seem to be a <silence> quicker way to <silence>
            do that because we have to go get the oranges <silence> and back again
utt70 :     right no that looks like the fastest
utt71 : B:  okay <silence> so
utt72 :     but do we still need one more boxcar
utt73 : A:  uh
utt74 : B:  I think
utt75 : A:  no I think we have <silence> two with the first engine <silence>
            + with the <silence> + the we have the orange juice in two <breathe>
            oh how many did we + need +
utt76 : B:  + uh- +
utt77 :     + we needed + five so + well +
utt78 : A:  + oh we needed + five in <silence> + uh total +
utt79 : B:  + yeah well +
utt80 : A:  oh we need one more yes
utt81 : B:  so
utt82 : A:  oh what we
utt83 : B:  couldn't we just u- <silence> attach that <silence>
            you said three boxcars + were for one engine + ]3
utt84 : A:  + right we could've just taken + three from Dansville
utt85 : B:  okay <silence> then <silence> the problem's finished
utt86 : A:  right <silence> okay ]
```

Figure 1. Excerpt from a booklet with a participant's indications of task-relevant text

Discussion

In Groen & Noyes (2007) it was reported that in English, Dutch and Mandarin-Chinese the words 'so', 'well' and 'but' (and their respective equivalences in Dutch and Mandarin-Chinese), appeared to aid the information searcher to locate task-relevant information across disparate language communities, topics and work domains. In this paper, experimental results are reported that indicate that these markers appear to exert their influence in recordings of conversations too when a human is required to locate goal-relevant information. Overall, the results indicate that the relevancy markers appear to be structural elements that people use in dialogues to highlight and locate the presence of task-relevant information and that are not specific to particular domains, topics, languages and media formats.

There is ample evidence of the low ability of humans to locate relevant information and of their problems with the search task itself. In order to support users to locate task-relevant information, it is suggested that they could benefit from support in which the information deemed relevant is automatically highlighted. A series of experiments have been conducted to test a possible provision of this functionality. The results are sufficiently consistent to present an opportunity to design this relevant information search support. This finding is very relevant to the design of support via human-computer user interfaces. The provision of generic, task-independent search support is preferred, as the determination of a person's context is notoriously difficult and problematic (Faraday & Sutcliffe, 1997; Grice, 1989; Mayer & Moreno, 2003). In addition, searchers become often incrementally aware (Langley et al., 2005; Lave & Wenger, 1991; Wieth & Burns, 2000) of their task goals.

Due to the problems with context determination and goal setting, it is important to avoid making assumptions about the task or its setting. Therefore, a bottom-up approach is adopted, focusing on low-level structural elements of information streams. As participants appear to orient on, or are influenced in their orientation by, the candidate relevancy markers, it could prove to be effective to assist searchers in finding task-relevant information by highlighting the candidate relevancy markers in some way.

There is some evidence that highlighting task-relevant information could be effective (Faraday & Sutcliffe, 1997; Grant & Spivey, 2003; Mayer & Moreno, 2003). In Grant and Spivey (2003), for example, a study is reported that showed that the performance of participants on an insight problem solving task can be improved by highlighting the task-relevant information. In a two-step study, first the eye-movements of participants working on a task were measured and analysed. Successful participants appeared to focus on different parts of the display than unsuccessful ones. In a second experiment the elements of the display focussed on by the successful participants were highlighted, assuming that this would assist the new participants in realising the task. Compared to other conditions in which this support was not provided, the participants performed significantly better (66% correct as opposed to 33%). Faraday and Sutcliffe (1997) reported a study in which eye movements of participants were investigated to examine the order in which visual information in a multimedia presentation is processed. Scan paths of the eye movements of participants over the displays were analysed and design recommendations were generated. Although not the primary focus of Faraday and Sutcliffe's study, the design recommendations seemed to centre on supporting the recognition of goal-relevant information. The design recommendations were used to re-author the multimedia presentation and an experiment was conducted to measure the effect of this redesign on recall. The results indicated that recall significantly improved when participants used the re-authored version as opposed to the original version of the multimedia presentation. It also appeared that highlighting the relevant information with text *and* images is important to establish the improved recall.

Mayer and Moreno (2003) investigated how instructional material can be adapted to allow mitigation of cognitive overload. *Cognitive load* is the result of the substantial cognitive processing that is required when a human encounters new or unknown material. *Cognitive overload* occurs when the processing demands exceed the cognitive resources of the human. Based on a 12-year research programme, Mayer and Moreno put forward a number of recommendations for the provision of support. One of the recommendations, which pertains to the theme of this paper, is to reduce cognitive load by providing cues about how to select and organise the material, called signalling. The results reported indicate that participants who worked with an information source embellished with signals (e.g., by stressing key words in speech, adding red and blue arrows to images, adding outlines and headings) performed significantly better on a problem solving transfer task, that is, a task where the solution could be accomplished by applying the acquired knowledge from a previous session. In the current study, similar task-relevant information elements have been found. The next step is to study whether highlighting them in a user interface (as detailed by Mayer & Moreno and Grant & Spivey) could aid task performance in a similar manner.

Acknowledgements

The authors would like to acknowledge funding and support for this work from GCHQ in Cheltenham, UK and the companies Achmea and Nuon in the Netherlands. Thanks to Meng Zhang and Dr Timothy Teo for their assistance with the Mandarin-Chinese study.

References

Bangerter, A., & Clark, H.H. (2003). Navigating joint projects with dialogue. *Cognitive Science, 27*, 195-225.
Brand-Gruwel, S., Wopereis, I., & Vermetten, Y. (2005). Information problem solving by experts and novices: Analysis of a complex cognitive skill. *Computers in Human Behavior, 21*, 487-508.
British National Corpus, Version 2 (BNC World). (2000). Distributed by Oxford University Computing Services on behalf of the BNC Consortium. http://www.natcorp.ox.ac.uk/.
Clark, H.H. (1996). *Using Language.* New York: Cambridge University Press.

Colburn, T.R. (1995). Heuristics, justification, and defeasible reasoning. *Minds and Machines, 5*, 467-487.
Dennett, D. (1978). *Brainstorms*. Cambridge, MA: MIT Press.
Dustdar, S., & Gall, H. (2003). Architectural concerns in distributed and mobile collaborative systems. *Journal of Systems Architecture, 49*, 457-473.
Faraday, P., & Sutcliffe, A. (1997). Designing effective multimedia presentations. In S. Pemberton (Ed.), *Proceedings of the SIGCHI conference on Human factors in computing systems* (pp. 272-278). New York: ACM Press.
Fodor, J.A. (1983). *The modularity of mind*. Cambridge, MA: MIT Press.
Fraser, B. (1999). What are discourse markers? *Journal of Pragmatics, 31*, 931-952.
Fu, W.T., & Gray, W.D. (2006). Suboptimal tradeoffs in information seeking. *Cognitive Psychology, 52*, 195-242.
Grant, E.R., & Spivey, M.J. (2003). Eye movements and problem solving: guiding attention guides thought. *Psychological Science, 14*, 462-466.
Grice, H.P. (1989). *Studies in the way of words*. Cambridge: Harvard University Press.
Groen, M. & Noyes, J. (2007). Locating task-objective relevant information in text. In D. de Waard, G.R.J. Hockey, P. Nickel, and K. Brookhuis (Eds.), *Human Factors Issues in Complex System Performance* (pp. 387-398). Maastricht, the Netherlands: Shaker.
Groen, M., Noyes, J., & Verstraten, F. (submitted). The effect of substituting discourse markers on their role in dialogue. Manuscript submitted for publication.
Haynes, R.B., & Wilczynski, N.L. (2004). Optimal search strategies for retrieving scientifically strong studies of diagnosis from Medline: analytical survey. *BMJ, 328*, 1040.
Hommel, B., Pratt, J., Colzato, L., & Godijn, R. (2001). Symbolic control of visual attention. *Psychological Science, 12*, 360-365.
Jenkins, C., Corritore, C.L., & Wiedenbeck, S. (2003). Patterns of information seeking on the Web: A qualitative study of domain expertise and Web expertise. *IT & Society, 1*(3), 64-89.
Langley, P., Choi, D., & Rogers, S. (2005). *Interleaving learning, problem solving, and execution in the Icarus architecture* (Technical Report). Stanford, CA: Computational Learning Laboratory, Stanford University.
Lave, J., & Wenger, E. (1991). *Situated learning: Legitimate Peripheral Participation*. New York: Cambridge University Press.
Lazonder, A.W. (2000). Exploring novice users' training needs in searching information on the WWW. *Journal of Computer Assisted Learning, 16*, 326-335.
Lewis, D.K. (1969). *Convention: A philosophical study*. Cambridge, MA: Harvard University Press.
Mayer, R.E., & Moreno, R. (2003). Nine ways to reduce cognitive load in multimedia learning. *Educational Psychologist, 38*, 43-52.
Newell, A. (1990). *Unified theories of cognition*. Cambridge, MA: Harvard University Press.
Sacks, H. (1995). *Lectures on Conversation*. Malden, NJ: Blackwell.

Schacter, J., Chung, G.K.W.K., & Dorr, A. (1998). Children's internet searching on complex problems: Performance and process analyses. *Journal of the American Society for Information Science, 49*, 840-849.

Schiffrin, D. (1987). *Discourse Markers*. New York: Cambridge University Press.

Schourup, L. (1999). Discourse markers. *Lingua, 107*, 227-265.

Shneiderman, B. (1997). Designing information-abundant web sites: issues and recommendations. *International Journal of Human Computer Studies, 47*, 5-29.

Smith, P., Newman, I., & Parks, L. (1997). Virtual hierarchies and virtual networks: some lessons from hypermedia usability research applied to the World Wide Web. *International Journal of Human-Computer Studies, 47*, 67-95.

Van Merriënboer, J.J.G. (1997). *Training complex cognitive skills*. Englewood Cliffs, NJ: Educational Technology Publications.

Waldman, M. (2003). Freshmen's use of library electronic resources and self-efficacy. *Information Research, 8*, 8-2.

Wieth, M., & Burns, B. (2000). Motivation in insight versus incremental problem solving. In L. Gleitman and A. Joshi (Eds.), *Proceedings of the 22nd Annual Meeting of the Cognitive Science Society* (pp. 1288-1304). Philadelphia, PA: Erlbaum.

Virtual Reality in HMI research: attentional and motor performance in 3D-space

Claudia Armbrüster[1], Marc Wolter[2], Torsten Kuhlen[2], Will Spijkers[1], & Brunno Fimm[3]
[1]Department of Psychology, RWTH Aachen
[2]Virtual Reality Group, RWTH Aachen
[3]Neuropsychology, Department of Neurology, RWTH Aachen
Germany

Abstract

Virtual reality (VR) offers the possibility to simulate reality and is highly controllable. Therefore, VR provides much potential for HMI (Human-Machine Interaction) research. Bülthoff, Foese-Mallot and Mallot (2000) describe the benefit of VR on three dimensions: control, realism and interactivity. One aspect of interactivity is the direct manipulation of virtual objects with our hands. This includes touching virtual objects without haptic feedback.

This paper presents an experimental study in which pointing movements were performed by 20 participants in VR. Psychomotor performance in three-dimensional space, as well as reaction time in two- and three-dimensional space was recorded. The virtual environment was displayed on a stereo rear projection screen. Hand and head movements were recorded with an optical tracking system. The influence of movement direction and attentional processes on virtual pointing movements as well as effects of validity and activation were analysed. Furthermore, attentional performance in a two-dimensional computer based test was compared with the attentional performance in virtual space (three-dimensional). Results show typical patterns for the orientation of attention in VR, which are comparable to results from standardised tests. Motor behaviour is affected by movement direction and it can be shown that pointing and touching is a very efficient manipulation possibility in virtual applications.

Introduction

Attention and motor performance

Attentional processes and psychomotor performance characterise human behaviour. In daily life we move our body in three-dimensional space and our movements depend mainly on information from our environment. All information has to be selected and combined to control our behaviour. Attention is considered to be an important aspect of those selection and control processes. We focus on information

In D. de Waard, F.O. Flemisch, B. Lorenz, H. Oberheid, and K.A. Brookhuis (Eds.) (2008), *Human Factors for assistance and automation* (pp. 457 - 468). Maastricht, the Netherlands: Shaker Publishing.

which is important for us, e.g. if we intent to grasp an object we have to direct our attention to it. On the physical side, that means that we move our eyes to fixate the object and prepare our motor system to initiate the grasp movement. On the attentional side, we shift our attention to the object to be able to fulfil a reasonable movement. It is known that there exists a direct relationship between motor performance and the orientation of attention (Mattingley et al., 1998). This suggests that visuo-spatial attentional networks can be influenced by the activation of brain areas being associated with movement planning and movement execution and vice versa.

Therefore the motivation for this empirical study is the examination of the influence of visuo-motor attentional processes on the planning and execution of pointing movements in three-dimensional space. In particular this study focuses on the temporal interdependence of attentional and motor processes, i.e. if movements can be influenced by the manipulation of the visuo-motor orientation of attention.

For HMI research the results of this study are important to generalize findings from VR applications to the real world. Human behaviour in simulations (e.g. cars, aircrafts, etc.) has to mirror human behaviour in reality, otherwise results from virtual applications can not be interpreted correctly. Therefore especially the correspondence of attentional and motor behaviour in the two different environments should be given.

Virtual Reality as research method

Since the last decade, virtual reality (VR) has found its way into psychological laboratories, and virtual environments seem to establish multifaceted experimental possibilities. Advantages of the use of VR are high ecological validity, experimental control, generalizability of experimental findings, experimental realism, the use of 'impossible' manipulations and the ease of implementation and conduction of experiments. Disadvantages are imperfection and high complexity of hardware and software, the difficulty of setting up high-quality virtual environments, high costs and side effects (Loomis, Blaskovich & Beall, 1999). The possibility to simulate three-dimensional environments, which are highly controllable, was the crucial aspect for this study. The disadvantages were ruled out because a state-of-the-art VR system was available for experimental purposes.

Method

Apparatus and materials

TAP. The Test for Attentional Performance (TAP, Zimmermann & Fimm, 2002) was developed for the assessment of attentional deficits. The core of the test are different reaction time tasks of low complexity allowing the evaluation of very specific deficiencies. The tasks consist of simple and easily distinguishable stimuli that the participants react to by a simple motor response. With few exceptions, the stimuli make no demand on verbal capacities. The test battery is a computerised procedure running on normal PC under MS-DOS. There are norms for 13 procedures from healthy persons. The procedures included in the test battery are: Alertness, Covert shift of attention, Crossmodal Integration, Divided attention, Eye movement, Flexibility, Go/no-go, Incompatibility, Sustained attention, Vigilance, Visual field examination / Neglect, Visual scanning and Working memory. For the purposes of this study the sub-test "Covert shift of attention" was selected to compare attentional performance in this standardized two-dimensional computer based test with attentional performance in three-dimensional virtual space. A covert shift of attention is considered to be a preliminary process to an eye movement by which the new visual target is determined. According to Posner and others this covert shift of attention can be decomposed in three components: (1) disengaging of the attentional focus, (2) moving of the focus, (3) engaging of the focus (Posner, 1980; Posner & Petersen, 1990). The examination consists of a simple reaction time task with a preceding cue, an arrow in the centre of the screen pointing with high probability to the side where the target stimulus will appear (valid cues: 80% of the presentations) or in rare cases to the opposed side (invalid cues: 20% of the presentations). The ability to shift the attentional focus is evaluated by the RT with valid cues and the difference of RT between the trials with valid and invalid cues.

Visual test. The subtests 1, 4 and 9 of the TITMUS Vision Tester® were used to test visual acuity. Test No. 1 measures myopia. Test No. 4 measures the ability to judge relative distances when all cues except binocular triangulation are eliminated. The level of difficulty is expressed as a percentage of the theoretical maximum stereopsis according to the Shepard-Fry formula and results are given in angle of stereopsis in seconds of arc and Shepard-Fry percentages. Test No. 9 measures hyperopia.

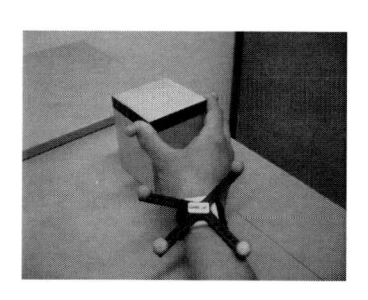

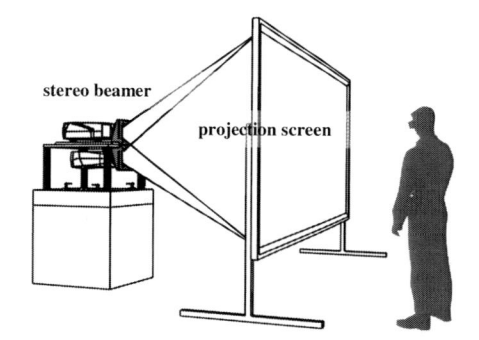

Figure 1. Left: example for a hand marker configuration; right: simplified illustration of the hardware set-up

Motion capture. For capturing the hand movements an optical tracking system was used. Three infrared cameras from Qualisys recorded the movements of the hand, which was marked with a single marker on the index finger and a 6DOF marker at the wrist (compare Figure 1, left).

VR hardware. The virtual environment was displayed on a rear projection screen (240 x 180 cm) with a stereo projector (resolution 1024x768 pixels). For stereoscopy, participants wore passive Infitec™ filter glasses (Figure 1, right). The participants' head motion was tracked using the optical tracking system, which was also used to register the hand movements (Qualisys).

VR software. The ReactorMan software was used to program the experimental paradigm. It is based on the VR-toolkit ViSTA and its multimedia extensions. Experimental set-ups are defined by scripts, which contain the basic experimental structure of sessions, blocks, trials and scenes. Chronological information about the user's reaction to events and the user's movement is saved with specific timing characteristics (Valvoda, Assenmacher, Kuhlen & Bischof, 2004; Wolter, Armbrüster, Valvoda & Kuhlen, 2007).

Post-questionnaire. The post-questioning in written form was divided into three sections. First, participants were asked about physical complaints. They had to indicate whether general discomfort occurred and then rate dizziness, malaise, eyestrain and headache on scale with 3 answering alternatives (little, medium, severe). The second part comprised statements about the virtual reality itself. Four dimensions were used: presence/immersion, external awareness, quality and enjoyment (Table 1). Answers were given on a scale with 5 answering alternatives (not applicable, rather not applicable, neither nor, rather applicable and applicable).

Table 1. Eleven statements on four dimensions of the post-questionnaire concerning the virtual experience

Dimension	Statement
Presence/ immersion	(1) I had the feeling to be in a virtual room/space. (2) I had the feeling to see pictures only, like in the cinema or on TV. (3) I could imagine the virtual space. (4) The virtual environments and the displayed objects seem to be realistic. (5) I had the feeling that I could reach into the virtual world and touch the objects.
External awareness	(6) I was not aware of the real world – the laboratory. (7) I knew all the time that I was in a real room – in the laboratory.
Quality	(8) I could see the virtual world clearly. (9) The quality of the graphical presentation was satisfying.
Enjoyment	(10) I was disappointed of the experience in the virtual reality. (11) The virtual experience was fascinating.

The last part of the questionnaire contained six specific statements about the experiment with the same scale with 5 answering alternatives as before. Statements were (1) 'I was able to touch the virtual objects.', (2) 'The visual feedback was direct and without delay.', (3) 'I could easily fixate the objects.', (4) 'My hand destroyed the 3D-impression', (5) 'In reality I would have touched the object in a different way.', (6) 'Haptic feedback was missing in the application.'

Participants

10 female and 10 male participants between 21 and 44 years of age (mean age M=26.9) took part in the study. They all had a normal or corrected to normal visual acuity. Visual test results of the TITMUS Vision Tester® revealed normal myopia and hyperopia values. Furthermore all participants were able to see binocular which is essential for the correct perception of three-dimensional virtual environments.

Task

The task was tapping at the top of a virtual log of wood as fast as possible after it appeared on the right or the left side of body midline. The log was displayed in a grey virtual room (Figure 2, 540 x 800 x 380 cm). Different conditions were applied. The movement direction (i.e. the side where the log would appear) was known or unknown (knowledge of movement direction), participants were activated by a preceding cue or not (activation), and the preceding cue was valid or not (validity).

Independent and dependent variables. Side (left versus right), knowledge of movement direction (known versus unknown, i.e. participants had the information where the log would appear), activation (activated versus not-activated, i.e. a cue activated the reaction or no cue was shown) and validity (valid or invalid, i.e. the cue showed in the correct or the incorrect direction) were used as independent variables. Reaction time (RT) and movement time (MT) were treated as dependent variables; both were measured in milliseconds (ms). Reaction time was defined as the time between the target appearance and the first movement of the index finger that had to be kept on a detection sensor in-between the trials. Movement time was defined as the time between the first movement of the index finger and the effective tap on the wooden log, which was registered by a bounding box, i.e. an invisible box around the log which is operating as a light barrier. When the hand breaks through the box the system knows that the log was touched.

Procedure

First, participants executed the TAP and the three subtests of the TITMUS Vision Tester®. Furthermore they filled out the Edinburgh inventory to test their handedness (only right handers were allowed to participate). The experiment started accordingly. During the experimental task participants sat 60 cm in front of the rear projection screen wearing passive stereo glasses. They were asked to move as little as possible during the trials. Minor head movements were registered with the optical tracking system and used for the update of the displayed user-centred visual perspective.

The task was to tap with the index finger on a virtual log which was displayed approximately 30 cm in front of the participant. All together 216 movements were conducted, which were divided in three blocks. Block 1: 48 trials without activation and a systematic variation of knowledge and side (Figure 2). Block 2: 48 with activation and a systematic variation of the two other variables (knowledge and side). Activation was induced by an unspecific cue (double arrow, see Figure 3, second from left) in the conditions without knowledge of the movement direction and with specific cues (arrow left, arrow right) in the conditions with knowledge. In the conditions without activation a red fixation point was shown for a randomised duration of 800, 1000, 1200 and 1400 ms (Figure 3, left) followed by the appearance of the log of wood (Figure 1, second from left). After a successful tap, which was registered by the system through a bounding box, the log's colour changed to provide visual feedback (Figure 3, right). In the conditions with activation the fixation point was shown for 1000 ms and the preceding cue (double, left or right arrow) for 100 ms. Then the target appeared after a randomised interval of 50, 150 or 250 ms.

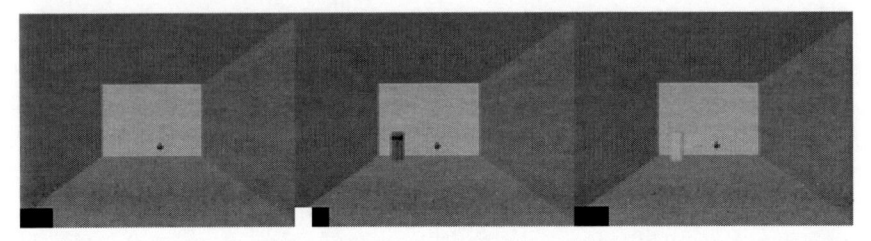

Figure 2. Virtual environment with fixation point (left), log of wood (middle) and coloured log of wood providing feedback (right), condition: without activation (i.e. without preceding cue)

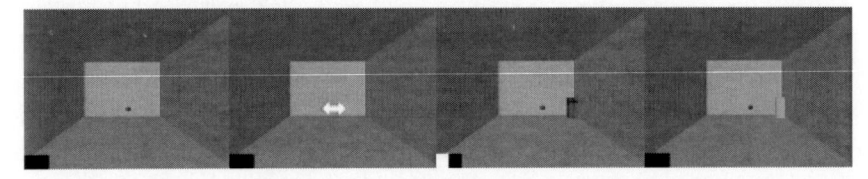

Figure 3. Virtual environment with fixation point (first from left), double arrow (second from left), log of wood (second from right) and coloured log of wood providing feedback (first from right), condition: with activation (i.e. with preceding cue)

In the last block (block 3) the targets location was unknown to the participants and a specified cue (arrow) activated the reactions. Either the cue was valid (96 trials) to the movement direction or not (24 trials). For all trials reaction times and movement times were recorded.

After the experiment participants filled out the post-questionnaire to rate physical complaints, the experience in the virtual reality and the experimental set-up itself.

Results

The result part is divided in two subsections. First the results of the TAP (two-dimensional) will be compared to the results of the experiment (three-dimensional). Subsequently attentional and motor behaviour in virtual reality will be examined. For statistical analysis reaction and movement times were used and analysed with repeated measures ANOVAs with the factors side, activation, knowledge and validity depending on the research question.

Attentional behaviour: 2D versus 3D

First reaction times in the two different environments were compared (2D-TAP versus 3D-VR). Therefore the average reaction times from the covert shift of attention task of the TAP (left valid, left invalid, right invalid, right valid) were compared with the corresponding times from the VR-experiment. All reactions were activated by a preceding cue. T-test results showed significant differences between the 2D and the 3D testing. Reaction times are displayed in Figure 4. In the left valid condition (cue to the left, target on the left) the difference is 102,8 ms, in the left invalid condition 126,2 ms (cue to the right, target on the left), in the right invalid condition 96,5 ms (cue to the left, target on the right) and in the right valid condition 106,1 ms(cue to the right, target on the right).

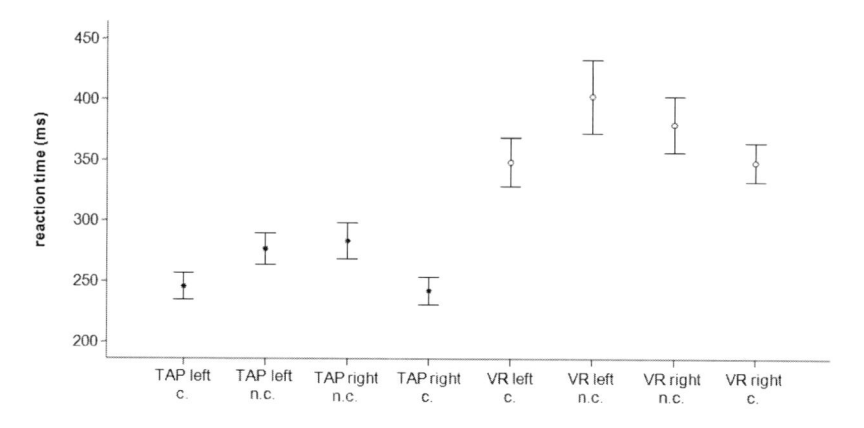

Figure 4. Influence of side and compatibility on reaction times - TAP 2-dimensional (left), VR 3-dimensional (right), condition: with activation

In contrast to the differences in absolute reaction times, the relative pattern is the same as can be seen in Figure 4. ANOVA results for the two testing environments confirm this. In 2D (TAP) as well as in 3D (VR) the main effect side is not significant (TAP: $F(1,19)=5.6$, NS, VR: $F(1,19)=2.7$, $p=0.12$) in contrast to the main effect validity (TAP: $F(1,19)=5.7$, $p<0.05$; VR: $F(1,19)=6.5$, $p<0.05$) and the interaction side x validity (TAP: $F(1,19)=69.2$, $p<0.01$; VR: $F(1,19)=42.0$, $p<0.01$) being significant. That means that covert shifts of attention occur in both environments.

Attentional versus motor behaviour in 3D space (virtual reality)

In a second step it was analysed whether the factors side and validity have a comparable influence on attentional and motor behaviour in the virtual environment in the activated condition, i.e. with preceding cues. Therefore main effects and interactions of the factors side and validity were looked at for the parameters reaction time and movement time recorded in the virtual experiment (Figure 5).

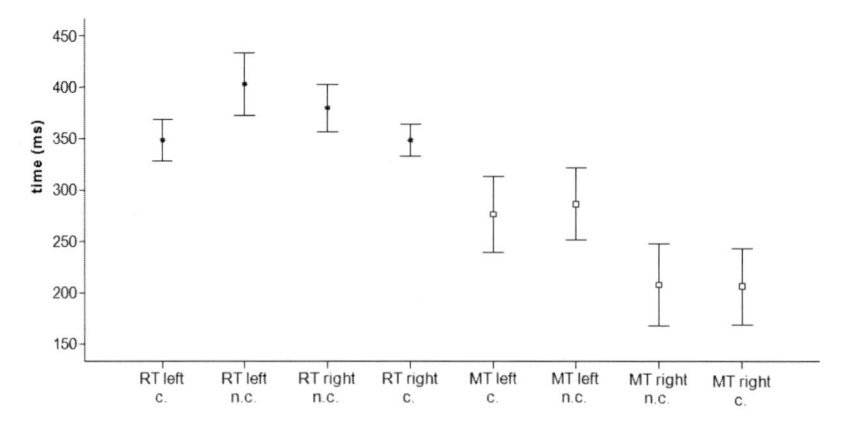

Figure 5. Influence of side and compatibility on attentional (left: reaction times) and motor (right: movement times) behaviour in VR, condition: with activation

As depicted in Figure 5 the pattern for reaction times (left) and movement times (right) were different. Whereas the invalid cue produced longer reaction times (main effect validity: $F(1,19)=41.9$, $p<0.01$) independent from the side ($F(1,19)=2.7$, $p=0.12$) on which the target occured, movement times were distinctly influenced by the factor side ($F(1,9)=4.3$, $p<0.01$) and not by the factor validity ($F(1,9)=1.9$, NS). Furthermore, interactions were significant for both parameters (reaction time: $F(1,19)=6.5$, $p<0.05$; movement time: $F(1,9)=1.5$, $p<0.05$). That means reaction time benefits from validity and movement time does not. Besides, movements to the left are faster then movements to the right independent from the preceding orientation of attention.

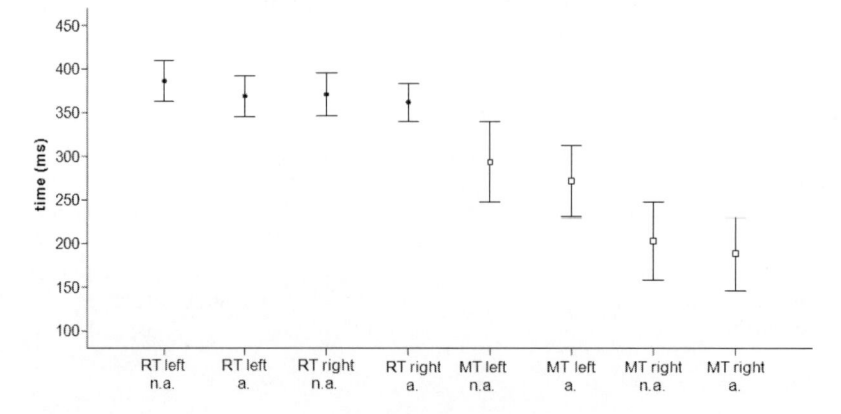

Figure 6. Influence of side and activation on attentional (left: reaction times) and motor (right: movement times) behaviour in VR, condition: without acquaintance

A further aspect is the influence of side and activation on attentional and motor behaviour in virtual reality. Here two conditions can be distinguished. Either participants did not know where the target would appear (Figure 6) or they knew (Figure 7). ANOVA results show significant side effects for reaction time and movement time in the condition with knowledge for reaction time (F(1,19)=4.3, p<0.05) as well as movement time (F(1,9)=13.3, p<0.01).

Overall attentional and motor processes are slower when a reaction and movement to the left is required (Figure 6) and participants did not know the movement direction in advance. If they knew the movement direction (Figure 7) activation has an influence on reaction time (F(1,19)=14.7, p<0.01) but not on movement time (F(1,9)=2.7, p=0.13). Whereas the factor side only influences the parameter movement time (F(1,9)=5.0, p<0.05) and not the reaction time (F(1,19)<1, NS). Interactions were not significant for either condition (without and with knowledge).

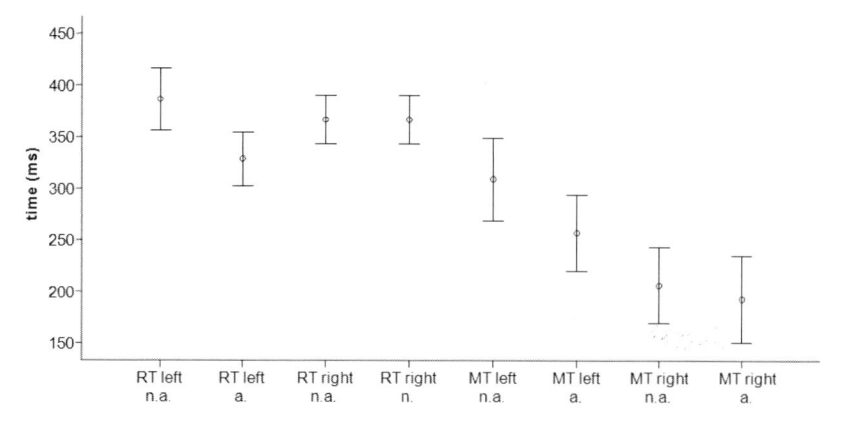

Figure 7. Influence of side and activation on attentional (left: reaction times) and motor (right: movement times) behaviour in VR, condition: with acquaintance

Post-questionnaire

The first part of the post-questionnaire asked whether general discomfort occurred and participants had to rate dizziness, malaise, eyestrain and headache on a 3-ary scale (little, medium, severe). Only 2 out of 20 participants felt discomfort after the virtual experience, one reported dizziness (little) and the other one dizziness (little) and malaise (little). On the basis of these results, it could be concluded that the virtual experience did not cause severe physical problems and that the objective data are not confounded by physical discomfort. The most frequent answers concerning the questions from Table 1 are listed in Table 2 including non-parametric test values of the conducted Chi^2-tests. Overall subjective ratings about the virtual reality itself were quite good (Table 2). Regarding the dimension 'presence/immersion', it can be stated that participants had the feeling of being in the virtual world (statements 1-5).

Table 2. Post-questionnaire: eleven statements on four dimensions concerning the virtual experience with the most frequent answer and Chi^2 – and p-values

Dimension	Statement	Most frequent answer	$Chi^2 (\chi^2_4)$ p-value
Presence/immersion	(1) I had the feeling to be in a virtual room/space.	Rather applicable	22 <.01
	(2) I had the feeling to see pictures only, like in the cinema or on TV.	Not applicable	15.5 <.01
	(3) I could imagine the virtual space.	applicable	21.5 <.01
	(4) The virtual environments and the displayed objects seem to be realistic.	Rather applicable	10.5 0.033
	(5) I had the feeling that I could reach into the virtual world and touch the objects.	applicable	26.5 <.01
External awareness	(6) I was not aware of the real world – the laboratory.	Not applicable	14.0 <.01
	(7) I knew all the time that I was in a real room – in the laboratory.	applicable	40.5 <.01
Quality	(8) I could see the virtual world clearly.	applicable	40.5 <.01
	(9) The quality of the graphical presentation was satisfying.	Rather applicable	14.5 <.01
Enjoyment	(10) I was disappointed of the experience in the virtual reality.	Not applicable	20.5 <.01
	(11) The virtual experience was fascinating.	Rather applicable	16 <.01

They did not loose external awareness (statement 6-7), what was expected because in the experimental setting the external environment (projection screen, etc.) was present all the time. The ratings on the dimensions quality (statement 8-9) and enjoyment (statements 10-11) were also satisfying. The quality was rated good and the participants were fascinated rather than disappointed.

The last six specific statements about the experiment itself were also answered in favour for the virtual environment. Participants were able to touch the virtual objects ($\chi^2_4=23.0$, $p<0.01$), the visual feedback came direct, without delay ($\chi^2_4=18.5$, $p<0.01$), they could fixate the objects ($\chi^2_4=70.5$, $p<0.01$), their own hand did not destroy the 3D-impression ($\chi^2_4=16.0$, $p<0.01$) and in reality they would have not touched the object in a different way ($\chi^2_4=33.5$, $p<0.01$). Only the last question about the haptic feedback was answered in a different way. 55% did not miss the haptic feedback whereas 30% missed it ($\chi^2_4=2.5$, $p=0.65$).

Discussion

The motivation for this study was to examine the influence of movement direction and attentional processes on virtual pointing movements. Furthermore the effects of the factors activation, knowledge and validity were analysed.

The comparison of attentional performance in a two-dimensional computer based test (TAP, Zimmermann & Fimm, 2002) with the attentional performance in virtual space showed typical patterns for the orientation of attention in VR, which are comparable to results from the standardised test. Altogether reaction times in VR are slower than in the standardized attention test (TAP). This could be caused by the higher complexity of the virtual environment being displayed. More visual information had to be processed. In the simple two-dimensional environment of the TAP only white signs on a black screen were visible. In the three dimensional environment a room, colour and depth were implemented.

Regarding the motor behaviour in virtual reality it can be stated that pointing movements are affected by movement direction but not by attentional processes. Movements to the left side were faster than movements to the right. This phenomenon is known from movements in reality, where movements towards the body-midline are easier and faster than outward movements (Wyke, 1969; Reed & Smith, 1961). Visuo-spatial attentional processes in advance of the executed movement do not influence the movement time.

Future research must show if those patterns can be found in other VR applications as well. Hardware and software, as well as task characteristics may play an important role for the generalizability of results. For the set-up that was used in this study the results are promising in regard of the use of VR applications in HMI research. Attentional (Heber et al., 2008) and motor behaviour in virtual reality seem to mirror real behaviour and therefore it can be recommended to use VR applications to simulate real environments and to conduct HMI experiments including the measurements of attentional and motor parameters.

Finally, with regard to tapping as an interaction method in virtual applications this study could show that this kind of manipulation with our own hands is effective and efficient in VR set-ups. In future this technique can be used to press virtual buttons, touch virtual control panels, etc.

Acknowledgement

This research was supported by a grant from the Interdisciplinary Center for Clinical Research "BIOMAT." within the faculty of Medicine at the RWTH Aachen University (TV N57).

References

Bülthoff, H.H., Foese-Mallot, B.M., & Mallot, H.A. (2000). Virtuelle Realität als Methode der Hirnforschung. In H. Krapp & T. Wagenbaur (Hrsg.), *Künstliche Paradiese, Virtuelle Realitäten* (pp. 241-260), München: Wilhelm Fink Verlag.

Heber, I.A., Valvoda, J.T., Kuhlen, T. & Fimm, B. (2008). Low arousal modulates visuo-spatial attention in three-dimensional virtual space. *Journal of the International Neuropsychological Society, 14*, 309-317.

Loomis, J. M., Blaskovich, J. J. & Beall A.C. (1999). Immersive virtual environment technology as a basic research tool in psychology. *Behavior Research Methods, Instruments, & Computers, 31*, 557-564.

Mattingley, J.B., Robertson, I.H., & Driver, J., (1998). Modulation of covert visual attention by hand movement: Evidence from parietal extinction after right-hemisphere damage. *Neurocase, 4*, 245–253.

Posner, M.I. (1980). Orienting of attention. *Quarterly Journal of Experimental Psychology, 32*, 3-25.

Posner, M.I. & Petersen, S.E. (1990). The attention system of the human brain. *Annual Review of Neuroscience, 13*, 25-42.

Reed, G.F. & Smith, A.C. (1961). Laterality and directional preferences in a simple perceptual-motor task. *Quarterly Journal of Experimental Psychology, 13*, 122-124.

Valvoda, T., Assenmacher, I., Kuhlen, T., & Bischof, C.H. (2004). Reaction-Time Measurement and Real-Time Data Acuisition for Neuroscientific Experiments in Virtual Environments. *Studies in Health Technology & Informatics, 98*, 391-393.

Wolter, M., Armbrüster, C., Valvoda, J. & Kuhlen, T. (2007). High Ecological Validity and Accurate Stimulus control in VR-based Psychological Experiments. In B. Fröhlich, R. Blach, and R. van Liere (Eds.), *Proceedings of EGVE 2007* (pp. 25-32). The Eurographics Association.

Wyke, M. (1969). Influence of direction on the rapidity of bilateral arm movements. *Neuropsychologia, 7*, 189-194.

Zimmermann, P. & Fimm, B. (2002). A test battery for attentional performance. In: Leclercq, M, Zimmermann, P, (Eds.): *Applied Neuropsychology of Attention. Theory, Diagnosis and Rehabilitation* (pp. 110-151). London: Psychology Press.

Evaluation of task-oriented package design to control attention in out-of-box experience on internet services

Masaru Miyamoto & Momoko Nakatani
NTT Corporation, Kanagawa
Japan

Abstract

We examine the out-of-the-box experience (OOBE) associated with setting up a small Internet hardware system consisting of three devices. OOBE is critical because it sets the first impression of the company and impacts the costs associated with call-centre operation. We found that many users called to ask for installation guidance without making any trial by themselves. One reason is that they could not find which setup guide was to be read first. Therefore, to direct the user to the correct guide, we design a task-oriented package tailored to the setup task and compare it to the current component-oriented package approach, which boxes each component separately. In tests, only 30% of participants using the component-oriented approach could succeed in starting the wiring procedure following the setup guide within 15 minutes, while the task-oriented approach yielded a 90% success rate. We discuss the methodology needed to control the user's initial attention.

Introduction

OOBE, which is the initial experience of the user when attempting to setup hardware or software, decides the first impression of the company and significantly influences the costs of call-centre operation. A company can, however, ensure that the users are encouraged to setup the service, which lowers overall cost, by providing an OOBE that ensures that novices find setup easy to initiate

Internet service traditionally provides one of the worst OOBEs. The Internet does not function as a stand-alone device. All devices and all functions must work at the same time. This connectivity is the reason for the difficulties experienced by novices. Therefore, calls to Internet service providers (ISP) and telecommunication companies continue to increase. The Ease of Use Roundtable surveyed user dissatisfaction for home and small business users with current network solutions. They identified both immediate and long-term improvement opportunity (The Ease of Use Roundtable, 1999).

The Japanese broadband internet market continues to become more competitive. Market participants seek any and all service advantages. Some companies provide free setup service, which is very expensive for the company, as a competitive advantage. In order to realize easy setup, with its low attendant costs, it is necessary to research OOBE to encourage self-installation. Therefore we discuss here a case study on OOBE targeting Internet Service installation.

We conducted preliminary experiments to explore the issues raised by OOBE with the task being to install an Internet Service. Many problems were discovered. The problems roughly parallel the installation steps. Ketola (2005) introduced the key elements of OOBE and related research.

When the user orders the Internet service, the company usually delivers a number of components (e.g. broadband router, wireless LAN card). Fig. 1a shows a sample of the package for Internet service. This includes two wireless LAN card and a broadband router. Usually each component is boxed separately and sent in one package to the user. When the user opens the internal router box, he sees the setup guide, the first document to be read. Preliminary experiments showed that many participants could not discover this target setup guide. Therefore in this paper we focus on the first step, the "packaging" problem. It is considered this packaging problem is one of the reasons why many users cannot understand what to do first and call to get installation hints without performing any trial by themselves. Any failure on this first step spills over onto the following steps.

To improve this situation, it is necessary to control the user's attention immediately upon opening the package and help the user locate the target setup guide. OOBE would be improved by designing the package according to the order of the tasks that should be performed just after opening the package.

In this paper our purpose is to compare the two extremes; the task-oriented package and the current component-oriented package and to consider a methodology for designing packages that appropriately control the user's initial attention.

The following text introduces experiments that compared the both packages. Next, we discuss the methodology to design packages that well control the user's attention. There has been some research on OOBE for Network services (Pirhonen, 2005, Schneider-Hufschmidt, 2006), but no work has adequately assessed the impact of package design.

Method

Independent variable

We selected package design type as the independent variable, it consists of the component-oriented arrangement (conventional approach) and the task-oriented arrangement (proposed approach).

In this experiment, we simulated the very common situation in which the user has ordered the IP phone service with wireless LAN option; in response, a broadband router with IP phone function and two wireless LAN cards are delivered, one card for the PC and one for the router.

In the component-oriented arrangement, the package sent from the telecommunication company contains three internal boxes as explained in the previous section (each device has its own box, Figure 1a). The packaging of each box is optimized but the overall packing arrangement is not. When the user opens the router box, set under the two LAN card boxes, he sees the setup guide, but the router itself is hidden by opaque packaging material.

In the task-oriented arrangement, the devices were packed without opaque covers and their order in the package was tailored to match the setup task. The setup steps were 1) wiring the router, 2) ISP user ID configuration, and 3) wireless LAN board installation and configuration. When the user opens the package, the setup guide to be read first was placed at the top of the package; immediately below the guide was the device to be treated first: the router (Figure 1b). The wireless LAN cards, which are not needed in the first step, wiring, were placed under the router. The documentation not needed at first was set on the underside of the package's lid.

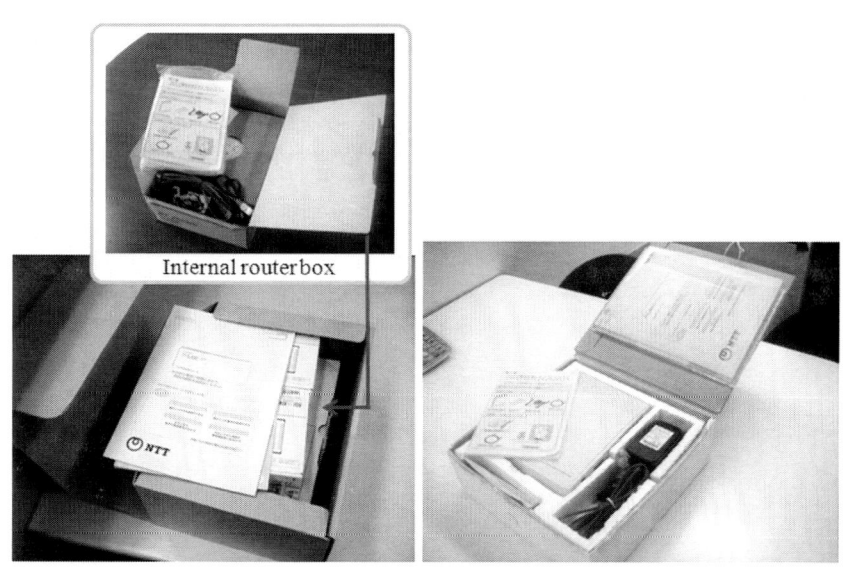

Figure 1a (left). Component-oriented type; Figure 1b (right) Task-oriented type

Dependent variable

We wanted to identify the design that makes it clear to the user what must be done first. We took the wiring start time from the beginning to open the box as the dependent variable. The experimental task was considered to be completed when the participant started the wiring process following the setup guide.

Participants

We asked a manpower company to recruit 20 paid participants. Their ages ranged from 50 to 70, and only computer novices were accepted. Because our goal is to develop a package design that supports novices, only older people were used since they have the least amount of self-confidence with regard to the Internet and PCs. To eliminate the learning effect, we designed a between-subject experiment; 10 participants were assigned to each package design approach.

Apparatus

A home environment was simulated in the laboratory. An FTTH (Fiber to the home) connection was prepared. Each participant was observed through a half-mirror. The participant's behaviour was videotaped for later analysis.

The participant was asked to sit at the desk on which there was the package, notebook PC with FTTH connection, and an envelope. The package was either selected component-oriented type or task oriented type. The envelope was mailed from the ISP and enclosed a printed User Account Sheet and some brochures.

Procedure

At first, an experimenter briefly explained the prior conditions and experimental task. As the prior condition, the participant was told that he/she was already using the internet via FTTH and had ordered the IP phone service and a Wireless LAN; in response, the package was delivered from a telecommunications company. The task was to open the package and configure the IP phone. We asked the participants to make a test call after completing the configuration process.

After the participant discovered the setup guide of the router, which was to be read first, and started wiring following the guide, we decided that setup had been started and the experimental task had been completed. We defined this elapsed time as wiring start time. We set 15 minutes as the time limit.

Results

Success rate

If the participant could start wiring following the setup guide within 15 minutes, we defined the trial as successful. 30% of the participants who used component-oriented type were successful while 90% of the participants who used task-oriented type could succeed (Figure 2). It is considered that task-oriented package design yields significantly better performance.

Wiring start time

We compared the wiring start times of successful participants (Figure 3). The average wiring start time of the participants who used component-oriented type was 7 min 19 s (n=3). That of the participants who used task-oriented type was 3 min 37 s (n=9). Because the success rate was low and the population of succeeded data is rather small, we made no statistical evaluation, but it is considered that the task-oriented package design yields significantly shorter wiring start times.

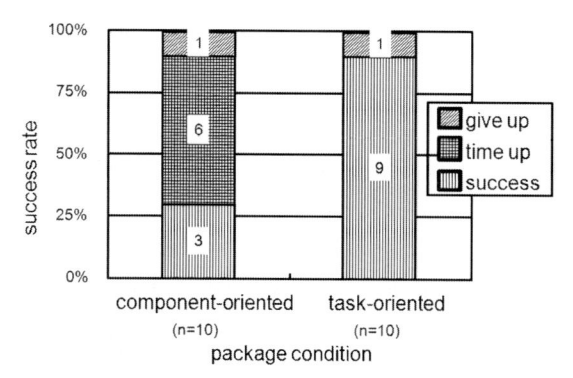

Figure 2. Success rate

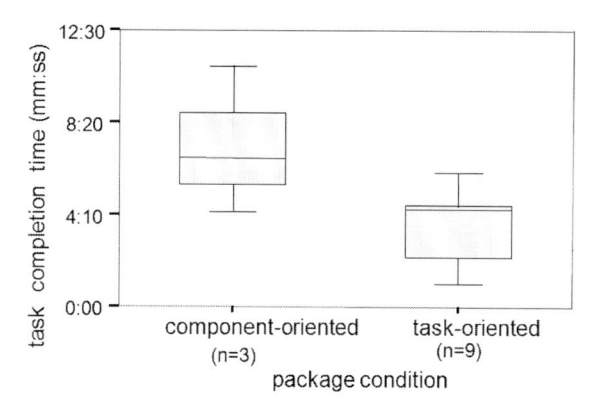

Figure 3. Wiring start time

Attention transition range

We developed a participant attention transition chart to analyze the participants' behaviour. Figure 4 shows a sample chart of one participant who used the component-oriented package. The vertical axis shows the devices/objects that were looked at by the participant. The horizontal axis plots time. The devices/objects were decided by the experimenter by visual observation of the participants as captured on video.

This chart is useful for visualizing participant behaviour. The shape of the line indicates attention transitions and is useful in categorizing the behaviour types. This participant exhibited a narrow view and could not reach the target setup guide. We made this chart for all 20 participants and divided them according to whether they reached the setup guide or not. Furthermore, in each category the participants were identified as either success or failure to start wiring (Table 1).

With the component-oriented package, five of the ten participants did not reach the setup guide, while all participants with t task-oriented package did reach it.

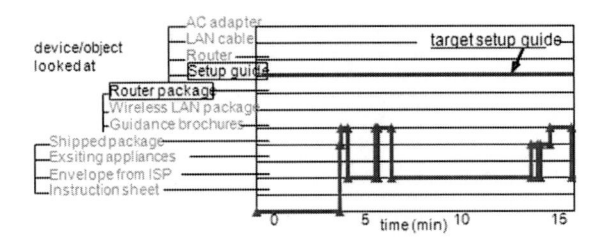

Figure 4. A sample of attention transition chart (Component-oriented type)

Table 1. Attention transition range

Package	Class	Success	Failure	Total
Component-Oriented	Not Reached setup guide	-	5	10
	Reached setup guide	3	2	
Task-oriented	Reached setup guide	9	1	10

Discussion

The critical package design factor to control user's initial attention

1) Navigation to the setup guide
One of the reasons why the task-oriented package could better control initial attention than the component-oriented package is its positioning of the setup guide in the hierarchy of the components. Figure 5 compares the positioning of the setup guide for the two packages. In the task-oriented arrangement, when the user opens the package, the setup guide to be read first was at the top of the package. Figure 6 shows that the participant focused on the setup guide immediately after opening the package. The documentation not needed at first was set on the underside of the package's lid. Therefore, there are no obstacles on the user's path to the setup guide.

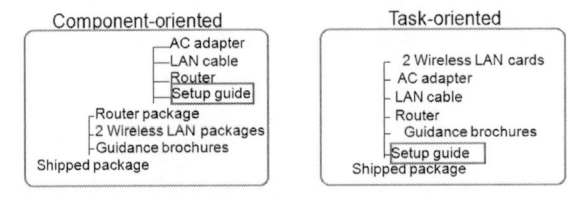

Figure 5. Comparison of the setup guide positioning

On the other hand, the component-oriented package contained three internal boxes and some initially extraneous documentation. The user had to open the router box to reach the setup guide. Figure 4 shows the participant remained focused on the Wireless LAN package until the time limit expired. Navigation to the setup guide should be very simple and obvious.

2) Focus affordance

Physical navigation to the setup guide is not enough to control initial attention because some participants reached the setup guide but did not use it. In order to focus on the setup guide two steps are needed, 1) physical acquisition and 2) identification of importance. In this paper, focus affordance means every element of the environment with the potential to provide a spontaneous clue that guides the user to focus on the setup guide.

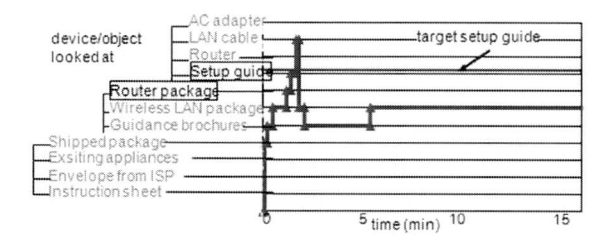

Figure 6. A sample of attention transition chart (Task-oriented type)

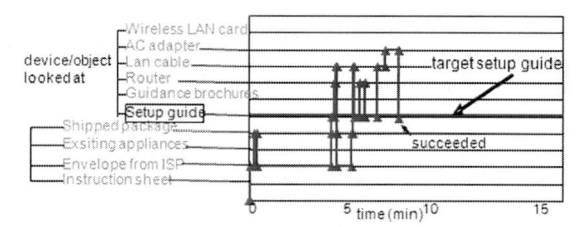

Figure 7. Another sample of attention transition chart (Component-oriented type)

Figure 6 shows the response of a participant with the component-oriented package; he reached the setup guide but then focused other brochures and the Wireless LAN packages. On the other hand, Figure 7 shows the response of a participant with the task-oriented package; he focused on the envelope from the ISP and then switched his attention to the setup guide. Both packages used the same setup guide. Both participants started with one document and then switched to the final target. It is considered that the document that has more focus affordance tends to become the target of attention.

Our results indicate that while physical proximity is important, it is not sufficient. Even though the setup guide was on the top of the box in the task-oriented package, some participants failed to recognize its importance. Some attribute or attributes of the guide, such as fonts, colour, or size, should be optimized to enhance the focus affordance of the guide.

Another example of focus affordance is provided by the physical appearance of the devices. With the component-oriented package, the illustration of the router printed on the setup guide did not match what the user immediately saw since the router was wrapped in opaque packaging material. With the task-oriented package, on the other hand, the illustration exactly matched what the user saw; the logical argument is that a strong immediate association is critical to creating strong focus affordance. Given the paucity of data, the factor of focus affordance should be examined more precisely in the future work.

This behaviour is similar to banner blindness which means that people tend to ignore big, flashy, colourful banners at the top of web pages (Benway, 1998, Norman, 1999). This banner blindness tendency shows the reason of first attention focus is not only relate to access from the start point and eye-catching appearance but also schemas, frameworks and expectations when users setup the electronic devices.

User behaviour model

In many cases it is difficult for the user to escape from the initial goal selected (Miyamoto *et al.*, 2004). All seven participants that used the component-oriented package failed, they continued to read the initially selected brochure, which was not the target setup guide. The participants of Figure 7 continued to focus on the Wireless LAN package for about ten minutes and so exceeded the time limit. This initial selection constrained the range of information explored. Therefore, it is very important to ensure that the user initially focuses on the target setup guide.

Conclusion

We conducted an experiment to compare the component-oriented package to the task-oriented package. The results of this experiment allowed us to consider the methodology needed to design packages that appropriately control the user's initial attention. It is considered that the critical factors controlling initial attention include navigation to the setup guide and affordance to focus. Because the initial selection constrain the range of information subsequently explored, it is very important to ensure that the user initially focuses on the target setup guide.

Future work

From this study the reproducible of this result and user behaviour tendency is limited. In order to generalize these findings, the study should be re-conducted using larger sample size so that a statistical analysis could be performed on the results.

Because this research is based on real internet business and simulated natural home setting and situation, the findings can impact on current and future internet and electronic industry. In the future, home network must be complicated more and more. From the practical research point of view, we will continue suggest theme about user experience about home network and try to contribute for user-centred network environment with not only industry community but also academic community.

Acknowledgements

We would like to give special thanks to Ms. Nakazawa for her support in conducting the experiment and analyzing the data.

References

Benway, J.P. (1998). Banner blindness: the irony of attention grabbing on the World Wide Web. *Proceedings of the Human Factors and Ergonomics Society 42nd Annual Meeting* (pp. 463-467). Santa Monica, CA: Human Factors and Ergonomics Society

Ketola, P. (2005). Special issue on out-of-box experience and consumer devices. *Personal and Ubiquitous Computing, 9*, 187-190.

Miyamoto, M Nakatani, M., Watanabe, M., Yonemura, S. & Ogawa, K. (2004). A comparison of Internet connection troubleshooting strategies by experts and novices. *Proceedings of the Human Factors and Ergonomics Society 48th Annual Meeting* (pp. 841-846). Santa Monica, CA: HFES.

Norman, D. A. (1999). Banner Blindness, Human Cognition and Web Design. Retrieved 01.03.2008 from http://www.jnd.org/dn.mss/banner_blindnes.html.

Pirhonen, A. (2005). Supporting a user facing a novel application: learnability in OOBE. *Personal and Ubiquitous Computing, 9* (4), 218-226.

Schneider-Hufschmidt, M. Williams, D., Bocker, M., Flygt, M., Ketola, P., Von Niman, B., & Tate, M. (2006). The right to information: setup of mobile terminals and services. *Proceedings of the 8th conference on Human-computer interaction with mobile devices and services, 159* (pp. 199-202). Espoo, Finland: ACM.

The Ease of Use Roundtable (1999). Ease-of-Use/PC Quality Roundtable: Home and Small Business Networking. Retrieved 01.03.2008 from http://www.eouroundtable.com/Files/NW_WP_final.pdf.

Human factors involved in container terminal ship-to-shore crane operator tasks: operator fatigue and performance analysis at Cagliari Port

Gianfranco Fancello, Gianmarco D'Errico, & Paolo Fadda
CIREM – University of Cagliari
Italy

Abstract

This paper is concerned with active safety and human factor aspects in maritime transport, in the specific case container terminals. In container handling safety optimization the human operator continues to play a central role and there are numerous sources of job stress that can lead to the serious risk of accidents.

This paper aims to define, for the first time, a tool that is able to understand the cause of human error in port operations most exposed to fatigue-related risk. The first experimental performance curves have been constructed for crane operators in container terminals.

A function has been constructed for crane operator performance under actual working conditions comparing subjective/perceived fatigue curves, developed using data drawn from a survey conducted among port workers at Cagliari Port, with performance curves for a number of crane operators at the container terminal, developed on the basis of productivity data. Furthermore it has been possibile to highlight the deviation between the productivity trend recorded by a container terminal operator and the actual performance curve disregarding idle time.

Introduction

Fatigue, and impaired performance in general, is regarded as a significant factor in the majority of accidents occurring in transport systems (Fadda, 1984). In air transport, the FAA (Federal Aviation Administration) states that 21% of reports submitted to the Aviation Safety Reporting System (ASRS) deal with fatigue related issues. As for road transport, the National Highway Traffic Safety Administration (NHTSA) estimates that in the U.S.A roughly 100,000 accidents are caused by drivers falling asleep at the wheel, resulting each year in 1,500 fatalities and 71,000 injuries. In the maritime sector an analysis carried out in 1996 by the US Coast Guard (USCG) showed that out of 279 accidents, fatigue accounted for 16% of no-injury accidents and 33% of accidents involving injuries. In railway transport, an analysis for the ten-year, period January 1990 - February 1999 carried out by the

Federal Railroad Administration (FRA) Safety Board showed as a contributing factor, 18 cases classified as *"operator fell asleep on the job"* (Ji et al., 2006).

In spite of the high degree of automation achieved over the last few years in container handling equipment in general (AGV, Shuttle Carrier etc.), for the ship-to-shore gantry crane, the real "bottleneck" in container terminals, the human operator continues to play a prominent role. Though on the one hand this maintains high levels of employment, on the other it increases the possibility of operator error causing an accident resulting in injury and/or damage.

This study falls within the area of "active safety", defined as the discipline that analyses factors influencing operator performance and that studies those processes underlying human error. It encompasses human factors and other branches such as anthropometry and ergonomics.

The causal factors of accidents in transportation systems can commonly be divided into *human*, *technical* and *environmental*. Human error can be the result of improper task and/or equipment design and ineffective training.

Over the last few years container traffic has registered an exponential growth that has resulted in

- the gigantism trend with container ships of increasingly larger size able to transport increasingly larger loads
- the effects of this phenomenon especially on the design and functionality of gantry cranes, on which container terminal productivity depends;
- the need to reduce the accident risk among gantry crane operators.

This has resulted in the need for gantry crane operators with increasing levels of specialization, but this has been accompanied by an increase in workplace stress levels creating situations where accidents are likely to occur. Thus research has focused on human factor issues involved in the occurrence of accidents in the maritime transport sector.

Accident reports tend to identify human factors as the cause of accidents without however accurately specifying the type of error involved. For example, a report issued by Maritime New Zealand (2004) describes an accident involving a ship-to-shore crane when the spreader collapsed as a result of the support cables snapping. It was not possible to identify the exact cause of the accident though it was perceived to be an *"error of judgement"* by the crane operator. This type of human error comprises work overload, physiological factors, fatigue, etc.

In this context, periodic refresher training and the analysis and assessment of performance and fatigue levels of crane operators can contribute to reducing the risk of workplace accidents.

Fatigue is usually analysed, also in relation to transportation systems, using a variety of tools. One of the most useful research tools is the arousal/performance curve that evaluates fatigue by means of task-based performance assessment.

Numerous laboratory tests have shown fatigue to be a complex phenomenon and no readily applicable methods are currently available for its evaluation. These tests include physiological, behavioural, visual and performance measurements conducted using medical instruments designed for diagnosis and not with research in mind, that produce data often not readily interpretable. Using the analytical methods typically adopted for handling complex data, such as neural networks (NN), Bayesian networks (BN) or fuzzy logic it is possible to treat and properly interpret incomplete, often partial information (Ji *et al.*, 2006).

As far as the authors are aware the specific issue of crane operator fatigue is not dealt with in the scientific literature and the present study is the first contribution of its kind.

Theory and previous research

Fatigue - Definition and Measurement

It has been ascertained that fatigue, alertness, vigilance, stress and performance all have physiological roots. Two physiological factors, in particular sleep and circadian rhythms, are the primary determinants of arousal state (Ji *et al.*, 2006).

Lack of sleep leads to deterioration in the main psychophysical functions which include cognitive processes, alertness, visual and physical coordination, judgement and decision making, communication, etc. Recently a definition of the factors causing workplace stress has been provided (Seck et al., 2005), distinguishing two macro areas, physical and mental, in turn divided into:

- physical: *environmental* (heat, cold, noise, vibrations, etc.) and *physiological* (lack of sleep, dehydration, muscle fatigue, etc.);
- mental: *cognitive* (too much or too little information, judgement difficulties, etc.) and *emotional* (pressure, frustration, boredom/inactivity etc.).

Past research conducted on human fatigue prevention has focused on both the physiological mechanism of and on methods for measuring fatigue levels (Sherry, 2000 & Ji *et al.*, 2002). The most common physiological measurements for determining the extent and length of reduced alterness, considered as an indicator of increased fatigue and/or drowsiness, employ the electroencephalogram (EEG) (Lal S.K.L. & Criag, A., 2002). A training manual prepared by the Transportation Development Centre of Canada in November 2002, provides guidelines for analysing fatigue, drowsiness and the resulting performance deterioration of Canadian navy personnel, combining EEG, EOG (eye movement), ECG (heartbeat) and EMG (muscle tone).

Behavioural measurements have also been gaining credibility over the last few years: These consist of recording frequency of a body movement during task performance over a specific time interval and can be significantly correlated with the EEG. Fatigue can also be readily detected by observing facial behaviour: changes of facial expression, eye and head movements, gaze are all indicators of fatigue.

However real time measurements of fatigue are plagued with numerous problems and shortcomings. The positioning of equipment and the real-time monitoring of psycho-physical parameters inside a crane cab in the specific case, interferes with the operator during his work, who already has to cope with the difficulties created by the continuous movement of the cab.

For these reasons, for some years now simulators are increasingly used for researching human factors in transport as these devices allow to reproduce and evaluate, also singly, all those factors contributing to fatigue. Recent studies conducted by the Universities of Taiwan and San Diego on the assessment of driver performance interpreting EEGs using fuzzy neural networks (Wu *et al.*, 2004), have shown that accidents caused by sleepy drivers involve a high percentage of fatalities due to a marked impairment in driver ability to control the vehicle.

It emerges from the above overview that no detailed analyses or significant studies are reported in the literature concerning fatigue in gantry crane operators. The only data published by Italian (Colombini, 2006) and Dutch researchers (Huysmans *et al.*, 2006) are concerned with posture of ship-to-shore crane operators.

Because of the lack of pertinent literature data, it was decided to adopt an original approach for the assessment of gantry crane operator performance, suitable for the type of job task but in any case drawn from analyses conducted in other sectors.

Methodological approach

The importance of the arousal/performance curve, already mentioned above, emerges from a recent study (Pang *et al.*, 2005) that by means of a subjective evaluation shows the performance curve to be an efficacious tool for graphically representing mental fatigue. These workers propose a simple method, the Auditory Vigilance Task (AVT) for determining the performance of a group of individuals subjected to a test session during which they were presented with acoustic messages and arousal was subjectively estimated through selecting a button. The fatigue severity scale (FSS), used by the subjects to indicate and evaluate that moment in time when fatigue was perceived and to what extent, is compared with objective measurements of brain activity (and hence performance) recorded with the EEG. This comparison showed that subjective assessment cannot be used by itself to gauge fatigue.

In the 1980s some workers (Daniel *et al.*, 1985), performed a subjective analysis to demonstrate the effect of shift work on industrial workers combining subjective evaluation with objective fatigue measurements, using FLIM (Fluorescence Lifetime Imaging Microscopy). This tool assesses performance by analysing flicker fusion frequency, an indicator of central nervous system (SNC) arousal. During this process an intermittent light source is perceived by the human eye as a flickering light changing to continuous with the onset of fatigue. Their results showed that the subjective answers underestimated the onset/duration of overwork load compared to the objective FLIM measurements.

Other examples of subjective contributions in fatigue analysis are the drowsiness/alertness perception scales One of the difficulties with this approach is that individuals need to have considerable intuition to be able to distinguish drowsiness from the other factors that contribute to defining performance For instance, the Stanford (Hoddes et al., 1972) and Karolinska (Akerstedt, 1996), sleepiness scales assess the momentary degree of drowsiness/alertness. These two scales are usually used to detect symptoms at a given point in time, rather than over a longer time interval. From this standpoint, the Epworth sleepiness scale proposes a more appropriate method (Johns, 1991).

As can be seen, subjective evaluations can be used in fatigue analysis but they need to be integrated with objective measurements. Indeed a subjective response to a perceived message evaluated by an individual by means of a rating scale is not, if taken by itself, reliable and unequivocal.

For this reason here the technique used combines subjective and objective evaluation.

The method used (shown in Fig. 1), can be summarised as follows:

1. <u>subjective evaluation</u>: based on self-assessment of peak performance and subsequent peak perceived fatigue (questionnaires administered to ship-to-shore crane operators);
2. <u>objective evaluation</u>: based on a performance indicator of ship-to-shore crane operators (number of containers handled per hour).

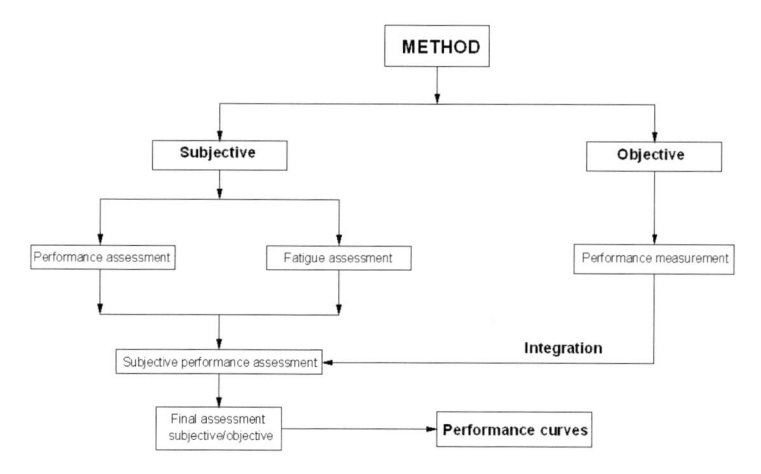

Figure 1. Methodological approach

The model for performance curves used here is based on the Yerkes-Dodson (Y-D) function, widely used for graphically representing performance trends. Recent simulation studies of human behaviour (Seck et al., 2005) adopted a dynamic stress model obtaining the Yerkes-Dodson performance curves shown in Figure 2.

The Yerkes Dodson law (Fadda, 1984) states that there exists an inverted-U relationship between arousal (that can be translated into task duration) and behavioural performance. The performance level is measured or plotted along the y-axis the *emotional arousal* (taskload) along the x-axis. Optimum arousal level and quality of performance will vary with task complexity. Initially, low workloads (early in shift) will result in poor performance, performance level then increasing in a directly proportional manner to arousal level until the optimum level has been attained. Viceversa, as workload increases (late in shift) so performance deteriorates, all the more so the more complex the task.

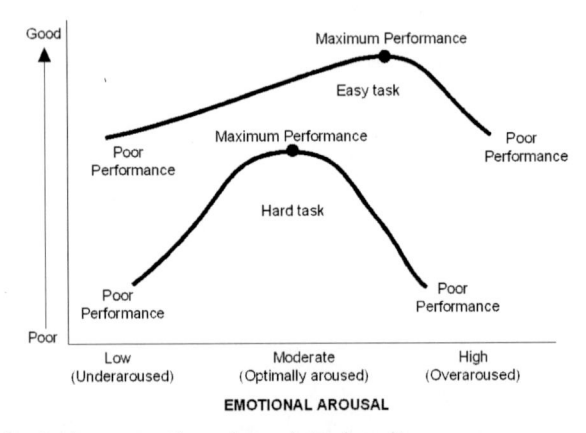

Figure 2. Yerkes-Dodson curves (complex and simple task)

Because of its simplicity of representation for analysing performance trends, here the Yerkes-Dodson function was chosen for providing a first contribution to the performance-fatigue assessment of gantry crane operators in a container terminal.

Crane operator study

Application field

Containers are loaded/unloaded from ship to shore using spreaders that are mechanically connected to the hoist motors via a beam suspended from cables and electrically connected to the crane (Fig. 3). The container is hooked/unhooked by means of four corner flippers on the spreader. The containers are transferred from ship to shore through a combination of two movements: the spreader-container system is hoisted to the maximum clearance height, and the crane then travels with its load along the bridge rails to the buffer area. This operation is generally repeated at least 20 times an hour, the gantry crane continuously travelling back and forth between the ship and the yard.

Thus throughout the six hour shift the crane operator is exposed both to high vibration, due to cab movements, and to high noise levels generated by the very nature of the operation,. Added to this, is the discomfort caused by the bent forward posture and awkward head/neck positions that the operator is forced to assume to follow the movement of the container some 40 m below. These conditions create psychophysical stress that, over time, can lead to serious health problems and in terms of operational efficiency impair operator performance, to the detriment of container terminal productivity.

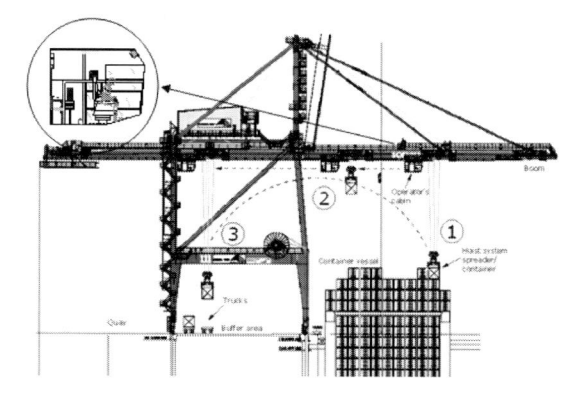

Figure 3. Quay crane unloading sequence from ship to truck

Method

Participants

The investigation is based on the results of a self-assessment questionnaire survey distributed to the 36 ship-to-shore crane operators at Cagliari's industrial port (Sardinia – Italy) between June 8 and 21 November 2006. The interviewer-assisted questionnaires were completed by groups of 3-4 operators at a time at the end of their shift.

The responses provided a subjective evaluation of average performance and fatigue perceived by the operators. These data were then integrated, given the difficulties in their correct interpretation (as mentioned in the section on methodological approach) with the objective evaluation obtained from container handling data recorded in the terminal's end-of-shift reports for each crane in operation. It was then discovered that crane driver performance is in actual fact distorted by a phenomenon that had not been accounted for. During the work cycle it often happens that the vehicle onto which the container is released is not immediately available (usually the truck-trailer, see Fig. 3). In this event, the container remains suspended from the spreader until the truck-trailer arrives. These idle times, that also depend on other operations coordination problems, were detected for two significant shifts, the two operators remaining inside the crane cab, and were used to correct the container handling data for the two shifts extrapolated from the reports shown in Fig. 5. The work shifts were purposely chosen so as to include both a novice and an experienced operator (among the crew who had responded to the questionnaire administered a year before the survey and handling data were collected) working on two cranes operating simultaneously.

Material

The questionnaire was divided into three sections: one section contained personal information including sex, age, qualifications, years of experience in task performed, years of employment in specific sector; in the second section the interviewee was asked to describe loading/unloading procedures (suggesting any possible improvements) and the degree of difficulty experienced in adverse weather conditions (wind, rough sea) while the third section aimed to identify the disturbing factors associated with operator visibility. Two identical shift assessment scales were included in this questionnaire which the respondent used to indicate at what time during the work shift he performed best and experienced the greatest physical and mental fatigue. (Fig. 4; the operators could indicate more than one hour).

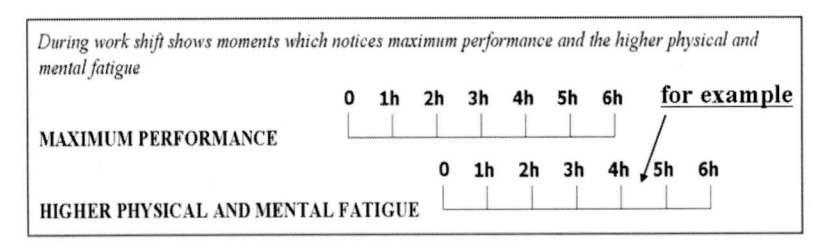

Figure 4. Rating scale showing the 6 hours for which the crane operators had to indicate greatest fatigue

The handling reports indicate the number of containers moved each hour at the end of each shift. Figure 5 shows the report for the two shifts examined, from which the total number of containers handled for each hour by the two operators (novice and experienced) were extrapolated.

Figure 5. Hourly handling reports

For the shifts examined wind speed and direction indicated in the crane cab instruments were also recorded each hour.

Procedure

A self-assessment questionnaire was administered to the 36 ship-to-shore crane operators at Cagliari port between June 8 and 21 November 2006. Idle time was determined for two different shifts corresponding to the two most significant ones in a container terminal, as they can also be affected by operator psychophysical stress and experience as well as environmental conditions. The data concerning hourly container handling were extrapolated for the same shifts.

1. 3rd shift (13:00-18:45) on 4 June 2007 for two gantry cranes, simultaneously unloading containers from a mothership (medium-low degree of difficulty), the first manoeuvered by an experienced (4 years), the second by a less experienced operator (1 year), with steady wind speed ranging from 15 to 26 km/h from the 1st to the 4th hour and gusts (9-33 km/h) in the 5th and 6th hours;
2. 4th shift (19:00-00:45) on 13 June 2007 again for two gantry cranes simultaneously unloading containers from a mothership, the first manoeuvered by an experienced (5 years), the second by a less experienced operator (1 year), with steady wind speed ranging from 5 to 23 km/h and artificial lighting from the second hour onwards.

Data analysis

Subjective evaluation (questionnaire)
From the responses to the questionnaire administered to crane operators at Cagliari port concerning years of experience in the specific task, the average number of years of experience was calculated over the entire crew (2.5 years). The data were then divided into two groups: more (16 - longest experience on the job 6 years) or less (20 shortest experience on the job 6 months) than 2½ years work experience. At this point the number of responses for each hour of the 6-hour shift indicated by operators as when they felt most fatigued was plotted on the Cartesian plane. The fitting curve was then plotted on the histogram. The x-axis of the fatigue curves obtained shows the number of hours per shift, the y-axis the response frequency for the time operators felt most fatigued. In fact the y-axis, was transformed from "number of responses" to "response frequency". For example, 15 out of 16 operators with above average experience indicated greatest fatigue between the 5th and 6th hours, i..e. 95% of the responses.

Actual operating performance assessment (terminal container reports)
Similarly to the Yerkes-Dodson model for the fatigue curves, the performance curves were plotted for the 4 crane operators (2 novices and 2 experienced) along the same Cartesian plane. The total number of containers handled per hour was considered as performance indicator and plotted along the y-axis for each time interval of the x axis (time on work shift).

Corrected operator performance assessment: values recalculated subtracting idle time
The idle time determined for each operator during the two shifts was then taken into account and the number of containers handled corrected accordingly. For example

on the 2nd hour of his shift on 13/6/07 the experienced operator moved 24 containers in 35 minutes, the other 25 minutes being idle time, so the figure was corrected to 41 for the whole hour.

In this way it was possible to better assess performance and fatigue, intended as actual work load of the crane operator during the shift.

Results: performance curves analysis

Three different curves were constructed:

- fatigue curve, obtained from crane operators response frequency to the question when (time interval) they felt most fatigued on their shift;
- actual performance, based on the number of containers handled per hour, indicated in the operators' end-of-shift reports;
- "corrected" performance considering the actual number of containers that the operators would have handled during the six hour shift subtracting idle time.

The 6 hours per shift are plotted along the x axis: wind speed and direction are also shown for each hour. The main features of the curves can be summarized as follows:

- the peak (peak performance of the Y-D function) of the corrected curve for number of containers handled occurs before the peak of the actual number of containers handled, indicating that the crane operator performs better during the first hours of the shift, but because of yard management problems is unable to move more than a certain number of containers;
- idle times are longer at the beginning of the shift, indicating that at least one hour is required to get the quayside crane-truck-yard crane system running regularly;
- the fatigue curve shows a more gradual trend for the less experienced operators in the crew, the more experienced workers tending to suffer fatigue only near the end of the shift (no fatigue reported during the first three hours);
- the peak of the fatigue curve obtained from analysis of the questionnaires always coincides with the decrease in number of containers handled by operators in the shifts examined (fifth and last hour);
- the peak of the actual performance curves always coincides with the operator unloading same size containers (20 ft or 40 ft containers), whereas performance deteriorates when containers are of different size (probably handled in a less orderly fashion forcing the operator to continuously adjust the spreader, increasing stress and disrupting unloading rhythms, Fig. 5);
- by superimposing the corrected operator performance curves for both the surveys conducted (Figs. 6/7), the correlation between the greatest number of containers handled and the time on shift emerges more clearly. On the third shift (13:00 to 19:00) both experienced and less experienced operators performed best at mid shift (3rd and 4th hours respectively), whereas on the fourth shift (19:00 to 01:00), peak performance was recorded early in the shift (during the 1st hour for less experienced and 2nd hour for experienced operators). This

suggests that during the afternoon shift the crane operator remains more alert and energetic throughout the 6 hours, actually improving performance with time, reaching a mid-afternoon peak. Conversely, operators working the late shift exhibit peak performance early in the shift.

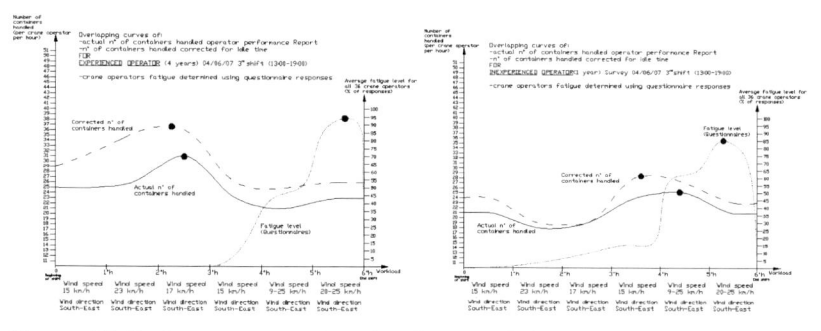

Figures 6/7. Performance curves (actual and corrected) for the third shift superimposed on fatigue curves based on questionnaire (experienced and less experienced operators)

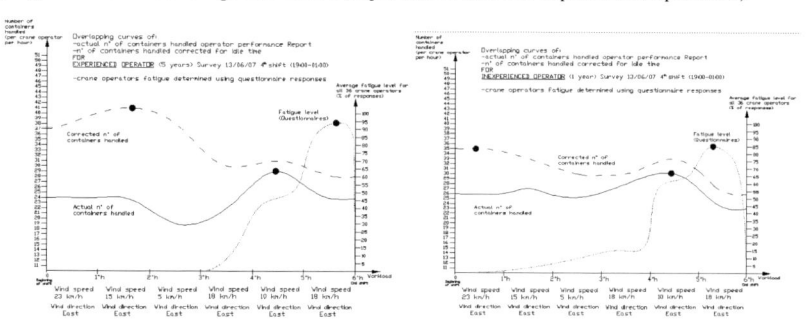

Figures 8/9. Performance curves (actual and corrected) for the fourth shift superimposed on fatigue curves based on questionnaire (experienced and less experienced operators)

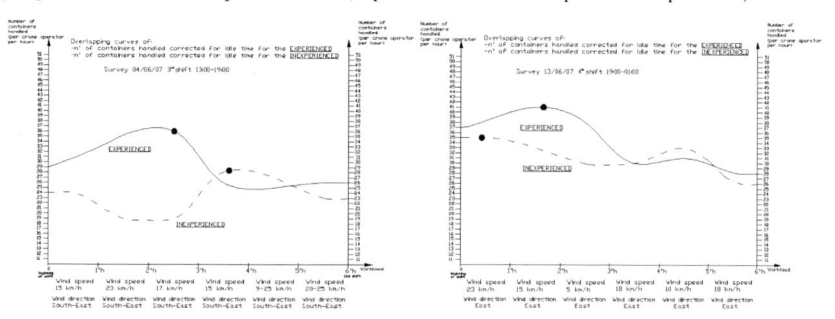

Figures 10/11. Corrected performance curves (superimposing experienced and less experienced operator) for the afternoon and night shifts

As can be seen (Figs. 6/7) the experienced operators working the afternoon shift attain peak performance levels long before their less experienced counterparts, unloading rhythms then decreasing though still to acceptable levels. By contrast, performance of the less experienced operators peaks in the 4th and 5th hours, deteriorating in the last hour.

There is less idle time on the afternoon shift than on the night shift (fewer organizational problems) generally a larger number of containers being handled, the experienced operators outperforming the less experienced ones. This suggests that the latter do not perform as well under pressure (note the poor performance in third hour, Fig. 7).

As far as late shift is concerned (Figs. 8/9), no significant differences were observed. Taking into account idle time, both experienced and less experienced operators handled a larger number of containers (Figs. 10/11). This can probably be explained by a less strenuous shift compared to the afternoon shift due to two factors associated with unloading procedures and to weather conditions:

- More hatch covers had to be removed on the afternoon shift (Fig. 5) and most questionnaire respondents indicated this operation to be one of the most time consuming;
- The majority of containers moved during the night shift were of the same size (Fig. 5);
- Wind speed was higher on the afternoon shift, especially during the second hour, and this is reflected in the performance of the less skilled crane operators in particular, who also experienced difficulties during wind gusts in the fifth and sixth hours of the shift.

Discussion

The following considerations can be drawn from the results of this investigation:

- this is the first contribution to the assessment of ship-to-shore crane operator performance in a transhipment container terminal (Port of Cagliari);
- the analysis of performance curves are undeniably an important aid in the design of workloads and shifts for the job task examined;
- a combination of subjective and objective evaluation has been used, a technique recognised in the literature.

Nonetheless, with this methodological approach it was not possible for example, to determine the stress levels in crane operators due to increased workload or to determine whether idle time resulted in loss of concentration that then repercusses negatively on productivity as a whole.

To draw more precise conclusions the psychophysical parameters of crane operators will need to be monitored during their tasks using electromedical instruments.

Because of the difficulties in obtaining accurate and reliable measurements of these parameters inside the gantry crane cab while the operator is actually working (confined spaces, difficulties in installing the instruments, conflicts with terminal operability, etc), and the numerous interference interactions, future research directions will focus on simulator studies. CIREM at the University of Cagliari, in collaboration with the COSMOLAB Consortium, is currently in the process of constructing a gantry crane simulator. One of the objectives of this project is to set up a permanent laboratory for monitoring fatigue by measuring parameters such as ECG, EEG, S.N.C., flicker fusion point, heartbeat, which provide a measure of the operator's psychophysical conditions under different workloads.

A simulator is a useful tool for obtaining objective undisturbed measurements of the workload (over- or underwork) and of any psychophysical stress caused by the complexity of the task, as well as of exposure to strain and vibrations and external stimuli (reproduced in the virtual environment) associated with the onset of fatigue.

References

Akerstedt T. (1996) *Wide awake at odd hours,* Stockholm, Swedish Council for Work Life Research

Colombini, D. (2006) *Results for an ergonomic re-design of a control station for containers crane for prevention of musculoskeletal disorders during container handling in harbour.* Proceedings of Symposium n° 9 "Prevention of WMSD", IEA TC on MSD Plans for IEA 06 Conference, Maastricht: Netherlands.

Daniel, J., Fabry, R., & Fickova, E. (1985) Changes in activation level during circadian rhythm in shift workers. *Studia Psychologica, 27,* 211-217.

Fadda, P. (1984). Interazione uomo-macchina. In *Introduzione al fattore "uomo" nella fenomenologia dei trasporti,* (pp. 155-179). Università degli Studi dell'Aquila: Italy

Fadda, P. (1984). Biomeccanica e Antropometria. In *Introduzione al fattore "uomo" nella fenomenologia dei trasporti,* (pp. 7-10). Università degli Studi dell'Aquila: Italy

Hoddes E., Dement W., & Zarcone V. (1972) The development and use of the Stanford sleepiness scale. *Psychophysiology, 9,* 150.

Huysmans M., de Looze M., Hoozemans M., van der Beek A., & van Dieën J. (2006) The effect of joystick handle size and gain at two levels of required precision on performance and physical load on crane operators. *Ergonomics, 49,* 1021-1035.

Johns, M.W. (1991) A new method for measuring daytime sleepiness: the Epworth sleepiness scale, *Sleep, 14,* 540–545.

Lal S.K.L. & Criag, A. (2002) Driver fatigue: Electroencephalography and psychological assessment, *Psychophysiology, 39,* 313-321.

Maritime New Zealand (2004) Accident Report. Port of Bluff, New Zealand

Pang Y.Y., Li X.P., Shen K.Q., Zheng H., Zhou W. & Wilder-Smith E.P.V. (2005) *An Auditory Vigilance Task for Mental Fatigue Detection.* Proceedings of the 2005 IEEE Engineering in Medicine and Biology 27th Annual Conference. Shanghai: China.

Qiang Ji, Lan, P., & Looney, C. (2006). A Probabilistic Framework for Modeling and Real-Time Monitoring Human Fatigue, *IEEE Transations on systems, Man and Cybernetics A, 36,* 746-754.

Seck, M., Frydman, C., Giambiasi, N., Ören, T.I., & Yilmaz, L. (2005). *Use of a Dynamic Personality Filter in Discrete Event Simulation of Human Behavior under Stress and Fatigue*, 1st International Conference on Augmented Cognition. Las Vegas: Nevada.

Sherry, P. (2000). *Fatigue Countermeasures in the Railroad Industry: Past and Current Developments*, Washington, D.C.: AAR Press

Wu, R.-C., Lin, C.-T., Liang. S.-F., & Huang, T.-Y. (2004) EEG-Based Fuzzy Neural Network Estimator for Driving Performance, *IEEE Transations on systems, Man and Cybernetics, 4,* 4034- 4040.

The different human factor in automation: the developer behind versus the operator in action

Hartmut Wandke & Jens Nachtwei
Humboldt University Berlin
Germany

Abstract

The rapid technological improvements in computer-based control systems make any direct comparison of performance between human operators and machines obsolete. Recent studies of automation do not concentrate on performance oriented comparisons but rather emphasise unwanted side-effects of function allocation, such as loss of competence, deficiencies in situation awareness, complacency, mistrust, etc. We focus on a "forgotten" question of function allocation: How well can future events be prospectively handled by developers compared with the operator's handling of the same events when they really happen? Developers of automation systems are able to program suitable algorithms for predictable problems. Human operators are needed for unpredictable situations. But analyses of accidents reveal that developers of automation are unable to consider all possible combinations of disturbances, while human operators are unable to act as quickly and accurately as needed. This represents a serious problem. A Socially Augmented Microworld ("SAM") was developed for the experimental comparison of developers' anticipation and operators' actions. We analyse the state-of-the-art concepts of function allocation from the described perspective. Developers and operators have different resources in anticipated vs. real-time process control. These resources are discussed and the methodological approach of "SAM" is explained. First laboratory studies are reviewed.

Introduction: a different perspective on function allocation

Recent technological developments have made one thing clear: as machines get smarter they still cannot consider all factors of human decision making (Norman, 2007). So, despite the exponential improvement of the technological basis for automation, it seems that Price's promising, more than 20 years old statement has not been culminated in an always reliable automation: "capabilities of machines to perform "intelligent" acts such as automation and decision-support are ever improving." (Sheridan, 2000, p. 204). Many implicit procedures and explicit principles of function allocation have been discussed: left-over and compensatory strategy had been replaced by a complementary approach (Hollnagel & Bye, 2000). Yet all have one thing in common: the comparison or integration of man and machine, i.e. their capabilities. Our approach could be described as human-centred in a wider sense. Human-centred automation is seen as a "romantic ideal" by some authors (Sheridan, 2000, p. 211). We however focus on the two "flesh and blood" protagonists of function allocation: the operator and the developer of a human-machine system. It is important to mention, that this paper describes a conceptual approach. First experiments are reviewed only to support the understanding of our approach but not especially conducted or reported in detail.

Thus function allocation can be examined without being dependent on technological developments and derived taxonomies. This can be seen as a huge advantage since "any proposed taxonomy (or fixed allocation table based on machine capabilities, the authors) is likely to be superceded by technological developments" (Parasuraman et al., 2000, p. 294). The historic dependency is even true for the term automation itself. "We define automation as the execution by a machine agent (usually a computer) of a function previously carried out by a human. What is considered automation will be therefore change with time." (Parasuraman & Riley, 1997, p. 231).

The unseen developer

Since technology and automation are rapidly changing, it seems reasonable not to be fixed to the machine part in human-machine systems, but to turn towards the human developer of the machine. The idea of having a closer look on the developer instead of the machine has been formulated before. Hoc (2001) puts it like this: "Rather than speaking of human-machine cooperation, it would be preferable to speak of designer-operator cooperation" (p. 535). And Norman (2007) describes the mind of the designer as the actual source of intelligence.

The success of the machine "Boss" in the DARPA Urban Challenge 2007 is not only a victory of this machine over other machines, but also a victory of the designer team. However it is easy to imagine a manned car (or the operator/ driver) winning if manned cars would participate. In contrast to playing chess (see Kramnik vs. Deep Fritz in 2006), driving a car through an urban area is an ill-structured task. The designers have to anticipate all possible obstacles and external influences onto the car.

Like in other cases, the successful computer programs are crystallized intelligence of the designers. Even if the programs were adaptable and could learn new ways to process information, this feature would be based on human intelligence. Therefore we suggest not comparing humans and machines. Instead we propose an empirical comparison between operators and designers. So one of the basic ideas of our approach is to create a system which can be used to study empirically the design and the operation of *one and the same* system (for a detailed explanation see section "Methodological approach of "SAM"").

Developer vs. operator

We aim at an experimental examination of developers' and operators' resources in programming vs. supervising human-machine-systems. Combining the resources of both the system's operators and developers "can be used to optimise the function allocation in man-machine systems on a long-term basis." (Gross & Nachtwei, 2007, p. 345). Hence, the research question is: Which functions should be performed by an operator who can react directly to unpredicted events, and which functions should be executed by automation designed by developers who have to anticipate all kinds of situations (Gross & Nachtwei, 2006). Beside reaction vs. anticipation there is another aspect to be considered: operators usually have much less time than developers for their decisions. The analysis of accidents has shown that lack of time (and the threat of danger in the case of inaction) can lead to catastrophic consequences. Thus, there is a trade-off between insight into the system's processes and time to plan actions and counteractions.

Our approach covers two key aspects. Firstly, the field of software engineering and design problem solving is examined in order to get insight into the resources and limitations of those who develop automation. Here, we can base our research on findings made in investigations of following research disciplines: psychology of programming (e.g. Hoc *et al.*, 1990), psychological research in construction process (e.g. Hacker *et al.*, 1998), research in empirical software-engineering (e.g. Subramanian *et al.*, 2006), and design research (e.g. Cross, 2004). Secondly, the prerequisites have to be found for successful operators' control, located on various levels as defined by Sheridan (2000). In the following section we will discuss our perspective on function allocation in the light of existing theories. After that we will describe the resources of developers and operators to be investigated. Afterwards our methodological approach and first studies will be presented briefly. We finish with a discussion of our conclusions so far and the work to come.

Analysis of state-of-the-art concepts of function allocation from the described perspective

After Fitts' list was recognized as futile in practice (Sheridan, 1998) more holistic approaches emerged. The notion of balanced work (Hollnagel & Bye, 2000) is one of these socio-technical perspectives that focus on the whole working system. Resources are seen in humans (e.g. semantic information processing), technology (e.g. storage capacity), and organisation (e.g. norms, procedures), whereas performance demands in safety and efficiency. Function allocation has direct implications for the balance of resources and demands (Hollnagel & Bye, 2000). Man and machine form a so-called joint system and are defined as natural vs. artificial cognitive systems (Bye *et al.*, 1999).

Our perspective is directed to the natural cognitive system only. Within this system we differentiate operator and developer respectively their resources. In contrast, the artificial cognitive system is considered in a way which allows generalisation of relevant variables to other domains in order to include a wide range of human-machine systems in our approach. Resources of technology and organisation are not considered, because of their dependence on the state of the technological development (e.g. speed and capacity of hardware components and power of software tools) and context of application (e.g. organisational rules or safety culture).

Linking threads in automation research

There are two main threads in human-automation literature, which are not yet directly linked: the conceptual one proposing general guidelines and principles (like socio-technological approach, human-centred automation, keeping operators in the loop, user centred design), providing *design variables* and the experimental one dealing with various factors of systems and their consequences to operator behaviour (*outcome variables* like performance, awareness, complacency, compliance, being-out-of-the-loop). We try to create this missing link in an empirical approach (see below).

The developer-operator comparison is also suited to handle the concept of dynamic function allocation. The concept of dynamic function allocation is closely related to the above mentioned dependence of automation on context of application. If the context changes, then the allocation of functions should also change. Instead of defining allocations for any kind of situation a priori, the automation should be flexible (Wright *et al.*, 2000). The question is: are there boundaries to the extent to which the automation should be flexible, and how are these boundaries defined? On the other hand, dynamic function allocation also depends on the assumptions of the developer. The question remains, when and how to reallocate and particularly whether the developer can anticipate this. To sum up, each protagonist (developer or operator) has to know the potentials and shortcomings of his counterpart in order to make human-machine interaction work. If two or more versions of function allocation (e.g. in terms of Sheridan's degrees) are available, in a given situation the question is who will decide when, under which condition and which version will be used? In adaptive automation, the developer will decide (and implement this decision in the program, to be executed later by the system), in adaptable automation the operator will decide. A recent approach replaces this either-or-logic with negotiation on function allocation. It has become known as the H-metaphor (Flemisch *et al.*, 2003). Norman (2007) describes it as two sentient systems, horse and rider, communicating with each other. But the question is whether the rider, i.e. the operator, has enough skill to exert control. Among other things, the acquisition of such skills can be explored with our approach.

Resources of developers and operators in anticipated vs. real-time process control

When we investigate an asynchronous division of labour between developers/designers and operators/users of human-computer systems, we need an appropriate unit of measurement, which is the resources these people have. Does the operator's situation awareness beat the anticipation capabilities of the developer? The term "division of labour" recalls the working system described above. So performance demands and resources are compared. The demands on developers are usually defined by their clients' product specifications, whereas the operator has to keep certain system's parameters under or above pre-defined thresholds. The challenge is to find the conditions which allow them to meet these demands.

The following table contrasts the heuristically derived resources of operators and designers. A plus (+) stands for an assumed positive effect on performance whereas a minus (-) stands for negative effects. The table serves as an initial point for a task analysis concerning operator's and developer's work. A further analysis of the literature and expert evaluation will specify the table in the future.

Table 1. Hypothetical resources of operators and developers and their effect on performance

	Operators' resources		Developers' resources	
	+	-	+	-
Time of decision	Real-time access to process parameters	Bad user interface narrow bandwidth		Difficult to anticipate future events
Time span for decision process		Need to decide and act within minutes or seconds.	Development of hard- and software within weeks/ years	Economic constraints
Feedback	Direct and real feedback to actions	Poor user interface, information overload		Feedback only imagined
Knowledge of the context (Expertise)	Direct perception of the situation	Limited capacities, situation awareness, information overload	Systematic gathering of potential context factors	Impracticality to consider all potential context factors
Access to Knowledge (Expertise)		Memory of personal experiences only	Databases, literature	Look for a needle in a haystack
Teamwork (Inter-disciplinarity)	Co-Operator, remote advisers	Often single operator only	Involvement of external experts possible	
Testing, Simulation		Not possible	Simulation of possible events	Ethical boundaries for simulation
Tools for analysing	Handbook, Help System	Not enough time for extensive analysis	Methods, measures for prospective evaluation	Bad practicability of evaluation methods
Emotional involvement	High involvement improves motivation/ effort	Stress impairs information processing and action regulation	Cool minds improve generation of suitable actions	

Different but comparable resources: developer vs. operator

Some relevant resources of developers are: advance information/ expertise concerning the system, time available for problem solving, interdisciplinarity of the team, and anticipation range regarding disturbances or unwanted side-effects. How can anticipation range be defined? Anticipation can be fixed to a time-line, i.e. there are different distances between the anticipation of a disturbance and the disturbance itself. For a developer this distance is quite immense with a direct implication for human-machine interaction as Norman (2007) describes the "system" as making guesses based on the designer's hunches. Lowering the anticipation range therefore means better hunches, so that range will be examined in the forthcoming research. The anticipation range bridges the gap between developer and operator since this concept can be applied to the operator too. The range is different and mostly the operator will react upon perception of a disturbance instead of proactively preventing them from happening. But nevertheless, since operators are able to anticipate disturbances minutes or seconds before they occur, the dualism of developer and operator is principally diminished.

Finally, a comment is needed on the resource "time for problem solving". We are well aware that design teams are obliged to deliver program code for example under time pressure determined by economic constraints. But in comparison to operators, who sometimes have to react in a matter of seconds, developers have all time in the world to decide when and how someone should be technically assisted.

With the resources of operators there are some parallels to the developers (see above). An intrinsic resource can be expertise/ training level as well. Extrinsic resources include the time for problem solving (which in contrast is extremely limited) and anticipation range, which is short. A very special resource is the information an operator has about the ongoing process. Many efforts in Human Factors research aim at this aspect: Augmented Reality, Ecological Interface Design, or Situation Awareness deal with providing this kind of information in order to improve operators' resources. Hence our methodological approach will also focus on it in the following section. Immediate feedback is not available to designers. In contrast to an operator, who should be "in the loop" in the sense of continually receiving information and feedback about actions, the designer will at best receive mediated feedback about the anticipated process. One can differentiate between the amount and quality of information, i.e. how many insights an operator gets and how well prepared or integrated these are. Some implications concerning the amount of information can be found in the literature. Sheridan (2000) puts it like this: "Too much information will overwhelm. We have seen this in aircraft cockpits, nuclear power plants and many other complex systems" (p. 210). So that's a question of dosage, which depends on different aspects, e.g. the state of the operator or his workload.

As for the aspect "quality of information" the old question of signal characteristics (bandwidth, timing) is addressed and recently picked up by Sheridan (2000). On the other hand, quality of information could be defined in a broader sense like modes of information integration etc.

The resources described above have to be specified through embedding them into our experimental setting. We focus on extrinsic resources (e.g. information about the process) because only these can be manipulated in an experimental setting. The next section will describe it in more detail.

Methodological approach of "SAM"

For our research we propose a Socially Augmented Microworld (SAM), an experimental setting with the following characteristics:

- quick-and-easy design, administration and evaluation
- illustration of the key features of a complex and dynamic technical environment
- independency of a specific domain (like process industry, aviation etc.)
- applicability for research on both operators and developers
- suitability for experimental research (the manipulation of independent variables, the controllability of confounders and measurability of a variety of dependent variables like time, error, behavioural and task load measures).

DiFonzo et al. (1998) name these last aspects for microworlds too. But classical approaches have a problem: "Either the functions are deterministic and events can be foreseen in principle - even if they are extremely complex and interdependent, or they contain stochastic parts, which makes prediction impossible." (Gross & Nachtwei, 2007, p. 345). An examination of mere determinism vs. chance would give predictable (dis-)advantages to developer and operator and thus would make an investigation obsolete. We therefore augment the classic microworld (see Brehmer & Dörner, 1993) by including microworld inhabitants as a social factor. Two participants perform a pursuit tracking task presented as a road environment either alone or cooperatively using a standard PC. Joysticks are used as input devices and 50% input (vertically and horizontally) are each assigned to them. The participants are part of the microworld (just as e.g. microprocessors are part of a real technical system). Especially their cooperation adds measurable complexity to this environment, because they are selected (by a questionnaire) and instructed differently concerning their tracking style. One has the tendency to track fast but makes more mistakes whereas the other microworld inhabitant shows the opposite tendency. As a result the process seems stochastic at first glance but is explainable afterwards. We can observe their behaviour and interview them, comparable to a supervisor in an aircraft cockpit.

When cooperation makes things difficult

This cooperative tracking task is the challenge for the developer and operator, who have to cope with it with their very own means. They have to deal, either directly or by means of automatic functions, with the conflicts and errors of these microworld inhabitants, which can be defined as system disturbances. The connection between cooperation and complexity is derived from a specially developed "Cooperation Conflict Complexity Model". It emphasizes personal features of the microworld inhabitants as a prerequisite of complexity (see above). This model combines theories and empirical results from cognitive, personality and social psychology and is explained by Gross & Nachtwei (2006). In addition, the track's composition is another important influence on the complexity of "SAM". The next section will show, that "SAM" can serve as an appropriate means to answer our research questions. However, there is still work to be done regarding its complexity.

The object-oriented SQUEAK environment (based on the programming language Smalltalk) is used for the implementation.

First laboratory studies

The long way to complexity

The first efforts were dedicated to the development of "SAM", beginning with solo tracking. Finally cooperative tracking was introduced. The aim was to induce an adequate support requirement for developers and operators similarly. As the cooperative tracking can be interpreted as a complex and dynamic process, significant variation can be expected. The microworld inhabitants differ in their ability and motivation to perform the tracking task accurately and quickly (Gross & Nachtwei, 2006). Without this variability neither an automation system designed by a team of developers nor an operator would need to supervise and intervene in order to keep the process running smoothly. Nevertheless, additional means to manipulate the complexity of "SAM" will be introduced.

What they know and who they are: expertise and interdisciplinarity of design teams

The next step was to take a closer look at some of the listed resources (see table 1). Concerning developers the analysis touches the field of system engineering and its conditions as well as processes. Two kinds of resources have been investigated in their effect on the results of a development process by now: expertise concerning the system and interdisciplinarity of the team. Krinner (2007) manipulated expertise through different forms of contact with the system and its users. The type of automation developed was used as the dependent variable. The taxonomy of assistance systems developed by Wandke (2005) was used to categorise the proposed systems. It was found that teams with most information about the system wanted to implement cooperative decision support, in contrast to teams under a less enriched representation of the system.

A further study by Krinner & Henkel (2007) dealt with the effects of maximising resources and especially the influence of interdisciplinarity of the design team on the developing process and outcome. The authors used an extreme group design regarding the resources which design teams had for the development of an assistance system. The experimental group could use all sources of information available (see Krinner, 2007). Additionally this group consisted of three students belonging to different fields of study (psychology, interface design, engineering/ computer science). In contrast, the control group could only read the system's description and just two students – both from engineering/ computer science – worked on the design problem. The authors report that the experimental group produced more comprehensive solutions, and more aspects of the design problem were discussed. Although in such an experimental approach the contributions of single factors remain undistinguishable, the authors reasonably claim that interdisciplinarity could be an important resource with an effect on the design process and outcome.

Ongoing studies, concluding remarks & further research plans

With regard to the perspective of an operator supervising the described processes in SAM, data of a recent experiment are just being analysed. After three years of development, the laboratory setting has been refined to a level that allows a smooth start into the next generation of experiments. SAM will be further elaborated and the manipulation of developers' and operators' resources implemented. The approach of using humans in order to augment a classical microworld has provided promising results and will be continued.

Obviously, not every potential resource could be investigated. The challenge is to select the right resources that meet the demands of experimental research and will help to clarify which tasks in which situations should remain with the operator or should be performed by an automatic function developed by designers. A combination of both is also possible in order to investigate the role of operators and developers in flexible/ adaptive automation. But regarding the rather pessimistic views on function allocation methods in general (see Fuld, 2000) and engineering methods in particular (see Sheridan, 2000) we think it is time to complement the theoretical approach with a strictly empirical one. This should be more promising than common sense judgement or human factors theory as rejected by Fuld (2000). Nevertheless we have to accept the challenge of bringing together the broad fields of software design *for* and supervisory control *in* complex and dynamic systems.

References

Brehmer, B. & Dörner, D. (1993). Experiments with Computer-simulated Microworlds: Escaping both the narrow straits of the laboratory and the deep blue sea of the field study. *Computers in Human Behavior, 9*, 171-184.

Bye, A. Hollnagel, E., & Brendeford, T.S. (1999). Human-machine function allocation: A functional modelling approach. *Reliability Engineering and Systems Safety, 64*, 291-300.

Cross, N. (2004). Creative Thinking by Expert Designers. *Journal of Design Research, 4*.

DiFonzo, N., Hantula, D., & Bordia, P. (1998). Microworlds for experimental Research: Having your (control and collection) cake, and realism too. *Behavior Research Methods, Instruments, & Computers, 30*, 278-286.

Flemisch, F.O., Adams, C.A., Conway, S.R., Goodrich, K.H., Palmer, M.T., & Schutte, P.C. (2003). The H-Metaphor as a Guideline for Vehicle Automation and Interaction. Technical Report NASA/TM 2003-212672. Available: http://techreports.larc.nasa.gov/ltrs/PDF/2003/tm/NASA-2003-tm212672.pdf

Fuld, R.B. (2000). The Fiction of Function Allocation, Revisited. *International Journal of Human-Computer Studies, 52*, 217-233.

Gross, B. & Nachtwei, J. (2006). Assistenzsysteme effizient entwickeln und nutzen - Die Mikrowelt als Methode zur Wissensakquisition für Entwickler und Operateure. In M. Grandt and A. Bauch (Eds.), *Cognitive Systems Engineering in der Fahrzeug- und Prozessführung*. (pp. 75-88). Bonn: Deutsche Gesellschaft für Luft- und Raumfahrt e.V. (DGLR-Bericht 2006-02/07).

Gross, B. & Nachtwei, J. (2007). How to develop and use assistance systems efficiently - Using the microworld to acquire knowledge for developers and operators. In D. de Waard, G.R.J. Hockey, P. Nickel, and K.A. Brookhuis (Eds.), *Human Factors Issues in Complex System Performance* (pp. 345-350). Maastricht, the Netherlands: Shaker Publishing.

Hacker, W., Sachse, P., & Schroda, F. (1998). Design thinking - Possible ways to successful solutions in product development. In H. Birkhofer, P. Badke-Schaub, and E. Frankenberger (Eds.), *Designers - The key to successful Product Development* (pp. 205 - 216). London: Springer.

Hoc, J-M. (2001). Towards a cognitive approach to human-machine cooperation in dynamic situations. *International Journal of Human-Computer Studies, 54*, 509-540.

Hoc, J-M., Green, T.R.G., Samurcay, R., & Gilmore, D.J. (Eds.) (1990). *Psychology of Programming.* London: Academic Press Ltd.

Hollnagel, E. & Bye, A. (2000). Principles for modelling function allocation. *International Journal of Human-Computer Studies, 52*, 253-265.

Krinner, C. (2007). How developers anticipate user behavior in the design of assistance systems. In D. Harris (Ed.), *Engineering psychology and cognitive ergonomics: Proceedings of the 12th International Conference on Human-Computer Interaction* (LNAI Bd. 4562, pp. 98-107). Heidelberg: Springer.

Krinner, C. & Henkel, S. (2007). Entwicklung von Assistenzkonzepten unter verschieden ressourcenreichen Bedingungen. In M. Rötting, G. Wozny, A. Klostermann, and J. Huss (Eds.), *Prospektive Gestaltung von Mensch-Technik-Interaktion.* 7. Berliner Werkstatt Mensch-Maschine-Systeme (pp. 489-494). Düsseldorf: VDI.

Norman, D.A. (2007). *The Design of Future Things.* New York: Basic Books.

Parasuraman, R. & Riley, V. (1997). Humans and Automation: Use, Misuse, Disuse, Abuse. *Human Factors, 39*, 230-253.

Parasuraman, R., Sheridan, T.B., & Wickens, C.D. (2000). A Model for Types and Levels of Human Interaction with Automation. *IEEE Transactions on Systems, Man, and Cybernetics - Part A: Systems and Humans, 30*, 286 – 297.

Sheridan, T.B. (1998). Allocating functions rationally between humans and machines. *Ergonomics in Design, 6*, 20–25.

Sheridan, T.B. (2000). Function allocation – algorithm, alchemy or apostasy? *International Journal of Human-Computer Studies, 52*, 203-216.

Subramanian, G.H., Pendharkar, P.C., & Wallace, M. (2006). An empirical study of the effect of complexity, platform, and program type on software development effort of business applications. *Empirical Software Engineering, 11*, 541-553.

Wandke, H. (2005). Assistance in human-machine interaction: A conceptual framework and a proposal for a taxonomy. *Theoretical Issues in Ergonomics Science, 6*, 129–155.

Wright, P.C., Dearden, A., & Fields, B. (2000). Function allocation: a perspective from studies of work practice. *International Journal of Human-Computer Studies, 52*, 335-355.

Acknowledgement to reviewers

The editors owe debt to the following colleagues who helped to review the manuscripts for this book:

Jennifer Elin Bahner, Technical University Berlin, Berlin, Germany
Mike Barnett, Southampton Solent University, Southampton, UK
Jan-Willem Bolderdijk, University of Groningen, Groningen, The Netherlands
Rino Brouwer, TNO Defense and Security, and Safety, Soesterberg, The Netherlands
Richard Bye, Network Rail, London, UK
Peter Burns, Transport Canada, Ottawa, Canada
Gary Burnett, University of Nottingham, Nottingham, UK
Jeff Caird, University of Calgary, Calgary, Canada
Sue Cobb, University of Nottingham, Nottingham, UK
Judy Edworthy, University of Plymouth, Plymouth, UK
Stephen Fairclough, Liverpool John Moores University, Liverpool, UK
Scott Galster, Wright-Patterson Air Force Base, Ohio, USA
Christhard Gelau, University of Regensburg, Regensburg, Germany
Carmen Hagemeister, Dresden University of Technology, Dresden, Germany
Daniel Hausmann, University of Zürich, Zürich, Switzerland
Geertje Hegeman, DHV, Amersfoort, The Netherlands
Bob Hockey, The University of Sheffield, Sheffield, UK
Martin Joosse, NLR Dutch Aerospace, Amsterdam, The Netherlands
Wolfgang Kallus, University of Graz, Graz, Austria
Sabine Langois, Renault Technocentre, Guyancourt, France
Claus Marberger, Fraunhofer IAO, University of Stuttgart, Stuttgart, Germany
Christoph Möhlenbrink, German Aerospace Center (DLR), Braunschweig, Germany
Ben Mulder, University of Groningen, Groningen, The Netherlands
Anjum Naweed, The University of Sheffield, Sheffield, UK
Alan Pope, NASA Langley Research Center, Hampton, VA, USA
Selma de Ridder, TNO Defense and Security, and Safety, Soesterberg, The Netherlands
Matthew Rizzo, University of Iowa, Iowa, USA
Anna Schieben, DLR, Braunschweig, Germany
Caroline Schießl, DLR, Braunschweig, Germany
Dirk Schulze-Kissing, DLR, German Aerospace Center (DLR), Hamburg, Germany
Tacha Serif, Brunel University, Uxbridge, UK
Sarah Sharples, University of Nottingham, Nottingham, UK
Steve Shladover, PATH, Berkeley University, Richmond, CA, USA
Leon Urbas, Technical University Berlin, Berlin, Germany
Mark Vollrath, Technical University Braunschweig, Braunschweig, Germany
Mark Young, Brunel University, Uxbridge, UK

In D. de Waard, F.O. Flemisch, B. Lorenz, H. Oberheid, and K.A. Brookhuis (Eds.) (2008), *Human Factors for assistance and automation* (p. 503). Maastricht, the Netherlands: Shaker Publishing.